BIRKHÄUSER

Modern Birkhäuser Classics

Many of the original research and survey monographs in pure and applied mathematics published by Birkhäuser in recent decades have been groundbreaking and have come to be regarded as foundational to the subject. Through the MBC Series, a select number of these modern classics, entirely uncorrected, are being re-released in paperback (and as eBooks) to ensure that these treasures remain accessible to new generations of students, scholars, and researchers.

Hans Triebel

Theory of Function Spaces

Reprint of the 1983 Edition

Birkhäuser

Prof. Dr. Hans Triebel
Mathematisches Institut
Friedrich-Schiller-Universität Jena
07743 Jena
Germany
hans.triebel@uni-jena.de

ISBN 978-3-0346-0415-4 e-ISBN 978-3-0346-0416-1
DOI 10.1007/978-3-0346-0416-1

Library of Congress Control Number: 2010924183

2000 Mathematics Subject Classification 46E35, 46-02, 46E15, 35J40

Originally printed in German Democratic Republic

Published with kind permission of Akademische Verlagsgesellschaft Geest & Portig, K.-G., Leipzig for sales in non-socialist countries in 1983 as volume 78 in the Monographs in Mathematics series by Birkhäuser Verlag, Switzerland, ISBN 978-3-7643-1381-1
Reprinted 2010 by Springer Basel AG

Printed on acid-free paper

Springer Basel AG is part of Springer Science+Business Media

www.birkhauser.ch

Preface

Function spaces have a long history. They play an important part in both classical and modern mathematics. Spaces whose elements are continuous, or differentiable, or p-integrable functions are of interest for their own sake. They are also useful tools for the study of ordinary and partial differential equations (although they have some shortcomings for that purpose). Since the thirties more sophisticated function spaces have been used in the theory of partial differential equations, in the first place the Hölder spaces and the Sobolev spaces. Later on (culminating in the fifties and in the sixties) many new spaces were created and investigated, e. g. Besov spaces (Lipschitz spaces), Bessel-potential spaces (Liouville spaces, Lebesgue spaces), Zygmund classes, Hardy spaces (real n-dimensional version) and the space *BMO*. In the sixties and seventies, deep new tools were discovered: interpolation theory and the methods of the socalled "hard analysis" (Fourier analysis, maximal inequalities, etc.). The main aim of this book is to present a unified theory of function spaces of the above-mentioned types from the standpoint of Fourier analysis. (In [I] we tried to give a corresponding representation of some of the above types of spaces on the basis of interpolation theory.) Although many results have been proved very recently, the theory seems now to be at a stage which justifies the writing of a book about this subject. The two books [F] and [S] of the author may be considered as forerunners. But they are more or less research reports written at a moment when some of the most important problems of the above theory were unsolved (in contrast to the present situation). However, we do incorporate some material from [F] and [S]. In this sense, this book also may be considered as a new and thoroughly revised edition of some parts of [F] and [S].

The heart of the book is Part I, in particular Chapter 2. In this chapter we give a systematic study of the spaces $B^s_{p,q}(R_n)$ and $F^s_{p,q}(R_n)$, where $s\in R_1$, $0<p\leqq\infty$, and $0<q\leqq\infty$, on the n-dimensional Euclidean space R_n, which is based on methods of Fourier analysis. These two scales $B^s_{p,q}(R_n)$ and $F^s_{p,q}(R_n)$ contain as special cases all the above-mentioned classical spaces (local versions in the case of the Hardy spaces and the space *BMO*). The preceding Chapter 1 deals with entire analytic functions on R_n and sequences of those functions which belong to $L_p(R_n)$ and $L_p(R_n, l_q)$, respectively. This theory is of interest in itself, but it is also useful for Chapter 2 (the technique of maximal functions is used from the very beginning). Chapter 3 concerns the theory of the spaces $B^s_{p,q}(\Omega)$ and $F^s_{p,q}(\Omega)$ on bounded smooth domains Ω in R_n. Finally, in Chapter 4 we extend the well-known theory of boundary value problems for regular elliptic differential equations in Hölder spaces and in L_p-spaces (where $1<p<\infty$) to the spaces $B^s_{p,q}(\Omega)$ and $F^s_{p,q}(\Omega)$ on domains. In Part II we discuss briefly further types of spaces: homogeneous spaces $\dot{B}^s_{p,q}(R_n)$ and $\dot{F}^s_{p,q}(R_n)$ with the Hardy spaces (n-dimensional real version) and the space *BMO* as special cases; weighted spaces of the above type on R_n and on domains; and corresponding spaces on the n-torus. We sketch a few applications to degenerate elliptic differential equations and to some problems for multiple trigonometric series. In connection with weighted spaces we give a brief introduction to the theory of the Beurling ultra-distributions.

The styles of Part I and Part II are different. In Part I we give detailed proofs of the main assertions, whereas we restrict ourselves in Part II to formulations, hints and references. In many cases we remark that with the help of the indicated modifications one can carry over corresponding proofs from Part I. The main assertions of Part I are those which are needed for the study of the spaces $B^s_{p,q}(\Omega)$ and $F^s_{p,q}(\Omega)$ on domains Ω and their applications to regular elliptic boundary value problems. In many cases proofs of other assertions (also in Part I) are only outlined or omitted. It is clear that Part II is chiefly directed to specialists who work in that field of research. On the other hand, we hope that Part I is not only of interest to specialists in the theory of function spaces, Fourier analysis, and partial differential equations, but also to mathematicians, physicists, and graduate students, who use function spaces as a tool or who wish to have some information about this subject.

Of course, it is not the aim of the book to give an exhaustive treatment of the theory of function spaces in the widest sense of the word. We are primarily interested in demonstrating the power of Fourier analysis (in particular of Fourier multiplier theorems and of the technique of maximal functions) in connection with the above-mentioned spaces. This branch of the theory of function spaces was established at the end of the sixties and the beginning of the seventies. Above all we have in mind the papers by C. Fefferman, E. M. Stein [2] and by J. Peetre [2, 4, 6] and Peetre's book [8] (of course, these papers have their own forerunners: in particular, maximal functions have a long history, cf. e. g., E. M. Stein [1]).

The reader is expected to have a working knowledge of functional analysis as presented in the classical textbooks (including the standard facts of the theory of distributions). A familiarity with the basic results of the spaces of differentiable functions, L_p-spaces and Sobolev spaces would be helpful. In this connection we refer to A. Kufner, O. John, S. Fučik [1].

In recent years several books have appeared which deal with function spaces of the above type from diverse points of view. Beside the classical book by S. L. Sobolev [4] and the cited books by E. M. Stein [1], J. Peetre [8], A. Kufner, O. John, S. Fučik [1] and the author [I, F, S], there are books by R. A. Adams [1], S. M. Nikol'skij [3], O. V. Besov, V. P. Il'jin, S. M. Nikol'skij [1], W. Mazja [1], and A. Kufner [1]. Furthermore, we refer to L. Hörmander [2], J.-L. Lions, E. Magenes [2] and E. M. Stein, G. Weiss [2].

The book is organized on the decimal system. "Theorem 2.2.2/1" refers to Theorem 1 in Subsection 2.2.2., "Theorem 2.2.3." means the theorem in 2.2.3., etc. All unimportant positive constants will be denoted by c (with additional indices if there are several c's in the same estimate). [I], [F], and [S] refer to the books of the author, cf. the list of references.

I take the opportunity to express my gratitude to Akademische Verlagsgesellschaft Geest & Portig K.-G. in Leipzig for producing this book and for giving it a perfect typographical format.

Jena, Summer 1982 Hans Triebel

Contents

1. Spaces of Entire Analytic Functions

1.1. Introduction

The Chapters 1 and 6 deal with analytic functions $f(x)$, defined on the Euclidean n-space R_n, in the framework of Fourier analysis. The basis is the Paley-Wiener-Schwartz theorem, which says that any tempered distribution f in R_n, whose Fourier transform has compact support, is an analytic function (more precisely: entire analytic function of exponential type). Let S' be the set of all complex-valued tempered distributions on R_n, and let Ff be the Fourier transform of $f \in S'$. If μ is an arbitrary Borel measure on R_n, then

$$\|f \mid L_{p,\mu}\| = \Big(\int_{R_n} |f(x)|^p \,\mathrm{d}\mu\Big)^{\frac{1}{p}} \tag{1}$$

has the standard meaning, $0<p\leqq\infty$ (usual modification if $p=\infty$). If Ω is a compact subset of R_n, then we shall be concerned with the spaces

$$L^{\Omega}_{p,\mu} = \{f \mid f\in S',\ \operatorname{supp} Ff\subset\Omega,\ \|f \mid L_{p,\mu}\|<\infty\}\,. \tag{2}$$

Let $\|f \mid L_p\| = \|f \mid L_{p,\mu}\|$ and $L^{\Omega}_p = L^{\Omega}_{p,\mu}$ if μ is the Lebesgue measure. If $f\in L^{\Omega}_p$, then inequalities of the following types are known:
(i) If $\{x^k\}_{k=1}^{\infty}$ is an appropriate set of points in R_n (e.g. lattice points, where the length of the mesh is sufficiently small) then

$$\Big(\sum_{k=1}^{\infty} |f(x^k)|^p\Big)^{\frac{1}{p}} \sim \|f \mid L_p\|, \quad 0<p\leqq\infty\,, \tag{3}$$

(modification if $p=\infty$),
(ii) If $0<p\leqq q\leqq\infty$ and if α is an arbitrary multi-index then

$$\|D^{\alpha} f \mid L_q\| \leqq c\|f \mid L_p\|\,. \tag{4}$$

The admissible sets $\{x^k\}_{k=1}^{\infty}$ in (3) and the constant c in (4) depend on Ω (and p, q, α) but not on f. Inequalities of type (3) and (4) have a long history and played (and play) a central role in the theory of function spaces. (3) is essentially due to M. Plancherel, G. Polya [1], cf. also R. P. Boas [1, pp. 197–199]. (4) has been extensively used by S. M. Nikol'skij [1, 3] (at least if $p\geqq 1$). Of course, (3) and (4) are special cases of the estimate

$$\|D^{\alpha} f \mid L_{q,\nu}\| \leqq c\|f \mid L_{p,\mu}\|, \quad f\in L^{\Omega}_{p,\mu}\,, \quad 0<p\leqq q\leqq\infty\,, \tag{5}$$

where ν and μ are Borel measures on R_n. Such estimates will be denoted as inequalities of Plancherel-Polya-Nikol'skij type. The problem is the following: If Ω, p and q are given, under which conditions for μ and ν does there exist a constant c (depending on α) such that (5) holds for all $f\in L^{\Omega}_{p,\mu}$?

Chapters 1 and 6 are concerned with inequalities of type (5) and with the study of the spaces $L^{\Omega}_{p,\mu}$. If one looks for natural conditions for ν and μ such that (5) holds, then one is faced from the very beginning with a serious problem. Let, for example, $\mathrm{d}\nu = \mathrm{d}\mu = \varrho(x)\,\mathrm{d}x$ where $\varrho(x)$ is a weight and $\mathrm{d}x$ refers to the Lebesgue measure. The theory presented in this chapter is applicable to weights $\varrho(x)$ of at most

H. Triebel, *Theory of Function Spaces*, Modern Birkhäuser Classics,
DOI 10.1007/978-3-0346-0416-1_1, © Birkhäuser Verlag 1983

polynomial growth, e.g. $\varrho(x) \sim (1+|x|)^{\varkappa}$ where $\varkappa$ is an arbitrary real number. On the other hand, inequalities of type (5) and related spaces $L^{\Omega}_{p,\mu}$ do not make sense if e.g. $\varrho(x)=e^{|x|}$ or $\varrho(x)=e^{-|x|}$. What about weights $\varrho(x)$ which grow more rapidly than any polynomial but less than $e^{|x|}$, e.g. $e^{|x|^\beta}$ with $0<\beta<1$? As we shall see in Chapter 6, $\varrho(x)=e^{|x|^\beta}$ or $\varrho(x)=e^{-|x|^\beta}$ with $0<\beta<1$ are admissible weights, and a natural assumption for general admissible weights is closely related to the socalled Denjoy-Carleman non-quasi-analyticity condition. But the theory of the usual distributions is inadequate to weights for type $e^{|x|^\beta}$ with $0<\beta<1$. One needs the (in many respects more natural) Beurling ultra-distributions. However we prefer not to develop the (natural but complicated) theory of ultra-distributions in Part I of this book, and postpone this task to Chapter 6. As justification for this strategy we recall that Chapters 2, 3, and 4 (and this is the heart of the book) are concerned with unweighted spaces and for that purpose, the usual distributions are quite sufficient. Although there are no problems in dealing with weights of at most polynomial growth in the framework of the (usual) tempered distributions, we shall concentrate our attention on the (unweighted) Lebesgue measure and related Borel measures, e.g. atomic measures in the sense of (3).

The periodic counterpart of (4) deals with inequalities for trigonometric polynomials. In Chapter 9 we shall be concerned briefly with this subject. Finally we mention that inequalities of type (4) have been considered extensively for the classical orthogonal expansions (Jacobi, Hermite, Laguerre polynomials). A unified treatment (from an abstract point of view) and many references may be found in R. J. Nessel, G. Wilmes [2]. Cf. also S. M. Nikol'skij [3], R. J. Nessel, G. Wilmes [1], J. Peetre [8] and [F]. Finally we wish to mention that a brief survey of spaces of type $L^{\Omega}_{p,\mu}$ (in particular with $0<p\leqq 1$) is also given in J. Peetre [8, Chapter 11].

1.2. Preliminaries

1.2.1. Distributions

It is expected that the reader is familiar with the theory of distributions, cf. e.g. L. Schwartz [1], K. Yosida [1] or L. Hörmander [2]. We recall some notations and results.

Let R_n be n-dimensional real Euclidean space. The general point in R_n is denoted by $x=(x_1, \ldots, x_n)$. Let $S=S(R_n)$ be the Schwartz space of all complex-valued rapidly decreasing infinitely differentiable functions on R_n (usually we write S instead of $S(R_n)$, there being no danger of confusion because all functions and distributions in this chapter are defined on R_n, and n is fixed once and for all. Similarly for the other spaces below.) The topology in the complete locally convex space S is generated by the norms

$$p_N(\varphi)=\sup_{x\in R_n}(1+|x|)^N \sum_{|\alpha|\leqq N} |D^\alpha\varphi(x)|, \quad N=1,2,3,\ldots \tag{1}$$

where $\varphi\in S$. Furthermore, $D^\alpha=\dfrac{\partial^{|\alpha|}}{\partial x_1^{\alpha_1}\ldots\partial x_n^{\alpha_n}}$ has the usual meaning: $\alpha=(\alpha_1,\ldots,\alpha_n)$ is a multi-index, $\alpha_j\geqq 0$ are integers and $|\alpha|=\sum_{j=1}^n \alpha_j$. Let $S'=S'(R_n)$ be the set of all tempered distributions on R_n, i.e. the topological dual of S, equipped with the strong topology. If $\varphi \in S$, then

$$(F\varphi)(x) = (2\pi)^{-\frac{n}{2}} \int_{R_n} e^{-ix\xi} \varphi(\xi)\, d\xi, \quad x \in R_n, \tag{2}$$

denotes the Fourier transform $F\varphi$ of φ. Here $x\xi = \sum_{j=1}^{n} x_j \xi_j$ is the scalar product in R_n of $x = (x_1, \ldots, x_n) \in R_n$ and $\xi = (\xi_1, \ldots, \xi_n) \in R_n$. The inverse Fourier transform $F^{-1}\varphi$ of φ is given by

$$(F^{-1}\varphi)(x) = (2\pi)^{-\frac{n}{2}} \int_{R_n} e^{ix\xi} \varphi(\xi)\, d\xi, \quad x \in R_n. \tag{3}$$

One extends F and F^{-1} in the usual way from S to S'. Recall that F yields an isomorphic mapping from S onto itself and from S' onto itself. Similarly F^{-1}. Furthermore, F (more precisely: the restriction of F to $L_2(R_n)$) is a unitary operator from $L_2(R_n)$ onto itself. Here, $L_2(R_n)$ is the usual L_2-space on R_n with respect to the Lebesgue measure.

Finally we recall the famous Paley-Wiener-Schwartz theorem. An analytic function f of n complex variables will be denoted by $f(z)$, where $z = x + iy$, $x \in R_n$ and $y \in R_n$. If $f(x)$ with $x \in R_n$ can be extended to an analytic function of n complex variables, then this extension will also be denoted by $f(z)$.

Theorem 1. The following two assertions are equivalent:
(i) $\varphi \in S$ and $\operatorname{supp} F\varphi \subset \{y \mid |y| \leqq b\}$,
(ii) $\varphi(z)$ is an entire analytic function of n complex variables and for any $\lambda > 0$ and any $\varepsilon > 0$ there exists a constant $c_{\lambda,\varepsilon}$ such that

$$|\varphi(z)| \leqq c_{\lambda,\varepsilon} (1 + |x|)^{-\lambda} e^{(b+\varepsilon)|y|} \tag{4}$$

holds for all $z = x + iy$ with $x \in R_n$ and $y \in R_n$.

Theorem 2. The following two assertions are equivalent:
(i) $f \in S'$ and $\operatorname{supp} Ff \subset \{y \mid |y| \leqq b\}$,
(ii) $f(z)$ is an entire analytic function of n complex variables and for an appropriate real number λ and for any $\varepsilon > 0$ there exists a constant c_ε such that

$$|f(z)| \leqq c_\varepsilon (1 + |x|)^{\lambda} e^{(b+\varepsilon)|y|} \tag{5}$$

holds for all $z = x + iy$ with $x \in R_n$ and $y \in R_n$.

Remark. It is clear how to understand these two theorems: If (i) of Theorem 2 holds, then $f \in S'$ is a regular distribution $f(x)$, which can be extended to an analytic function $f(z)$, which satisfies (5). Vice versa, if (ii) of Theorem 2 holds, then the restriction of $f(z)$ to R_n satisfies (i). Similarly for Theorem 1. Proofs may be found in L. Schwartz [1, p. 272], L. Hörmander [2, Theorem 1.7.7] and K. Yosida [1, VI. 4]. The formulation in Hörmander's book looks more handsome but the above version is the direct counterpart of a corresponding theorem for ultra-distributions, cf. 6.1.4.

1.2.2. L_p-Spaces and Quasi-Banach Spaces

If μ is a Borel measure on R_n and $0 < p < \infty$, then we put

$$\|f \mid L_{p,\mu}\| = \left(\int_{R_n} |f(x)|^p \, d\mu\right)^{\frac{1}{p}} \tag{1}$$

and

$$\|f \mid L_{\infty,\mu}\| = \mu - \operatorname*{ess\,sup}_{x \in R_n} |f(x)|. \tag{2}$$

If $0<p\leqq\infty$, then $L_{p,\mu}=L_{p,\mu}(R_n)$ denotes the set of all Borel measurable complex-valued functions on R_n, such that $\|f \mid L_{p,\mu}\|<\infty$. Again, we write $L_{p,\mu}$ instead of $L_{p,\mu}(R_n)$ if there is no danger of confusion. If μ is Lebesgue measure, then we put $L_p=L_{p,\mu}$ and $\|f \mid L_p\|=\|f \mid L_{p,\mu}\|$.

Let A be a (real or complex) linear vector space. $\|a \mid A\|$ is said to be a quasi-norm if $\|a \mid A\|$ satisfies the usual conditions of a norm with the exception of the triangle inequality, which is replaced by

$$\|a_1+a_2 \mid A\|\leqq c\,(\|a_1 \mid A\|+\|a_2 \mid A\|)\,. \tag{3}$$

More precisely: There exists a constant c such that (3) holds for all $a_1\in A$ and all $a_2\in A$. Of course $c\geqq 1$. (If $c=1$ is admissible, then A is a normed space.) A quasi-normed space is said to be a quasi-Banach space if it is complete (i.e. any fundamental sequence in A with respect to $\|\cdot \mid A\|$ converges).

It is well-known that $L_{p,\mu}$ with $0<p\leqq\infty$ is a quasi-Banach space (Banach space if $p\geqq 1$).

If $0<q\leqq\infty$, then l_q is the set of all sequences $\{b_k\}_{k=0}^{\infty}$ of complex numbers such that

$$\|b_k \mid l_q\|=\left(\sum_{k=0}^{\infty}|b_k|^q\right)^{\frac{1}{q}}<\infty \tag{4}$$

(modification if $q=\infty$). Of course, l_q is a special $L_{q,\mu}$-space with respect to an appropriate atomic measure μ. In particular, l_q is a quasi-Banach space (if $q\geqq 1$, a Banach space). For the sake of simplicity we write $\|b_k \mid l_q\|$ instead of the more correct notation $\|\{b_k\}_{k=0}^{\infty} \mid l_q\|$.

Let $0<p\leqq\infty$ and $0<q\leqq\infty$. If $\{f_k(x)\}_{k=0}^{\infty}$ is a sequence of complex-valued Borel measurable functions on R_n, then

$$\|f_k \mid L_p(l_q)\|=\|\,\|f_k(\cdot) \mid l_q\|\;L_p\,\|=\left(\int_{R_n}\left(\sum_{k=0}^{\infty}|f_k(x)|^q\right)^{\frac{p}{q}}\mathrm{d}x\right)^{\frac{1}{p}}, \tag{5}$$

$$\|f_k \mid l_q(L_p)\|=\|\,\|f_k(\cdot) \mid L_p\|\;l_q\,\|=\left(\sum_{k=0}^{\infty}\left(\int_{R_n}|f_k(x)|^p\,\mathrm{d}x\right)^{\frac{q}{p}}\right)^{\frac{1}{q}} \tag{6}$$

(modification if $p=\infty$ and/or $q=\infty$). Let $L_p(l_q)=L_p(R_n, l_q)$ be the set of all sequences $\{f_k\}$ such that $\|f_k \mid L_p(l_q)\|<\infty$. Similarly $l_q(L_p)=l_q(L_p(R_n))$ (again we omit R_n if there is no danger of confusion). $L_p(l_q)$ and $l_q(L_p)$ are quasi-Banach spaces (Banach spaces if $p\geqq 1$ and $q\geqq 1$).

1.2.3. Maximal Inequalities

If $f(x)$ is a complex-valued locally Lebesgue-integrable function on R_n, then

$$(Mf)\,(x)=\sup\frac{1}{|B|}\int_B|f(y)|\,\mathrm{d}y \tag{1}$$

is the Hardy-Littlewood maximal function, where the supremum is taken over all balls B centered at x. We have

$$(Mf)\,(x)\geqq|f(x)|\quad\text{for almost every } x\in R_n \tag{2}$$

(with respect to the Lebesgue measure). If B in (1) stands for arbitrary balls with $x\in B$, or for cubes with $x\in B$, or for cubes centered at x, then the corresponding maximal functions can be mutually estimated from above and from below with the help of constants which are independent of $f(x)$. In other words, the theorem below holds for any of these maximal functions. Assertions concerning maximal functions and historical remarks can be found in E. M. Stein [1, § 1].

Theorem. Let $1<p<\infty$ and $1<q<\infty$. There exists a constant c such that

$$\|Mf_k \mid L_p(l_q)\| \leqq c \, \|f_k \mid L_p(l_q)\| \tag{3}$$

holds for all sequences $\{f_k(x)\}_{k=0}^{\infty}$ of complex-valued locally Lebesgue-integrable functions on R_n.

Remark. This fundamental theorem is due to C. Fefferman, E. M. Stein [1]. If $f_0(x)=f(x)$ and $f_j(x)\equiv 0$ for $j=1, 2, \ldots$, then one obtains the scalar case of (3),

$$\|Mf \mid L_p\| \leqq c\|f \mid L_p\|, \quad \text{where} \quad 1<p\leqq\infty . \tag{4}$$

The case $p=\infty$ is trivial, because $(Mf)(x)\leqq\|f \mid L_\infty\|$. On the other hand, (4) is not valid if $p=1$. As for proofs and further comments we refer again to E. M. Stein [1] and C. Fefferman, E. M. Stein [1].

1.2.4. Admissible Borel Measures

As has been said in 1.1., the study of the general spaces $L^{\Omega}_{p,\mu}$ from (1.1/2) is postponed to Chapter 6. In the present chapter we restrict ourselves to the Lebesgue measure and related Borel measures. In this subsection we explain what kind of Borel measures we have in mind.

Let Z_n be the lattice of all points $k=(k_1, \ldots, k_n)\in R_n$, where the components $k_1, \ldots, k_n$ are integers. If $h>0$ and $k\in Z_n$, then we put

$$Q^h_k=\{x \mid x\in R_n,\ hk_j\leqq x_j<h\,(k_j+1);\quad (j=1, \ldots, n)\} . \tag{1}$$

Of course, $x=(x_1, \ldots, x_n)$ in (1). Obviously, $\{Q^h_k\}_{k\in Z_n}$ yields a decomposition of R_n.

Definition. If $h>0$ is given, then K_h denotes the set of all Borel measures μ in R_n for which there exist two positive numbers c_1 and c_2 such that

$$c_1\leqq\mu(Q^h_k)\leqq c_2 \tag{2}$$

holds for every cube Q^h_k, where $k\in Z_n$.

Remark. The definition shows that $\mu\in K_h$ has a lattice structure. Besides Lebesgue measure, atomic measures are of special interest:

$$\mu(Q^h_k)=\mu(\{x^k\})=1, \quad \text{where} \quad x^k\in Q^h_k \tag{3}$$

is a fixed point in Q^h_k, e.g. $x^k=hk$. If μ is given by $d\mu=\varkappa(x)\,dx$, where dx indicates Lebesgue measure and $\varkappa(x)$ stands for a non-negative weight function, then (2) reads as follows:

$$0<c_1\leqq \int_{Q^h_k} \varkappa(x)\,dx\leqq c_2<\infty \quad \text{for all} \quad k\in Z_n . \tag{4}$$

In particular, weights of type $\varkappa(x)=\min(1, |x|^{\lambda})$ with $\lambda\geqq 0$ and $\varkappa(x)=\max(1, |x|^{-\sigma})$ with $0\leqq\sigma<n$ are admissible.

1.3. Inequalities of Plancherel-Polya-Nikol'skij Type

1.3.1. A Maximal Inequality

Section 1.3. follows closely [F, 1.3]. If Ω is a compact subset of R_n, then we put

$$S^{\Omega}=\{\varphi \mid \varphi\in S,\quad \operatorname{supp} F\varphi\subset\Omega\} . \tag{1}$$

Recall that $S=S(R_n)$. Furthermore, M is the maximal function from 1.2.3.

Theorem. Let Ω be a compact subset of R_n and let $0<r<\infty$. There exist two positive numbers c_1 and c_2 such that

$$\sup_{z\in R_n} \frac{|\nabla\varphi(x-z)|}{1+|z|^{\frac{n}{r}}} \leqq c_1 \sup_{z\in R_n} \frac{|\varphi(x-z)|}{1+|z|^{\frac{n}{r}}} \leqq c_2 \left[(M|\varphi|^r)(x)\right]^{\frac{1}{r}} \tag{2}$$

holds for all $\varphi\in S^\Omega$ and all $x\in R_n$.

Proof. Step 1. Let $\psi\in S$ such that $(F\psi)(x)=1$ if $x\in\Omega$. If $\varphi\in S^\Omega$ then we have $F\varphi=F\varphi\cdot F\psi$ and $\varphi=F^{-1}(F\varphi\cdot F\psi)$. By well-known properties of the Fourier transform it follows that

$$\varphi(x)=c\int_{R_n}\varphi(y)\,\psi(x-y)\,dy\,. \tag{3}$$

If one replaces x in (3) by $x-z$, then differentiation yields

$$\begin{aligned}\left|\frac{\partial\varphi}{\partial x_1}(x-z)\right| &= c\left|\int_{R_n}\varphi(y)\frac{\partial\psi}{\partial x_1}(x-z-y)\,dy\right| \\ &\leqq c'\int_{R_n}|\varphi(y)|\,(1+|x-y-z|)^{-\lambda}\,dy\,.\end{aligned} \tag{4}$$

Here c and c' are positive numbers, and λ may be chosen arbitrarily large. Of course, one can replace $\frac{\partial\varphi}{\partial x_1}$ on the left-hand side of (4) by $\nabla\varphi$. We divide both sides of this modified estimate by $1+|z|^{\frac{n}{r}}$ and use the inequality

$$\frac{1+|x-y|^{\frac{n}{r}}}{1+|z|^{\frac{n}{r}}} \leqq c\,(1+|x-y-z|^{\frac{n}{r}}),\quad x\in R_n,\quad y\in R_n,\quad z\in R_n\,.$$

Then it follows that

$$\sup_{z\in R_n}\frac{|\nabla\varphi(x-z)|}{1+|z|^{\frac{n}{r}}} \leqq c\sup_{z\in R_n}\int_{R_n}\frac{|\varphi(y)|}{1+|x-y|^{\frac{n}{r}}}(1+|x-y-z|)^{-\varkappa}\,dy\,,$$

where $\varkappa$ can be chosen arbitrarily large. Then we have

$$\sup_{z\in R_n}\frac{|\nabla\varphi(x-z)|}{1+|z|^{\frac{n}{r}}} \leqq c\sup_{z\in R_n}\frac{|\varphi(x-z)|}{1+|z|^{\frac{n}{r}}}\,. \tag{5}$$

Step 2. We need an auxiliary assertion. Let $g(z)$ be a complex-valued continuously differentiable function in the closed unit ball $B=\{y\,|\,|y|\leqq 1\}$ in R_n. If $0<r<\infty$, then we claim that there exists a constant c such that

$$|g(z)|\leqq c\left[\sup_{w\in B}|\nabla g(w)|+\left(\int_B|g(w)|^r\,dw\right)^{\frac{1}{r}}\right],\quad z\in B\,, \tag{6}$$

holds for all functions $g(z)$ with the above properties. But (6) is an easy consequence of the mean value theorem:

$$\begin{aligned}|g(z)| &\leqq c\,[\min_{w\in B}|g(w)|+\sup_{w\in B}|\nabla g(w)|] \\ &\leqq c\left(\int_B|g(w)|^r\,dw\right)^{\frac{1}{r}}+c\sup_{w\in B}|\nabla g(w)|\,.\end{aligned}$$

If B_δ is an arbitrary closed ball of radius $\delta>0$ in R_n, then there exists a constant c such that for all $\delta>0$ and for all functions $g(z)$, which are continuously differentiable in B_δ,

$$|g(z)|\leq c\delta \sup_{w\in B_\delta} |\nabla g(w)|+c\delta^{-\frac{n}{r}} \left(\int_{B_\delta} |g(w)|^r \,\mathrm{d}w\right)^{\frac{1}{r}}, \quad z\in B_\delta, \tag{7}$$

holds. We assume that $B_\delta=\{y \mid |y|\leq\delta\}$. If we replace $g(z)$ in (6) by $g(\delta z)$, then (7) follows from (6).

Step 3. If $g(z)=\varphi(x-z)$, where φ is as in Step 1, then (7) yields

$$|\varphi(x-z)|\leq c\delta \sup_{|y|\leq\delta} |\nabla\varphi(x-z-y)|+c\delta^{-\frac{n}{r}} \left(\int_{|y|\leq\delta} |\varphi(x-z-y)|^r \,\mathrm{d}y\right)^{\frac{1}{r}}. \tag{8}$$

Let $0<\delta\leq 1$. The the last integral can be estimated from above by

$$\left(\int_{|u|\leq|z|+1} |\varphi(x-u)|^r \,\mathrm{d}u\right)^{\frac{1}{r}} \leq c\,(1+|z|^{\frac{n}{r}})\,[(M\,|\varphi|^r)(x)]^{\frac{1}{r}},$$

where M is the Hardy-Littlewood maximal function from (1.2.3/1). We put this estimate into (8) and divide both sides by $1+|z|^{\frac{n}{r}}$. Taking the supremum with respect to $z\in R_n$, we obtain

$$\sup_{z\in R_n} \frac{|\varphi(x-z)|}{1+|z|^{\frac{n}{r}}} \leq c\delta \sup_{z\in R_n} \frac{|\nabla\varphi(x-z)|}{1+|z|^{\frac{n}{r}}} + c\delta^{-\frac{n}{r}}[(M\,|\varphi|^r)(x)]^{\frac{1}{r}}. \tag{9}$$

Recall that $\delta\leq 1$ and that c in (9) is independent of δ, φ and x. Substituting (9) in (5) and choosing δ sufficiently small, we obtain

$$\sup_{z\in R_n} \frac{|\nabla\varphi(x-z)|}{1+|z|^{\frac{n}{r}}} \leq c\,[(M\,|\varphi|^r)(x)]^{\frac{1}{r}}, \quad x\in R_n, \tag{10}$$

where c is independent of φ and x. Now (2) follows from (5), (9) and (10).

Remark. The proof is a simplified version of a corresponding assertion in [F, 1.3.2]. However we wish to emphasize that some key ideas in the above proof are due to J. Peetre [6]. Cf. also C. Fefferman, E. M. Stein [2], where the technique of maximal functions of the above type has been developed.

1.3.2. Inequalities for the Lebesgue-Measure

Our goal is the proof of inequalities of type (1.1/5) for admissible Borel measures ν and λ in the sense of Subsection 1.2.4. In this section we restrict ourselves to functions $f\in S^\Omega$ (in order to avoid technical difficulties). The extension to $f\in L^\Omega_{p,\mu}$ will be treated in Section 1.4. Furthermore, it is convenient to begin with Lebesgue measure. Recall that D^α stands for $\frac{\partial^{|\alpha|}}{\partial x_1^{\alpha_1}\ldots\partial x_n^{\alpha_n}}$, where $\alpha=(\alpha_1,\ldots,\alpha_n)$ is a multi-index, i.e. $\alpha_1,\ldots,\alpha_n$ are non-negative integers, and $|\alpha|=\sum_{j=1}^{n}\alpha_j$.

Proposition. Let Ω be a compact subset of R_n and let $0<p\leq q\leq\infty$. If α is a multi-index, then there exists a positive constant c such that

$$\|D^\alpha\varphi \mid L_q\|\leq c\,\|\varphi \mid L_p\| \tag{1}$$

holds for all $\varphi\in S^\Omega$.

Proof. Step 1. We prove that

$$\|\varphi \mid L_\infty\| \leqq c \, \|\varphi \mid L_p\| , \quad 0<p\leqq\infty . \tag{2}$$

If $1\leqq p\leqq\infty$, then (2) is an immediate consequence of (1.3.1/3) and Hölder's inequality. If $0<p<1$, then (1.3.1/3) yields

$$|\varphi(x)| \leqq c \, (\sup_{y\in R_n} |\varphi(y)|)^{1-p} \int_{R_n} |\varphi(y)|^p \, dy .$$

Taking the supremum with respect to $x\in R_n$, we obtain (2). Let $0<p<q<\infty$. By (2) we have

$$\begin{aligned} \|\varphi \mid L_q\| &= \Big(\int_{R_n} |\varphi(x)|^q \, dx\Big)^{\frac{1}{q}} \\ &\leqq (\sup_{x\in R_n} |\varphi(x)|)^{1-\frac{p}{q}} \Big(\int_{R_n} |\varphi(x)|^p \, dx\Big)^{\frac{1}{q}} \leqq c \, \|\varphi \mid L_p\| . \end{aligned} \tag{3}$$

This proves (1) for $\alpha=(0, \ldots, 0)$.

Step 2. First we remark that $D^\alpha\varphi\in S^\Omega$ holds for $\varphi\in S^\Omega$ and an arbitrary α. Consequently, iteration of (1.3.1/2) yields

$$|D^\alpha\varphi(x)| \leqq \sup_{z\in R_n} \frac{|D^\alpha\varphi \, (x-z)|}{1+|z|^{\frac{n}{r}}} \leqq c_1 \sup_{z\in R_n} \frac{|\varphi \, (x-z)|}{1+|z|^{\frac{n}{r}}} \leqq c_2 \, [(M|\varphi|^r) \, (x)]^{\frac{1}{r}} ,$$

where $0<r<\infty$. If $0<p\leqq\infty$, then we choose $0<r<p$ (with $r=1$ if $p=\infty$). It follows that

$$\|D^\alpha\varphi \mid L_p\| \leqq c \, \|(M \, |\varphi|^r)^{\frac{1}{r}} \mid L_p\| = c \, \| \, M \, |\varphi|^r \mid L_{\frac{p}{r}} \, \|^{\frac{1}{r}} .$$

The maximal inequality (1.2.3/4) yields

$$\|D^\alpha\varphi \mid L_p\| \leqq c \, \||\varphi|^r \mid L_{\frac{p}{r}}\|^{\frac{1}{r}} = c \, \|\varphi \mid L_p\| . \tag{4}$$

This proves (1) with $p=q$. The general case follows from (4) and (3).

Remark 1. The constant c in (1) depends on Ω, p, q, α and the dimension n. The following consequence of (1) will be of interest later on: Let $0<p\leqq q\leqq\infty$ and let α be a multi-index. There exists a constant c such that

$$\|D^\alpha\varphi \mid L_q\| \leqq c b^{|\alpha|+n\left(\frac{1}{p}-\frac{1}{q}\right)} \|\varphi \mid L_p\| \tag{5}$$

holds for all $b>0$ and all $\varphi\in S^{B_b}$ with $B_b=\{y \mid |y|\leqq b\}$. The proof of (5) is simple. If $\varphi(x)\in S^{B_b}$ then $\varphi(b^{-1}x)\in S^{B_1}$, because

$$(F\varphi(b^{-1}\cdot)) \, (\xi) = b^n (F\varphi) \, (b\xi) . \tag{6}$$

If we put $\varphi(b^{-1}x)$ instead of $\varphi(x)$ into (1), then this modified estimate yields (5). More sophisticated estimates of c in (1) have been given by R. J. Nessel, G. Wilmes [1].

Remark 2. If $1\leqq p\leqq q\leqq\infty$, then (5) is the famous Nikol'skij inequality, which has been extensively used in the theory of function spaces by S. M. Nikol'skij and the Russian school, cf. S. M. Nikol'skij [3] and O. V. Besov, V. P. Il'jn, S. M. Nikol'skij [1]. There exist many modifications and generalizations: Nikol'skij inequalities for trigonometric polynomials and for orthogonal polynomials, extensions to $0<p\leqq q\leqq\infty$ and to general measures and weights, abstract generalizations etc. References and detailed descriptions may be found in S. M. Nikol'skij [3] and R. J. Nessel, G. Wilmes [1, 2].

1.3.3. Inequalities for Atomic Measures

Let μ be an appropriate atomic measure in R_n. Our aim is to prove the counterpart of (1.3.2/1), where L_p and L_q are replaced by $L_{p,\mu}$ and $L_{q,\mu}$ respectively. For that purpose it is sufficient to show that the quasi-norms $\|\varphi \mid L_p\|$ and $\|\varphi \mid L_{p,\mu}\|$ are equivalent, $\varphi \in S^\Omega$ (recall that $D^\alpha \varphi \in S^\Omega$ if $\varphi \in S^\Omega$). The lattice Z_n and the cubes Q_k^h have the same meaning as in 1.2.4.

Proposition. Let Ω be a compact subset of R_n and let $0 < p \leqq \infty$. There exist three positive numbers h_0, c_1 and c_2 such that

$$c_1 \Big(\sum_{k \in Z_n} |\varphi(x^k)|^p \Big)^{\frac{1}{p}} \leqq h^{-\frac{n}{p}} \|\varphi \mid L_p\| \leqq c_2 \Big(\sum_{k \in Z_n} |\varphi(x^k)|^p \Big)^{\frac{1}{p}} \tag{1}$$

holds for all h with $0 < h < h_0$, all sets $\{x^k\}_{k \in Z_n}$ with $x^k \in Q_k^h$, and all $\varphi \in S^\Omega$ (modification if $p = \infty$).

Proof. Step 1. We prove the right-hand side of (1). We may assume that $h \leqq 1$. If $x \in Q_k^h$ and $\varphi \in S^\Omega$, then

$$|\varphi(x)| \leqq |\varphi(x^k)| + ch \sup_{z \in Q_k^h} |\nabla \varphi(z)| \leqq |\varphi(x^k)| + ch \sup_{|x-z| \leqq c'} |\nabla \varphi(z)| , \tag{2}$$

where c and c' are independent of h and φ. If $p = \infty$, then (1.3.2/1) with $p = q = \infty$ yields

$$\|\varphi \mid L_\infty\| = \sup_k |\varphi(x^k)| + ch \|\varphi \mid L_\infty\| ,$$

where c is independent of h and φ (but depends on Ω). If h is small, then we get the right-hand side of (1). If $p < \infty$, then (2) yields

$$\int_{R_n} |\varphi(x)|^p \, dx \leqq ch^n \sum_k |\varphi(x^k)|^p + ch^p \int_{R_n} \sup_{|x-z| \leqq c'} |\nabla \varphi(z)|^p \, dx . \tag{3}$$

Let $0 < r < p$. By (1.3.1/2) and the maximal inequality (1.2.3/4),

$$\begin{aligned} \int_{R_n} |\varphi(x)|^p \, dx &\leqq ch^n \sum_k |\varphi(x^k)|^p + ch^p \int_{R_n} |(M |\varphi|^r)(x)|^{\frac{p}{r}} \, dx \\ &\leqq ch^n \sum_k |\varphi(x^k)|^p + c'h^p \, \| |\varphi|^r \mid L_{\frac{p}{r}} \|^{\frac{p}{r}} \\ &\leqq ch^n \sum_k |\varphi(x^k)|^p + c'h^p \int_{R_n} |\varphi(x)|^p \, dx . \end{aligned} \tag{4}$$

If h is small, then we obtain the right-hand side of (1).

Step 2. In order to prove the left-hand side of (1) with $p < \infty$ we start with

$$|\varphi(x^k)| \leqq |\varphi(x)| + ch \sup_{|x-z| \leqq c'} |\nabla \varphi(z)| , \quad x \in Q_k^h ,$$

instead of (2). Then we obtain (3) and (4) with $h^n \sum_k |\varphi(x^k)|^p$ on the left-hand sides and $\int_{R_n} |\varphi(x)|^p \, dx$ instead of $h^n \sum_k |\varphi(x^k)|^p$ on the right-hand sides. This proves the left-hand side of (1). The left-hand side of (1) with $p = \infty$ is trivial.

Remark. Of interest is the number h_0. It depends on Ω and p. The restriction $h_0 \leqq 1$ is unimportant. One can prove the following assertion: if

$$\Omega \subset Q_b = \{ y \mid |y_j - y^0| \leqq b ; \quad (j = 1, \ldots, n) \}, \quad y^0 = (y_1^0, \ldots, y_n^0) \in R_n ,$$

then $h_0 \leqq \frac{\pi}{b}$, cf. [F, pp. 31, 33, 48] and 1.3.5.

1.3.4. Inequalities for Admissible Borel Measures

The lattice Z_n and the set K_h of Borel measures in R_n have the same meaning as in 1.2.4. Special elements of K_h are the Lebesgue measure and the atomic measures of the last subsection, cf. (1.2.4/3). Let $h_0 = h_0(\Omega, p)$ be the positive number from Proposition 1.3.3 and Remark 1.3.3. Let $h(\Omega, p, q) = \min(h_0(\Omega, p), h_0(\Omega, q))$, where $0 < p \leqq q \leqq \infty$.

Theorem. Let Ω be a compact subset of R_n and let $0 < p \leqq q \leqq \infty$. Let $\nu \in K_h$ and $\mu \in K_h$, where $0 < h < h(\Omega, p, q)$. If α is a multi-index then there exists a constant c such that

$$\|D^\alpha \varphi \mid L_{q,\nu}\| \leqq c \, \|\varphi \mid L_{p,\mu}\| \tag{1}$$

holds for all $\varphi \in S^\Omega$.

Proof. Step 1. If $\varphi \in S^\Omega$ is given, then we choose $x^k \in Q_k^h$ and $y^k \in Q_k^h$ such that

$$|\varphi(x^k)| = \max_{z \in Q_k^h} |\varphi(z)|, \quad |\varphi(y^k)| = \min_{z \in Q_k^h} |\varphi(z)| .$$

Let $p < \infty$. Then (1.2.4/2) and Proposition 1.3.3 yield

$$\begin{aligned} \|\varphi \mid L_{p,\nu}\|^p &= \int_{R_n} |\varphi(x)|^p \, d\nu \leqq c_1 \sum_{k \in Z_n} |\varphi(x^k)|^p \\ &\leqq c_2 \sum_{k \in Z_n} |\varphi(y^k)|^p \leqq c_3 \int_{R_n} |\varphi(x)|^p \, d\mu = c_3 \|\varphi \mid L_{p,\mu}\|^p . \end{aligned} \tag{2}$$

Of course, the trick in this estimate is the possibility of switching over from $\{x^k\}_{k \in Z_n}$ to $\{y^k\}_{k \in Z_n}$ on the basis of Proposition 1.3.3 (equivalent quasi-norms for different choices of $\{x^k\}_{k \in Z_n}$ in (1.3.3/1)). Similarly if $p = \infty$.

Step 2. In particular, (2) yields that $\|\varphi \mid L_{p,\mu}\|$ and $\|\varphi \mid L_p\|$ are equivalent on S^Ω (i.e. $\|\varphi \mid L_{p,\mu}\|$ can be estimated from above and from below by $\|\varphi \mid L_p\|$ with the help of constants which are independent of $\varphi \in S^\Omega$). Similarly, $\|D^\alpha \varphi \mid L_{q,\nu}\|$ and $\|D^\alpha \varphi \mid L_q\|$ are equivalent. Now, (1) follows from Proposition 1.3.2.

Remark 1. As has been mentioned in Step 2, we obtain as a special case of (1) the following improvement of Proposition 1.3.3: Let Ω be a compact subset of R_n and let $0 < p \leqq \infty$. If $\mu \in K_h$, where $0 < h < h_0(\Omega, p)$, then there exist two positive numbers c_1 and c_2 such that

$$c_1 \|\varphi \mid L_{p,\mu}\| \leqq \|\varphi \mid L_p\| \leqq c_2 \|\varphi \mid L_{p,\mu}\| \tag{3}$$

holds for all $\varphi \in S^\Omega$ (equivalent quasi-norms on S^Ω).

Remark 2. Formula (1) coincides with (1.1/5) if $f \in S^\Omega$, provided that ν and μ satisfy the above hypotheses. Later on, we extend (1) to $L^\Omega_{p,\mu} = L^\Omega_p$, cf. Theorem 1.4.1.

1.3.5. A Representation Formula

In order to shed some light upon the lattice structure of the preceding inequalities, cf. (1.3.3/1), (1.3.4/3), and Theorem 1.3.4, we prove a representation formula for $\varphi \in S^\Omega$. We specialize Ω by considering

$$\Omega = Q_b = \{y \mid |y_j| \leqq b; \quad (j = 1, \ldots, n)\}, \quad b > 0 . \tag{1}$$

Furthermore, the characteristic function of Q_b will be denoted by $\chi_b(x)$. Let Z_n be the lattice from 1.2.4 with the lattice-points $k = (k_1, \ldots, k_n)$.

Proposition. If $\Omega=Q_b$ is given by (1) then any $\varphi\in S^{\Omega}$ can be represented by

$$\varphi(x)=\left(\frac{\pi}{2}\right)^{\frac{n}{2}} b^{-n} \sum_{k\in Z_n} \varphi\left(\frac{\pi}{b}k\right)(F^{-1}\chi_b)\left(x-\frac{\pi}{b}k\right) \tag{2}$$

$$=\sum_{k\in Z_n}\varphi\left(\frac{\pi}{b}k\right)\prod_{j=1}^{n}\frac{\sin(bx_j-k_j\pi)}{bx_j-k_j\pi},\quad x\in R_n,$$

(absolutely convergent series).

Proof. Step 1. If $x=(x_1,\ldots,x_n)\in R_n$ and $k=(k_1,\ldots,k_n)\in Z_n$, then $xk=\sum_{j=1}^{n}x_jk_j$ is the usual scalar product in R_n. As a smooth function, $F\varphi$ can be developed in Q_b in an absolutely convergent multiple trigonometric series. Hence,

$$(F\varphi)(x)=\chi_b(x)\sum_{k\in Z_n}a_k e^{-i\frac{\pi}{b}kx}\quad\text{with}$$

$$a_k=(2b)^{-n}\int_{Q_b}(F\varphi)(x)\,e^{i\frac{\pi}{b}kx}\,dx$$

$$=(2b)^{-n}(2\pi)^{\frac{n}{2}}\left[F^{-1}(F\varphi)\right]\left(\frac{\pi}{b}k\right)=\left(\frac{\pi}{2}\right)^{\frac{n}{2}}b^{-n}\varphi\left(\frac{\pi}{b}k\right),\quad k\in Z_n.$$

This proves the first representation in (2) if one takes into consideration that

$$F^{-1}\left(\chi_b(x)\,e^{-i\frac{\pi}{b}kx}\right)(y)=(F^{-1}\chi_b)\left(y-\frac{\pi}{b}k\right).$$

Step 2. In the one-dimensional case, i.e. $Q_b=[-b,b]$, we have

$$(F^{-1}\chi_b)(x)=(2\pi)^{-\frac{1}{2}}\int_{-b}^{b}e^{ixy}dy=\left(\frac{2}{\pi}\right)^{\frac{1}{2}}b\,\frac{\sin bx}{bx},\quad x\in R_1.$$

In the n-dimensional case it follows by multiplication that

$$(F^{-1}\chi_b)\left(x-\frac{\pi}{b}k\right)=\left(\frac{2}{\pi}\right)^{\frac{n}{2}}b^n\prod_{j=1}^{n}\frac{\sin(bx_j-k_j\pi)}{bx_j-k_j\pi}, \tag{3}$$

where $x=(x_1,\ldots,x_n)\in R_n$ and $k=(k_1,\ldots,k_n)\in Z_n$. Now the second representation follows from the first one and (3). The absolute convergence of the series in (2) is clear, because $\sum_{k\in Z_n}\left|\varphi\left(\frac{\pi}{b}k\right)\right|<\infty$.

Remark 1. Formula (2) shows that the function $\varphi(x)$ is known if one knows its values at the lattice points $\frac{\pi}{b}k$, where $k\in Z_n$. This provides a better insight into the lattice structure of the inequalities in the preceding subsections. Furthermore, one can understand that $h_0\leq\frac{\pi}{b}$ in Remark 1.3.3: Let $\frac{\pi}{b}<h<h_0$. We assume that (1.3.3/1) holds for $\varphi=F^{-1}\chi_b$; at least for $1<p<\infty$ this is true if (by our assumption); (1.3.3/1) holds for all functions of S^{Q_b} and $h>\frac{\pi}{b}$. Then we can choose x^k in such a way that $\varphi(x^k)=0$ for all $k\in Z_n$, cf. (3) with $k=0$. This is a contradiction. More careful arguments, which cover all $0<p<\infty$, may be found in the references given in Remark 1.3.3: mollification of $F^{-1}\chi_b$ is the key-word.

Remark 2. If $1<p<\infty$ then we have the following addendum to (1.3.3/1): The norms $\|\varphi \mid L_p\|$ and $\left(\sum_{k\in Z_n}\left|\varphi\left(\frac{\pi}{b}k\right)\right|^p\right)^{\frac{1}{p}}$ are mutually equivalent on S^Ω with $\Omega=Q_b$. A proof may be found in [F, 1.4.6]. As far as the lattice-constant $\frac{\pi}{b}$ is concerned, this is the best possible result, cf. Remark 1.

1.4. L_p-Spaces of Analytic Functions

1.4.1. Definition and Main Inequalities

Let K_h be the set of Borel measures defined in 1.2.4.

Definition. Let $\mu\in K_h$ and $0<p\leqq\infty$. If Ω is a compact subset of R_n then

$$L^\Omega_{p,\mu}=\{f \mid f\in S',\ \mathrm{supp}\ Ff\subset\Omega,\ \|f \mid L_{p,\mu}\|<\infty\}\,. \tag{1}$$

If μ is Lebesgue measure, then we put $L^\Omega_{p,\mu}=L^\Omega_p$.

Remark 1. The definition of $\|f \mid L_{p,\mu}\|$ has been given in 1.2.2. Furthermore, we recall that $S'=S'(R_n)$. These are the spaces from (1.1/2). As we shall see $\|f \mid L_{p,\mu}\|$ is a quasi-norm and $L^\Omega_{p,\mu}$ is a quasi-Banach space if $h>0$ is sufficiently small.

Theorem. (i) Let Ω be a compact subset of R_n and let $0<p\leqq\infty$. If $0<r<p$ then there exists a positive constant c such that

$$\left\|\sup_{z\in R_n}\frac{|f(\cdot-z)|}{1+|z|^{\frac{n}{r}}}\ \middle|\ L_p\right\|\leqq c\,\|f \mid L_p\| \tag{2}$$

holds for all $f\in L^\Omega_p$.
(ii) Let Ω be a compact subset of R_n and let $0<p\leqq q\leqq\infty$. There exists a positive number H, which depends on Ω, p, and q, with the following property: if $\nu\in K_h$ and $\mu\in K_h$ with $0<h<H$, and if α is a multi-index, then there exists a positive constant c such that

$$\|D^\alpha f \mid L_{q,\nu}\|\leqq c\,\|f \mid L_{p,\mu}\| \tag{3}$$

holds for all $f\in L^\Omega_{p,\mu}$.

Proof. Step 1. We prove (2). If $p=\infty$, then (2) is obvious. Let $0<r<p<\infty$. Let $\varphi\in S$ with $\varphi(0)=1$ and $\mathrm{supp}\ F\varphi\subset\{y \mid |y|\leqq 1\}$, cf. Remark 2 below. Let $f_\delta(x)=\varphi(\delta x)f(x)$ with $0<\delta<1$ and $f\in L^\Omega_p$. By the two Paley-Wiener-Schwartz theorems from 1.2.1. we have $f_\delta(x)\in S^B$, where B is a closed ball, centered at the origin, such that

$$\{y \mid \exists x \text{ with } x\in\Omega \quad\text{and}\quad |x-y|\leqq 1\}\subset B\,.$$

We apply Theorem 1.3.1 to $f_\delta(x)$ and obtain

$$\frac{|f_\delta(x-z)|}{1+|z|^{\frac{n}{r}}}\leqq c[(M\,|f_\delta|^r)(x)]^{\frac{1}{r}}\leqq c'[(M\,|f|^r)(x)]^{\frac{1}{r}}\,,\quad x\in R_n\,,\quad z\in R_n\,, \tag{4}$$

where c and c' are independent of x and z. We have $\varphi(\delta x)\,\psi(x)\to\psi(x)$ in S if $\delta\downarrow 0$ and $\psi\in S$. Consequently, $f_\delta\to f$ in S' if $\delta\downarrow 0$. Proposition 1.3.2 with $q=\infty$, $\alpha=(0,\ldots,0)$ and $\varphi=f_\delta-f_{\delta'}$ shows that $\{f_\delta\}_{0<\delta<1}$ is fundamental in L_∞. By standard arguments it follows that $f\in L_\infty$ and $f_\delta(x)\Rightarrow f(x)$ (uniform convergence) if $\delta\downarrow 0$. Hence letting $\delta\downarrow 0$ we see that it is meaningful to take the supremum in (4) over all $z\in R_n$. We have

$$\left\| \sup_{z\in R_n} \frac{|f(\cdot - z)|}{1+|z|^{\frac{n}{r}}} \,\middle|\, L_p \right\| \leqq c \|[(M\,|f|^r)(\cdot)]^{\frac{1}{r}} \mid L_p\| = c\,\|M\,|f|^r \mid L_{\frac{p}{r}}\|^{\frac{1}{r}}$$

where $\frac{p}{r}>1$. By (1.2.3/4) it follows that

$$\left\| \sup_{z\in R_n} \frac{|f(\cdot - z)|}{1+|z|^{\frac{n}{r}}} \,\middle|\, L_p \right\| \leqq c\,\||f|^r \mid L_{\frac{p}{r}}\|^{\frac{1}{r}} = c\,\|f \mid L_p\| .$$

This proves (2).

Step 2. We prove (3). If B is the above ball, then we choose $H=h(B, p, q)$, where $h(B, p, q)>0$ has the same meaning as in Theorem 1.3.4. If $f\in L^\Omega_{p,\mu}$, then $f_\delta\in S^B$ has the above meaning. We apply (1.3.4/1) to $\varphi=f_\delta$ to obtain

$$\|D^\alpha f_\delta \mid L_{q,\nu}\| \leqq c\,\|f_\delta \mid L_{p,\mu}\| \leqq c'\,\|f \mid L_{p,\mu}\| . \tag{5}$$

Similarly as in the first step, it follows that $D^\alpha f_\delta(x)\to D^\alpha f(x)$ if $\delta\downarrow 0$ (pointwise convergence) and $D^\alpha f\in L_\infty$. Now (3) is a consequence of (5).

Remark 2. There exist functions $\varphi\in S$ with $\varphi(0)=1$ and $\operatorname{supp} F\varphi\subset\{y \mid |y|\leqq 1\}$: Let $(F\varphi)(x)>0$ if $|x|<1$ and $(F\varphi)(x)=0$ if $|x|\geqq 1$. Then

$$\varphi(0)=(F^{-1}F\varphi)(0)=(2\pi)^{-\frac{n}{2}}\int_{R_n}(F\varphi)(x)\,dx>0 .$$

A normalization of φ yields a function with the desired properties.

Remark 3. Let $p=q$ and $\alpha=(0, \ldots, 0)$ in (3). Then it follows that (1.3.4/3) holds for all $\varphi\in L^\Omega_{p,\mu}$ (equivalent quasi-norms). In particular, we have

$$L^\Omega_{p,\mu}=L^\Omega_{p,\nu}=L^\Omega_p,\quad 0<p\leqq\infty,\quad \nu\in K_h,\quad \mu\in K_h,\quad 0<h<H, \tag{6}$$

where $H=h(B, p, p)$ has the above meaning.

Remark 4. By the above arguments, (1.3.2/5) holds for all $\varphi\in L^\Omega_p$ with $\Omega=B_b$.

1.4.2. Basic Properties

If S^Ω is given by (1.3.1/1), then the topology in S^Ω is generated by the norms (1.2.1/1).

Theorem. Let Ω be a compact subset of R_n and let $0<p\leqq\infty$. Let $0<h<H$, where H has the meaning of Theorem 1.4.1 (ii) (with $q=p$). If $\nu\in K_h$ and $\mu\in K_h$, then

$$L^\Omega_{p,\mu}=L^\Omega_{p,\nu}=L^\Omega_p \quad \text{(equivalent quasi-norms)} \tag{1}$$

holds. Furthermore, L^Ω_p is a quasi-Banach space and

$$S^\Omega\subset L^\Omega_p\subset S' \tag{2}$$

(topological embedding).

Proof. Step 1. Formula (1) coincides with Remark 1.4.1/3, in particular, $\|f \mid L^\Omega_{p,\mu}\|$ is a quasi-norm. We prove (2). If $\varphi\in S^\Omega$ then

$$\|\varphi \mid L_p\| \leqq \|(1+|x|)^{\frac{n+1}{p}}\varphi \mid L_\infty\| \leqq p_N(\varphi),\quad N=1+\left[\frac{n+1}{p}\right].$$

cf. (1.2.1/1). This proves the left-hand side of (2). For the right-hand side, note that if $f \in L_p^\Omega$ and $\varphi \in S$, then

$$|f(\varphi)| = |\int_{R_n} f(x)\, \varphi(x)\, dx| \leq \|f \mid L_\infty\| \, \|\varphi \mid L_1\|$$
$$\leq c \, \|f \mid L_p\| \, \|(1+|x|)^{n+1} \varphi \mid L_\infty\| \leq c \, \|f \mid L_p\| \, p_{n+1}(\varphi) \,,$$

where c is independent of f (we used (1.4.1/3) with $q=\infty$ and $\alpha=(0,\ldots,0)$). This proves the right-hand side of (2).

Step 2. In order to prove that L_p^Ω is a quasi-Banach space we must show that any fundamental sequence in L_p^Ω converges. Let $\{f_j\}_{j=1}^\infty$ be a fundamental sequence in L_p^Ω. By (1.4.1/3) with $q=\infty$ and $\alpha=(0,\ldots,0)$, it follows that $\{f_j\}_{j=1}^\infty$ is a fundamental sequence in L_∞. Hence $f_j \to f$ in L_∞ and, consequently, $f_j \to f$ in S'. Hence $Ff_j \to Ff$ in S' and this proves that $\operatorname{supp} Ff \subset \Omega$. On the other hand, it follows by standard arguments that $f \in L_p$ and $\|f_j - f \mid L_p\| \to 0$ if $j \to \infty$. Hence $f \in L_p^\Omega$ and $f_j \to f$ in L_p^Ω. The proof is complete.

1.4.3. Further Properties

If $1 \leq p \leq \infty$, then L_p^Ω from Definition 1.4.1 is a Banach space. Hence, one can study standard problems in the framework of the Banach space theory: the topological dual, separability, Schauder bases, isomorphism properties, density of subsets etc. In [F, pp. 39–49] we dealt with some of these problems. We recall here some results without proof.

Dual Spaces. Let Ω be a compact subset of R_n and let $-\Omega = \{x \mid -x \in \Omega\}$. Let $1 < p < \infty$ and $\frac{1}{p} + \frac{1}{p'} = 1$. We ask for a representation of $g \in (L_p^\Omega)'$ by

$$g(f) = \int_{R_n} f(x)\, g(x)\, dx \,, \quad \|g \mid (L_p^\Omega)'\| \sim \|g \mid L_{p'}\| \,. \tag{1}$$

Here, "$\sim$" means that the two norms are equivalent to each other. We have

$$g(f) = F^{-1}g(Ff) \quad \text{and} \quad \operatorname{supp} F^{-1}g = -\operatorname{supp} Fg \,. \tag{2}$$

This shows that only the restriction of Fg to $-\Omega$ is of interest in (1). Hence, we ask whether

$$(L_p^\Omega)' = L_{p'}^{-\Omega} \,, \quad 1 < p < \infty, \quad \frac{1}{p} + \frac{1}{p'} = 1 \tag{3}$$

in the sense of (1). This is a rather sophisticated problem, cf. [F, 1.4.5]. We quote two results: (i) If Ω is a rectangle in R_n and if $2 \leq p \leq \infty$, then (3) holds; (ii) If Ω is a ball and $1 < p < 2$, then (3) is not valid. The latter negative result is based on the following well-known assertion by C. Fefferman [2]: the characteristic function of a ball is not a Fourier multiplier in L_p if $1 \leq p \leq \infty$ and $p \neq 2$.

Density. In connection with (1.4.2/2) the problem arises whether S^Ω is dense in L_p^Ω. In some cases, one can give an affirmative answer, cf. [F, 1.4.4]. Let ω be a bounded domain (bounded open set) in R_n, and let $\Omega = \bar{\omega}$ be the closure of ω. We say that ω has the segment property if there exist N open balls K_j and N vectors $y^j \in R_n$ such that

$$\bigcup_{j=1}^N K_j \supset \Omega \,, \quad (\Omega \cap K_j) + ty^j \subset \omega \,, \quad 0 < t \leq 1 \,, \quad j = 1, \ldots, N \,.$$

If ω is a bounded domain with the segment property, $\Omega = \bar{\omega}$, and $0 < p < \infty$, then S^ω (and hence also S^Ω) is dense in L_p^Ω.

Isomorphism Properties and Schauder Bases. In connection with Proposition 1.3.5 one can ask whether (1.3.5/2) can be extended to $\varphi \in L_p^\Omega$ with $\Omega = Q_b$. We recall some results from [F, 1.4.6]: if $1 < p < \infty$ and $\Omega = Q_b$, then (1.3.5/2) holds for all $\varphi \in L_p^\Omega$ (convergence in L_p^Ω); furthermore, L_p^Ω is isomorphic to l_p, and $\left\{(F^{-1}\chi_b)\left(x-\frac{\pi}{b}k\right)\right\}_{k\in Z_n}$ is an unconditional Schauder basis of analytic functions in L_p^Ω. That L_p^Ω is isomorphic to l_p has also been proved in [I, pp. 239/240]. The last assertion is true for arbitrary compact subsets Ω of R_n.

1.5. Fourier Multipliers for L_p-Spaces

1.5.1. Definition and Criterion

Fourier multipliers for function spaces play an important role in this book. If Ω is a compact subset of R_n and if $0<p\leqq\infty$, then L_p^Ω has the above meaning. In [F, 1.5] we studied Fourier multipliers for L_p^Ω and discussed several interpretations of the term "Fourier multiplier". Here we restrict ourselves to a simplified version, which is completely sufficient for our later purposes.

Definition. Let Ω be a compact subset of R_n and let $0<p\leqq\infty$. Let $M\in S'$ and $F^{-1}M\in L_1$. Then M is said to be a Fourier multiplier for L_p^Ω if there exists a constant c such that

$$\|F^{-1}MFf \mid L_p\|\leqq c\,\|f \mid L_p\| \quad \text{for all } f\in L_p^\Omega . \tag{1}$$

Remark 1. For the sake of simplicity we write $F^{-1}MFf=F^{-1}[M\cdot Ff]$. If $f\in L_p^\Omega$ and $F^{-1}M\in L_1$ $(=L_1(R_n))$, then $F^{-1}MFf$ is well-defined: by (1.4.1/3) we have $f\in L_\infty$, and hence

$$(F^{-1}MFf)(x)=c\int_{R_n}(F^{-1}M)(y)\,f(x-y)\,dy\,,\quad x\in R_n\,, \tag{2}$$

makes sense, where c is an appropriate positive number. One can criticize the above definition, because we assume that a Fourier multiplier for L_p^Ω belongs from the very beginning to FL_1, the Fourier image of L_1. However in the framework of the theory of the spaces L_p^Ω, this restriction is not very severe. A more detailed discussion of this somewhat sophisticated question may be found in [F, 1.5.1 and 1.5.6].

Proposition. Let Ω and Γ be compact subsets of R_n. Let $0<p\leqq\infty$ and $\tilde{p}=\min(1,p)$. There exists a constant c such that

$$\|F^{-1}MFf \mid L_p\|\leqq c\,\|F^{-1}M \mid L_{\tilde{p}}\|\,\|f \mid L_p\| \tag{3}$$

holds for all $f\in L_p^\Omega$ and all $F^{-1}M\in L_{\tilde{p}}^\Gamma$.

Proof. By (1.4.1/3) we have $F^{-1}M\in L_1^\Gamma$. Hence, (2) makes sense. If $1\leqq p\leqq\infty$, then (3) follows immediately from (2). Let $0<p<1$ and $f\in L_p^\Omega$. We claim that

$$|(F^{-1}MFf)(x)|^p\leqq c\int_{R_n}|(F^{-1}M)(y)|^p\,|f(x-y)|^p\,dy\,, \tag{4}$$

where c is independent of $x\in R_n$. We have $\operatorname{supp} M\subset\Gamma$ and, by (1.4.1/3), $F^{-1}M\in L_\infty$. Hence, for any fixed $x\in R_n$,

$$(F^{-1}M)(y)\,f(x-y)\in L_p \quad (\text{as a function of } y) \tag{5}$$

and

$$\begin{aligned}\operatorname{supp} F\,((F^{-1}M)(\cdot)\,f(x-\cdot))&\subset\operatorname{supp} M+\operatorname{supp} Ff\,(x-\cdot)\\ &=\operatorname{supp} M-\operatorname{supp} Ff\subset\Gamma-\Omega\,.\end{aligned} \tag{6}$$

Here we used

$$\operatorname{supp} Ff\,(x-\cdot)=-\operatorname{supp} Ff\,(\cdot-x)=-\operatorname{supp} Ff$$

and the notation $-A=\{\zeta \mid -\zeta\in A\}$, where A is a set in R_n. Hence, by (5) and (6), we have

$$(F^{-1}M)(\cdot)f(x-\cdot)\in L_p^{\Gamma-\Omega}, \quad \text{where } x\in R_n .$$

Now (4) is a consequence of (2) and (1.4.1/3) with $q=1$ and $0<p<1$. Finally, (3) follows from (4).

Remark 2. By (1.4.1/3) we have $F^{-1}M\in L_1$ in any case. Hence $M(x)=FF^{-1}M$ is a continuous bounded function on R_n with supp $M(x)\subset\Gamma$. Because supp $Ff\subset\Omega$, it is clear that only the values of $M(x)$ with $x\in\Gamma\cap\Omega$ are of interest. This justifies the restriction to Fourier multipliers $M(x)$ with compact support.

1.5.2. A Multiplier Theorem

If s is a real number, then

$$H_2^s=\{f \mid f\in S',\ \|f \mid H_s^2\|=\|(1+|x|^2)^{\frac{s}{2}}(Ff)(x) \mid L_2\|<\infty\}\,. \tag{1}$$

Again, we write H_2^s instead of $H_2^s(R_n)$, because n is fixed once and for all in this chapter. By the well-known fact that F is a unitary operator in L_2 we have

$$\|f \mid H_2^s\|=\|F^{-1}(1+|x|^2)^{\frac{s}{2}}Ff \mid L_2\|\,. \tag{2}$$

We recall that

$$F^{-1}(1+|x|^2)^{\frac{s}{2}}Ff=F^{-1}[(1+|x|^2)^{\frac{s}{2}}\cdot Ff(x)]\,.$$

Furthermore, H_2^s are special Bessel-potential spaces (Liouville spaces, Lebesgue spaces), which will be studied later on. Sometimes these spaces are also denoted by $W_2^s=H_2^s$ (Sobolev-Slobodeckij spaces, at least if $s\geqq 0$). If $s=m=0, 1, 2, \ldots$, then W_2^m is the usual Sobolev space,

$$W_2^m=\left\{f \mid f\in S',\ \|f \mid W_2^m\|=\Big(\sum_{|\alpha|\leqq m}\|D^\alpha f \mid L_2\|^2\Big)^{\frac{1}{2}}<\infty\right\}. \tag{3}$$

We shall not discuss the properties of these spaces because they are very special (almost classical) cases of the spaces which are treated in Chapter 2. But we wish to mention that s in H_2^s is a smoothness index: as s increases, then the functions belonging to H_2^s become smoother, cf. (3). Of course, $H_2^{s_0}\subset H_2^{s_1}$ if $s_0>s_1$ and $H_2^0=L_2$.

Theorem. Let Ω be a compact subset of R_n and let $0<p\leqq\infty$. Let $\sigma_p=n\left(\frac{1}{\min(p,1)}-\frac{1}{2}\right)$. If $s>\sigma_p$, then there exists a constant c such that

$$\|F^{-1}MFf \mid L_p\|\leqq c\,\|M \mid H_2^s\|\,\|f \mid L_p\| \tag{4}$$

for all $f\in L_p^\Omega$ and all $M\in H_2^s$.

Proof. Step 1. Let us assume that (4) is valid for all $M\in H_2^s$ with

$$\text{supp}\, M\subset\Omega^1=\{y \mid \exists x\in\Omega \quad \text{with} \quad |x-y|\leqq 1\}\,. \tag{5}$$

Let $\psi\in S$ with supp $\psi\subset\Omega^1$ and $\psi(x)=1$ if $x\in\Omega$. If $M\in H_2^s$ is an arbitrary function, then

$$F^{-1}MFf=F^{-1}\psi MFf, \quad f\in L_p^\Omega\,. \tag{6}$$

By our assumption, (4) with ψM instead of M is valid. Finally we remark that

$$\|\psi M \mid H_2^s\| \leqq c \, \|M \mid H_2^s\| , \tag{7}$$

where c depends only on ψ (and not on M). This is a well-known fact, which is a special case of more general assertions proved in Chapter 2. But (4) with ψM instead of M, (6) and (7) prove (4) for all $M \in H_2^s$.

Step 2. In this step we prove the following assertion: if $0 < p \leqq 1$ and $s > \sigma_p$, then there exists a constant c such that

$$\|FM \mid L_p\| \leqq c \, \|M \mid H_2^s\| \tag{8}$$

holds for all $f \in H_2^s$. Let

$$K_0 = \{x \mid |x| \leqq 1\}, \quad K_j = \{x \mid 2^{j-1} \leqq |x| \leqq 2^j\} \quad \text{for} \quad j = 1, 2, 3, \ldots \tag{9}$$

Then we have

$$\begin{aligned} \|FM \mid L_p\|^p &= \sum_{j=0}^{\infty} \int_{K_j} |(FM)(x)|^p \, dx \\ &\leqq c \sum_{j=0}^{\infty} 2^{jn\frac{2-p}{2}} \left(\int_{K_j} |(FM)(x)|^2 \, dx \right)^{\frac{p}{2}} \\ &= c \sum_{j=0}^{\infty} \left[2^{j\sigma_p} \left(\int_{K_j} |(FM)(x)|^2 \, dx \right)^{\frac{1}{2}} \right]^p \\ &\leqq c' \left[\sum_{j=0}^{\infty} 2^{2js} \int_{K_j} |(FM)(x)|^2 \, dx \right]^{\frac{p}{2}} \leqq c'' \, \|f \mid H_2^s\|^p . \end{aligned} \tag{10}$$

This proves (8).

Step 3. Let $M \in H_2^s$ with $s > \sigma_p$ and supp $M \subset \Omega^1$, cf. (5). If $\tilde{p} = \min(1, p)$, then (8) yields $F^{-1}M \in L_{\tilde{p}}^{\Omega^1}$. Now (4) follows from Proposition 1.5.1 and (8). The proof is complete.

Remark 1. In Step 2 we proved a bit more than stated: If $0 < p \leqq 2$, then

$$\|FM \mid L_p\| \leqq c \left(\sum_{j=0}^{\infty} \left[2^{jn\left(\frac{1}{p} - \frac{1}{2}\right)} \left(\int_{K_j} |(FM)(x)|^2 \, dx \right)^{\frac{1}{2}} \right]^p \right)^{\frac{1}{p}}, \tag{11}$$

cf. 1.5.4.

Remark 2. The above theorem is comparatively simple, but it reminds us of the famous Michlin-Hörmander multiplier theorem for the L_p-spaces, $1 < p < \infty$. If $m_p = [\sigma_p] + 1$, then (3) and (4) yield

$$\|F^{-1}MFf \mid L_p\| \leqq c \left(\sum_{|\alpha| \leqq m_p} \|D^\alpha M \mid L_2\|^2 \right)^{\frac{1}{2}} \|f \mid L_p\| , \tag{12}$$

where c is independent of M and f.

Remark 3. As in Remark 1.3.2/1 we add a homogeneity consideration. Let $0 < p \leqq \infty$ and let

$$\Omega = B_b = \{y \mid |y| \leqq b\} \quad \text{with} \quad b > 0.$$

If $f \in L_p^\Omega$ then we have $f(b^{-1}x) \in L_p^\omega$, where $\omega = B_1$, cf. (1.3.2/6) (which holds for any $\varphi \in S'(R_n)$) and the argument given there. Let $M(x) \in H_2^s$ with $s > \sigma_p$, where σ_p has the same meaning as in the theorem. Iteration of (1.3.2./6) (extended to $S'(R_n)$) yields

$$(F^{-1}MFf)(x) = \{F^{-1}[M(b\,\cdot)\,(Ff(b^{-1}\,\cdot))(\cdot)]\}(bx) .$$

Now, by (4) with ω instead of Ω and $f(b^{-1}\cdot)$ instead of f we have

$$\|F^{-1}MFf \mid L_p\| = b^{-\frac{n}{p}} \|F^{-1}[M(b\,\cdot)\,(Ff(b^{-1}\cdot))\,(\cdot)] \mid L_p\| \tag{13}$$
$$\le cb^{-\frac{n}{p}} \|M(b\,\cdot) \mid H_2^s\| \cdot \|f(b^{-1}\cdot) \mid L_p\| = c\|M(b\,\cdot) \mid H_2^s\|\,\|f \mid L_p\|,$$

where c is independent of b and $f \in L_p^\Omega$.

1.5.3. Convolution Algebras

If $f \in L_1$ and $g \in L_1$ then the convolution $f*g$ is defined by

$$(f*g)(x) = \int_{R_n} f(x-y)\,g(y)\,dy\,. \tag{1}$$

If c is an appropriate positive number, then (1) can be rewritten as $f*g = cF^{-1}[Ff \cdot Fg]$, and one can define $f*g$ by the last formula for more general distributions f and g if $F^{-1}[Ff \cdot Fg]$ makes sense.

Proposition. Let Ω be a compact subset of R_n and let $0<p\le 1$. Then $f*g$, with $f \in L_p^\Omega$ and $g \in L_p^\Omega$, is well-defined. Furthermore, there exists a constant c such that

$$\|f*g \mid L_p\| \le c\,\|f \mid L_p\|\,\|g \mid L_p\| \quad \text{for all} \quad f \in L_p^\Omega \quad \text{and all} \quad g \in L_p^\Omega\,. \tag{2}$$

Proof. If $M = Ff$ and $\Gamma = \Omega$, then (2) is an immediate consequence of Proposition 1.5.1.

Remark 1. If the multiplication is defined by the convolution, then all the usual properties of an algebra are satisfied. In other words: L_p^Ω with $0<p\le 1$ is a quasi-normed convolution algebra. Assertions of this type have been proved in J. Peetre [8, p. 237] and [F, 1.5.3].

Remark 2. Let $\Omega = B_b = \{y \mid |y| \le b\}$ with $b>0$. If $f \in L_p^\Omega$ and $g \in L_p^\Omega$, then we apply (2) to $f(b^{-1}x)$ and $g(b^{-1}x)$ instead of f and g, respectively, cf. Remark 1.5.2/3. Elementary transformations of coordinates, cf. also (1), show that there exists a constant c such that

$$\|f*g \mid L_p\| \le cb^{n\left(\frac{1}{p}-1\right)} \|f \mid L_p\|\,\|g \mid L_p\| \tag{3}$$

holds for all $b>0$, all $f \in L_p^\Omega$ and all $g \in L_p^\Omega$.

1.5.4. Further Multiplier Assertions

Improvement of Theorem 1.5.2. In 2.3.1. we define the spaces $B_{u,v}^s = B_{u,v}^s(R_n)$, where $s \in R_1$, $0<u\le\infty$, and $0<v\le\infty$. If $u=2$, then

$$\|F^{-1}\varphi_j Ff \mid L_2\| = \|\varphi_j Ff \mid L_2\|,$$

where φ_j has the same meaning as in 2.3.1., and it is not hard to see that the right-hand ide of (1.5.2/11) is an equivalent quasi-norm in $B_{2,p}^{n\left(\frac{1}{p}-\frac{1}{2}\right)}$. Hence (1.5.2/11) can be re-formuated as follows. If $0<p\le 2$, then

$$\|FM \mid L_p\| \le c\|f \mid B_{2,p}^{n\left(\frac{1}{p}-\frac{1}{2}\right)}\|\,. \tag{1}$$

Estimates of this type are known, cf. J. Peetre [1, pp. 9–11] and [F, pp. 60/61]. With the help of this inequality, Theorem 1.5.2 can be strengthened: if $0<p\le\infty$, and $\tilde{p} = \min(1,p)$, then there exists a constant c such that

$$\|F^{-1}MFf \mid L_p\| \le c\|M \mid B_{2,\tilde{p}}^{\sigma_p}\|\,\|f \mid L_p\|$$

for all $f \in L_p^\Omega$ and all $M \in B_{2,\tilde{p}}^{\sigma_p}$.

Discrete Multiplier Assertions. Until now we have not used the lattice structure of the spaces L_p^Ω, cf. (1.4.1/6) and (1.3.3/1). If Ω is a compact subset of R_n and $0<p\leqq\infty$, then $h_0=h_0(\Omega, p)$ has the meaning of Proposition 1.3.3. Let $a>\frac{\pi}{h_0}$ and

$$Q_a=\{y \mid |y_j|\leqq a\,;\quad (j=1,\ldots,n)\}\,.$$

Let $M\in L_\infty$ and $M_k=(F^{-1}M)\left(\frac{\pi}{a}k\right)$ with $k\in Z_n$. Then $M(x)$ is a Fourier multiplier in L_p^Ω (in a somewhat wider sense as in Definition 1.5.1, cf. [F, pp. 51/52]) if and only if there exists a constant c such that

$$\left(\sum_{k\in Z_n}\left|\sum_{l\in Z_n} M_{k-l}f\left(\frac{\pi}{a}l\right)\right|^p\right)^{\frac{1}{p}}\leqq c\left(\sum_{k\in Z_n}\left|f\left(\frac{\pi}{a}k\right)\right|^p\right)^{\frac{1}{p}} \tag{3}$$

(modification for $p=\infty$) holds for all $f\in L_p^\Omega$. Proofs and explanations may be found in [F, 1.5.2]. In [F] we started with (3) and proved (1.5.1/3) and (1.5.2/4) afterwards.

Non-Multipliers. The question arises whether (1.5.2/4) and (2) are sharp. If $1\leqq p\leqq\infty$, then $\sigma_p=\frac{n}{2}$. In [F, 1.5.7] we proved the following assertion: if $1<p<\infty$, $1<q<\infty$, and $\Omega=Q_b$ with $b>0$, then there exists a function $M\in B^{\frac{n}{2}}_{2,q}$ which is not a Fourier multiplier for L_p^Ω.

1.6. $L_p(l_q)$-Spaces of Analytic Functions

1.6.1. Definition and Basic Properties

We recall the abbreviations introduced in 1.2.2. In particular, $\|f_k \mid L_p(l_q)\|$ has the meaning of (1.2.2/5).

Definition. Let $0<p\leqq\infty$ and $0<q\leqq\infty$. If $\Omega=\{\Omega_k\}_{k=0}^\infty$ is a sequence of compact subsets of R_n, then

$$L_p^\Omega(l_q)=\{f \mid f=\{f_k\}_{k=0}^\infty\subset S'\,,\quad \operatorname{supp} Ff_k\subset\Omega_k \quad\text{if}\quad k=0,1,2,\ldots \text{ and } \|f_k \mid L_p(l_q)\|<\infty\}\,. \tag{1}$$

Remark. As in the preceding sections we omit R_n in the notation of the spaces, i.e. we write $L_p^\Omega(l_q)$ instead of $L_p^\Omega(R_n, l_q)$. In contrast to Definition 1.4.1 we restrict ourselves from the very beginning to Lebesgue measure.

Theorem. Under the hypotheses of the above definition, $L_p^\Omega(l_q)$ is a quasi-Banach space with the quasi-norm $\|f_k \mid L_p(l_q)\|$.

Proof. It is obvious that $\|f_k \mid L_p(l_q)\|$ is a quasi-norm. Furthermore, by Theorem 1.4.2, any space $L_p^{\Omega_k}$ is complete. Then it follows by standard arguments that $L_p^\Omega(l_q)$ is complete too.

1.6.2. Maximal Inequalities

Let $\Omega=\{\Omega_k\}_{k=0}^\infty$ be a sequence of compact subsets of R_n. Let $d_k=\sup_{x\in\Omega_k,\, y\in\Omega_k}|x-y|$ be the diameter of Ω_k. We always assume that $d_k>0$, i.e. Ω_k contains at least two different points.

Theorem. Let $0<p<\infty$ and $0<q\leq\infty$. Let $\Omega=\{\Omega_k\}_{k=0}^\infty$ be a sequence of compact subsets of R_n. Let $d_k>0$ be the diameter of Ω_k. If $0<r<\min(p, q)$, then there exists a constant c such that

$$\left\| \sup_{z\in R_n} \frac{|f_k(\cdot-z)|}{1+|d_k z|^{\frac{n}{r}}} \,\Big|\, L_p(l_q) \right\| \leq c\,\|f_k \mid L_p(l_q)\| \tag{1}$$

holds for all $f\in L_p^\Omega(l_q)$, where $f=\{f_k\}$.

Proof. Step 1. Let $y^k\in\Omega_k$ and $h_k(x)=e^{-ixy^k}f_k(x)$, where xy^k is the scalar product of $x\in R_n$ and $y^k\in R_n$. Furthermore, $\{f_k\}\in L_p^\Omega(l_q)$. Then we have $(Fh_k)(x)=(Ff_k)(x+y^k)$. Hence,

$$\operatorname{supp} Fh_k\subset\Omega_k-y^k\,.$$

If (1) holds for $\{f_k\}\in L_p^\Omega(l_q)$, then we have a corresponding inequality for $\{h_k\}$, where Ω is replaced by $\{\Omega_k-y^k\}_{k=0}^\infty$ and vice versa. Hence, we may assume that $0\in\Omega_k$. Then it is sufficient to consider the case $\Omega_k=D_k=\{y\mid |y|\leq d_k\}$.

Step 2. We must prove (1) with $\Omega_k=D_k=\{y\mid |y|\leq d_k\}$ and $d_k>0$. If $\{f_k\}\in L_p^\Omega(l_q)$, then $f_k\in L_p^{D_k}$. If $g_k(x)=f_k(d_k^{-1}x)$, then

$$(Fg_k)(\xi)=d_k^n(Ff_k)(d_k x) \quad\text{and}\quad \operatorname{supp} Fg_k\subset\{y\mid |y|\leq 1\}\,.$$

Hence, by the arguments in the first step of the proof of Theorem 1.4.1, we have

$$\frac{|g_k(x-z)|}{1+|z|^{\frac{n}{r}}}\leq c[(M|g_k|^r)(x)]^{\frac{1}{r}},\quad x\in R_n,\quad z\in R_n\,,$$

and

$$\frac{|f_k(x-z)|}{1+|d_k z|^{\frac{n}{r}}}\leq c[(M|f_k|^r)(x)]^{\frac{1}{r}},\qquad x\in R_n,\quad z\in R_n\,, \tag{2}$$

where c is independent of x, z, and $k=0, 1, 2, \ldots$ Here we used the fact that

$$(Mh(c\,\cdot))(\xi)=(Mh)(c\xi),\quad c>0\,.$$

If $q<\infty$, then it follows that

$$\left\| \sup_{z\in R_n} \frac{|f_k(\cdot-z)|}{1+|d_k z|^{\frac{n}{r}}} \,\Big|\, L_p(l_q) \right\| \leq c\|[(M|f_k|^r)(\cdot)]^{\frac{1}{r}}\mid L_p(l_q)\| = c\|M|f_k|^r\mid L_{\frac{p}{r}}(l_{\frac{q}{r}})\|^{\frac{1}{r}}\,.$$

Because $0<r<\min(p, q)$ we have $\frac{p}{r}>1$ and $\frac{q}{r}>1$. We apply Theorem 1.2.3 to obtain

$$\left\| \sup_{z\in R_n} \frac{|f_k(\cdot-z)|}{1+|d_k z|^{\frac{n}{r}}} \,\Big|\, L_p(l_q) \right\| \leq c\||f_k|^r\,| L_{\frac{p}{r}}(l_{\frac{q}{r}})\|^{\frac{1}{r}} = c\|f_k\mid L_p(l_q)\|\,. \tag{3}$$

This proves (1) with $\Omega_k=D_k=\{y\mid |y|\leq d_k\}$ and $q<\infty$. If $q=\infty$ then (2) yields

$$\sup_k \frac{|f_k(x-z)|}{1+|d_k z|^{\frac{n}{r}}}\leq c\,\sup_k\,[(M\mid f_k(\cdot)|^r)(x)]^{\frac{1}{r}}$$

$$\leq c[(M\mid \sup_k |f_k(\cdot)|\,|^r)(x)]^{\frac{1}{r}},\quad x\in R_n,\quad z\in R_n\,.$$

Hence,

$$\left\| \sup_k \sup_{z\in R_n} \frac{|f_k(\cdot - z)|}{1+|d_k z|^{\frac{n}{r}}} \,\Big|\, L_p \right\| \leqq c \left\| M \,\Big|\, \sup_k |f_k(\cdot)|^r \,\Big|\, L_{\frac{p}{r}} \right\|^{\frac{1}{r}}.$$

Now we apply (1.2.3/4) with $\sup_k |f_k(x)|^r$ instead of $f(x)$ and $\frac{p}{r}>1$ instead of p. This proves (3) with $q=\infty$ and the proof is complete.

Remark. In contrast to 1.6.1. we restricted p in the above theorem to $p<\infty$. The case $p=q=\infty$ is trivial. What about the case $p=\infty$, $0<q<\infty$? If the above theorem would be valid for $p=\infty$, $0<q<\infty$, then the Fourier multiplier theorem which will be proved in the following subsection, would be valid, also for $p=\infty$ and $0<q<\infty$. But in [S, 2.1.4] we proved that this is impossible. In other words, the above theorem cannot be extended to $p=\infty$, $0<q<\infty$. As a consequence, one obtains that the vector-valued maximal inequality from Theorem 1.2.3 cannot be extended to $p=\infty$, $1<q<\infty$. The last assertion is well-known, cf. C. Fefferman, E. M. Stein [1].

1.6.3. A Multiplier Theorem

The spaces H_2^s have been defined in 1.5.2.

Theorem. Let $0<p<\infty$, $0<q\leqq\infty$. Let $\Omega=\{\Omega_k\}_{k=0}^\infty$ be a sequence of compact subsets of R_n. Let $d_k>0$ be the diameter of Ω_k. If $\varkappa>\frac{n}{2}+\frac{n}{\min(p,q)}$, then there exists a constant c such that

$$\|F^{-1}M_kFf_k \mid L_p(l_q)\| \leqq c \sup_l \|M_l(d_l\,\cdot) \mid H_2^\varkappa\| \cdot \|f_k \mid L_p(l_q)\| \tag{1}$$

holds for all systems $\{f_k(x)\}_{k=0}^\infty \in L_p^\Omega(l_q)$ and all sequences $\{M_k(x)\}_{k=0}^\infty \subset H_2^\varkappa$.

Proof. First we remark that $F^{-1}M_kFf_k=F^{-1}[M_k\cdot Ff_k]$ is well-defined: By (1.5.2/8) we have $F^{-1}M_k\in L_1$. Furthermore, $f_k\in L_p^{\Omega_k}$, and hence by (1.4.1/3), $f_k\in L_\infty$. Remark 1.5.1/1 shows that $F^{-1}M_kFf_k$ makes sense. Next we want to prove that

$$\sup_{z\in R_n} \frac{|F^{-1}M_kFf_k(x-z)|}{1+|d_kz|^{\frac{n}{r}}} \leqq c \sup_{z\in R_n} \frac{|f_k(x-z)|}{1+|d_kz|^{\frac{n}{r}}} \| M_k(d_k\cdot) \mid H_2^\varkappa \| \tag{2}$$

holds where $0<r<\min(p,q)$ and $\varkappa>\frac{n}{2}+\frac{n}{r}$. Here c is independent of x and k (and the functions M_k and f_k). By (1.5.1/2) and the above remarks we have

$$|(F^{-1}M_kFf_k)(x-z)| \leqq c \int_{R_n} |(F^{-1}M_k)(x-z-y)|\,|f_k(y)|\,dy \tag{3}$$

$$\leqq c \sup_{u\in R_n} \frac{|f_k(u)|}{1+|d_k(x-u)|^{\frac{n}{r}}} \int_{R_n} |(F^{-1}M_k)(x-z-y)|\,(1+|d_k(x-y)|^{\frac{n}{r}})\,dy\,.$$

But

$$1+|d_k(x-y)|^{\frac{n}{r}} \leqq c\,(1+|d_k(x-y-z)|^{\frac{n}{r}})\,(1+|d_kz|^{\frac{n}{r}})\,.$$

Substituting this estimate into (3) and dividing both sides by $1+|d_kz|^{\frac{n}{r}}$, we find that

$$\sup_{z\in R_n} \frac{|(F^{-1}M_kFf_k)(x-z)|}{1+|d_kz|^{\frac{n}{r}}} \leqq c \sup_{z\in R_n} \frac{|f_k(x-z)|}{1+|d_kz|^{\frac{n}{r}}} \int_{R_n} |(F^{-1}M_k)(y)|\,(1+|d_ky|^{\frac{n}{r}})\,dy\,. \tag{4}$$

Using $(F^{-1}M_k(d_k\cdot))(y)=d_k^{-n}(F^{-1}M_k)(d_k^{-1}y)$ and $\varkappa>\frac{n}{r}+\frac{n}{2}$, we then obtain

$$\int_{R_n}|(F^{-1}M_k)(y)|\,(1+|d_ky|^{\frac{n}{r}})\,dy=\int_{R_n}|(F^{-1}M_k(d_k\cdot))(y)|\,(1+|y|^{\frac{n}{r}})\,dy$$
$$\leqq c\,[\int_{R_n}(1+|y|^{\varkappa})^2\,|\,(F^{-1}M_k(d_k\cdot))(y)|^2\,dy]^{\frac{1}{2}}\leqq c'\|M_k(d_k\cdot)\mid H_2^{\varkappa}\|\,.$$

Putting this estimate into (4), we obtain (2). Now (1) is an immediate consequence of (2), Theorem 1.6.2 and the trivial estimate

$$|(F^{-1}M_kFf_k)(x)|\leqq\sup_{z\in R_n}\frac{|(F^{-1}M_kFf_k)(x-z)|}{1+|d_kz|^{\frac{n}{r}}}.$$

Remark. The scalar case of (1) is related to (1.5.2/13). However (1.5.2/13) holds also for $p=\infty$ and the restrictions for $s=\varkappa$ are more natural, cf. also the following subsection.

1.6.4. Further Multiplier Assertions

The number σ_p in Theorem 1.5.2 and in (1.5.4/2) is natural. The vector-valued counterpart is

$$\sigma_{p,q}=n\left(\frac{1}{\min(1,p,q)}-\frac{1}{2}\right). \tag{1}$$

This number is smaller than the critical number $n\left(\frac{1}{2}+\frac{1}{\min(p,q)}\right)$ in Theorem 1.6.3. The reason for this awkward difference is the use of the technique of maximal functions: it spoils some constants and this has the above effect, cf. the proof of Theorem 1.6.3. However, a combination of Theorem 1.6.3, a Fourier multiplier theorem of Michlin-Hörmander type for $L_p(l_q)$ with $1<p<\infty$, $1<q<\infty$, and complex interpolation yields the desired result: *the assertion of Theorem* 1.6.3 *is valid for all* $\varkappa>\sigma_{p,q}$, *cf.* 2.4.9.

2. Function Spaces on R_n

2.1. Introduction

As we mentioned in the preface, this chapter is the heart of the book. We give a thorough description of the two scales $B^s_{p,q}(R_n)$ and $F^s_{p,q}(R_n)$ of spaces on R_n. On this basis we switch over later on from R_n to smooth bounded domains Ω in R_n, study the spaces $B^s_{p,q}(\Omega)$ and $F^s_{p,q}(\Omega)$ (Chapter 3), and apply the results to regular elliptic boundary value problems (Chapter 4). The spaces $B^s_{p,q}$ and $F^s_{p,q}$ contain many well-known spaces as special cases: Besov-Lipschitz, Hölder, Zygmund, Sobolev, Bessel-potential, Lebesgue and Hardy spaces, and also spaces of functions of bounded mean oscillation. All these spaces have their own definitions, but we give for them (and their generalizations $B^s_{p,q}$ and $F^s_{p,q}$) a unified approach via decomposition methods in the framework of Fourier analysis. The advantage is clear: one deals with all these apparently rather diversified spaces from a common point of view. The disadvantage is that the formal definition of $B^s_{p,q}$ and $F^s_{p,q}$ looks somewhat complicated at first glance. In order to provide a better understanding of what happens we give in Section 2.2. some (more or less heuristic) motivations. We also add a few historical comments and list the above-mentioned spaces. The rigorous definition of $B^s_{p,q}(R_n)$ and $F^s_{p,q}(R_n)$ is given in 2.3., where the chapter starts in earnest. The rest of the chapter (which is the main part) deals with properties of these spaces.

2.2. The Historical Background, Motivations, and Principles

2.2.1. On the History of Function Spaces

In this Section 2.2. "function space" always means a normed or quasi-normed space of functions and distributions on R_n, which is (with the usual technical terms) unweighted, isotropic and non-homogeneous. The historical remarks below and the list of spaces given in 2.2.2. deal mostly with spaces of such type. There is hardly any doubt that this type of space was and is the backbone (in the widest sense of the word) of the theory of function spaces from the very beginning up to our time. Corresponding spaces, defined on bounded smooth domains in R_n, will be considered in Chapter 3. Also, in Part II, we deal briefly with their homogeneous, periodic and weighted counterparts. A lot of other spaces, which have attracted much attention in recent years, remain outside the scope of this book, e. g. Lorentz spaces, Campanato-Morrey spaces, Orlicz spaces, Orlicz-Sobolev spaces etc. An excellent survey of the realm of function spaces in this wider sense has been given by A. Kufner, O. John, S. Fučik [1]. Cf. also the books mentioned in the Preface.

The history of function spaces (in the above restricted sense) can be divided into three periods. The first (basic) period starts at the beginning of our century and ends in the mid-thirties. During that time, the classical basic spaces L_p of p-integrable functions and C^m of m times differentiable functions were thoroughly investigated (here $m=0, 1, 2, \ldots$; we set $C^0=C$; definitions of all the spaces mentioned here may be found in 2.2.2.) The Hölder spaces C^s with $0<s\neq$ integer and the Hardy spaces H_p with $0<p<\infty$ belong to that period and anticipate the second (constructive) period. Originally, the Hardy spaces were introduced

H. Triebel, *Theory of Function Spaces*, Modern Birkhäuser Classics,
DOI 10.1007/978-3-0346-0416-1_2, © Birkhäuser Verlag 1983

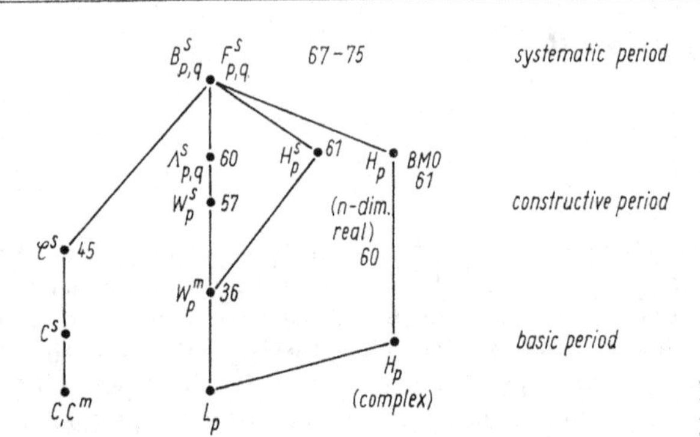

Fig. 2.2.1 (45 means the year 1945 etc.)

in order to characterize the boundary values of analytic functions in a disk in the complex plane. The close connection between the Hardy spaces and the methods of Fourier analysis was quite clear from the beginning. The second (constructive) period starts with the papers by S. L. Sobolev [1–3] in 1935–1938, where the spaces that are nowadays called the Sobolev spaces W^m_p, with $m=0, 1, 2, \ldots$, were introduced. A new tool, the theory of distributions, was discovered and new techniques and results (e. g. embedding theorems) were used successfully in order to investigate partial differential equations. The second period is characterized by more or less speculative constructions of many new spaces with the help of explicit norms (and quasi-norms) on the basis of the above classical spaces. This period culminated at the end of the fifties and in the sixties, and flourishes also in our time. First of all we must note the direct descendants of the above-mentioned classical spaces (which include the Hölder spaces, Hardy spaces and Sobolev spaces):
(i) *The Zygmund spaces* (*or classes*) $\mathcal{C}^s$. These are extensions of the Hölder spaces C^s to values $s=1, 2, 3, \ldots$ The beginning is marked by A. Zygmund [1], who discovered the remarkable fact that sometimes second differences $\Delta^2_h f$ of functions f (cf. 2.2.2. for the definition) are much more useful than first differences $\Delta^1_h f$.
(ii) *The Slobodeckij spaces* W^s_p with $0<s\neq$ integer. These constitute an attempt to fill the gaps between the Sobolev spaces L_p, W^1_p, $W^2_p, \ldots$ in the spirit of the Hölder spaces: use of the first differences $\Delta^1_h f$. A second, and even more convincing, reason for the introduction of the spaces W^s_p with fractional s is the fact that these spaces are necessary for the description of the exact traces on hyper-planes in R_n of functions belonging to Sobolev spaces. These spaces were introduced simultaneously by N. Aronszajn [1], L. N. Slobodeckij [1] and E. Gagliardo [1].
(iii) *The Besov* (*or Lipschitz*) *spaces* $\Lambda^s_{p,q}$. The combination of the above ideas (Zygmund's proposal to use second differences $\Delta^2_h f$ instead of first differences $\Delta^1_h f$, and the replacement of the sup-norm in the Hölder spaces by the L_p-norm in the Slobodeckij spaces) yields essentially the spaces $\Lambda^s_{p,q}$, defined by O. V. Besov [1, 2]. However two points should be emphasized. Firstly, there was from the very beginning a close connection with problems in approximation theory. We shall return to that point in 2.2.5. Secondly, the spaces $\Lambda^s_{p,\infty}$ (sometimes called Nikol'skij spaces) had been investigated in the early fifties by S. M. Nikol'skij [1].
(iv) *The Bessel-potential (or Lebesgue or Liouville) spaces* H^s_p, where s is real. This is another attempt to fill the gaps between the Sobolev spaces L_p, W^1_p, $W^2_p, \ldots$, because $H^m_p = W^m_p$ if $m=1, 2, 3, \ldots$ Instead of derivatives and (first and higher) differences (as in the above spaces) the spaces H^s_p are defined via Fourier transforms. This seemed to be reasonable after L. Schwartz discovered in the late forties the importance of the Fourier transform in the framework of the theory of distributions.

The spaces H^s_p were introduced by N. Aronszajn, K. T. Smith [1] and A. P. Calderón [1].
(v) Finally we mention *the space BMO* of functions of bounded mean oscillation, cf. F. John, L. Nirenberg [1], and the n-dimensional real version of the Hardy spaces H_p due to E. M. Stein, G. Weiss [1].

Besides these spaces, many extensions, generalizations, and modifications have been treated extensively during the last 20 years: anisotropic spaces, weighted spaces, spaces with dominating derivatives, Lorentz spaces, Campanato-Morrey spaces, Orlicz spaces, Orlicz-Sobolev spaces etc. The third (systematic) period begins in the sixties and overlaps heavily with the constructive period. This inflation of the number of spaces demands simple and far-reaching methods which will enable us to deal with function spaces (either in the above restricted or in a wider sense) from the point of view of a few general principles. The first remarkable success was achieved in the framework of the abstract interpolation theory of Banach spaces and its applications to function spaces (E. Gagliardo, J. L. Lions, E. Magenes, J. Peetre, A. P. Calderón, 1960–1964). A thorough description of the theory of function spaces from the standpoint of interpolation theory and detailed references have been given in [I], in particular in Chapter 2. Starting with simple spaces (e. g. L_p, C^m, W^m_p) one obtains on the basis of general abstract procedures (interpolation methods) a lot of new spaces (e. g. $\Lambda^s_{p,q}$ and H^s_p) in a systematic way. The disadvantage of this method is that one needs "endpoint" spaces. At the end of the sixties and the beginning of the seventies, new methods (and essentially improved old methods) of Fourier analysis and of the theory of the so-called maximal inequalities gave a new impetus in the theory of function spaces (J. Peetre, E. M. Stein, C. Fefferman in 1967–1975, detailed references will be given later on). With the help of these new far-reaching powerful techniques one can deal with all the above-mentioned function spaces in (i)–(v) (cf. also Fig. 2.2.1) from unified point of view and on the basis of very few general principles. All these spaces are contained in the two scales $B^s_{p,q}$ and $F^s_{p,q}$, which are defined in the framework of methods of Fourier analysis. (In the case of the Hardy spaces H_p and BMO one needs the so-called "local" versions, h_p and bmo, respectively. H_p and BMO are special homogeneous spaces of the type $\dot{F}^s_{p,q}$, cf. Chapter 5.)

Remark. References concerning the spaces $B^s_{p,q}$ and $F^s_{p,q}$ will be given in 2.3.5. Of course, the above historical remarks about the "constructive" function spaces are very rough. Definitions of these spaces are given in the following subsection. In 2.3.5. we describe how the "basic" and "constructive" spaces are related with $B^s_{p,q}$ and $F^s_{p,q}$. More detailed historical references concerning the "basic" and "constructive" spaces may be found in the note sections of S. M. Nikol'skij [3], A. Kufner, O. John, S. Fučik [1, in particular Chapter 8] and [I, in particular Chapter 2]. The theory of the "constructive" spaces (and also of the "basic" spaces) has been treated extensively from several points of view in the books by R. A. Adams [1], O. V. Besov, V. P. Il'in, S. M. Nikol'skij [1], A. Kufner, O. John, S. Fučik [1], S. M. Nikol'skij [3], E. M. Stein [1], E. M. Stein, G. Weiss [2] and [I]. Systematic treatments in the framework of interpolation theory or Fourier analysis may be found in J. Peetre [8], [I], [F] and [S].

2.2.2. The Constructive Spaces

We summarize the spaces from (i)–(v) of the preceding subsection as constructive spaces. We use the notations introduced in 1.2.1. and 1.2.2., in particular R_n, $S'(R_n)$, $L_p(R_n)$ with $\|f \mid L_p(R_n)\|$ and $0<p\leq\infty$ (for the sake of clarity we add now the suffix "R_n"). Furthermore, as in those subsections, D^α stands for derivative, F and F^{-1} denote the n-dimensional Fourier transform and its inverse on $S'(R_n)$, respectively.

First we introduce $C(R_n)$ as the set of all complex-valued bounded and uniformly continuous functions on R_n. If $m=1, 2, 3, \ldots$, we define

$$C^m(R_n)=\{f \mid D^\alpha f \in C(R_n) \quad \text{for all} \quad |\alpha| \leqq m\}. \tag{1}$$

Let $C^0(R_n)=C(R_n)$. Here D^α are classical derivatives. Of course, $C^m(R_n)$ endowed with the norm

$$\|f \mid C^m(R_n)\| = \sum_{|\alpha| \leqq m} \|D^\alpha f \mid L_\infty(R_n)\|$$

is a Banach space. Concerning the space $C(R_n)$ (and consequently also for the spaces $C^m(R_n)$) one has at least two other candidates: (i) the space of all bounded and continuous (but not necessarily uniformly continuous) functions on R_n and (ii) the completion of $S(R_n)$ in $C(R_n)$. The first space is larger and the second one is smaller than $C(R_n)$. There are some reasons (which will not be explained in detail at this moment) that $C(R_n)$ is the best choice. After some "basic" spaces in the sense of 2.2.1. have been defined, we come to the constructive spaces and the Hölder spaces. First we describe the spaces and the (in Remark 1, below) we give the necessary explanations.

(i) *The Hölder spaces* $C^s(R_n)$. If s is a real number, then

$$s=[s]+\{s\} \text{ with } [s] \text{ integer and } 0 \leqq \{s\} < 1. \tag{2}$$

If $s>0$ is not an integer, then

$$C^s(R_n)=\Big\{f \mid f \in C^{[s]}(R_n), \quad \|f \mid C^s(R_n)\| = \|f \mid C^{[s]}(R_n)\| + \sum_{|\alpha|=[s]} \sup_{\substack{x \neq y \\ x \in R_n, y \in R_n}} \frac{|D^\alpha f(x) - D^\alpha f(y)|}{|x-y|^{\{s\}}} < \infty\Big\}. \tag{3}$$

(ii) *The Zygmund spaces* $\mathscr{C}^s(R_n)$. If s is a real number, then

$$s=[s]^- + \{s\}^+ \text{ with } [s]^- \text{ integer and } 0 < \{s\}^+ \leqq 1. \tag{4}$$

Furthermore, if $f(x)$ is an arbitrary function on R_n and $h \in R_n$, then

$$(\Delta_h^1 f)(x) = f(x+h) - f(x), \quad (\Delta_h^l f)(x) = \Delta_h^1(\Delta_h^{l-1} f)(x) \tag{5}$$

where $l=2, 3, \ldots$ and $x \in R_n$. If $s>0$ then

$$\mathscr{C}^s(R_n)=\{f \mid f \in C^{[s]^-}(R_n), \ \|f \mid \mathscr{C}^s(R_n)\| = \|f \mid C^{[s]^-}(R_n)\| + \sum_{|\alpha|=[s]^-} \sup_{0 \neq h \in R_n} |h|^{-\{s\}^+} \|\Delta_h^2 D^\alpha f \mid C(R_n)\| < \infty\}. \tag{6}$$

(iii) *The Sobolev spaces* $W_p^m(R_n)$. If $1<p<\infty$ and $m=1, 2, 3, \ldots$, then

$$W_p^m(R_n)=\{f \mid f \in L_p(R_n), \|f \mid W_p^m(R_n)\| = \sum_{|\alpha| \leqq m} \|D^\alpha f \mid L_p(R_n)\| < \infty\}. \tag{7}$$

Furthermore, $W_p^0(R_n)=L_p(R_n)$.

(iv) *The Slobodeckij spaces* $W_p^s(R_n)$. If $1 \leqq p < \infty$ and $0<s \neq$ integer, then

$$W_p^s(R_n)=\Big\{f \mid f \in W_p^{[s]}(R_n), \|f \mid W_p^s(R_n)\| = \|f \mid W_p^{[s]}(R_n)\| + \sum_{|\alpha|=[s]} \Big(\int_{R_n \times R_n} \frac{|D^\alpha f(x) - D^\alpha f(y)|^p}{|x-y|^{n+\{s\}p}} \, dx dy \Big)^{\frac{1}{p}} < \infty\Big\}. \tag{8}$$

(v) *The Besov (or Lipschitz) spaces* $\Lambda^s_{p,q}(R_n)$. If $s>0$, $1\leqq p<\infty$ and $1\leqq q<\infty$, then

$$\Lambda^s_{p,q}(R_n)=\left\{f \mid f\in W_p^{[s]^-}(R_n),\ \|f \mid \Lambda^s_{p,q}(R_n)\|=\|f \mid W_p^{[s]^-}(R_n)\| \right. \\ \left. +\sum_{|\alpha|=[s]^-}\left(\int_{R_n}|h|^{-\{s\}^+q}\|\Delta_h^2 D^\alpha f \mid L_p(R_n)\|^q \frac{dh}{|h|^n}\right)^{\frac{1}{q}}<\infty\right\}. \tag{9}$$

If $s>0$ and $1\leqq p<\infty$, then

$$\Lambda^s_{p,\infty}(R_n)=\{f \mid f\in W_p^{[s]^-}(R_n),\ \|f \mid \Lambda^s_{p,\infty}(R_n)\|=\|f \mid W_p^{[s]^-}(R_n)\| \\ +\sum_{|\alpha|=[s]^-}\sup_{0\neq h\in R_n}|h|^{-\{s\}^+}\|\Delta_h^2 D^\alpha f \mid L_p(R_n)\|<\infty\}. \tag{10}$$

(vi) *The Bessel-potential (or Lebesgue or Liouville) spaces* $H^s_p(R_n)$. If s is a real number and $1<p<\infty$, then

$$H^s_p(R_n)=\{f \mid f\in S'(R_n),\ \|f \mid H^s_p(R_n)\|=\|F^{-1}(1+|x|^2)^{\frac{s}{2}} Ff \mid L_p(R_n)\|<\infty\}. \tag{11}$$

(vii) *The local Hardy spaces* $h_p(R_n)$. If $\varphi(x)$ is a test function on R_n (i.e. an infinitely differentiable function on R_n with compact support) with $\varphi(0)=1$, then we put $\varphi_t(x)=\varphi(tx)$ with $x\in R_n$ and $t>0$. If $0<p<\infty$, then

$$h_p(R_n)=\{f \mid f\in S'(R_n),\ \|f \mid h_p(R_n)\|^\varphi=\|\sup_{0<t<1}|F^{-1}\varphi_t Ff|\ |L_p(R_n)\|<\infty\}. \tag{12}$$

(viii) *The space* $bmo(R_n)$. If $f(x)$ is a locally Lebesgue-integrable function on R_n and if Q is a cube in R_n, then

$$f_Q=\frac{1}{|Q|}\int_Q f(x)\,dx \tag{13}$$

is the mean value of $f(x)$ with respect to Q.

$$bmo(R_n)=\left\{f \mid f(x) \text{ locally Lebesgue-integrable on } R_n,\ \|f \mid bmo(R_n)\| \right. \\ \left. =\sup_{|Q|\leqq 1}\frac{1}{|Q|}\int_Q|f(x)-f_Q|\,dx+\sup_{|Q|>1}\frac{1}{|Q|}\int_Q|f(x)|dx<\infty\right\}. \tag{14}$$

Remark 1 (Technical explanations). $D^\alpha f$ in (3) and (6) are classical derivations. On the other hand $D^\alpha f$ in (7)–(10) must be interpreted in the sense of distributions. In other words, $f\in W_p^m(R_n)$ means that the distribution f and all its derivatives $D^\alpha f$ with $|\alpha|\leqq m$ are regular and that they belong (interpreted as functions) to $L_p(R_n)$. Similarly in (8)–10). We recall that $F^{-1}(1+|x|^2)^{\frac{s}{2}} Ff$ in (11) means $F^{-1}[(1+|x|^2)^{\frac{s}{2}} Ff(x)](\cdot)$, cf. 1.5.2. and 1.5.1. If $f\in S'(R_n)$ then $F^{-1}(1+|x|^2)^{\frac{s}{2}} Ff\in S'(R_n)$ and one asks in (11) whether this is a regular distribution belonging to $L_p(R_n)$. Similarly, $F^{-1}\varphi_t Ff$ in (12) means $(F^{-1}[\varphi_t\cdot Ff])(\xi)$. If $\xi\in R_n$ is fixed, then one takes in (12) the supremum of $|(F^{-1}[\varphi_t\cdot Ff])(\xi)|$ with respect to t. Then one asks whether the result, considered as a function of $\xi\in R_n$, belongs to $L_p(R_n)$. Finally, $\sup_{|Q|\leqq 1}$ in (14) means that the supremum is taken over all cubes in R_n with $|Q|\leqq 1$, where $|Q|$ is the volume of Q (similarly for $\sup_{|Q|>1}$).

Remark 2. The definition of $h_p(R_n)$ is somewhat complicated. If $f\in S'(R_n)$ then $F^{-1}\varphi_t Ff$ satisfies the hypotheses of the Paley-Wiener-Schwartz theorem, cf. Theorem 1.2.1/2. Hence, $(F^{-1}\varphi_t Ff)(\xi)$ is an analytic function with respect to $\xi\in R_n$ for every fixed $t>0$. Hence $\sup_{0<t<1}|(F^{-1}\varphi_t Ff)(\xi)|$ makes sense for any $\xi\in R_n$ and any $f\in S'(R_n)$. Furthermore, because

$\varphi(0)=1$ we have

$$F^{-1}\varphi_t Ff \to f \quad \text{in} \quad S'(R_n) \quad \text{if} \quad t\downarrow 0. \tag{15}$$

This shows that $F^{-1}\varphi_t Ff$ is an approximation of $f\in S'(R_n)$ by entire analytic functions. We return to problems of this type in 2.2.4. As we shall see later on, $h_p(R_n)$ is independent of the chosen test function $\varphi(x)$. If φ and ψ are two test functions with $\varphi(0)=\psi(0)=1$, then the corresponding quasi-norms (norms if $p\geqq 1$) are equivalent to each other (the notion of a quasi-norm has been explained in 1.2.2.). The definition (12) is due to D. Goldberg [1, 2]. It is the local (or better, non-homogeneous, in our language) version of the Hardy spaces $H_p(R_n)$. C. Fefferman, E. M. Stein [2] have shown that the Hardy spaces $H_p(R_n)$ (n-dimensional real version) which have been mentioned in 2.2.1. can be defined by (12) if one replaces $\sup\limits_{0<t<1}$ by $\sup\limits_{t>0}$. But $H_p(R_n)$ is a homogeneous space and, as has been mentioned in 2.2.1., we wish to work with non-homogeneous spaces. This is the reason why we prefer $h_p(R_n)$ (later on, in Chapter 5, we return briefly to $H_p(R_n)$ and homogeneous spaces, cf. 5.2.4.). Concerning BMO and bmo we have the same situation. bmo is a non-homogeneous space and BMO is a homogeneous one. The definition of $BMO(R_n)$ is the same as in (14) if one replaces $\|f \mid bmo(R_n)\|$ by

$$\|f \mid BMO(R_n)\| = \sup_Q \frac{1}{|Q|} \int_Q |f(x)-f_Q| \, dx\,, \tag{16}$$

where the supremum is taken over all cubes Q in R_n.

Remark 3. We mention some results which will be proved later on:

$$h_p(R_n)=L_p(R_n) \quad \text{if} \quad 1<p<\infty\,. \tag{17}$$

If $1\leqq p<\infty$ and $0<s\neq$ integer then

$$W_p^s(R_n)=\Lambda_{p,p}^s(R_n) \quad \text{and} \quad C^s(R_n)=\mathcal{C}^s(R_n)\,. \tag{18}$$

If $1<p<\infty$ and $m=0, 1, 2, \ldots$, then

$$W_p^m(R_n)=H_p^m(R_n)\,. \tag{19}$$

This justifies some assertions which we have made above: $\mathcal{C}^s(R_n)$ extends $C^s(R_n)$ from $0<s\neq$ integer to all values $s>0$ (it should be mentioned that $C^m(R_n)\neq\mathcal{C}^m(R_n)$ if $m=1, 2, \ldots$). Furthermore, the scale $H_p^s(R_n)$ contains the Sobolev spaces $W_p^m(R_n)$ as special cases. Finally, the Slobodeckij spaces $W_p^s(R_n)$ are special Besov spaces. This justifies the corresponding lines drawn in Fig. 2.2.1.

Remark 4. We restricted p in (9) and (10) to $1\leqq p<\infty$. This is convenient at the moment, but not necessary. Later on we include also values $p<1$.

2.2.3. The Criterion

First we recall of the notion of a quasi-norm and a quasi-Banach space, explained in 1.2.2. The spaces which have been described in (i)–(viii) in 2.2.2. are Banach spaces, with the exception of $h_p(R_n)$ if $0<p<1$. The local Hardy spaces $h_p(R_n)$ with $0<p<1$ are only quasi-Banach spaces. All these spaces (and also $C^m(R_n)$ with $m=0, 1, 2, \ldots$, as well as $L_p(R_n)$ with $1\leqq p\leqq\infty$) are continuously embedded in $S'(R_n)$. Our aim is to develop a unified theory of function spaces, which contain the above spaces as special cases. Hence, it seems reasonable to restrict ourselves to

quasi-normed spaces of distributions which are continuously embedded in $S'(R_n)$. (1)

This means that we exclude from the very beginning the spaces $L_p(R_n)$ with $0<p<1$, which cannot be interpreted as subspaces of $S'(R_n)$. Our task is to find simple and convincing criteria and methods in order to select a sufficiently rich family of subspaces of $S'(R_n)$ in the sense of (1). Hence, we can pose the question:

Which of the spaces (1) *are good, how do we find them and how do we define them?* (2)

First of all, one must say what is meant by "good" spaces. This is a matter of taste. There are at least two different points of view concerning a criterion for a "good" space:

(i) Simple definition.
(ii) "Good" properties.

What is meant by "simple definition" is quite clear. $L_p(R_n)$ and $C^m(R_n)$ (maybe also $W_p^m(R_n)$) have a simple definition. On the other hand, the definitions of $\Lambda^s_{p,q}(R_n)$ and $h_p(R_n)$ look (at least at the first glance) complicated. Which properties are "good" ones again depends on the task under consideration. Our point of view here is that we are looking for spaces which are useful in the theory of partial differential equations (in particular, elliptic differential equations). For instance, the spaces $L_p(R_n)$ with $1<p<\infty$ have this property, but the spaces $L_1(R_n)$, $L_\infty(R_n)$ and $C(R_n)$ (and also $C^m(R_n)$ with $m=1, 2, 3, \ldots$) give rise to trouble. In order to give the last remarks a more precise meaning, we consider a (normed or quasi-normed) space A, which is continuously embedded in $S'(R_n)$. Let

$$A^m=\{f \mid D^\alpha f\in A \quad \text{if} \quad 0\leq|\alpha|\leq m\},$$
$$\|f \mid A^m\|=\sum_{|\alpha|\leq m}\|D^\alpha f \mid A\|.$$

Of course, $D^\alpha f$ must be understood in the sense of distributions. Let Δ be the Laplacian and let E be the identity mapping. From the standpoint of elliptic differential operators it seems to be reasonable to ask whether $-\Delta+E$ yields an isomorphic mapping from A^2 onto A (and more generally, from A^{m+2} onto A^m). For instance (as we shall see later on) $A=L_p(R_n)$ with $1<p<\infty$ has this property (in that case we have $A^m=W_p^m(R_n)$). On the other hand if $n\geq 2$ then $L_1(R_n)$, $L_\infty(R_n)$, $C(R_n)$, and more generally $C^k(R_n)$ with $k=0, 1, 2, \ldots$, do not, cf. Remark 2 below. If A has this property, then there exist positive constants c and c' such that

$$\sum_{|\alpha|\leq 2}\|D^\alpha f \mid A\|\leq c\|(-\Delta+E)f \mid A\|\leq c'\left(\|f \mid A\|+\sum_{j=1}^{n}\left\|\frac{\partial^2 f}{\partial x_j^2}\,\middle|\, A\right\|\right) \tag{3}$$

holds for all $f\in A^2$. In particular, "mixed" derivatives $D^\alpha f$ can be estimated by "pure" derivatives. If one uses other elliptic differential operators, then it follows that corresponding assertions should also be valid for higher derivatives. One could take this as a "good" property and as a criterion we are looking for. However, a more qualitative formulation is desirable. If $\xi=(\xi_1, \ldots, \xi_n)\in R_n$ and if $\alpha=(\alpha_1, \ldots, \alpha_n)$ is a multi-index, then $\xi^\alpha=\prod_{j=1}^{n}\xi_j^{\alpha_j}$ by definition. It is not hard to see that the above-mentioned generalization of (3) yields

$$\|F^{-1}\xi^\alpha Ff \mid A\|\leq c\left\|F^{-1}\left(1+\sum_{j=1}^{n}\xi_j^{2m}\right)Ff \,\middle|\, A\right\| \tag{4}$$

for every α with $|\alpha|\leq 2m$ and every $f\in A^{2m}$. The positive number c in (4) is independent of $f\in A^{2m}$ (and α). Let $m_\alpha(\xi)=\xi^\alpha\left(1+\sum_{j=1}^{n}\xi_j^{2m}\right)^{-1}$. Then (4) can be reformulated in the following way: there exists a positive constant c such that

$$\|F^{-1}m_\alpha Ff \mid A\|\leq c\|f \mid A\| \tag{5}$$

holds for every α with $|\alpha| \leqq 2m$ and every $f \in A$. If $m(\xi) = m_\alpha(\xi)$, then we have

$$\sup_{\xi \in R_n} (1+|\xi|)^{|\beta|} |D^\beta m(\xi)| < \infty \tag{6}$$

for any multi-index β. This describes the qualitative behaviour of $m_\alpha(\xi)$, which fits very well with the topology in $S(R_n)$, cf. (1.2.1./1). Thus it seems to be reasonable to take the class of all infinitely differentiable functions $m(x)$ which satisfy (6) for every multi-index β as the qualitative generalization of the above functions $m_\alpha(\xi)$.

Definition. The quasi-normed space A is said to have the weak Fourier multiplier property if, for any complex-valued infinitely differentiable function $m(\xi)$ on R_n, which satisfies (6) for every multi-index β, there exists a positive constant c such that

$$\|F^{-1} m F f \mid A\| \leqq c \|f \mid A\| \tag{7}$$

holds for any $f \in A$.

Remark 1. A function $m(\xi)$ with (7) is called a Fourier multiplier for A. The above definition is close to the well-known Michlin-Hörmander multiplier theorem for $L_p(R_n)$ with $1<p<\infty$. In this case it is sufficient to know (6) for every multi-index with $|\beta| \leqq n$. In the above definition we assume that (6) is satisfied for every multi-index β, and this is the reason why we denote this Fourier multiplier property as "weak". (A brief survey on Fourier multipliers for $L_p(R_n)$ with $1<p<\infty$ may be found in [I, 2.2.4.], cf. also R. E. Edwards, G. I. Gaudry [1].)

Criterion. The space A has the weak Fourier multiplier property.

Remark 2. This is the criterion we are looking for. As we shall see, all "constructive" spaces listed in (i)–(viii) in 2.2.2. satisfy this criterion (including $L_p(R_n)$ with $1<p<\infty$). On the other hand the remaining spaces $L_1(R_n)$, $L_\infty(R_n)$ and $C^k(R_n)$ with $k=0, 1, 2, \ldots$ do not have the weak Fourier multiplier property. A proof of this assertion may be found in [S, pp. 21–23]. What about the weaker property (3) for these exceptional spaces? If $n \geqq 2$, then again the answer is negative. We refer to J. Boman [1], K. DeLeeuw, H. Mirkil [1], O. V. Besov [3] and D. Ornstein [1]. Cf. also the footnote of the editor of the Russian edition of [I, p. 407].

2.2.4. Decomposition Method, the Principle

In the preceding subsection we have given a criterion for "good" spaces (from our point of view). The next task is to find a general method for the construction of such spaces. On the one hand such method should be as simple as possible, while on the other it should be sophisticated enough to cover all the spaces from 2.2.1. and 2.2.2. as far as they satisfy the Criterion 2.2.3 (cf. Remark 2.2.3/2).

Spectral Decomposition. In order to provide a feeling for what happens we start with some considerations in the framework of spectral theory of self adjoint operators in Hilbert spaces. Let H be a complex separable Hilbert space endowed with the norm $\|\cdot\|$. Let T be a positive-definite self-adjoint operator in H with the spectral decomposition

$$Tf = \int_0^\infty t \, dE(t) \, f,$$

where $\{E(t)\}_{t>0}$ stands for the spectral measure of T. Then

$$T^s f = \int_0^\infty t^s \, dE(t) \, f \quad \text{with} \quad -\infty < s < \infty$$

are the usual fractional powers of T. More generally,

$$\varphi(T)\, f = \int_0^\infty \varphi(t)\, \mathrm{d}E(t)\, f\,.$$

If $a>1$, then

$$\|T^s f\|^2 = \int_0^\infty t^{2s} \mathrm{d}\|E(t)\, f\|^2 \sim \|E(1)\, f\|^2 + \sum_{j=1}^\infty a^{2js} \|\chi_{[a^{j-1}, a^j)}(T)\, f\|^2\,, \tag{1}$$

where $\chi_{[a^{j-1}, a^j)}$ is the characteristic function of the interval $[a^{j-1}, a^j)$. Furthermore, "$\sim$" means that the right-hand side of (1) can be estimated from above and from below by the left-hand side with the aid of positive constants which are independent of f (belonging to the domain of definition of T^s). In this sense, (1) is a characterization of $\|T^s f\|$ by means of spectral decompositions. Let us consider two examples. Let

$$H = L_2(R_n) \quad \text{and} \quad (Tf)(x) = (1+|x|^2)\, f(x)\,. \tag{2}$$

Then it is easy to see that $(\varphi(T)\, f)(x) = \varphi\,(1+|x|^2)\cdot f(x)$. An immaterial modification of (1) yields

$$\|T^s f \mid L_2(R_n)\|^2 \sim \sum_{j=0}^\infty a^{4js} \|\chi_j^{(a)}(\cdot)\, f(\cdot) \mid L_2(R_n)\|^2\,, \tag{3}$$

where $\chi_j^{(a)}(x)$ is the characteristic function of $\{x \mid a^{j-1} \leqq |x| < a^j\}$ if $j=1, 2, 3, \ldots$ and $\chi_0^{(a)}(x)$ is the characteristic function of $\{x \mid |x|<1\}$. Roughly speaking, (3) is a decomposition of $f(x)$ with respect to its "weight properties", expressed by the weight $1+|x|^2$. We consider a second example. Let

$$H = L_2(R_n) \quad \text{and} \quad (Vf)(x) = (-\Delta + E)\, f(x)\,, \tag{4}$$

where Δ is again the Laplacian and E stands for the identity. The operators T and V in (2) and (4), respectively, and their functions φ are related to each other by

$$V = F^{-1} T F\,, \quad \varphi(V) = F^{-1} \varphi(T)\, F\,, \tag{5}$$

where F and F^{-1} have the same meaning as before, cf. 1.2.1. In particular, the counterpart of (3) is given by

$$\begin{aligned} \|V^s f \mid L_2(R_n)\|^2 &\sim \sum_{j=0}^\infty a^{4js}\, \|F^{-1}\chi_j^{(a)} Ff \mid L_2(R_n)\|^2 \\ &= \left\| \left(\sum_{j=0}^\infty a^{4js} \mid (F^{-1}\chi_j^{(a)} Ff)\,(\cdot)|^2 \right)^{\frac{1}{2}} \mid L_2(R_n) \right\|^2. \end{aligned} \tag{6}$$

This is a decomposition of f with respect to its "differentiability properties" (expressed by $-\Delta + E$). In particular, s is a smoothness index.

Now we take the idea of a decomposition of smoothness in the sense of (6) as motivation and return to our original task. Because we wish to include all spaces from 2.2.1. and 2.2.2. (as far as they fulfil Criterion 2.2.3.) we are forced to generalize the above L_2-setting. (6) involves an l_2-norm (with respect to j) and an $L_2(R_n)$-norm, where one can change the order (first $L_2(R_n)$ and then l_2 or vice versa). If one replaces l_2 by l_q and L_2 by L_p, then one obtains the following two generalizations of (6):

$$\|a^{js} F^{-1}\chi_j^{(a)} Ff \mid l_q(L_p)\| \quad \text{and} \tag{7}$$

$$\|a^{js} F^{-1}\chi_j^{(a)} Ff \mid L_p(l_q)\|\,, \tag{8}$$

where we used the abbreviations introduced in (1.2.2/5) and (1.2.2/6). Furthermore, we again recall that $F^{-1}\chi_j^{(a)} Ff = F^{-1}[\chi_j^{(a)} Ff]$. Both (7) and (8) are quasi-norms.

Now one would try to define a space A consisting of all $f\in S'(R_n)$ such that the quasi-norm (7) is finite. Similarly with (8) instead of (7). In order to discuss such a possibility we ask some questions:
(i) Since the Fourier transform is involved it seems reasonable that at least the Schwartz space $S(R_n)$ be a subspace of A (continuous embedding).
(ii) The number $a>1$ in the above arguments has been chosen arbitrarily. In particular, the spaces in (6) are independent of a. Thus it seems reasonable to ask whether A is also independent of $a>1$ (of course. one would expect A to depend on s, p, q; similarly if one replaces (7) by (8)).

In order to discuss question (i) we take for sake of simplicity $n=1$ (one-dimensional case). If $f\in S(R_1)$ such that $(Ff)(x)=1$ for $|x|\leqq 1$ then we have

$$(F^{-1}\chi_0^{(a)}Ff)(x)=(F^{-1}\chi_0^{(a)})(x)=c\,\frac{\sin x}{x}\,,\quad c\neq 0\,.$$

This shows that $f\notin A$ if $0<p\leqq 1$ (also if one replaces (7) by (8)). This calculation shows that it would be reasonable to replace $\chi_0^{(a)}$, and more generally $\chi_j^{(a)}$, by smooth functions φ_j. Such a replacement has several advantages. Let us assume that φ_j is a test function (i.e. a function from $S(R_n)$ with compact support). In that case $F^{-1}\varphi_jFf$ makes sense for any $f\in S'(R_n)$ (and belongs also to $S'(R_n)$). Furthermore, by the Paley-Wiener-Schwartz theorem (cf. Theorem 1.2.1/2) $(F^{-1}\varphi_jFf)(x)$ is an analytic function. In other words, if one replaces $\{\chi_j^{(a)}\}_{j=0}^{\infty}$ in (7) by an appropriate sequence $\{\varphi_j\}_{j=0}^{\infty}$ of test functions then the question of whether the so-modified quasi-norm is finite makes sense for any $f\in S'(R_n)$ (similarly for (8)). A detailed discussion of the above problem (ii) yields the same result: the characteristic functions $\chi_j^{(a)}$ are not smooth enough to guarantee that A is independent of a in some cases. Hence it will be reasonable to replace the resolution of unity

$$1=\sum_{j=0}^{\infty}\chi_j^{(a)}(x)\,,\quad x\in R_n\,, \tag{9}$$

by a smooth resolution of unity,

$$1=\sum_{j=0}^{\infty}\varphi_j(x)\,,\qquad x\in R_n\,. \tag{10}$$

The problem is the following: on the one hand we wish to preserve the above decomposition of differentiability properties of $f\in S'(R_n)$, cf. (6) and the attempted generalizations (7) and (8). On the other hand, we are forced to replace the non-smooth resolution of unity (9) by the smooth resolution (10). The best possible choice which fits with both requirements is the following. Let $\Phi^{(a)}$ (with $a>1$) be the set of all systems $\{\varphi_j(x)\}_{j=0}^{\infty}\subset S(R_n)$ satisfying (10) such that

$$\begin{cases}\operatorname{supp}\varphi_0\subset\{x\mid |x|\leqq a\}\,,\\ \operatorname{supp}\varphi_j\subset\{x\mid a^{j-1}\leqq|x|\leqq a^{j+1}\}\quad\text{if}\quad j=1,2,3,\ldots,\end{cases} \tag{11}$$

$$\sup_{\substack{x\in R_n\\ j=0,1,2,\ldots}} a^{j|\alpha|}\,|D^\alpha\varphi_j(x)|<\infty\quad\text{for every multi-index }\alpha. \tag{12}$$

There exist systems with these properties. If $\{\varphi_j\}\in\Phi^{(a)}$ is a fixed system then we replace $\chi_j^{(a)}$ in (7) and (8) by φ_j, respectively, and define spaces

$$\{f\mid f\in S'(R_n)\,,\ \|a^{js}F^{-1}\varphi_jFf\mid l_q(L_p)\|<\infty\}\,, \tag{13}$$

$$\{f\mid f\in S'(R_n)\,,\ \|a^{js}F^{-1}\varphi_jFf\mid L_p(l_q)\|<\infty\}\,. \tag{14}$$

It is easy to see that $p=q=2$ yields essentially (6) with $2s$ instead of s. In other words: (13) and (14) are generalizations of (6) which have the advantage of circumventing the troubles raised by the characteristic functions in (7) and (8). Again one would ask the above questions (i) and (ii), and furthermore, one would ask whether the spaces in (13) and (14) are independent of $\{\varphi_j\}\in\Phi^{(a)}$. One has affirmative answers for all $-\infty<s<\infty$, $0<p\leqq\infty$, $0<q\leqq\infty$ with the exception of $p=\infty$, $0<q<\infty$ in the case of the quasi-norm (14). Furthermore, all these spaces (with the above-mentioned exception) satisfy Criterion 2.2.3. This will be proved later on. For the moment we take the underlying method as the basis of our further considerations.

Principle. The space A from the Criterion 2.2.3. will be constructed by measuring the decomposition $\{F^{-1}\varphi_j Ff\}_{j=0}^{\infty}$ of the differentiability properties of $f\in S'(R_n)$ via the quasi-norms (1.2.2/5) and (1.2.2/6) in the sense of (13) and (14).

Thus we have established one criterion (the weak Fourier multiplier property from 2.2.3.) and one principle (decomposition of differentiability properties of tempered distributions $f\in S'(R_n)$) in order to construct a sufficiently rich family of function spaces which contains the spaces from 2.2.1. and 2.2.2. as special cases (as far as they satisfy Criterion 2.2.3.).

Remark 1. Perhaps at first glance it seems a little astonishing that (13) and (14) make sense even if $p<1$. We recall that $L_p(R_n)$ with $p<1$ is a bad space from the standpoint of Fourier analysis. However the situation is similar to that in (2.2.2/12), where we defined the local Hardy spaces $h_p(R_n)$: Both $F^{-1}\varphi_t Ff$ in (2.2.2./12) and $F^{-1}\varphi_j Ff$ in (13), (14) are analytic functions in R_n, and hence they do not have the nasty local singularities that a function $f(x)\in L_p(R_n)$ with $p<1$ may have (e. g. $f(x)=|x|^{-n}$ near the origin). The basic idea in (2.2.2/12) and in (14) is similar: By (2.2.2/15), $F^{-1}\varphi_t Ff$ is an approximation of $f\in S'(R_n)$ by analytic functions, measured in an $L_p(L_\infty)$-quasi-norm. On the other hand, (14) is a decomposition of $f\in S'(R_n)$ by analytic functions measured in an $L_p(l_q)$-quasi-norm.

Remark 2. What about the exceptional case $p=\infty$, $0<q<\infty$ in (14)? In that case, one must modify (14) somewhat, cf. 2.3.4. After this modification, the corresponding spaces with $p=\infty$ and $1<q<\infty$ satisfy Criterion 2.2.3. Mostly we shall be concerned with $p<\infty$, $0<q<\infty$ in (14). But we add a series of remarks concerning that modified limiting case (14) with $p=\infty$ and $1<q<\infty$, because the interesting space $bmo(R_n)$ from (2.2.2/14) is a special case.

Remark 3. We mention an interesting special case of (14). If $s=0$, $1<p<\infty$ and $q=2$, then the norm in (14) is equivalent to $\|f \mid L_p(R_n)\|$ and the space in (14) with $s=0$, $1<p<\infty$ and $q=2$ coincides with $L_p(R_n)$. This is essentially the famous Littlewood-Paley theorem for $L_p(R_n)$. On the other hand, this assertion sheds also new light upon (14), which may also be considered as a generalization of this theorem.

Remark 4. The spaces in (13) and (14) are described as non-homogeneous (or local) spaces. In Chapter 5 we deal with their homogeneous counterparts (including the modifications in the sense of Remark 2), which contain as special cases the usual Hardy spaces $H_p(R_n)$ and $BMO(R_n)$.

2.2.5. Approximation Procedures

In 2.2.2. we characterized smoothness of functions (as elements of function spaces) by the classical means of derivatives D^α and differences Δ_h^l (at least in the spaces (i)–(v) in 2.2.2.). Since the fundamental thesis of D. Jackson in 1911 it has beena well-known fact that smoothness of functions can also be expressed by approximation procedures. We discuss briefly this possibility because it offers new insights into the structure of (2.2.4/13) and (2.2.4/14), which hold a key position in our approach.

Let $T=\{x \mid -\pi\leqq x\leqq\pi\}$ be the 1-torus where the endpoints $-\pi$ and π are identified. In other words, we deal temporarily with 2π-periodic functions. Let $\mathfrak{T}_N$ be the set of all trigonometric polynomials

$$P_N(x)=\sum_{k=-N}^{N} a_k e^{ikx}, \quad x\in T. \tag{1}$$

Of course, $N>0$ is an integer. If $0<p<\infty$, then we ask the question: What about the approximation numbers

$$d_N=\inf_{P_N\in\mathfrak{P}_N} \left(\int_T |f(x)-P_N(x)|^p\,dx\right)^{\frac{1}{p}}? \tag{2}$$

We assume that $f(x)$ is p-integrable on T. By Weierstrass' theorem it follows that $d_N\downarrow 0$ if $N\to\infty$. Of course, d_N is monotonically decreasing. Of interest is how rapidly d_N decreases. What we have in mind is a characterization of functions which guarantees that

$$\sum_{N=1}^{\infty} N^{sq} d_N^q<\infty, \quad s\geqq 0, \quad 0<q<\infty. \tag{3}$$

We wish to have a one-to-one relation between smoothness properties of functions and assertions of type (3).

Now we try to translate the one-dimensional periodic case to R_n. What is the counterpart of a trigonometric polynomial? A trigonometric polynomial in T can be characterized as an integrable function on T with only a finite number of its Fourier coefficients being different from zero. In other words, an integrable function on T is a trigonometric polynomial if and only if the set of its non-vanishing Fourier coefficients is compact. But this formulation makes sense in R_n, too, if one replaces the Fourier coefficients by the Fourier transform. Hence, by the Paley-Wiener-Schwartz theorem, cf. Theorem 1.2.1/2, we are looking for approximations by functions

$$g\in S'(R_n), \quad \operatorname{supp} Fg\subset\{y \mid |y|\leqq N\}. \tag{4}$$

The counterpart of (2) reads as follows:

$$d_N=\inf \|f-g \mid L_p(R_n)\|, \tag{5}$$

where $f\in S'(R_n)\cap L_p(R_n)$ with $0<p<\infty$ is given and the infimum is taken over all $g\in S'(R_n)\cap L_p(R_n)$ satisfying (4). Again we ask for the class of all functions satisfying (3) (where d_N is given by (5)). It is easy to see that it is sufficient (and convenient) to deal only with $N=2^j$ in (5), where $j=0, 1, 2, \ldots$ A second reason for that "dyadic" procedure is the structure of (2.2.4/13) and (2.2.4/14) (this will be clearer just now). Then (3) (with (5)) can be reformulated (and slightly generalized) in the following way. If $0<p\leqq\infty$, then we introduce

$$\mathfrak{A}_p(R_n)=\{a \mid a=\{a_j(x)\}_{j=0}^{\infty}, \quad a_j(x)\in S'(R_n)\cap L_p(R_n), \quad \operatorname{supp} Fa_j\subset\{y \mid |y|\leqq 2^{j+1}\}\}. \tag{6}$$

If $s>0$, $0<p\leqq\infty$ and $0<q\leqq\infty$ then one obtains as a refinement of (3) (with (5))

$$\{f \mid f\in S'(R_n)\cap L_p(R_n), \quad \exists a\in\mathfrak{A}_p(R_n) \text{ with } a_k(x)\to f \text{ in } S'(R_n) \text{ if } k\to\infty \text{ and } \|a_0 \mid L_p(R_n)\|+\|2^{sj}(f(x)-a_j(x)) \mid l_q(L_p)\|<\infty\}. \tag{7}$$

We used the abbreviation (1.2.2/6). This is a typical approximation procedure. It resembles (2.2.4/13), at least in the structure. As we shall see later on, (2.2.4/13) and (7) coincide under the restrictions $0<p\leqq\infty$, $0<q\leqq\infty$, $s>\max\left(0, n\left(\frac{1}{p}-1\right)\right)$, cf. Theorem 2.5.3. The approximation procedure (7) is meaningless if $s<0$, in contrast to (2.2.4/13) which makes sense for all $-\infty<s<\infty$, $0<p\leqq\infty$, $0<q\leqq\infty$. We recall that (2.2.4/13) stands for a decomposition. In this sense we modify (6) by

$$\mathfrak{B}_p(R_n)=\{b \mid b=\{b_j(x)\}_{j=0}^{\infty}, \quad b_j\in S'(R_n)\cap L_p(R_n), \quad \operatorname{supp} Fb_0\subset\{y \mid |y|\leqq 2\}, \quad \operatorname{supp} Fb_j\subset\{y \mid 2^{j-1}\leqq|y|\leqq 2^{j+1}\} \text{ if } j=1, 2, 3, \ldots\}, \tag{8}$$

where $0<p\leqq\infty$. Of course, $\mathfrak{B}_p(R_n)\subset\mathfrak{A}_p(R_n)$. The counterpart of (7) reads as follows,

$$\{f \mid f\in S'(R_n)\cap L_p(R_n), \quad \exists b\in\mathfrak{B}_p(R_n) \text{ with } \sum_{k=0}^{\infty} b_k=f \text{ in } S'(R_n), \quad \|2^{sj}b_j \mid l_q(L_p)\|<\infty\}. \tag{9}$$

If $a=2$ in (2.2.4/11) and (2.2.4/12) then $b_k=F^{-1}\varphi_k Ff$ is a special decomposition. Now it does not come as a surprise that (2.2.4/13) and (9) coincide for all parameters $-\infty<s<\infty$, $0<p\leqq\infty$, $0<q\leqq\infty$. What about (2.2.4/14) in the sense of the above considerations? One obtains similar results if one replaces $\|\cdot\mid l_q(L_p)\|$ in (7) and (9) by $\|\cdot\mid L_p(l_q)\|$, cf. (1.2.2/5). Detailed formulations and proofs will be given later on.

2.3. Definition and Fundamental Properties

2.3.1. Definition

In Section 2.3. we give the definition of the spaces $B^s_{p,q}$ and $F^s_{p,q}$ in the sense of the Principle 2.2.4 and the formulas (2.2.4/13) and (2.2.4/14) (this subsection), and we prove some fundamental properties, including an affirmative answer concerning the problem raised in Criterion 2.2.3. However the main result of this section is the maximal inequality proved in 2.3.6., which plays a key role in the further considerations of this chapter.

We use the notions from 1.2.1. and 1.2.2. For the sake of clarity we now write $S(R_n)$ and $S'(R_n)$ (instead of S and S') for the Schwartz space and the space of tempered distributions on R_n, respectively. Furthermore, we replace $\|f_k\mid L_p(l_q)\|$ from (1.2.2/5) by $\|f_k\mid L_p(R_n,l_q)\|$ and $\|f_k\mid l_q(L_p)\|$ from (1.2.2/6) by $\|f_k\mid l_q(L_p(R_n))\|$.

Definition 1. Let $\Phi(R_n)$ be the collection of all sytems $\varphi=\{\varphi_j(x)\}_{j=0}^\infty\subset S(R_n)$ such that

$$\begin{cases}\operatorname{supp}\varphi_0\subset\{x\mid |x|\leqq 2\}\,,\\ \operatorname{supp}\varphi_j\subset\{x\mid 2^{j-1}\leqq|x|\leqq 2^{j+1}\}\quad\text{if}\quad j=1,2,3,\ldots,\end{cases}\tag{1}$$

for every multi-index α there exists a positive number c_α such that

$$2^{j|\alpha|}\,|D^\alpha\varphi_j(x)|\leqq c_\alpha\quad\text{for all}\quad j=0,1,2,\ldots\quad\text{and all}\quad x\in R_n\tag{2}$$

and

$$\sum_{j=0}^\infty\varphi_j(x)=1\quad\text{for every}\quad x\in R_n\,.\tag{3}$$

Remark 1. First we remark that $\Phi(R_n)$ coincides with $\Phi^{(a)}$ from 2.2.4. if $a=2$. In this sense, $\varphi\in\Phi(R_n)$ is a smooth resolution of unity. Next we prove that $\Phi(R_n)$ is not empty. Let $0\leqq\psi(x)\in S(R_n)$ with

$$\operatorname{supp}\psi\subset\left\{x\;\middle|\;\frac{1}{2}\leqq|x|\leqq 2\right\}\quad\text{and}\quad\psi(x)>0\quad\text{if}\quad\frac{1}{\sqrt{2}}\leqq|x|\leqq\sqrt{2}\,.\tag{4}$$

Let $\varphi_j(x)=\psi(2^{-j}x)\left(\sum_{k=-\infty}^{\infty}\psi(2^{-k}x)\right)^{-1}$ if $j=1,2,3,\ldots$ and $\varphi_0(x)=1-\sum_{j=1}^\infty\varphi_j(x)$. Then it is easy to see that $\varphi=\{\varphi_j(x)\}_{j=0}^\infty\in\Phi(R_n)$.

Definition 2. Let $-\infty<s<\infty$ and $0<q\leqq\infty$. Let $\varphi=\{\varphi_j(x)\}_{j=0}^\infty\in\Phi(R_n)$.

(i) If $0<p\leqq\infty$, then

$$B^s_{p,q}(R_n)=\{f\mid f\in S'(R_n),\ \|f\mid B^s_{p,q}(R_n)\|^\varphi=\|2^{sj}F^{-1}\varphi_jFf\mid l_q(L_p(R_n))\|<\infty\}\tag{5}$$

(ii) If $0<p<\infty$, then

$$F^s_{p,q}(R_n)=\{f\mid f\in S'(R_n),\ \|f\mid F^s_{p,q}(R_n)\|^\varphi=\|2^{sj}F^{-1}\varphi_jFf\mid L_p(R_n,l_q)\|<\infty\}\,.\tag{6}$$

Remark 2. We recall that $F^{-1}\varphi_jFf=F^{-1}[\varphi_jFf](x)$ is an analytic function on R_n, cf. the discussion in front of formula (2.2.4/9). Hence, (5) and (6) make sense. Of course, the quasi-norms $\|f\mid B^s_{p,q}(R_n)\|^\varphi$ and $\|f\mid F^s_{p,q}(R_n)\|^\varphi$ depend on the chosen system $\varphi\in\Phi(R_n)$. But this is not so for the spaces $B^s_{p,q}(R_n)$ and $F^s_{p,q}(R_n)$ itself (this will be proved in the next subsection). This justifies our notation.

Remark 3. In comparison with the discussion in 2.2.4. we fixed $a=2$ from the very beginning. On the basis of the considerations in 2.3.2. it is not hard to see that the above space $B^s_{p,q}(R_n)$ can also be defined via (2.2.4/13) with $a>1$ and $\{\varphi_j(x)\}_{j=0}^{\infty}\in\Phi^{(a)}$. Similarly for $F^s_{p,q}(R_n)$. In this sense the spaces defined by (2.2.4/13) with $-\infty<s<\infty$, $0<p\leqq\infty$, $0<q\leqq\infty$ and the spaces defined by (2.2.4/14) with $-\infty<s<\infty$, $0<p<\infty$, $0<q\leqq\infty$, are independent of $a>1$.

Remark 4. What about $p=\infty$ in (6)? One can show that in this case (6) does not make sense. In particular, a corresponding space is not independent of $\varphi\in\Phi(R_n)$. However the proof of the last assertion is complicated and we refer for details to [S, 2.1.4], where we proved a "non-multiplier" theorem. Later on, we introduce spaces $F^s_{\infty,q}(R_n)$ with $1<q<\infty$ on the basis of a modification of (6), cf. 2.3.4.

2.3.2. Equivalent Quasi-Norms and Elementary Embeddings

The notion of a quasi-norm has been described in 1.2.2. We say that two quasi-norms $\|a\|_1$ and $\|a\|_2$ on a given linear vector space A are equivalent to each other if there exist two positive numbers c_1 and c_2 such that

$$\|a\|_1\leqq c_1\|a\|_2 \quad\text{and}\quad \|a\|_2\leqq c_2\|a\|_1 \quad\text{for every}\quad a\in A.$$

Proposition 1. Let $\varphi=\{\varphi_j(x)\}_{j=0}^{\infty}\in\Phi(R_n)$ and $\psi=\{\psi_j(x)\}_{j=0}^{\infty}\in\Phi(R_n)$.
(i) If $-\infty<s<\infty$, $0<p\leqq\infty$ and $0<q\leqq\infty$, then $\|f\mid B^s_{p,q}(R_n)\|^{\varphi}$ and $\|f\mid B^s_{p,q}(R_n)\|^{\psi}$ are equivalent quasi-norms on $B^s_{p,q}(R_n)$.
(ii) If $-\infty<s<\infty$, $0<p<\infty$ and $0<q\leqq\infty$, then $\|f\mid F^s_{p,q}(R_n)\|^{\varphi}$ and $\|f\mid F^s_{p,q}(R_n)\|^{\psi}$ are equivalent quasi-norms on $F^s_{p,q}(R_n)$.

Proof. Step 1. We prove (i). It is easy to see that both $\|f\mid B^s_{p,q}(R_n)\|^{\varphi}$ and $\|f\mid B^s_{p,q}(R_n)\|^{\psi}$ are quasi-norms. In order to prove the equivalence of these two quasi-norms we apply (1.5.2/13). We recall that $H^m_2=W^m_2(R_n)$ are the usual Sobolev spaces if $m=1, 2, 3, \ldots$, normed by (1.5.2/3). If $\psi_{-1}\equiv 0$, then we have $\varphi_j(x)=$ $=\varphi_j(x)\sum_{r=-1}^{1}\psi_{j+r}(x)$, where $j=0, 1, 2, \ldots$ Consequently,

$$F^{-1}\varphi_j Ff=\sum_{r=-1}^{1}F^{-1}\varphi_j FF^{-1}\psi_{j+r}Ff\,. \tag{1}$$

If we replace f, M and b in (1.5.2/13) by $F^{-1}\psi_{j+r}Ff$, φ_j and $d2^j$, respectively, where $d>0$ is an appropriate number, then we obtain

$$\|F^{-1}\varphi_j FF^{-1}\psi_{j+r}Ff\mid L_p(R_n)\|\leqq c\|\varphi_j(d2^j\cdot)\mid W^m_2(R_n)\|\,\|F^{-1}\psi_{j+r}Ff\mid L_p(R_n)\|\,, \tag{2}$$

where c is independent of j. Of course, we assume that m is sufficiently large. Now it follows from (2.3.1/2) and (2.3.1/1) that

$$\|\varphi_j(d2^j\cdot)\mid W^m_2(R_n)\|\leqq c \quad\text{for}\quad j=0, 1, 2, \ldots \tag{3}$$

Hence there exists a positive constant c such that

$$\|F^{-1}\varphi_j\psi_{j+r}Ff\mid L_p(R_n)\|\leqq c\|F^{-1}\psi_{j+r}Ff\mid L_p(R_n)\| \tag{4}$$

for $j=0, 1, 2, \ldots$ and $r=-1, 0, 1$. But this proves that $\|f\mid B^s_{p,q}(R_n)\|^{\varphi}$ can be estimated from above by $c\|f\mid B^s_{p,q}(R_n)\|^{\psi}$. Hence these two quasi-norms are equivalent to each other.

Step 2. The proof of part (ii) is the same if one replaces (1.5.2/13) by (1.6.3/1).

Remark. If one extends (2.3.1/6) to $p=\infty$ and $0<q<\infty$, then the corresponding assertion of part (ii) of the above proposition (with $p=\infty$ and $0<q<\infty$) is not valid, cf. [S, 2.1.4] and Remark 2.3.1/4.

In the following proposition, $A_1 \subset A_2$ always means that the quasi-normed space A_1 is continuously embedded in the quasi-normed space A_2, i.e. there exists a constant c such that $\|a \mid A_2\| \leqq c\|a \mid A_1\|$ holds for all $a \in A_1$.

Proposition 2. (i) Let $0<q_0 \leqq q_1 \leqq \infty$ and $-\infty<s<\infty$. Then

$$B^s_{p,q_0}(R_n) \subset B^s_{p,q_1}(R_n) \quad \text{if} \quad 0<p\leqq\infty \tag{5}$$

and

$$F^s_{p,q_0}(R_n) \subset F^s_{p,q_1}(R_n) \quad \text{if} \quad 0<p<\infty\,. \tag{6}$$

(ii) Let $0<q_0\leqq\infty$, $0<q_1\leqq\infty$, $-\infty<s<\infty$ and $\varepsilon>0$. Then

$$B^{s+\varepsilon}_{p,q_0}(R_n) \subset B^s_{p,q_1}(R_n) \quad \text{if} \quad 0<p\leqq\infty \tag{7}$$

and

$$F^{s+\varepsilon}_{p,q_0}(R_n) \subset F^s_{p,q_1}(R_n) \quad \text{if} \quad 0<p<\infty\,. \tag{8}$$

(iii) If $0<q\leqq\infty$, $0<p<\infty$ and $-\infty<s<\infty$, then

$$B^s_{p,\min(p,q)}(R_n) \subset F^s_{p,q}(R_n) \subset B^s_{p,\max(p,q)}(R_n)\,. \tag{9}$$

Proof. Step 1. (5) and (6) are simple consequences of the monotinicity of the l_q-spaces, cf. (1.2.2/4). Similarly, (7) and (8) follow from

$$\left(\sum_{k=0}^{\infty} 2^{skq_1}\,|b_k|^{q_1}\right)^{\frac{1}{q_1}} \leqq \sup_k\, 2^{(s+\varepsilon)k}\,|b_k| \left(\sum_{l=0}^{\infty} 2^{-\varepsilon l q_1}\right)^{\frac{1}{q_1}} \leqq c\,\sup_k\, 2^{(s+\varepsilon)k}\,|b_k|$$

and (5) or (6) (modification if $q_1=\infty$).

Step 2. We prove (iii). If $f\in S'(R_n)$ and $\varphi=\{\varphi_k(x)\}_{k=0}^{\infty}\in\Phi(R_n)$ then we put $a_k(x)=2^{sk}(F^{-1}\varphi_k Ff)(x)$. We shall use the generalized triangle inequality for Banach spaces. If $0<q\leqq p<\infty$, then (9) follows from

$$\|a_k \mid l_p(L_p(R_n))\| \leqq \|a_k \mid L_p(R_n, l_q)\| = \left\| \sum_{k=0}^{\infty} |a_k|^q \mid L_{\frac{p}{q}}(R_n) \right\|^{\frac{1}{q}}$$

$$\leqq \left(\sum_{k=0}^{\infty} \|\, |a_k|^q \mid L_{\frac{p}{q}}(R_n)\|\right)^{\frac{1}{q}} = \|a_k \mid l_q(L_p(R_n))\|\,.$$

If $0<p<q\leqq\infty$, then (9) follows from

$$\|a_k \mid l_q(L_p(R_n))\| = \|\int_{R_n} |a_k|^p\,dx \mid l_{\frac{q}{p}}\|^{\frac{1}{p}}$$

$$\leqq \left(\int_{R_n} \|\,|a_k|^p \mid l_{\frac{q}{p}}\|\,dx\right)^{\frac{1}{p}} = \|a_k \mid L_p(R_n, l_q)\| \leqq \|a_k \mid l_p(L_p(R_n))\|\,.$$

We used the abbreviation (1.2.2/4).

2.3.3. Basic Properties

The notion of a quasi-Banach space has been introduced in 1.2.2. The topology of the locally convex space $S(R_n)$ is generated by (1.2.1/1). The topological dual $S'(R_n)$ of $S(R_n)$ is equipped with the strong topology. If A is a quasi-normed space, then $S(R_n)\subset A$ always means continuous (topological) embedding. Similarly $A\subset S'(R_n)$.

Theorem. (i) $B^s_{p,q}(R_n)$ is a quasi-Banach space if $-\infty<s<\infty$, $0<p\leqq\infty$ and $0<q\leqq\infty$ (Banach space if $1\leqq p\leqq\infty$ and $1\leqq q\leqq\infty$), and the quasi-norms $\|f \mid B^s_{p,q}(R_n)\|^\varphi$ with $\varphi\in\Phi(R_n)$ and $\|f \mid B^s_{p,q}(R_n)\|^\psi$ with $\psi\in\Phi(R_n)$ are equivalent. Furthermore,

$$S(R_n)\subset B^s_{p,q}(R_n)\subset S'(R_n) \tag{1}$$

for $-\infty<s<\infty$, $0<p\leqq\infty$, and $0<q\leqq\infty$. If $-\infty<s<\infty$, $0<p<\infty$ and $0<q<\infty$, then $S(R_n)$ is dense in $B^s_{p,q}(R_n)$.

(ii) $F^s_{p,q}(R_n)$ is a quasi-Banach space if $-\infty<s<\infty$, $0<p<\infty$ and $0<q\leqq\infty$ (Banach space if $1\leqq p<\infty$ and $1\leqq q\leqq\infty$), and the quasi-norms $\|f \mid F^s_{p,q}(R_n)\|^\varphi$ with $\varphi\in\Phi(R_n)$ and $\|f \mid F^s_{p,q}(R_n)\|^\psi$ with $\psi\in\Phi(R_n)$ are equivalent. Furthermore,

$$S(R_n)\subset F^s_{p,q}(R_n)\subset S'(R_n) \tag{2}$$

for $-\infty<s<\infty$, $0<p<\infty$, and $0<q\leqq\infty$. If $-\infty<s<\infty$, $0<p<\infty$ and $0<q<\infty$, then $S(R_n)$ is dense in $F^s_{p,q}(R_n)$.

Proof. Step 1. The equivalence of the quasi-norms in (i) as well as in (ii) has been proved in Proposition 2.3.2/1. Furthermore, in order to prove (1) and (2) we may restrict ourselves to (1) with $q=\infty$. This follows from Proposition 2.3.2/2.

Step 2. We prove the left-hand side of (1) with $q=\infty$. Let $f\in S(R_n)$ and $\varphi=\{\varphi_k(x)\}_{k=0}^\infty\in\Phi(R_n)$. We recall that F yields a one-to-one mapping from $S(R_n)$ onto itself, in particular $p_N(F\varphi)$ from (1.2.1/1) with $N=1, 2, 3, \ldots$ generates the topology of $S(R_n)$. If L, M and N are sufficiently large natural numbers, then

$$\begin{aligned}
\|f \mid B^s_{p,\infty}(R_n)\|^\varphi &= \sup_k 2^{ks}\|F^{-1}\varphi_k Ff \mid L_p(R_n)\| \\
&\leqq \sup_k 2^{ks}\,\|(1+|x|^2)^{2L}F^{-1}\varphi_k Ff \mid L_\infty(R_n)\| \\
&= \sup_k 2^{ks}\,\|F^{-1}\,(1+(-\Delta)^L)\,\varphi_k Ff \mid L_\infty(R_n)\| \\
&\leqq c\,\sup_k 2^{ks}\,\|(1+(-\Delta)^L)\,\varphi_k Ff \mid L_1(R_n)\| \\
&\leqq c'\,\|(1+|x|)^M\,(1+(-\Delta)^L)\,Ff \mid L_\infty(R_n)\| \leqq c' p_N(Ff)\,.
\end{aligned}$$

Here, c and c' are appropriate positive numbers.

Step 3. We prove the right-hand side of (1) with $q=\infty$. If $\varphi\in\Phi(R_n)$ is the above system, then we put $\chi_k(x)=\varphi_{k-1}(x)+\varphi_k(x)+\varphi_{k+1}(x)$ if $k=0, 1, 2, \ldots$ (with $\varphi_{-1}=0$). Then we have $\chi_k(x)=1$ if $x\in\operatorname{supp}\varphi_k$. If $f\in B^s_{p,\infty}(R_n)$ and $\psi\in S(R_n)$, then $f(\psi)$ denotes the value of the functional f of $S'(R_n)$ for the test function ψ. We obtain

$$\begin{aligned}
|f(\psi)| &= \left|\sum_{k=0}^\infty F^{-1}\varphi_k Ff(F\chi_k F^{-1}\psi)\right| \\
&\leqq \sum_{k=0}^\infty \|F^{-1}\varphi_k Ff \mid L_\infty(R_n)\|\cdot\|F\chi_k F^{-1}\psi \mid L_1(R_n)\|\,.
\end{aligned} \tag{3}$$

Recall that both $F^{-1}\varphi_k Ff$ and $F\chi_k F^{-1}\psi$ are analytic functions. Hence, the last estimate makes sense. We wish to use (1.3.2/5) with $\alpha=(0,\ldots,0)$, $q=\infty$ and $b=2^{k+1}$. The approximation procedure from the proof of Theorem 1.4.1 shows that (1.3.2/5) is applicable to $F^{-1}\varphi_k Ff$. Hence,

$$\|F^{-1}\varphi_k Ff \mid L_\infty(R_n)\|\leqq c2^{k\frac{n}{p}}\|F^{-1}\varphi_k Ff \mid L_p(R_n)\|\,.$$

Putting this estimate into (3) we find that

$$|f(\psi)|\leqq c\|f \mid B^s_{p,\infty}(R_n)\|^\varphi \sum_{k=0}^\infty 2^{-sk+k\frac{n}{p}}\|F\chi_k F^{-1}\psi \mid L_1(R_n)\|\,.$$

The last sum can be estimated from above by $\|\psi \mid B_{1,1}^{-s+\frac{n}{p}}(R_n)\|^{\varphi}$. By Step 2, the left-hand side of (1) holds for $B_{1,1}^{-s+\frac{n}{p}}(R_n)$. Consequently, if N is a sufficiently large natural number, then we have

$$|f(\psi)| \leqq c\|f \mid B^s_{p,\infty}(R_n)\|^{\varphi}\, p_N(\psi) \tag{4}$$

for any $\psi \in S(R_n)$. This proves that $B^s_{p,\infty}(R_n)$ is continuously embedded in $S'(R_n)$. Hence, (1) and (2) are proved.

Step 4. $B^s_{p,q}(R_n)$ is a quasi-normed space. We prove the completeness. Let $\{f_l\}_{l=1}^{\infty}$ be a fundamental sequence in $B^s_{p,q}(R_n)$ (with respect to a fixed quasi-norm in $B^s_{p,q}(R_n)$). Then (1) shows that $\{f_l\}_{l=1}^{\infty}$ is also a fundamental sequence in $S'(R_n)$. Because $S'(R_n)$ is complete we find a limit element $f \in S'(R_n)$. If $\varphi = \{\varphi_k(x)\}_{k=0}^{\infty} \in \Phi(R_n)$, then $F^{-1}\varphi_k F f_l$ converges to $F^{-1}\varphi_k F f$ in $S'(R_n)$ if $l \to \infty$. On the other hand $\{F^{-1}\varphi_k F f_l\}_{l=1}^{\infty}$ is a fundamental sequence in $L_p(R_n)$. By Theorem 1.4.1 it is also a fundamental sequence in $L_\infty(R_n)$. This shows that the limiting element of $\{F^{-1}\varphi_k F f_l\}_{l=1}^{\infty}$ in $L_p(R_n)$ (which is the same as in $L_\infty(R_n)$) coincides with $F^{-1}\varphi_k F f$. Now it follows by standard arguments that f belongs to $B^s_{p,q}(R_n)$ and that f_l converges in $B^s_{p,q}(R_n)$ to f. Hence, $B^s_{p,q}(R_n)$ is complete. Similarly for $F^s_{p,q}(R_n)$.

Step 5. We prove that $S(R_n)$ is dense in $F^s_{p,q}(R_n)$ if $-\infty < s < \infty$, $0 < p < \infty$ and $0 < q < \infty$. Let $f \in F^s_{p,q}(R_n)$. If $\varphi = \{\varphi_k(x)\}_{k=0}^{\infty} \in \Phi(R_n)$ then we put

$$f_N(x) = \sum_{k=0}^{N} F^{-1}\varphi_k F f\,.$$

Of course, $f_N \in F^s_{p,q}(R_n)$. Consequently,

$$\|f - f_N \mid F^s_{p,q}(R_n)\|^{\varphi} \leqq c \left\| \left(\sum_{k=N}^{\infty} \sum_{r=-1}^{1} 2^{skq} |F^{-1}\varphi_k \varphi_{k+r} F f|^q \right)^{\frac{1}{q}} \Big| L_p(R_n) \right\|,$$

and by Theorem 1.6.3. and (2.3.2/3)

$$\|f - f_N \mid F^s_{p,q}(R_n)\|^{\varphi} \leqq c \left\| \left(\sum_{k=N}^{\infty} 2^{skq} |F^{-1}\varphi_k F f|^q \right)^{\frac{1}{q}} \Big| L_p(R_n) \right\|. \tag{5}$$

Lebesgue's bounded convergence theorem proves that the right-hand side of (5) tends to zero if $N \to \infty$. Hence, f_N approximates f in $F^s_{p,q}(R_n)$. Next we apply the smoothness procedure from Step 1 of the proof of Theorem 1.4.1. Then $(f_N)_\delta \in S(R_n)$ approximates f_N in L_p^{Ω} with $\Omega = \{y \mid |y| \leqq 2^{k+2}\}$. But this is also an approximation of f_N in $F^s_{p,q}(R_n)$. This proves that $S(R_n)$ is dense in $F^s_{p,q}(R_n)$. A similar argument is valid for $B^s_{p,q}(R_n)$ with $-\infty < s < \infty$, $0 < p < \infty$, $0 < q \leqq \infty$.

Remark. It is not very complicated to prove that $S(R_n)$ is not dense in $B^s_{\infty,q}(R_n)$, $B^s_{p,\infty}(R_n)$ and $F^s_{p,\infty}(R_n)$ if $0 < q \leqq \infty$ and $0 < p < \infty$.

Convention. In the sequel we shall not distinguish between equivalent quasi-norms of a given quasi-Banach space. In this sense we shall write $\|f \mid B^s_{p,q}(R_n)\|$ instead of $\|f \mid B^s_{p,q}(R_n)\|^{\varphi}$ and $\|f \mid F^s_{p,q}(R_n)\|$ instead of $\|f \mid F^s_{p,q}(R_n)\|^{\varphi}$ with $\varphi \in \Phi(R_n)$.

2.3.4. The Spaces $F^s_{\infty,q}(R_n)$

We mentioned several times that an extension of (2.3.1/6) to $p=\infty$ does not make sense if $0<q<\infty$ (if $p=q=\infty$ then an extension of (2.3.1/6) yields the spaces $B^s_{\infty,\infty}(R_n)$ from (2.3.1/5)). This subsection deals with a modification of (2.3.1/6) which enables us to define $F^s_{\infty,q}(R_n)$ with $1<q<\infty$.

Definition. If $-\infty<s<\infty$, $1<q\leqq\infty$ and $\varphi=\{\varphi_k(x)\}_{k=0}^\infty\in\Phi(R_n)$, then

$$F^s_{\infty,q}(R_n)=\left\{f \mid f\in S'(R_n),\quad \exists\{f_k(x)\}_{k=0}^\infty\subset L_\infty(R_n)\quad \text{such that}\right.$$
$$\left. f=\sum_{k=0}^\infty F^{-1}\varphi_k F f_k \quad \text{in}\quad S'(R_n)\quad \text{and}\quad \|2^{sk}f_k \mid L_\infty(R_n,l_q)\|<\infty\right\}, \tag{1}$$

$$\|f \mid F^s_{\infty,q}(R_n)\|^\varphi=\inf\|2^{sk}f_k \mid L_\infty(R_n,l_q)\|, \tag{2}$$

where the infimum is taken over all admissible representations of f in the sense of (1).

Remark 1. Recall that $F^{-1}\varphi_k F f_k=F^{-1}[\varphi_k F f_k]$ and $\|\cdot \mid L_\infty(R_n,l_q)\|$ has the same meaning as in 2.3.1. and (1.2.2/5),

$$\|2^{sk}f_k \mid L_\infty(R_n,l_q)\|=\sup_{x\in R_n}\left(\sum_{k=0}^\infty 2^{skq}|f_k(x)|^q\right)^{\frac{1}{q}}. \tag{3}$$

Remark 2. There is no difficulty in extending (1) and (2) to $0<q\leqq1$. The restriction to $1<q\leqq\infty$ comes from Proposition 2, which in turn is the basis of a sufficiently rich theory of the spaces $F^s_{\infty,q}(R_n)$ with $1<q\leqq\infty$. Furthermore, one can try to replace $L_\infty(R_n)$ in (1) and (2) by $L_p(R_n)$. If $0<p<1$, then $F^{-1}\varphi_k F f_k$ does not make sense for certain systems $\{f_k(x)\}_{k=0}^\infty\subset L_p(R_n)$. Hence, $p\geqq1$ is necessary. On the other hand, in the following proposition we shall see that at least in the case $1<p<\infty$ and $1<q<\infty$ one obtains simply a new description of the above spaces $F^s_{p,q}(R_n)$.

Proposition 1. If $-\infty<s<\infty$, $1<p<\infty$, $1<q<\infty$ and $\varphi=\{\varphi_k(x)\}_{k=0}^\infty\in\Phi(R_n)$, then

$$F^s_{p,q}(R_n)=\left\{f \mid f\in S'(R_n),\quad \exists\{f_k(x)\}_{k=0}^\infty\subset L_p(R_n)\quad \text{such that}\right.$$
$$\left. f=\sum_{k=0}^\infty F^{-1}\varphi_k F f_k \quad \text{in}\quad S'(R_n)\quad \text{and}\quad \|2^{sk}f_k \mid L_p(R_n,l_q)\|^\varphi<\infty\right\}. \tag{4}$$

Furthermore,

$$\inf\|2^{sk}f_k \mid L_p(R_n,l_q)\|^\varphi \tag{5}$$

is an equivalent norm in $F^s_{p,q}(R_n)$, where the infimum is taken over all admissible representations of f in the sense of (4).

Proof. If $f\in F^s_{p,q}(R_n)$, we choose $f_k=\sum_{r=-1}^{1} F^{-1}\varphi_{k+r}Ff$ (with $\varphi_{-1}=0$). Then it follows that f belongs to the space on the right-hand side of (4) and that (5) can be estimated from above by $c\|f \mid F^s_{p,q}(R_n)\|^\varphi$. In order to prove the reverse statement we use the l_q-valued $L_p(R_n)$-Fourier multiplier theorem of Michlin-Hörmander, cf. (2.4.8/7) and Remark 2.4.8/1. If f is given by the right-hand side of (4) then it follows from that theorem that

$$\|2^{sk}F^{-1}\varphi_k Ff \mid L_p(R_n,l_q)\|\leqq c\|2^{sk}f_k \mid L_p(R_n,l_q)\|,$$

where c is independent of $\{f_k\}$. Taking the infimum on the right-hand side we obtain the desired assertion.

Hence (1) and its generalization to p instead of ∞ in the sense of (4) is only of interest for limiting cases, and ∞ seems to be the most interesting one. By Theorem 2.3.3 (ii) the Schwartz space $S(R_n)$ is dense in $F^s_{p,q}(R_n)$ if $-\infty<s<\infty$, $0<p<\infty$ and $0<q<\infty$. Furthermore, we have

(2.3.3/2). If $(F^s_{p,q}(R_n))'$ is the topological dual of $F^s_{p,q}(R_n)$ then we have by the usual interpretation (which will be explained in more detail in 2.11.1.)

$$S(R_n) \subset (F^s_{p,q}(R_n))' \subset S'(R_n) .$$

Proposition 2. If $-\infty < s < \infty$ and $1 \leq q < \infty$, then

$$(F^s_{1,q}(R_n))' = F^{-s}_{\infty,q'}(R_n) \tag{6}$$

with $\frac{1}{q}+\frac{1}{q'}=1$.

Remark 3. This is a special case of Theorem 2.11.2 (recall that $F^{-s}_{\infty,\infty}(R_n)=B^{-s}_{\infty,\infty}(R_n)$). Consequently, the spaces $F^s_{\infty,q}(R_n)$ from the above definition can be interpreted as the dual spaces of the Banach spaces $F^{-s}_{1,q'}(R_n)$ $\left(\text{again with } \frac{1}{q}+\frac{1}{q'}=1\right)$. In particular, $F^s_{\infty,q}(R_n)$ from(1) is a Banach space, endowed with the norm (2), it is independent of the choice of $\varphi \in \Phi(R_n)$, and the norms (2) are equivalent to each other for different φ's. Our method concerning the spaces $F^s_{\infty,q}(R_n)$ from (1) is the following: properties, which have been proved for the spaces $F^s_{p,q}(R_n)$ with $-\infty < s < \infty$, $0 < p < \infty$, $0 < q \leq \infty$ will be extended to the spaces (1) via a duality procedure.

2.3.5. An Orientation and Some Historical Remarks

As we mentioned in 2.1., the main aim of Chapter 2 is the thorough study of the spaces $B^s_{p,q}(R_n)$ and $F^s_{p,q}(R_n)$ from Definition 2.3.1/2 (and Definition 2.3.4). Our standpoint is to collect sufficiently many properties in order to deal successfully with regular elliptic boundary value problems in corresponding spaces $B^s_{p,q}(\Omega)$ and $F^s_{p,q}(\Omega)$ on smooth bounded domains Ω in R_n (Chapters 3 and 4). This was our point of view in 2.2.3. Concerning the spaces on R_n this led us to Criterion 2.2.3. An affirmative answer for the spaces $B^s_{p,q}(R_n)$ and $F^s_{p,q}(R_n)$ will be given in 2.3.7. Of course, for our task we need a lot of other properties for the spaces on R_n, which will be proved step by step in this chapter: several types of equivalent quasi-norms, interpolation assertions, embeddings, traces on hyperplanes, Fourier multipliers, pointwise multipliers, diffeomorphism properties etc. The maximal inequality from 2.3.6. plays a key role in all our later considerations.

The Spaces from 2.2.2. Another aim will be to prove that the "basic" and "constructive" spaces from 2.2.1. and 2.2.2. are special cases of the two scales $B^s_{p,q}$ and $F^s_{p,q}$, insofar as they satisfy Criterion 2.2.3. By Remark 2.2.3./2, Remark 2.2.2/3 and the list in 2.2.2. we must consider:

the Hölder-Zygmund spaces $\mathcal{C}^s(R_n)$ with $s>0$;
the Bessel-potential spaces $H^s_p(R_n)$ with $1<p<\infty$ and $-\infty<s<\infty$;
the Besov spaces $\Lambda^s_{p,q}(R_n)$ with $s>0$, $1\leq p<\infty$, and $1\leq q\leq\infty$;
the local Hardy spaces $h_p(R_n)$ with $0<p<\infty$;
the spaces $bmo(R_n)$.

In order to provide orientation we formulate some results which will be proved later on. We have:

$$\mathcal{C}^s(R_n)=B^s_{\infty,\infty}(R_n) \quad \text{if} \quad s>0 , \tag{1}$$

$$H^s_p(R_n)=F^s_{p,2}(R_n) \quad \text{if} \quad 1<p<\infty, \quad -\infty<s<\infty , \tag{2}$$

$$\Lambda^s_{p,q}(R_n)=B^s_{p,q}(R_n) \quad \text{if} \quad s>0, \quad 1\leq p<\infty, \quad 1\leq q\leq\infty , \tag{3}$$

$$h_p(R_n)=F^0_{p,2}(R_n) \quad \text{if} \quad 0<p<\infty , \tag{4}$$

$$bmo(R_n)=F^0_{\infty,2}(R_n) . \tag{5}$$

Concerning the proofs we refer to 2.5.6., 2.5.7., 2.5.8. and 2.5.12. In other words, after (1)–(5) and (2.2.2/17)–(2.2.2/19) have been established, we shall know that all "basic" and "constructive" spaces from 2.2.1. and 2.2.2. which satisfy Criterion 2.2.3. are special cases of the spaces $B^s_{p,q}(R_n)$ and $F^s_{p,q}(R_n)$.

Remark. $B^s_{p,q}(R_n)$ and $F^s_{p,q}(R_n)$ are non-homogeneous spaces. In Chapter 5 we deal briefly with their homogeneous counterparts $\dot{B}^s_{p,q}(R_n)$ and $\dot{F}^s_{p,q}(R_n)$. Then (4) and (5) read as follows,

$$H_p(R_n)=\dot{F}^0_{p,2}(R_n) \quad \text{if} \quad 0<p<\infty\,, \tag{6}$$

$$BMO(R_n)=\dot{F}^0_{\infty,2}(R_n)\,. \tag{7}$$

where $H_p(R_n)$ and $BMO(R_n)$ are the usual (real n-dimensional) Hardy spaces and the space of all functions of bounded mean oscillation in the sense of Remark 2.2.2/2, respectively.

Historical Remarks. We add a few historical remarks and references as far as the spaces $B^s_{p,q}(R_n)$ and $F^s_{p,q}(R_n)$ are concerned. The definition of $B^s_{p,q}(R_n)$ with $1\leq p\leq\infty$ is due to J. Peetre [2] (with a forerunner in L. Hörmander [2] if $p=q=2$). A systematic treatment of his results has been given in J. Peetre [8] (in this book, in particular in the notes at the end of the chapters, one can find further references and comments). The definition of the spaces $F^s_{p,q}(R_n)$ with $1<p<\infty$ and $1<q<\infty$ has been given in H. Triebel [2]. At the same time, P. I. Lizorkin [2,4] introduced spaces which are closely related to $F^s_{p,q}(R_n)$ with $1<p<\infty$ and $1<q<\infty$. In particular, $\overset{+}{\mathfrak{L}}{}^s_{p,q}=F^s_{p,q}(R_n)$, where $\overset{+}{\mathfrak{L}}{}^s_{p,q}$ are the spaces introduced in P. I. Lizorkin [4]. The surprising extension of $B^s_{p,q}(R_n)$ to $p<1$ is due to J. Peetre [4]. In J. Peetre [8, pp. 225–227] one can find a vivid description of how he discovered that possibility. In particular, he emphasizes that there was a forerunner by T. M. Flett [1] (based in turn on classical ideas of G. H. Hardy and J. E. Littlewood). The key paper by C. Fefferman, E. M. Stein [2] gave also a new impetus to the theory of the spaces $F^s_{p,q}(R_n)$. On the basis of the technique of maximal functions, J. Peetre [6] observed that the definition of the spaces $F^s_{p,q}(R_n)$ can be extended to $0<p<\infty$ and $0<q\leq\infty$. Formula (2) is essentially the classical Littlewood-Paley theorem. Furthermore, (3) was known from the very beginning (after the spaces $B^s_{p,q}(R_n)$ had been defined via Definition 2.3.1/2). Further references may be found in [I, pp. 179 and 190]. (1) has been proved in H. Triebel [9] (the homogeneous case is due to B. Grevholm [1], cf. also J. Peetre [3]). (6) is essentially due to J. Peetre [5, 6], cf. also H. Triebel [13] for a more general case and [S, p. 167]. The proof of (4) follows the same line, cf. also Bui Huy Qui [1]. Finally, (7) (and similarly (5)) is proved in H. Triebel [13] and [S, p. 170].

2.3.6. Maximal Inequalities

Maximal inequalities play a fundamental role in our later considerations. It is not complicated to re-formulate (1.4.1/2) and (1.6.2/1) in the language of the spaces $B^s_{p,q}(R_n)$ and $F^s_{p,q}(R_n)$, respectively. However the corresponding assertions are not sufficient for our later purposes. We need some essential generalizations.

Definition 1. If L is a natural number then $\mathcal{A}_L(R_n)$ is the collection of all systems $\varphi=\{\varphi_j(x)\}_{j=0}^\infty\subset S(R_n)$ of functions with compact supports such that

$$C(\varphi)=\sup_{x\in R_n}|x|^L\sum_{|\alpha|\leq L}|D^\alpha\varphi_0(x)|+\sup_{\substack{x\in R_n-\{0\}\\ j=1,2,3}}(|x|^L+|x|^{-L})\sum_{|\alpha|\leq L}|D^\alpha\varphi_j(2^jx)|<\infty\,. \tag{1}$$

Remark 1. $D^\alpha\varphi_j(2^jx)$ means that the derivative D^α (with respect to $x\in R_n$) is applied to the function $g_j(x)=\varphi_j(2^jx)$. An important example of such a system φ can be obtained if one starts with appropriate functions $\varphi_0(x)$ and $\psi(x)$ and puts $\varphi_j(x)=\psi(2^{-j}x)$ if $j=1, 2, 3, \ldots$ Of course, $\Phi(R_n)\subset\mathcal{A}_L(R_n)$ for every natural number L, cf. Definition 2.3.1/1.

Definition 2. If L is a natural number, $\varphi=\{\varphi_j(x)\}_{j=0}^{\infty}\in\mathcal{A}_L(R_n)$, $f\in S'(R_n)$ and $a>0$, then

$$(\varphi_j^* f)(x)=\sup_{y\in R_n}\frac{|(F^{-1}\varphi_j Ff)(x-y)|}{1+|2^j y|^a},\quad x\in R_n, \tag{2}$$

(maximal function), $j=0, 1, 2, \ldots$

Remark 2. In any case, $(F^{-1}\varphi_j Ff)(x)$ is an analytic function, cf. Theorem 1.2.1/2. Hence, (2) makes sense if one does not exclude $(\varphi_j^* f)(x)=\infty$. In order to discuss the last possibility we remark that for $x\in R_n$ and $z\in R_n$,

$$(\varphi_j^* f)\,(x-z)=\sup_{y\in R_n}\frac{|(F^{-1}\varphi_j Ff)(x-z-y)|}{1+|2^j(z+y)|^a}\cdot\frac{1+|2^j(z+y)|^a}{1+|2^j y|^a}.$$

The last factor can be estimated from above by $c(1+|2^j z|^a)$, where c depends only on a. Consequently,

$$(\varphi_j^* f)(x-z)\leqq c(1+|2^j z|^a)(\varphi_j^* f)\,(x)\,. \tag{3}$$

This shows that either $(\varphi_j^* f)(x)<\infty$ for all $x\in R_n$ or $(\varphi_j^* f)(x)=\infty$ for all $x\in R_n$. Of course, $(\varphi_j^* f)(x)$ depends on a. But in later considerations, the value of a is not of interest (a must be sufficiently large later on). This is the reason why we omitted the index a on the left-hand side of (2).

Remark 3. There is no difficulty in later considerations in replacing the supremum with respect to $x\in R_n-\{0\}$, resp., R_n, in (1) by an L_p-quasi-norm. In that case, $p=2$ is closely related to the class $\mathcal{A}$ in C. Fefferman, E. M. Stein [2, p. 183]. That class $\mathcal{A}$ plays a crucial role in the n-dimensional real variable theory for the Hardy spaces which has been developed in C. Fefferman, E. M. Stein [2]. The above class $\mathcal{A}_L(R_n)$ is the counterpart of $\mathcal{A}$ in our framework. Generalizations of $\mathcal{A}_L(R_n)$ for more general classes of $B^s_{p,q}$-$F^s_{p,q}$ type (including anisotropic spaces, homogeneous spaces, spaces with dominating mixed derivatives etc.) have been given in [F, 2.2.2]. $\{\varphi_j^* f\}_{j=0}^{\infty}$ is the counterpart of the "tangential" maximal function $u^{**}(x)$ in C. Fefferman, E. M. Stein [2, p. 166]. This maximal function $u^{**}(x)$ proved very useful in the real variable characterization of the Hardy spaces $H_p(R_n)$, cf. (2.2.2/12) and Remark 2.2.2/2. The above modification (2) of this idea in the framework of the homogeneous spaces of $F^s_{p,q}$ type is due to J. Peetre [6, p. 125]. Further modifications for more general spaces of $B^s_{p,q}$-$F^s_{p,q}$ type have been given in [F, 2.3.1].

We use the above abbreviations: $\|b_k \mid l_q\|$ has the meaning of (1.2.2/4) (with the usual modification if $q=\infty$) and $\|f_k \mid l_q(L_p(R_n))\|$ has been defined in (1.2.2/6) (including the limiting cases if $p=\infty$ and/or $q=\infty$), cf. also the beginning of 2.3.1. Finally, $C(\varphi)$ is given by (1).

Proposition. Let $a>0$ in (2) be fixed. Let $-\infty<s<\infty$ and let L be a natural number larger than $|s|+3a+n+2$.

(i) If $0<q\leqq\infty$, then there exists a positive number c such that

$$\|2^{ks}\sup_{0<\tau<1}(\varphi_k^{\tau *} f)(x)\mid l_q\|\leqq c\sup_{0<\tau<1}C(\varphi^\tau)\,\|2^{ks}(\varphi_k^* f)(x)\mid l_q\| \tag{4}$$

holds for all systems $\varphi=\{\varphi_k(x)\}_{k=0}^{\infty}\in\Phi(R_n)$, all families of systems $\varphi^\tau=\{\varphi_k^\tau(x)\}_{k=0}^{\infty}\in\mathcal{A}_L(R_n)$ with $0<\tau<1$, all $f\in S'(R_n)$ and all $x\in R_n$.

(ii) If $0<q\leqq\infty$ and $0<p\leqq\infty$, then there exists a positive number c such that

$$\|2^{ks}\sup_{0<\tau<1}(\varphi_k^{\tau *} f)\,(\cdot)\mid l_q(L_p(R_n))\|\leqq c\sup_{0<\tau<1}C(\varphi^\tau)\|\,2^{ks}(\varphi_k^* f)(\cdot)\mid l_q(L_p(R_n))\| \tag{5}$$

holds for all systems $\varphi=\{\varphi_k(x)\}_{k=0}^{\infty}\in\Phi(R_n)$, all families of systems $\varphi^\tau=\{\varphi_k^\tau(x)\}_{k=0}^{\infty}\in\mathcal{A}_L(R_n)$ with $0<\tau<1$ and all $f\in S'(R_n)$.

Proof. Step 1. First we prove a pointwise inequality for the maximal functions $\varphi_k^{\tau *}f$ and $\varphi_l^{*}f$. Let τ be fixed, $0<\tau<1$. Let $\{\psi_k(x)\}_{k=0}^{\infty}\in\mathcal{A}_L(R_n)$ with

$$\operatorname{supp}\psi_0\subset\{x\mid |x|\le 4\},\ \operatorname{supp}\psi_j\subset\{x\mid 2^{j-2}\le|x|\le 2^{j+2}\}\quad\text{if}\quad j=1,2,3,\ldots,\tag{6}$$

$$\psi_k(x)=1\quad\text{if}\quad x\in\operatorname{supp}\varphi_k\quad\text{and}\quad k=0,1,2,\ldots\tag{7}$$

We recall that $\{\varphi_k(x)\}_{k=0}^{\infty}\in\Phi(R_n)$, cf. Definition 2.3.1. It is not hard to see that there exist systems $\{\psi_k(x)\}_{k=0}^{\infty}$ with the required properties, where L is completely immaterial. We have

$$F^{-1}\varphi_k^{\tau}Ff=\sum_{l=0}^{\infty}F^{-1}[\varphi_k^{\tau}\psi_l FF^{-1}\varphi_l Ff],\quad f\in S'(R_n)\,,$$

where only finitely many summands on the right-hand side are different from zero. If $*$ indicates the convolution, then it follows that

$$\begin{aligned}&|(F^{-1}\varphi_k^{\tau}Ff)(x-z)|\\&\le c\sum_{l=0}^{\infty}\int_{R_n}|(F^{-1}\varphi_k^{\tau}*F^{-1}\psi_l)(y)|\,|(F^{-1}\varphi_l Ff)(x-z-y)|\,dy\,,\end{aligned}\tag{8}$$

where $x\in R_n$ and $z\in R_n$. The last estimate makes sense because $F^{-1}\varphi_k^{\tau}*F^{-1}\psi_l\in S(R_n)$ and $F^{-1}\varphi_l Ff$ is an entire analytic function and, consequently, a regular distribution of at most polynomial growth. By (2) we have

$$|(F^{-1}\varphi_l Ff)(x-z-y)|\le(1+|2^l(y+z)|^a)(\varphi_l^{*}f)(x)\,.$$

We note that $1+|2^l(y+z)|^a\le c\,(1+|2^l y|^a)(1+|2^l z|^a)$. Now (8) yields

$$2^{ks}\frac{|(F^{-1}\varphi_k^{\tau}Ff)(x-z)|}{1+|2^k z|^a}\le c\sum_{l=0}^{\infty}2^{(k-l)s}\frac{1+|2^l z|^a}{1+|2^k z|^a}\,2^{ls}(\varphi_l^{*}f)(x)\,I_{k,l}\tag{9}$$

with

$$I_{k,l}=\int_{R_n}|(F^{-1}\varphi_k^{\tau}*F^{-1}\psi_l)(y)|\,(1+|2^l y|^a)\,dy\,.\tag{10}$$

We estimate $I_{h,l}$:

$$\begin{aligned}I_{k,l}&=2^{-kn}\int_{R_n}|(F^{-1}\varphi_k^{\tau}*F^{-1}\psi_l)(2^{-k}y)|\,(1+|2^{l-k}y|^a)\,dy\\&\le 2^{-kn+a|k-l|}\int_{R_n}|(F^{-1}\varphi_k^{\tau}*F^{-1}\psi_l)(2^{-k}y)|\,(1+|y|^a)\,dy\\&\le c2^{-kn+a|k-l|}\|(1+|\cdot|^{a+\frac{n}{2}+\varkappa})(F^{-1}\varphi_k^{\tau}*F^{-1}\psi_l)(2^{-k}\cdot)\mid L_2(R_n)\|\,,\end{aligned}\tag{11}$$

where $\varkappa$ is an arbitrary positive number. Next we recall that

$$(F^{-1}g(2^k.))\,(y)=2^{-kn}(F^{-1}g)(2^{-k}y)\;;$$

furthermore, the convolution $g*h$ of two smooth functions is given by

$$(g*h)(x)=\int_{R_n}g(y)\,h(x-y)\,dy\,.$$

Now it is not hard to see that

$$\begin{aligned}&[F^{-1}(\varphi_k^{\tau}(2^k\cdot))*F^{-1}(\psi_l(2^k\cdot))](y)\\&=2^{-2kn}\int_{R_n}(F^{-1}\varphi_k^{\tau})(2^{-k}z)\cdot(F^{-1}\psi_l)(2^{-k}y-2^{-k}z)\,dz\\&=2^{-kn}(F^{-1}\varphi_k^{\tau}*F^{-1}\psi_l)(2^{-k}y)\,.\end{aligned}\tag{12}$$

We put this calculation into (11) and replace $a+\frac{n}{2}+\varkappa$ by an even number N with $a+\frac{n}{2}+2\geq N>a+\frac{n}{2}$. Then

$$\begin{aligned}I_{k,l}&\leq c2^{a|k-l|}\|(1+|\cdot|^N)\,[F^{-1}(\varphi_k^\tau(2^k\cdot))*F^{-1}(\psi_l(2^k\cdot))](\cdot)\mid L_2(R_n)\|\\&=c2^{a|k-l|}\|F\{(1+|\cdot|^N)\,[F^{-1}(\varphi_k^\tau(2^k\cdot))*F^{-1}(\psi_l(2^k\cdot))]\}(\cdot)\mid L_2(R_n)\|\\&\leq c'2^{a|k-l|}\|\varphi_k^\tau(2^k\cdot)\,\psi_l(2^k\cdot)\mid W_2^N(R_n)\|\,,\end{aligned}$$

where $W_2^N(R_n)$ is the Sobolev space, normed by (2.2.2/7). By (6) the function $\psi_l(2^kx)$ vanishes outside of $\{y\mid 2^{l-k-2}\leq|y|\leq 2^{l-k+2}\}$ (modification if $l=0$). On the other hand, if x belongs to that set then (1) yields

$$|D^\alpha\varphi_k^\tau(2^kx)|\leq c2^{-L|l-k|}C(\varphi^\tau)\,, \tag{13}$$

Assuming that $L\geq N$, we obtain

$$\begin{aligned}I_{k,l}&\leq cC(\varphi^\tau)\,2^{(a-L)|l-k|}\|\psi_l(2^k\cdot)\mid W_2^N(R_n)\|\\&=cC(\varphi^\tau)\,2^{(a-L)|l-k|}\|\psi_l(2^l\cdot 2^{k-l}\cdot)\mid W_2^N(R_n)\|\\&\leq c'C(\varphi^\tau)\,2^{\left(a+N+\frac{n}{2}-L\right)|l-k|}\|\psi_l(2^l\cdot)\mid W_2^N(R_n)\|\\&\leq c''C(\varphi^\tau)\,2^{\left(a+N+\frac{n}{2}-L\right)|l-k|}\,,\end{aligned}$$

where c'' is independent of l, k, φ and φ^τ. We return to (9). The factor in front of $2^{ls}(\varphi_l^*f)\,I_{k,l}$ can be estimated from above by $2^{(|s|+a)|k-l|}$. Now take the supremum with respect to $z\in R_n$ on the left-hand side of (9) to obtain

$$2^{ks}(\varphi_k^{\tau *}f)(x)\leq cC(\varphi^\tau)\sum_{l=0}^\infty 2^{\left(|s|+2a+N+\frac{n}{2}-L\right)|l-k|}2^{ls}(\varphi_l^*f)(x)\,, \tag{14}$$

where c is independent of $f\in S'(R_n)$, $x\in R_n$, k, τ, $\varphi\in\Phi(R_n)$ and $\varphi^\tau\in\mathcal{R}_L(R_n)$.

Step 2. We prove (4). Let $\sigma=|s|+2a+N+\frac{n}{2}$. Let $L>\sigma$. If $1\leq q\leq\infty$, then

$$2^{ks}\sup_{0<\tau<1}(\varphi_k^{\tau *}f)(x)\leq c\sup_{0<\tau<1}C(\varphi^\tau)\left(\sum_{l=0}^\infty 2^{(\sigma+\varepsilon-L)|l-k|q}2^{lsq}(\varphi_l^*f)^q(x)\right)^{\frac{1}{q}} \tag{15}$$

with $\varepsilon>0$ and $L>\sigma+\varepsilon$ (modification if $q=\infty$). Now (4) is an immediate consequence. If $0<q<1$, then (14) yields

$$2^{ksq}(\varphi_k^{\tau *}f)^q(x)\leq c^qC(\varphi^\tau)^q\sum_{l=0}^\infty 2^{q(\sigma-L)|l-k|}2^{lsq}(\varphi_l^*f)^q(x)\,. \tag{16}$$

Hence, we have the same situation as in (15). This proves (4) for $q<1$, too. Next we prove (5). If $1\leq p\leq\infty$, then the triangle inequality gives

$$\begin{aligned}&\|2^{ks}\sup_{0<\tau<1}(\varphi_k^{\tau *}f)(\cdot)\mid L_p(R_n)\|\\&\leq c\sup_{0<\tau<1}C(\varphi^\tau)\sum_{l=0}^\infty 2^{(\sigma-L)|l-k|}\|2^{ls}(\varphi_l^*f)(\cdot)\mid L_p(R_n)\|\,.\end{aligned} \tag{17}$$

The next steps are the same as in (15) and (16), respectively. If $0<p<1$, we begin as in (16) (with p instead of q). The rest is the same. This proves (4) and (5) provided

$$L>|s|+2a+\frac{n}{2}+a+\frac{n}{2}+2=|s|+3a+n+2\,. \tag{18}$$

Theorem. Let $-\infty<s<\infty$.
(i) Let $0<p\leq\infty$ and $0<q\leq\infty$. If $a>\frac{n}{p}$ in (2) and if the natural number L from (18) is larger than $|s|+3\frac{n}{p}+n+2$, then there exists a positive number c such that

$$\|2^{ks}\sup_{0<\tau<1}(\varphi_k^{\tau*}f)(\cdot)\mid l_q(L_p(R_n))\|\leq c\sup_{0<\tau<1}C(\varphi^\tau)\|f\mid B^s_{p,q}(R_n)\|^\varphi \tag{19}$$

holds for all systems $\varphi=\{\varphi_k(x)\}_{k=0}^\infty\in\Phi(R_n)$, all families of systems

$$\varphi^\tau=\{\varphi_k^\tau(x)\}_{k=0}^\infty\in\mathcal{A}_L(R_n) \text{ with } 0<\tau<1, \text{ and all } f\in B^s_{p,q}(R_n)\,.$$

(ii) Let $0<p<\infty$ and $0<q\leq\infty$. If $a>\frac{n}{\min(p,q)}$ in (2) and if the natural number L from (18) is larger than $|s|+3\frac{n}{\min(p,q)}+n+2$, then there exists a positive number c such that

$$\|2^{ks}\sup_{0<\tau<1}(\varphi_k^{\tau*}f)(\cdot)\mid L_p(R_n,l_q)\|\leq c\sup_{0<\tau<1}C(\varphi^\tau)\|f\mid F^s_{p,q}(R_n)\|^\varphi \tag{20}$$

holds for all systems $\varphi=\{\varphi_k(x)\}_{k=0}^\infty\in\Phi(R_n)$, all families of systems $\varphi^\tau=\{\varphi_k^\tau(x)\}_{k=0}^\infty\in\mathcal{A}_L(R_n)$ with $0<\tau<1$, and all $f\in F^s_{p,q}(R_n)$.

Proof. Step 1. We prove (ii). By (4) we have to show that

$$\|2^{ks}(\varphi_k^*f)(\cdot)\mid L_p(R_n,l_q)\|\leq c\|2^{ks}F^{-1}\varphi_kFf\mid L_p(R_n,l_q)\| \tag{21}$$

if $f\in F^s_{p,q}(R_n)$ and $a>\frac{n}{\min(p,q)}$ in (2), where c is independent of $\varphi\in\Phi(R_n)$. Recall that the right-hand side of (21) coincides with $\|f\mid F^s_{p,q}(R_n)\|^\varphi$, cf. (2.3.1/6). However, (21) is an immediate consequence of Theorem 1.6.2. with $d_k=2^{k+2}$ (cf. Definition 2.3.1/1), $a=\frac{n}{r}$ and $f_k=F^{-1}\varphi_kFf$.

Step 2. We prove (i). By (5) it is sufficient to show that

$$\|\varphi_k^*f\mid L_p(R_n)\|\leq c\|F^{-1}\varphi_kFf\mid L_p(R_n)\| \tag{22}$$

if $f\in B^s_{p,q}(R_n)$ and $a>\frac{n}{p}$ in (2), where c is independent of $\varphi\in\Phi(R_n)$ and $k=0,1,2,\ldots$ If $p=\infty$, then (22) is trivial:

$$\|\varphi_k^*f\mid L_\infty(R_n)\|=\|F^{-1}\varphi_kFf\mid L_\infty(R_n)\|\,.$$

If $0<p<\infty$, then (22) follows from the scalar case of Theorem 1.6.2 (e.g. with $q=p$).

Remark 1. We need an extension of the theorem. $C(\varphi)$ from (1) is a norm in a corresponding Banach space of sequences $\{\varphi_j(x)\}_{j=0}^\infty$ of functions $\varphi_j(x)$ which are defined in $R_n-\{0\}$ (with the exception of $\varphi_0(x)$, which is defined in R_n). For the sake of clarity we write temporarily $C(\varphi,L)$. If $C(\varphi,L+1)<\infty$, then it is easy to see that φ can be approximated in the norm $C(\varphi,L)$ by systems $\varphi_\varepsilon=\{\varphi_{\varepsilon,j}(x)\}_{j=0}^\infty\in\mathcal{A}_L(R_n)$. In particular,

$$C(\varphi,L)=\lim_{\varepsilon\downarrow 0}C(\varphi_\varepsilon,L),\quad C(\varphi_\varepsilon-\varphi_\delta,L)\to 0 \quad\text{if}\quad \varepsilon\downarrow 0 \text{ and } \delta\downarrow 0$$

(and similarly, if we start with a family φ^τ of systems such that $\sup_{0<\tau<1}C(\varphi^\tau,L+1)<\infty$). Now we assume that $\varphi\in\Phi(R_n)$ is as in the statement of the theorem and that $\{\varphi^\tau\}_{0<\tau<1}$ is an arbitrary family with $\sup_{0<\tau<1}C(\varphi^\tau,L+1)<\infty$. We apply the above approximation procedure to (14) with $\{\varphi^\tau_{\varepsilon,k}\}_{k=0}^\infty\in\mathcal{A}_L(R_n)$. This shows that the pointwise limit of $(\varphi^{\tau*}_{\varepsilon,k}f)(x)$ exists and hence so does the pointwise limit of $(F^{-1}\varphi^{\tau}_{\varepsilon,k}Ff)(x)$ (which by modulus is less than or equal to $(\varphi^{\tau *}_{\varepsilon,k}f)\ (x)$). By (3), this convergence is also valid in a weighted L_∞-space (the weight is an appropriate polynomial). Hence, $(F^{-1}\varphi^{\tau}_{\varepsilon,k}Ff)(x)$ converges also in $S'(R_n)$ if $\varepsilon\downarrow 0$ and we take the limit as the definition of $F^{-1}\varphi^{\tau}_{\varepsilon}Ff$. Then (14) holds for any family of systems $\varphi^\tau=\{\varphi^\tau_k(x)\}^\infty_{k=0}$ with $\sup_{0<\tau<1} C(\varphi^\tau, L+1)<\infty$. Now it is clear that both the proposition and the theorem can be extended to these more general families φ^τ.

2.3.7. A Fourier Multiplier Theorem

Our next aim is to prove that the spaces $B^s_{p,q}(R_n)$ and $F^s_{p,q}(R_n)$ from Definition 2.3.1/2 (and also from Definition 2.3.4) satisfy Criterion 2.2.3, cf. also the program which has been described in 2.3.5. We recall Convention 2.3.3, which will be used consistently in the sequel.

Definition. Let $m(x)$ be a complex-valued infinitely differentiable function on R_n. Let $-\infty<s<\infty$ and $0<q\leqq\infty$. Let A be either $B^s_{p,q}(R_n)$ with $0<p\leqq\infty$ or $F^s_{p,q}(R_n)$ with $0<p<\infty$. Then $m(x)$ is said to be a Fourier multiplier for A if there exists a positive constant c such that

$$\|F^{-1}mFf \mid A\|\leqq c\|f\mid A\| \tag{1}$$

holds for all $f\in A$.

Remark 1. $F^{-1}mFf=F^{-1}[mFf]$ makes sense for any $f\in S'(R_n)$ if $m(x)$ and all its derivatives have at most polynomial growth. In order to avoid technical difficulties we assume for the present that $m(x)$ alway has this property (later on we shall relax this assumption essentially via limiting procedures).

Remark 2. If $A=F^s_{\infty,q}(R_n)$ with $1<q\leqq\infty$ and $-\infty<s<\infty$, then (1) gives the definition of a Fourier multiplier for the spaces from Definition 2.3.4.

Let N be a natural number. For the sake of brevity we introduce

$$\|m\|_N=\sup_{|\alpha|\leqq N}\sup_{x\in R_n}(1+|x|^2)^{\frac{|\alpha|}{2}}\,|D^\alpha m(x)|\,. \tag{2}$$

Theorem. Let $-\infty<s<\infty$ and $0<q\leqq\infty$. Let A be either $B^s_{p,q}(R_n)$ with $0<p\leqq\infty$ or $F^s_{p,q}(R_n)$ with $0<p<\infty$. If the natural number N is sufficiently large, then there exists a positive number c such that

$$\|F^{-1}mFf\mid A\|\leqq c\|m\|_N\,\|f\mid A\| \tag{3}$$

holds for all infinitely differentiable functions $m(x)$ and all $f\in A$.

Proof. Let $A=F^s_{p,q}(R_n)$. We use (2.3.6/20) with $\varphi=\{\varphi_k(x)\}^\infty_{k=0}\in\Phi(R_n)$. We have

$$F^{-1}\varphi_kF[F^{-1}mFf]=F^{-1}\varphi_kmFf\,. \tag{4}$$

Now (2.3.6/20) with $\varphi^\tau_k=m\varphi_k$ and the obvious fact that

$$|(F^{-1}\varphi^\tau_kFf)(x)|\leqq(\varphi^{\tau *}_kf)(x)$$

yields (3) with $N>|s|+3\dfrac{n}{\min(p,q)}+n+2$. Similarly if $A=B^s_{p,q}(R_n)$.

Remark 3. The estimates for N are very rough:

$$N>|s|+3\frac{n}{p}+n+2 \qquad \text{for}\quad B^s_{p,q}(R_n)\,, \tag{5}$$

$$N>|s|+3\frac{n}{\min(p,q)}+n+2 \quad \text{for}\quad F^s_{p,q}(R_n)\,. \tag{6}$$

Later on we obtain substantial improvement on the basis of Theorem 1.5.2 and (1.5.4/2) for the spaces $B^s_{p,q}(R_n)$, and on the basis of Theorem 1.6.3 (and the interpolation theory which will be developed in Section 2.4.) for the space $F^s_{p,q}(R_n)$. Roughly speaking, instead of the unnatural right-hand sides of (5) and (6) we shall deal with the numbers σ_p from Theorem 1.5.2 (for $B^s_{p,q}(R_n)$) and $\sigma_{p,q}$ from (1.6.4/1) (for $F^s_{p,q}(R_n)$). For details we refer to Subsection 2.4.8. In particular, the parameter $|s|$ in (5) and (6) can be cancelled. The latter fact is also an easy consequence of the simple lifting property from 2.3.8. by which it follows that the set of all Fourier multipliers for $F^s_{p,q}(R_n)$ is independent of s (similarly for $B^s_{p,q}(R_n)$). On the other hand, the above simple proof of the theorem makes it clear that sufficiently general and strong Fourier multiplier assertions can be obtained via maximal inequalities. The right-hand sides of (5) and (6) (and also the numbers σ_p and $\sigma_{p,q}$) tend to infinity if $p\downarrow 0$. This effect (in connection with Fourier multipliers) is known, cf. J. Peetre [6] and (for more general spaces) [F, p. 96], cf. also C. Fefferman, E. M. Stein [2, p. 191] as far as the Hardy spaces are concerned, which by (2.3.5/4) and (2.3.5/6) are closely related to spaces of $F^s_{p,q}$-type.

Remark 4. Let $f\in F^s_{\infty,q}(R_n)$ with $-\infty<s<\infty$ and $1<q\leqq\infty$. Let $m(x)$ be an infinitely differentiable function in the sense of Remark 1. If $\varphi\in S(R_n)$, then we have

$$(F^{-1}mFf)(\varphi)=f(FmF^{-1}\varphi)\,. \tag{7}$$

Next we use $(F^{-s}_{1,q'}(R_n))'=F^s_{\infty,q}(R_n)$ with $\frac{1}{q}+\frac{1}{q'}=1$, cf. Proposition 2.3.4/2. By Theorem 2.3.3, the set $S(R_n)$ is dense in $F^{-s}_{1,q'}(R_n)$. Then it follows easily from (7) that any Fourier multiplier for $F^{-s}_{1,q'}(R_n)$ is also a Fourier multiplier for $F^s_{\infty,q}(R_n)$.

Corollary. The spaces

$B^s_{p,q}(R_n)$ with $-\infty<s<\infty,\quad 0<p\leqq\infty,\quad 0<q\leqq\infty$;

$F^s_{p,q}(R_n)$ with $-\infty<s<\infty,\quad 0<p<\infty,\quad 0<q\leqq\infty$; and

$F^s_{\infty,q}(R_n)$ with $-\infty<s<\infty,\quad 1<q\leqq\infty$;

satisfy Criterion 2.2.3.

Proof. This follows immediately from the above theorem and Remark 4.

Remark 5. In [S, p. 32] we proved the following assertion: if a Banach space A has the weak Fourier multiplier property in the sense of Definition 2.2.3 (hence, it satisfies Criterion 2.2.3), then there exists a natural number N and a positive number c, such that (3) holds for all $m(x)$ and all $f\in A$.

2.3.8. Lifting Property and Related Equivalent Quasi-Norms

In order to show the usefulness of Theorem 2.3.6 and Theorem 2.3.7 we prove the so-called lifting property and some related assertions. If σ is a real number, then the operator I_σ is defined by

$$I_\sigma f=F^{-1}\,(1+|x|^2)^{\frac{\sigma}{2}}\,Ff,\quad f\in S'(R_n)\,. \tag{1}$$

It is well-known that I_σ is a one-to-one mapping from $S'(R_n)$ onto itself. Furthermore, the restriction of I_σ to $S(R_n)$ is a one-to-one mapping from $S(R_n)$ onto itself. Obviously, $I_\sigma I_\varkappa=I_{\sigma+\varkappa}$. We shall say in the sequel that "I_σ maps A_0 onto A_1" instead of "the restriction of I_σ to A_0 maps A_0 onto A_1".

Theorem. Let $-\infty<s<\infty$, $-\infty<\sigma<\infty$, $m=1, 2, 3, \ldots$ and $0<q\leqq\infty$.
(i) Let $0<p\leqq\infty$. Then I_σ maps $B^s_{p,q}(R_n)$ isomorphically onto $B^{s-\sigma}_{p,q}(R_n)$ and

$\|I_\sigma f \mid B^{s-\sigma}_{p,q}(R_n)\|$ is an equivalent quasi-norm on $B^s_{p,q}(R_n)$. Furthermore,

$$\sum_{|\alpha|\leq m} \|D^\alpha f \mid B^{s-m}_{p,q}(R_n)\| \quad \text{and} \quad \|f \mid B^{s-m}_{p,q}(R_n)\| + \sum_{j=1}^n \left\| \frac{\partial^m f}{\partial x_j^m} \,\middle|\, B^{s-m}_{p,q}(R_n) \right\| \tag{2}$$

are equivalent quasi-norms on $B^s_{p,q}(R_n)$.

(ii) Let $0<p<\infty$. Then I_σ maps $F^s_{p,q}(R_n)$ isomorphically onto $F^{s-\sigma}_{p,q}(R_n)$ and $\|I_\sigma f \mid F^{s-\sigma}_{p,q}(R_n)\|$ is an equivalent quasi-norm on $F^s_{p,q}(R_n)$. Furthermore,

$$\sum_{|\alpha|\leq m} \|D^\alpha f \mid F^{s-m}_{p,q}(R_n)\| \quad \text{and} \quad \|f \mid F^{s-m}_{p,q}(R_n)\| + \sum_{j=1}^n \left\| \frac{\partial^m f}{\partial x_j^m} \,\middle|\, F^{s-m}_{p,q}(R_n) \right\| \tag{3}$$

are equivalent quasi-norms on $F^s_{p,q}(R_n)$.

Proof. Step 1. First we prove the mapping properties of I_σ. If $\varphi = \{\varphi_k(x)\}_{k=0}^\infty \in \Phi(R_n)$, then $\psi = \{\psi_k(x)\}_{k=0}^\infty \in \mathcal{R}_L(R_n)$, where L is an arbitrary natural number and $\psi_k(x) = 2^{-k\sigma}(1+|x|^2)^{\frac{\sigma}{2}} \varphi_k(x)$, cf. Definitions 2.3.1/1 and 2.3.6/1. If $f \in F^s_{p,q}(R_n)$, then Theorem 2.3.6 and the obvious estimate $|(F^{-1}\psi_k Ff)(x)| \leq \psi_k^* f(x)$ yield

$$\begin{aligned} \|I_\sigma f \mid F^{s-\sigma}_{p,q}(R_n)\| &= \|2^{(s-\sigma)k} F^{-1}(1+|\cdot|^2)^{\frac{\sigma}{2}} \varphi_k Ff \mid L_p(R_n, l_q)\| \\ &= \|2^{sk} F^{-1} \psi_k Ff \mid L_p(R_n, l_q)\| \leq c \|f \mid F^s_{p,q}(R_n)\| . \end{aligned} \tag{4}$$

This proves that I_σ maps $F^s_{p,q}(R_n)$ into $F^{s-\sigma}_{p,q}(R_n)$. If $g \in F^{s-\sigma}_{p,q}(R_n)$, then we have by the same argument $f = I_{-\sigma}\, g \in F^s_{p,q}(R_n)$ and

$$\|f \mid F^s_{p,q}(R_n)\| \leq c\|g \mid F^{s-\sigma}_{p,q}(R_n)\| = c\|I_\sigma f \mid F^{s-\sigma}_{p,q}(R_n)\| . \tag{5}$$

Because, I_σ is a one-to-one mapping in $S'(R_n)$, it is also a one-to-one mapping from $F^s_{p,q}(R_n)$ onto $F^{s-\sigma}_{p,q}(R_n)$. By (4) and (5), $\|I_\sigma f \mid F^{s-\sigma}_{p,q}(R_n)\|$ is an equivalent quasi-norm in $F^s_{p,q}(R_n)$. This proves the first part of (ii). The proof of the first part of (i) is the same.

Step 2. We prove that the quasi-norms in (3) are equivalent on $F^s_{p,q}(R_n)$. If $x = (x_1, \ldots, x_n) \in R_n$, we put $x^\alpha = \prod_{j=1}^n x_j^{\alpha_j}$, where $\alpha = (\alpha_1, \ldots, \alpha_n)$ is a multi-index. It is an easy consequence of Theorem 2.3.7 that $x^\alpha(1+|x|^2)^{-\frac{m}{2}}$ is a Fourier multiplier for any space $F^\varkappa_{p,q}(R_n)$, provided that $|\alpha| \leq m$. Consequently, if $f \in F^s_{p,q}(R_n)$, then we have

$$\begin{aligned} \sum_{|\alpha|\leq m} \|D^\alpha f \mid F^{s-m}_{p,q}(R_n)\| &= \sum_{|\alpha|\leq m} \|F^{-1} x^\alpha Ff \mid F^{s-m}_{p,q}(R_n)\| \\ &= \sum_{|\alpha|\leq m} \|F^{-1} x^\alpha (1+|x|^2)^{-\frac{m}{2}} FF^{-1} (1+|x|^2)^{\frac{m}{2}} Ff \mid F^{s-m}_{p,q}(R_n)\| \\ &\leq c\|I_m f \mid F^{s-m}_{p,q}(R_n)\| \leq c'\|f \mid F^s_{p,q}(R_n)\| . \end{aligned} \tag{6}$$

Next we assume that $f \in F^{s-m}_{p,q}(R_n)$ and $\frac{\partial^m f}{\partial x_j^m} \in F^{s-m}_{p,q}(R_n)$ for $j=1, \ldots, n$. We wish to show that f belongs to $F^s_{p,q}(R_n)$. In the next step we shall prove that there exist Fourier multipliers $\varrho_1(x), \ldots, \varrho_n(x)$ for $F^{s-m}_{p,q}(R_n)$ (in the sense of Theorem 2.3.7) and a positive number c such that

$$1 + \sum_{j=1}^n \varrho_j(x)\, x_j^m \geq c\,(1+|x|^2)^{\frac{m}{2}} \quad \text{for all} \quad x \in R_n \tag{7}$$

(if m is even, then $\varrho_j(x) \equiv 1$ has the desired properties; if m is odd, then the situation is a bit more complicated). Again by Theorem 2.3.7,

$$(1+|x|^2)^{\frac{m}{2}} \left[1 + \sum_{j=1}^n \varrho_j(x) x_j^m\right]^{-1} = M(x)$$

is also a Fourier multiplier for $F^{s-m}_{p,q}(R_n)$. Consequently,

$$\|f \mid F^s_{p,q}(R_n)\| \leqq c \, \|F^{-1}(1+|x|^2)^{\frac{m}{2}} Ff \mid F^{s-m}_{p,q}(R_n)\|$$
$$\leqq c' \left\| F^{-1} M(x) \, FF^{-1}\left[1+\sum_{j=1}^{n} \varrho_j(x) \, x_j^m\right] Ff \mid F^{s-m}_{p,q}(R_n)\right\|$$
$$\leqq c'' \|f \mid F^{s-m}_{p,q}(R_n)\| + c'' \sum_{j=1}^{n} \|F^{-1}\varrho_j(x) \, x_j^m Ff \mid F^{s-m}_{p,q}(R_n)\| \, .$$

We have $x_j^m Ff = cF \dfrac{\partial^m f}{\partial x_j^m}$. Hence, the multiplier property of $\varrho_j(x)$ yields

$$\|f \mid F^s_{p,q}(R_n)\| \leqq c\|f \mid F^{s-m}_{p,q}(R_n)\| + c \sum_{j=1}^{n} \left\| \frac{\partial^m f}{\partial x_j^m} \,\middle|\, F^{s-m}_{p,q}(R_n)\right\| . \tag{8}$$

Now (6) and (8) prove that the quasi-norms in (3) are equivalent in $F^s_{p,q}(R_n)$. In the same way one proves that the quasi-norms in (2) are equivalent in $B^s_{p,q}(R_n)$.

Step 3. We prove that there exist Fourier multipliers $\varrho_1(x), \ldots, \varrho_n(x)$ with the property (7). If m is even, we choose $\varrho_j(x) \equiv 1$. Let m be odd and $n=2$. Let

$$S_t^1 = \{x \in R_2 \mid x_1 > 0, \, |x_2| \leqq t x_1\}, \quad S_t^2 = \{x \in R_2 \mid x_1 < 0, \, |x_2| \leqq t|x_1|\} ,$$

$t>0$. Let $\varphi^j(x)$ be an infinitely differentiable function in $R_2 - \{0\}$ with $\varphi^j(x) \geqq 0$, $\varphi^j(x) = \varphi^j\left(\frac{x}{|x|}\right)$, $\varphi^j(x)=1$ if $x \in S_1^j$ and supp $\varphi \subset S_2^j$, where $j=1, 2$. Let $\psi(x)$ be a non-negative infinitely differentiable function in R_2 with $\psi(x)=1$ if $|x| \geqq 1$ and $0 \notin \operatorname{supp} \psi$. Let

$$\varrho_1(x) = \psi(x) \, [\varphi^1(x) - \varphi^2(x)] \, .$$

In a similar way one constructs a function $\varrho_2(x)$. It is not hard to see that $\varrho_1(x)$ and $\varrho_2(x)$ have the desired properties. This method can be extended to $n=3, 4, 5, \ldots$

If $h \in R_n$ then the translation operator T_h is defined by $(T_h f)(x) = f(x+h)$ for any regular distribution. T_h can be extended in the usual way to $S'(R_n)$: if $f \in S'(R_n)$ and $\varphi \in S(R_n)$, then we put $(T_h f)(\varphi(\cdot)) = f\,(\varphi(\cdot - h))$. Recall that

$$FT_h f = e^{ihx} Ff \quad \text{and} \quad F^{-1}(e^{ihx} f) = T_h F^{-1} f \, , \tag{9}$$

cf. (1.2.1/2) and (1.2.1/3) and the usual extension procedures from $S(R_n)$ to $S'(R_n)$. In particular, if $\varphi = \{\varphi_j(x)\}_{j=0}^\infty \in \Phi(R_n)$, then

$$(F^{-1}\varphi_j F T_h f)(x) = T_h(F^{-1}\varphi_j Ff)(x) = (F^{-1}\varphi_j Ff)(x+h) \, . \tag{10}$$

Proposition. Let $-\infty < s < \infty$ and $0 < q \leqq \infty$. If $0 < p \leqq \infty$, then T_h yields an isomorphic mapping from $B^s_{p,q}(R_n)$ onto itself. If $0 < p < \infty$, then T_h yields an isomorphic mapping from $F^s_{p,q}(R_n)$ onto itself.

Proof. The proof follows immediately from (10).

Remark 1. By (9) and the proposition it follows that e^{ixh} is a Fourier multiplier for $B^s_{p,q}(R_n)$ and $F^s_{p,q}(R_n)$. However this simple Fourier multiplier is not covered by 2.3.7.

Remark 2. Both the above theorem and the proposition can be extended to $F^s_{\infty,q}(R_n)$ with $1 < q \leqq \infty$. In the case of the proposition this follows simply by duality, cf. Proposition 2.3.4/2. As for the theorem, a proof can be based on Remark 2.3.7/4.

2.3.9. Diversity of the Spaces $B^s_{p,q}(R_n)$ and $F^s_{p,q}(R_n)$

The diversity of the spaces $\Lambda^s_{p,q}(R_n)$ and $H^s_p(R_n)$ from 2.2.2. attracted much attention in the history of these spaces, cf. O. V. Besov [2], K. K. Golovkin [2] and M. H. Taibleson [1, I, Lemma 2]. Cf. also S. M. Nikol'skij [3, 9.7] and [I, 2.12]. Thus it is natural to ask the same question in our more general framework.

Theorem. Let $-\infty<s_0<\infty$, $-\infty<s_1<\infty$, $0<q_0\leq\infty$ and $0<q_1\leq\infty$.
(i) Let $0<p_0\leq\infty$ and $0<p_1\leq\infty$. Then we have $B^{s_0}_{p_0,q_0}(R_n)=B^{s_1}_{p_1,q_1}(R_n)$ if and only if $s_0=s_1$, $q_0=q_1$ and $p_0=p_1$.
(ii) Let $0<p_0<\infty$ and $0<p_1<\infty$. Then we have $F^{s_0}_{p_0,q_0}(R_n)=F^{s_1}_{p_1,q_1}(R_n)$ if and only if $s_0=s_1$, $q_0=q_1$ and $p_0=p_1$.
(iii) Let $0<p_0<\infty$ and $0<p_1\leq\infty$. Then we have $F^{s_0}_{p_0,q_0}(R_n)=B^{s_1}_{p_1,q_1}(R_n)$ if and only if $s_0=s_1$ and $p_0=q_0=p_1=q_1<\infty$.

Proof. Step 1. The "if"-part of the theorem is obvious because $F^s_{p,p}(R_n)=B^s_{p,p}(R_n)$. We prove the "only if"-part.

Step 2. Let $\psi(x)\in S(R_n)$ be an appropriate function with supp $F\psi\subset\{x\mid 1\leq|x|\leq 2\}$. If $\psi_j(x)=\psi(2^jx)$, then supp $F\psi_j\subset\{x\mid 2^j\leq|x|\leq 2^{j+1}\}$, cf. (1.3.2/6). Then it follows from the Definitions 2.3.1/1 and 2.3.1/2 (and appropriate choices of ψ and $\{\varphi_j\}\in\Phi(R_n)$) that

$$\|\psi_j\mid B^s_{p,q}(R_n)\|=\|\psi_j\mid F^s_{p,q}(R_n)\|=\|\psi(2^j\cdot)\mid L_p(R_n)\|=2^{-\frac{jn}{p}}\|\psi\mid L_p(R_n)\| \tag{1}$$

if $j=0,-1,-2,\ldots$ and

$$\|\psi_j\mid B^s_{p,q}(R_n)\|=\|\psi_j\mid F^s_{p,q}(R_n)\|=2^{sj}\|\psi(2^j\cdot)\mid L_p(R_n)\|=2^{\left(s-\frac{n}{p}\right)j}\|\psi\mid L_p(R_n)\| \tag{2}$$

if $j=1,2,3,\ldots$ Now we assume that two spaces, on the one hand $B^{s_0}_{p_0,q_0}(R_n)$ or $F^{s_0}_{p_0,q_0}(R_n)$ and on the other hand $B^{s_1}_{p_1,q_1}(R_n)$ or $F^{s_1}_{p_1,q_1}(R_n)$ coincide. Then (1) yields $p_0=p_1$ and (2) (afterwards) $s_0=s_1$.

Step 3. Let $x^0=(1,0,0,\ldots,0)$ and let $\psi(x)\in S(R_n)$ be an appropriate function with supp $\psi\subset\{x\mid |x-x^0|\leq 1\}$. Let $x'=(x_2,\ldots,x_n)$. If f is defined by

$$(Ff)(x)=\sum_{j=1}^N a_j\psi(x_1-2^j,x'),\quad a_j \text{ complex numbers,}$$

then we have

$$\|f\mid F^s_{p,q}(R_n)\|=\|F^{-1}\psi\mid L_p(R_n)\|\left(\sum_{j=1}^N 2^{sjq}|a_j|^q\right)^{\frac{1}{q}}$$

(modification if $q=\infty$), cf. also formula (3) below. Consequently, $q_0=q_1$ if $F^s_{p,q_0}(R_n)=F^s_{p,q_1}(R_n)$. This assertion and Step 2 prove (ii).

Step 4. In order to prove (i) and (iii) we may assume that $s_0=s_1=0$ and $p_0=p_1=p$. This follows from Step 2 and the lifting property from Theorem 2.3.8. Let ψ be the same function as in the previous step. Let

$$(Ff)(x)=\sum_{j=1}^N e^{2^jx_1}\psi(x_1-2^j,x').$$

If $\varphi=\{\varphi_j(x)\}_{j=0}^\infty\in\Phi(R_n)$ is an appropriate system, then we have

$$(F^{-1}\varphi_jFf)(x)=F^{-1}(e^{2^jy_1}\psi(y_1-2^j,y'))(x)=e^{i2^j(x_1+2^j)}(F^{-1}\psi)(x_1+2^j,x'), \tag{3}$$

cf. (1.2.1/3). Consequently,

$$\|f\mid B^0_{p,q}(R_n)\|\sim\|F^{-1}\psi\mid L_p(R_n)\|\,N^{\frac{1}{q}} \tag{4}$$

and

$$\|f\mid F^0_{p,q}(R_n)\|\sim\|F^{-1}\psi\mid L_p(R_n)\|\,N^{\frac{1}{q}}. \tag{5}$$

Now (i) follows from (4). Finally, if $\overset{0}{F}_{p,q_0}(R_n)=\overset{0}{B}_{p,q_1}(R_n)$ then (4) and (5) yield $q_1=p$. Hence $\overset{0}{F}_{p,p}(R_n)=\overset{0}{B}_{p,p}(R_n)=\overset{0}{F}_{p,q_0}(R_n)$. Then we obtain from (ii) that $q_0=p$. This proves (iii).

Remark. The Besov spaces $\Lambda^s_{p,q}(R_n)$ and the Bessel-potential spaces $H^s_p(R_n)$ from 2.2.2. are always different (with the exception of $p=q=2$). This classical result, which has been treated in the above-cited papers, is a special case of the theorem, cf. (2.3.5/2) and (2.3.5/3). Furthermore, in [I, 2.12] we proved that the spaces $\Lambda^s_{p,q}(R_n)$ and $H^s_p(R_n)$ with $p\neq 2$ are not only different from a set-theoretical point of view, but they differ also in their isomorphic structure as Banach spaces. For details we refer to [I, 2.12]. A corresponding result for the more general spaces $B^s_{p,q}(R_n)$ and $F^s_{p,q}(R_n)$ would be desirable. Finally we mention that the above theorem can be extended to the spaces $F^s_{\infty,q}(R_n)$ with $-\infty<s<\infty$ and $1<q\leqq\infty$. This follows by duality arguments, cf. 2.3.4. and 2.11.2.

2.4. Interpolation

2.4.1. Preliminaries

Interpolation methods are of some interest for our later considerations. But the central ideas of this chapter are independent of the results of this section. On the other hand, we shall occasionally use some interpolation formulas which are proved in this section. A reader who is not interested in interpolation theory may skip this section, provided he is ready to consult the corresponding results later on, when they will be needed from time to time. Furthermore, we shall not develop the abstract interpolation theory (with the exception of the few remarks below about the real method in quasi-Banach spaces). We prefer to give direct proofs for the main assertions of this section which are not based on general interpolation assertions.

Interpolation methods. As far as concrete interpolation procedures are concerned there exist essentially two methods: The real interpolation method by J.-L. Lions and J. Peetre, and the complex interpolation method by J.-L. Lions, A. P. Calderón and S. G. Krejn. An extensive treatment of these abstract methods (and many references) has been given in [I], cf. also J. Bergh, J. Löfström [1]. Originally, both methods were developed in the framework of Banach space theory. However it is not difficult to see (and well-known nowadays) that the real method (or at least some of its crucial assertions) can be extended immediately to quasi-Banach spaces (concerning the notion of a quasi-Banach space cf. 1.2.2.). Below we give a brief description of the real method for quasi-Banach spaces and we prove the so-called interpolation property. In contrast to the real method, the complex method cannot be extended immediately from Banach spaces to quasi-Banach spaces. Recently, A. P. Calderón, A. Torchinsky [1] discovered a modification of the usual complex method, which can be applied to some concrete quasi-Banach spaces in the framework of Fourier analysis, in particular to the above spaces $B^s_{p,q}(R_n)$ and $F^s_{p,q}(R_n)$. This modified concrete complex method (and not the usual abstract complex method) will be described in 2.4.4.–2.4.7. (including a weak interpolation property).

The real method. The general background of an interpolation method is the following: let $\mathcal{H}$ be a linear complex Hausdorff space, and let A_0 and A_1 be two complex quasi-Banach spaces such that $A_0\subset\mathcal{H}$ and $A_1\subset\mathcal{H}$. Recall that "$\subset$" indicates a linear and continuous embedding. In particular, the algebraic operations (addition of elements and multiplication by complex numbers) are the same in

A_0, A_1, and $\mathcal{H}$ (as far as they make sense). Let A_0+A_1 be the set of all elements $a\in\mathcal{H}$ which can be represented as $a=a_0+a_1$ with $a_0\in A_0$ and $a_1\in A_1$. If $0<t<\infty$ and $a\in A_0+A_1$, then Peetre's celebrated K-functional is given by

$$K(t, a)=K(t, a; A_0, A_1)=\inf (\|a_0 \mid A_0\|+t\|a_1 \mid A_1\|), \tag{1}$$

where the infimum is taken over all representations of a of the form $a=a_0+a_1$ with $a_0\in A_0$ and $a_1\in A_1$.

Definition. Let $0<\Theta<1$. If $0<q<\infty$ then

$$(A_0, A_1)_{\Theta,q}=\left\{a \mid a\in A_0+A_1, \|a \mid (A_0, A_1)_{\Theta,q}\|=\left(\int_0^\infty [t^{-\Theta}K(t,a)]^q \frac{dt}{t}\right)^{\frac{1}{q}}<\infty\right\}. \tag{2}$$

If $q=\infty$, then

$$(A_0, A_1)_{\Theta,\infty}=\{a \mid a\in A_0+A_1, \|a \mid (A_0, A_1)_{\Theta,\infty}\|=\sup_{0<t<\infty} t^{-\Theta}K(t, a)<\infty\}. \tag{3}$$

Remark 1. These are the well-known real interpolation spaces. One can prove that $(A_0, A_1)_{\Theta,q}$ is always a quasi-Banach space. In our concrete cases, which will be treated in the following subsections, this will be clear from the very beginning, because we shall only deal with the spaces $B^s_{p,q}(R_n)$ and $F^s_{p,q}(R_n)$.

Remark 2. One can replace the complex spaces A_0, A_1 and $\mathcal{H}$ by corresponding real spaces. This is immaterial for the real interpolation method.

Proposition (Interpolation property). $\mathcal{H}$, A_0 and A_1 have the above meaning. Let T be a linear operator mapping $\mathcal{H}$ into itself such that its restriction to A_k yields a linear and bounded mapping from A_k into itself, where $k=0$ or 1. Then the restriction of T to $(A_0, A_1)_{\Theta,q}$ (where $0<\Theta<1$ and $0<q\leqq\infty$) yields a linear and bounded mapping from $(A_0, A_1)_{\Theta,q}$ into itself.

Proof. It is easy to see that $(A_0, A_1)_{\Theta,q}$ is a linear space. Hence, the restriction of T to $(A_0, A_1)_{\Theta,q}$ is a linear operator. Let

$$\|Ta_k \mid A_k\|\leqq\|T\|_k \|a_k \mid A_k\| \quad \text{for all} \quad a_k\in A_k \tag{4}$$

if $k=0$ or 1. If $a=a_0+a_1$ with $a_0\in A_0$ and $a_1\in A_1$, then $Ta=Ta_0+Ta_1$ and (1) gives

$$\begin{aligned} K(t, Ta)&\leqq\inf (\|Ta_0 \mid A_0\|+t\|Ta_1 \mid A_1\|) \\ &\leqq\|T\|_0 \inf (\|a_0 \mid A_0\|+t\|T\|_1 \|T\|_0^{-1} \|a_1 \mid A_1\|) \\ &=\|T\|_0 K(t\|T\|_1 \|T\|_0^{-1}, a)\,, \end{aligned} \tag{5}$$

where the infimum in (5) is taken over all representations $a=a_0+a_1$ (we assume that $\|T\|_0>0$). By a simple transformation of coordinates we have

$$\|Ta \mid (A_0, A_1)_{\Theta,q}\|\leqq\|T\|_0^{1-\Theta} \|T_1\|^{\Theta} \|a \mid (A_0, A_1)_{\Theta,q}\|\,. \tag{6}$$

In particular $Ta\in(A_0, A_1)_{\Theta,q}$. The proof is complete.

Remark 3. We proved slightly more than we stated. If

$$\|Ta \mid (A_0, A_1)_{\Theta,q}\|\leqq\|T\|_{\Theta,q} \|a \mid (A_0, A_1)_{\Theta,q}\|\,,$$

then (6) shows that

$$\|T\|_{\Theta,q}\leqq\|T\|_0^{1-\Theta} \|T\|_1^{\Theta}\,. \tag{7}$$

By a limiting argument this is also valid if $\|T\|_0=0$.

Remark 4. The following elementary remark will be useful later on. Let $\mathcal{H}$ be the above linear complex Hausdorff space and let $A_0\subset B_0\subset\mathcal{H}$ and $A_1\subset B_1\subset\mathcal{H}$, where A_0, A_1, B_0 and B_1

are quasi-Banach spaces. If $0<\Theta<1$ and $0<q\leqq\infty$, then

$$(A_0, A_1)_{\Theta,q}\subset(B_0, B_1)_{\Theta,q} \tag{8}$$

This follows easily from (1), (2) and (3).

2.4.2. Real Interpolation for the Spaces $B^s_{p,q}(R_n)$ and $F^s_{p,q}(R_n)$ with Fixed p

The basis for the real interpolation theory for the spaces $B^s_{p,q}(R_n)$ and $F^s_{p,q}(R_n)$ is Theorem 2.3.3, which shows that the Hausdorff space $\mathcal{H}$ from 2.4.1. can be identified with $S'(R_n)$.

Theorem. Let $0<\Theta<1$, $0<q_0\leqq\infty$, $0<q_1\leqq\infty$ and $0<q\leqq\infty$. Furthermore, let $-\infty<s_0<\infty$, $-\infty<s_1<\infty$, $s_0\neq s_1$ and $s=(1-\Theta)\,s_0+\Theta s_1$.
(i) If $0<p\leqq\infty$, then

$$(B^{s_0}_{p,q_0}(R_n), B^{s_1}_{p,q_1}(R_n))_{\Theta,q}=B^s_{p,q}(R_n)\,. \tag{1}$$

(ii) If $0<p<\infty$, then

$$(F^{s_0}_{p,q_0}(R_n), F^{s_1}_{p,q_1}(R_n))_{\Theta,q}=B^s_{p,q}(R_n)\,. \tag{2}$$

Proof. Step 1. First we prove that

$$(B^{s_0}_{p,\infty}(R_n), B^{s_1}_{p,\infty}(R_n))_{\Theta,q}\subset B^s_{p,q}(R_n)\,. \tag{3}$$

We may assume that $s_0>s_1$. Let $q<\infty$. We have

$$(0,\infty)=\bigcup_{k=-\infty}^{\infty}[2^{(k-1)(s_0-s_1)}, 2^{k(s_0-s_1)})$$

and

$$\int_0^\infty t^{-\Theta q}K^q(t, f; B^{s_0}_{p,\infty}(R_n), B^{s_1}_{p,\infty}(R_n))\frac{dt}{t} \\ \geqq c\sum_{k=0}^{\infty}2^{-\Theta qk(s_0-s_1)}K^q(2^{k(s_0-s_1)}, f; B^{s_0}_{p,\infty}(R_n), B^{s_1}_{p,\infty}(R_n))\,. \tag{4}$$

c is an appropriate positive constant and $f\in B^{s_0}_{p,\infty}(R_n)+B^{s_1}_{p,\infty}(R_n)$. Let $f=f_0+f_1$ with $f_0\in B^{s_0}_{p,\infty}(R_n)$ and $f_1\in B^{s_1}_{p,\infty}(R_n)$. If $\{\varphi_j(x)\}_{j=0}^{\infty}\in\Phi(R_n)$ then (2.3.1/5) yields

$$\begin{aligned}&2^{ks_0}\|F^{-1}\varphi_kFf\mid L_p(R_n)\| \\ &\leqq c2^{ks_0}\|F^{-1}\varphi_kFf_0\mid L_p(R_n)\|+c2^{k(s_0-s_1)}2^{ks_1}\|F^{-1}\varphi_kFf_1\mid L_p(R_n)\| \\ &\leqq c\|f_0\mid B^{s_0}_{p,\infty}(R_n)\|+c2^{k(s_0-s_1)}\|f_1\mid B^{s_1}_{p,\infty}(R_n)\|\,.\end{aligned} \tag{5}$$

The infimum with respect to all representations $f=f_0+f_1$ yields the K-functional in (4). Consequently,

$$\sum_{k=0}^{\infty}2^{qks}\|F^{-1}\varphi_kFf\mid L_p(R_n)\|^q$$

can be estimated from above by the left-hand side of (4). This proves (3). A small modification yields a corresponding result for $q=\infty$.

Step 2. If $0<r<q$, we prove that

$$B^s_{p,q}(R_n)\subset(B^{s_0}_{p,r}(R_n), B^{s_1}_{p,r}(R_n))_{\Theta,q}. \tag{6}$$

Again we assume that $s_0>s_1$. Because $s>s_1$, Proposition 2.3.2/2 gives

$$K(t, f; B^{s_0}_{p,r}(R_n), B^{s_1}_{p,r}(R_n))\leqq ct\|f\mid B^{s_1}_{p,r}(R_n)\|\leqq c't\|f\mid B^s_{p,q}(R_n)\|\,.$$

In particular,

$$\left(\int_0^1 t^{-\Theta q} K^q(t, f;\, B^{s_0}_{p,r}(R_n),\, B^{s_1}_{p,r}(R_n))\, \frac{dt}{t}\right)^{\frac{1}{q}} \leq c\|f \mid B^s_{p,q}(R_n)\| \tag{7}$$

(modification if $q=\infty$). Furthermore,

$$\begin{aligned} I &= \int_1^\infty t^{-\Theta q} K^q(t, f;\, B^{s_0}_{p,r}(R_n),\, B^{s_1}_{p,r}(R_n))\, \frac{dt}{t} \\ &\leq c \sum_{k=0}^\infty 2^{-\Theta q k(s_0-s_1)} K^q(2^{k(s_0-s_1)}, f;\, B^{s_0}_{p,r}(R_n),\, B^{s_1}_{p,r}(R_n)) \,. \end{aligned} \tag{8}$$

If again $\varphi=\{\varphi_j(x)\}_{j=0}^\infty \in \Phi(R_n)$, then we put (for the sake of brevity) $\xi_j = \|F^{-1}\varphi_j Ff \mid L_p(R_n)\|$. We use the representation $f=f_0+f_1$ with

$$f_0=\sum_{j=0}^k F^{-1}\varphi_j Ff \quad \text{and} \quad f_1=\sum_{j=k+1}^\infty F^{-1}\varphi_j Ff \,.$$

The properties of the system φ and the scalar case of Theorem 1.6.3 (cf. also Remark 1.5.2/3) show that

$$\|f_0 \mid B^{s_0}_{p,r}(R_n)\|^r \leq c \sum_{j=0}^{k+1} 2^{js_0 r}\xi_j^r, \quad \|f_1 \mid B^{s_1}_{p,r}(R_n)\|^r \leq c \sum_{j=k}^\infty 2^{js_1 r}\xi_j^r \,.$$

If we use this special decomposition, then (8) yields

$$I \leq c \sum_{k=0}^\infty 2^{qks} \left[\sum_{j=0}^{k+1} 2^{s_0 r(j-k)}\xi_j^r + \sum_{j=k}^\infty 2^{s_1 r(j-k)}\xi_j^r\right]^{\frac{q}{r}} . \tag{9}$$

Let $s_0>\varkappa_0>s>\varkappa_1>s_1$. By Hölder's inequality with $\frac{r}{q}+\frac{r}{\sigma}=1$ we have

$$\begin{aligned} I &\leq c \sum_{k=0}^\infty 2^{qk(s-s_0)} \left(\sum_{j=0}^{k+1} 2^{(s_0-\varkappa_0)\sigma j}\right)^{\frac{q}{\sigma}} \left(\sum_{j=0}^{k+1} 2^{\varkappa_0 qj}\xi_j^q\right) \\ &\quad + c \sum_{k=0}^\infty 2^{qk(s-s_1)} \left(\sum_{j=k}^\infty 2^{(s_1-\varkappa_1)\sigma j}\right)^{\frac{q}{\sigma}} \left(\sum_{j=k}^\infty 2^{\varkappa_1 qj}\xi_j^q\right) \\ &\leq c \sum_{j=0}^\infty 2^{\varkappa_0 qj}\xi_j^q \sum_{k=j-1}^\infty 2^{qk(s-\varkappa_0)} + c \sum_{j=0}^\infty 2^{\varkappa_1 qj}\xi_j^q \sum_{k=0}^j 2^{qk(s-\varkappa_1)} \\ &\leq c' \sum_{j=0}^\infty 2^{jqs}\xi_j^q = c'\|f \mid B^s_{p,q}(R_n)\|^q \,. \end{aligned} \tag{10}$$

(7), (8) and (10) prove (6) (if $q=\infty$, then the above calculations must be modified in the usual way).

Step 3. In order to prove (1), we use Remark 2.4.1/4 and the embeddings from Proposition 2.3.2/2. If $r\leq q_0$ and $r\leq q_1$, then (3) and (6) yield

$$\begin{aligned} B^s_{p,q}(R_n) &\subset (B^{s_0}_{p,r}(R_n),\, B^{s_1}_{p,r}(R_n))_{\Theta,q} \\ &\subset (B^{s_0}_{p,q_0}(R_n),\, B^{s_1}_{p,q_1}(R_n))_{\Theta,q} \\ &\subset (B^{s_0}_{p,\infty}(R_n),\, B^{s_1}_{p,\infty}(R_n))_{\Theta,q} \subset B^s_{p,q}(R_n) \,. \end{aligned}$$

This proves (1). Finally, (2) follows in the same way from (1) and (2.3.2/9).

Remark 1. The proof is essentially a modification of a corresponding proof in [I, pp. 124/125].

Remark 2. If $1<p<\infty$, $1\leqq q_0\leqq\infty$ and $1\leqq q_1\leqq\infty$, then (1) and (2) coincide essentially with [I, Theorem 2.4.2/2]. The cases with $p<1$ (and $q_0\leqq 1$ or $q_1\leqq 1$) are also more or less known, cf. J. Peetre [4 and 8, p. 245], B. Jawerth [1, 2] and [S, 2.2.10].

2.4.3. Real Interpolation for the Spaces $B^s_{p,p}(R_n)$

Theorem 2.4.2 is essentially sufficient for our later purposes (as far as the real method is concerned). The number p in Theorem 2.4.2 is fixed. One can raise the question: what about real interpolation formulas if spaces with different p's are involved? This problem is of interest for several reasons, but it is harder than the corresponding problem with fixed p's, which we treated in 2.4.2. In the Banach space case, i.e. $1<p<\infty$ and $1\leqq q\leqq\infty$, an extensive treatment has been given in [I, 2.4]. Concerning the general case we refer to [S, pp. 73/74]. We formulate here an interesting result and add some remarks concerning its proof.

Theorem. Let $-\infty<s_0<\infty$, $-\infty<s_1<\infty$, $0<p_0<\infty$ and $0<p_1<\infty$. If $0<\Theta<1$, and

$$s=(1-\Theta)s_0+\Theta s_1\,,\qquad \frac{1}{p}=\frac{1-\Theta}{p_0}+\frac{\Theta}{p_1}\,, \tag{1}$$

then

$$(B^{s_0}_{p_0,p_0}(R_n),\, B^{s_1}_{p_1,p_1}(R_n))_{\Theta,p}=B^s_{p,p}(R_n)\,. \tag{2}$$

Remark 1. In [S, pp. 72/73] we outlined a proof of (2) which is based on abstract interpolation methods and on

$$(H_{p_0}(R_n),\, H_{p_1}(R_n))_{\Theta,p}=H_p(R_n)\,, \tag{3}$$

where $H_p(R_n)$ are the Hardy spaces from Remark 2.2.2/2. The interpolation formula (3) is due to C. Fefferman, N. M. Riviere, Y. Sagher [1], cf. also N. M. Riviere, Y. Sagher [1] and W. R. Madych [1]. One can prove that a corresponding result holds for the local Hardy spaces $h_p(R_n)$ from (2.2.2/12), i.e.

$$(h_{p_0}(R_n),\, h_{p_1}(R_n))_{\Theta,p}=h_p(R_n)\,, \tag{4}$$

where again $0<p_0<\infty$, $0<p_1<\infty$, $0<\Theta<1$ and $\frac{1}{p}=\frac{1-\Theta}{p_0}+\frac{\Theta}{p_1}$. On the basis of (4) and the so-called retraction method of interpolation theory one can carry over many interpolation formulas from the Banach space case (considered in [I, 2.4]) to the more general case treated here. A description of this technique in the above framework may be found in [S, pp. 72–74].

Remark 2. However, in order to prove (2), it is not necessary to use the far-reaching retraction method of interpolation theory and the deep formulas (3) or (4). As in 2.4.2. one can give a direct proof of (2). For that purpose one modifies Step 1 of the proof of the theorem in [I, p. 121] in the same way as we modified the proof in [I, pp. 124/125] in connection with Theorem 2.4.2. But in contrast to 2.4.2. one needs a new interpolation method, the so-called L-method, cf. [I, 1.4]. We do not go into detail.

2.4.4. Complex Interpolation: Definitions

The rest of Section 2.4. deals with a complex interpolation method for the spaces $B^s_{p,q}(R_n)$ and $F^s_{p,q}(R_n)$, and an application to Fourier multipliers for $F^s_{p,q}(R_n)$. Furthermore, we discuss briefly corresponding problems for the spaces $L^\Omega_p(R_n, l_q)$ from 1.6.1. As has been mentioned in 2.4.1., the complex method under consideration here is not the classical method by J. L. Lions, A. P. Calderón and S. G. Krejn, which is restricted to Banach spaces.

Let $A=\{z\mid 0<\operatorname{Re} z<1\}$ be a strip in the complex plane. Its closure $\{z\mid 0\leqq \operatorname{Re} z\leqq 1\}$ is denoted by $\bar{A}$. We say that $f(z)$ is an $S'(R_n)$-analytic function in A if the following properties are satisfied:
(i) For every fixed $z\in\bar{A}$ we have $f(z)\in S'(R_n)$,
(ii) $(F^{-1}\varphi Ff)(x,z)$ is a uniformly continuous and bounded function in $R_n\times\bar{A}$ for every $\varphi\in S(R_n)$ with compact support in R_n,
(iii) $(F^{-1}\varphi Ff)(x,z)$ is an analytic function in A for every $\varphi\in S(R_n)$ with compact support in R_n and every fixed $x\in R_n$.
We recall that $(F^{-1}\varphi Ff)(x,z)$ is an entire analytic function in R_n with respect to x if $z\in\bar{A}$ is fixed, cf. 1.2.1.

Definition 1. Let $-\infty<s_0<\infty$, $-\infty<s_1<\infty$, $0<q_0\leqq\infty$ and $0<q_1\leqq\infty$.
(i) If $0<p_0\leqq\infty$ and $0<p_1\leqq\infty$, then

$$\begin{aligned}&F(B^{s_0}_{p_0,q_0}(R_n),\,B^{s_1}_{p_1,q_1}(R_n))\\&=\{f(z)\mid f(z)\ \text{ is an } S'(R_n)\text{-analytic function in } A,\\&\quad f(it)\in B^{s_0}_{p_0,q_0}(R_n)\,,\quad f(1+it)\in B^{s_1}_{p_1,q_1}(R_n)\quad\text{for every}\quad t\in R_1\,,\\&\quad \|f(z)\mid F(B^{s_0}_{p_0,q_0},\,B^{s_1}_{p_1,q_1})\|=\max_{l=0,1}\sup_{t\in R_1}\|f(l+it)\mid B^{s_l}_{p_l,q_l}(R_n)\|<\infty\}\,.\end{aligned}\tag{1}$$

(ii) If $0<p_0<\infty$ and $0<p_1<\infty$, then

$$\begin{aligned}&F(F^{s_0}_{p_0,q_0}(R_n),\,F^{s_1}_{p_1,q_1}(R_n))\\&=\{f(z)\mid f(z) \text{ is an } S'(R_n)\text{-analytic function in } A,\\&\quad f(it)\in F^{s_0}_{p_0,q_0}(R_n),\ f(1+it)\in F^{s_1}_{p_1,q_1}(R_n)\quad\text{for every}\quad t\in R_1,\\&\quad \|f(z)\mid F(F^{s_0}_{p_0,q_0},\,F^{s_1}_{p_1,q_1})\|=\max_{l=0,1}\sup_{t\in R_1}\|f(l+it)\mid F^{s_l}_{p_l,q_l}(R_n)\|<\infty\}\,.\end{aligned}\tag{2}$$

Remark 1. This is the substitute of the well-known construction by A. P. Calderón [2] in the case of Banach spaces, cf. also [I, p. 56]. In the classical case (which is restricted to Banach spaces) $f(z)$ is a vector-valued analytic function in A (with values in a Banach space). The idea of using $S'(R_n)$-analytic functions is due to A. P. Calderón, A. Torchinsky [1, II, p. 135]. Cf. also L. Päivärinta [1] and H. Triebel [15, 16]. Our treatment follows closely H. Triebel [19].

Definition 2. Let $-\infty<s_0<\infty$, $-\infty<s_1<\infty$, $0<q_0\leqq\infty$, $0<q_1\leqq\infty$ and $0<\Theta<1$.
(i) If $0<p_0\leqq\infty$ and $0<p_1\leqq\infty$, then

$$\begin{aligned}&(B^{s_0}_{p_0,q_0}(R_n),\,B^{s_1}_{p_1,q_1}(R_n))_\Theta\\&=\{g\mid \exists f(z)\in F(B^{s_0}_{p_0,q_0}(R_n),\,B^{s_1}_{p_1,q_1}(R_n))\quad\text{with}\quad g=f(\Theta)\}\,,\end{aligned}\tag{3}$$

$$\|g\mid (B^{s_0}_{p_0,q_0}(R_n),\,B^{s_1}_{p_1,q_1}(R_n))_\Theta\|=\inf\|f(z)\mid F(B^{s_0}_{p_0,q_0},\,B^{s_1}_{p_1,q_1})\|\,,\tag{4}$$

where the infimum is taken over all admissible functions $f(z)$ in the sense of (3).
(ii) If $0<p_0<\infty$ and $0<p_1<\infty$, then

$$\begin{aligned}&(F^{s_0}_{p_0,q_0}(R_n),\,F^{s_1}_{p_1,q_1}(R_n))_\Theta\\&=\{g\mid \exists f(z)\in F(F^{s_0}_{p_0,q_0}(R_n),\,F^{s_1}_{p_1,q_1}(R_n))\quad\text{with}\quad g=f(\Theta)\}\,,\end{aligned}\tag{5}$$

$$\|g\mid (F^{s_0}_{p_0,q_0}(R_n),\,F^{s_1}_{p_1,q_1}(R_n))_\Theta\|=\inf\|f(z)\mid F(F^{s_0}_{p_0,q_0},\,F^{s_1}_{p_1,q_1})\|\,,\tag{6}$$

where the infimum is taken over all admissible functions $f(z)$ in the sense of (5).

Remark 2. The classical complex method is denoted by $[\cdot,\cdot]_\Theta$, in contrast to our notation $(\cdot,\cdot)_\Theta$, cf. A. P. Calderón [2] or [I, p. 58/59]. If the above numbers p_0, p_1, q_0 and q_1 are larger than or equal to 1 (i.e. in the Banach space case) both methods makes sense and the question arises whether they coincide or not. In many cases the two methods yield the same spaces, but in some limiting cases the two methods yield different results, cf. Remark 2.4.7/2.

2.4.5. Complex Interpolation: Properties

In 2.4.7. we shall prove that the interpolation spaces from Definition 2.4.4/2 coincide with some of the spaces $B^s_{p,q}(R_n)$ and $F^s_{p,q}(R_n)$, respectively. Then it will be clear that the spaces from Definition 2.4.4/2 are quasi-Banach spaces. One can ask the same question for the spaces from Definition 2.4.4/1.

Proposition. Let $-\infty<s_0<\infty$, $-\infty<s_1<\infty$, $0<q_0\leqq\infty$ and $0<q_1\leqq\infty$.
(i) If $0<p_0\leqq\infty$ and $0<p_1\leqq\infty$, then $F(B^{s_0}_{p_0,q_0}(R_n), B^{s_1}_{p_1,q_1}(R_n))$ is a quasi-Banach space.
(ii) If $0<p_0<\infty$ and $0<p_1<\infty$, then $F(F^{s_0}_{p_0,q_0}(R_n), F^{s_1}_{p_1,q_1}(R_n))$ is a quasi-Banach space.

Remark. This is an interesting result, but we shall not need it later on. A (comparatively elementary) proof may be found in H. Triebel [19]. In this paper we have also given a more direct proof (without using the full theorem 2.4.7) ot thc fact that the interpolation space from Definition 2.4.4/2 are quasi-Banach spaces.

2.4.6. Some Preparations

Our main goal is the proof of Theorem 2.4.7. For that purpose it is convenient to start with some preparations.

Lemma 1. Let $\{a_k\}_{k=0}^\infty$ be a sequence of positive numbers. Let $b_k=a_k\left(\sum_{m=0}^k a_m\right)^{-\varkappa}$ where $k=0, 1, 2, \dots$ If $0\leqq\varkappa<1$, then

$$\sum_{k=0}^\infty b_k=2\,(2^{1-\varkappa}-1)^{-1}\left(\sum_{k=0}^\infty a_k\right)^{1-\varkappa}. \tag{1}$$

Proof. We may assume that $\sum_{k=0}^\infty a_k=1$, the general case following from a homogeneity argument. We choose a sequence of integers $K_1, K_2, \dots$ such that $\infty>K_1\geqq\geqq K_2\geqq\dots\geqq 0$ and

$$\sum_{l=0}^{K_m} a_l\geqq 2^{-m} \quad\text{and}\quad \sum_{l=0}^{K_m-1} a_l<2^{-m} \quad\text{if}\quad m=1, 2, \dots$$

(with the usual interpretation $\sum_{l=0}^{-1} a_l=0$). If $K_0-1=\infty$, then

$$\sum_{k=0}^\infty b_k=\sum_{m=1}^\infty\sum_{j=K_m}^{K_{m-1}-1} b_j\leqq\sum_{m=1}^\infty 2^{m\varkappa}\sum_{j=K_m}^{K_{m-1}-1} a_j\leqq\sum_{m=1}^\infty 2^{m\varkappa}2^{-m+1}.$$

This proves (1).

Lemma 2. Let $A=\{z\mid z \text{ complex}, 0<\operatorname{Re} z<1\}$, $\bar{A}=\{z\mid 0\leqq\operatorname{Re} z\leqq 1\}$ and $0<r<\infty$. Then there exist two positive functions $\mu_0(\Theta, t)$ and $\mu_1(\Theta, t)$ in $(0,1)\times R_1$ such that

$$|g(z)|^r\leqq\left(\frac{1}{1-\Theta}\int_{R_1}|g(it)|^r\,\mu_0(\Theta, t)\right)^{1-\Theta}\left(\frac{1}{\Theta}\int_{R_1}|g\,(1+it)|^r\,\mu_1(\Theta, t)\,dt\right)^{\Theta} \tag{2}$$

with $\Theta=\operatorname{Re} z$ for any analytic function $g(z)$ in A which is uniformly continuous and bounded in $\bar{A}$. Furthermore, if $0<\Theta<1$, then

$$\frac{1}{1-\Theta}\int_{R_1}\mu_0(\Theta, t)\,dt=\frac{1}{\Theta}\int_{R_1}\mu_1(\Theta, t)\,dt=1. \tag{3}$$

Proof. (Outline). If $r=1$, then (2) is known, cf. A. P. Calderón [2, 9.4 and 29.4] or [I, pp. 65 and 67]. $\mu_0(\Theta, t)$ and $\mu_1(\Theta, t)$ are the Poisson kernels for the strip A. The key inequality of that proof reads as follows,

$$\log |g(z)| \leq \int_{R_1} \log |g(\mathrm{i}t)|\, \mu_0(\Theta, t)\, \mathrm{d}t + \int_{R_1} \log |g(1+\mathrm{i}t)|\, \mu_1(\Theta, t)\, \mathrm{d}t\,. \tag{4}$$

We multiply (4) by r, in other words we replace $|g(\cdot)|$ in (4) by $|g(\cdot)|^r$. Now the same proof as in [I, p. 67] yields (2).

Lemma 3. (i) If all notations have the same meaning as in Definition 2.4.4/2 (i), then

$$\begin{aligned} &\|g \mid (B^{s_0}_{p_0,q_0}(R_n), B^{s_1}_{p_1,q_1}(R_n)_\Theta\| \\ &= \inf\,(\sup_{t\in R_1} \|f(\mathrm{i}t) \mid B^{s_0}_{p_0,q_0}(R_n)\|^{1-\Theta} \sup_{t\in R_1} \|f(1+\mathrm{i}t) \mid B^{s_1}_{p_1,q_1}(R_n)\|^{\Theta})\,, \end{aligned} \tag{5}$$

where the infimum is taken over all admissible functions $f(z)$ in the sense of (2.4.4/3).
(ii) If all notations have the same meaning as in Definition 2.4.4/2 (ii), then

$$\begin{aligned} &\|g \mid (F^{s_0}_{p_0,q_0}(R_n), F^{s_1}_{p_1,q_1}(R_n))_\Theta\| \\ &= \inf\,(\sup_{t\in R_1} \|f(\mathrm{i}t) \mid F^{s_0}_{p_0,q_0}(R_n)\|^{1-\Theta} \sup_{t\in R_1} \|f(1+\mathrm{i}t) \mid F^{s_1}_{p_1,q_1}(R_n)\|^{\Theta})\,, \end{aligned} \tag{6}$$

where the infimum is taken over all admissible functions $f(z)$ in the sense of (2.4.4/5).

Proof. We prove (5). The proof of (6) is the same. It is obvious that the right-hand side of (5) can be estimated from above by the left-hand side, cf. (2.4.4/4) and (2.4.4/1). In order to prove the reverse inequality we replace $f(z)$ in (2.4.4/3) and (2.4.4/4) by $a^{z-\Theta}f(z)$, where a is a positive number. Then it follows that

$$\|g \mid (B^{s_0}_{p_0,q_0}(R_n), B^{s_1}_{p_1,q_1}(R_n))_\Theta\| \leq \max_{l=0,1}\,[a^{l-\Theta} \sup_{t\in R_1} \|f(l+\mathrm{i}t) \mid B^{s_l}_{p_l,q_l}(R_n)\|]\,.$$

If $\|f(1+\mathrm{i}t) \mid B^{s_1}_{p_1,q_1}(R_n)\| \equiv 0$ then $a\to\infty$ yields the desired result. Otherwise, we choose

$$a = \sup_{t\in R_1} \|f(\mathrm{i}t) \mid B^{s_0}_{p_0,q_0}(R_n)\|\ (\sup_{t\in R_1} \|f(1+\mathrm{i}t) \mid B^{s_1}_{p_1,q_1}(R_n)\|)^{-1}\,.$$

2.4.7. Complex Interpolation for the Spaces $B^s_{p,q}(R_n)$ and $F^s_{p,q}(R_n)$

Theorem. Let $-\infty<s_0<\infty$, $-\infty<s_1<\infty$, $0<q_0\leq\infty$, $0<q_1\leq\infty$ and $0<\Theta<1$.
(i) If $0<p_0\leq\infty$, $0<p_1\leq\infty$ and

$$s=(1-\Theta)s_0+\Theta s_1\,,\quad \frac{1}{p}=\frac{1-\Theta}{p_0}+\frac{\Theta}{p_1}\,,\quad \frac{1}{q}=\frac{1-\Theta}{q_0}+\frac{\Theta}{q_1} \tag{1}$$

then

$$(B^{s_0}_{p_0,q_0}(R_n), B^{s_1}_{p_1,q_1}(R_n))_\Theta = B^s_{p,q}(R_n)\,. \tag{2}$$

(ii) If $0<p_0<\infty$ and $0<p_1<\infty$ and if (1) is satisfied, then

$$(F^{s_0}_{p_0,q_0}(R_n), F^{s_1}_{p_1,q_1}(R_n))_\Theta = F^s_{p,q}(R_n)\,. \tag{3}$$

Proof. We prove (3). The proof of (2) is essentially the same, but simpler in some details. Furthermore, we assume that $q_0<\infty$ and $q_1<\infty$. If either $q_0=\infty$ or $q_1=\infty$ then one has to modify in the usual way.

Step 1. We prove that

$$(F^{s_0}_{p_0,q_0}(R_n), F^{s_1}_{p_1,q_1}(R_n))_\Theta \subset F^s_{p,q}(R_n)\,. \tag{4}$$

Let $f(z)\in F(F^{s_0}_{p_0,q_0}(R_n), F^{s_1}_{p_1,q_1}(R_n))$ and $\{\varphi_k(x)\}_{k=0}^\infty \in \Phi(R_n)$, cf. 2.4.4. and 2.3.1. In the sense of Definition 2.3.1/2 and Definition 2.4.4/2 we put $g_k(x, z)=$

$=(F^{-1}\varphi_k Ff(z))(x)$, where $k=0, 1, 2, \ldots$ Let $0<r<\min(p_0, p_1, q_0, q_1)$. Then

$$I=\left[\int_{R_n}\left(\sum_{k=0}^{\infty}|2^{ks}g_k(x,\Theta)|^q\right)^{\frac{p}{q}}dx\right]^{\frac{1}{p}}=\left[\int_{R_n}\left(\sum_{k=0}^{\infty}(|2^{ks}g_k(x,\Theta)|^r)^{\frac{q}{r}}\right)^{\frac{r}{q}\cdot\frac{p}{r}}dx\right]^{\frac{r}{p}\cdot\frac{1}{r}}, \tag{5}$$

where the spaces involved $l_{\frac{q}{r}}$ and $L_{\frac{p}{r}}(R_n)$ are Banach spaces. We use (2.4.6/2) with $2^{ks}g_k(x, z)$ instead of $g(z)$ (where $x\in R_n$ is fixed). Let

$$a_k(x)=\frac{1}{1-\Theta}\int_{R_1}|g_k(x, it)|^r\,\mu_0(\Theta, t)\,dt \tag{6}$$

and

$$b_k(x)=\frac{1}{\Theta}\int_{R_1}|g_k(x, 1+it)|^r\,\mu_1(\Theta, t)\,dt\,. \tag{7}$$

Then (1) and elementary applications of Hölder's inequality yield

$$\left[\sum_{k=0}^{\infty}(|2^{ks}\,g_k(x,\Theta)|^r)^{\frac{q}{r}}\right]^{\frac{r}{q}}\leqq\left[\sum_{k=0}^{\infty}(2^{ksr}a_k^{1-\Theta}(x)\,b_k^{\Theta}(x))^{\frac{q}{r}}\right]^{\frac{r}{q}} \tag{8}$$
$$\leqq\left(\sum_{k=0}^{\infty}2^{ks_0q_0}\,a_k^{\frac{q_0}{r}}(x)\right)^{\frac{r}{q_0}(1-\Theta)}\left(\sum_{k=0}^{\infty}2^{ks_1q_1}b_k^{\frac{q_1}{r}}(x)\right)^{\frac{r}{q_1}\Theta}.$$

By (6) and a well-known inequality for integrals we have

$$\|\{2^{ks_0r}a_k(x)\}_{k=0}^{\infty}\mid l_{\frac{q_0}{r}}\|\leqq\frac{1}{1-\Theta}\int_{R_1}\|\{|2^{ks_0}g_k(x, it)|^r\}_{k=0}^{\infty}|\,l_{\frac{q_0}{r}}\|\mu_0(\Theta, t)\,dt$$
$$=\frac{1}{1-\Theta}\int_{R_1}\left(\sum_{k=0}^{\infty}2^{ks_0q_0}|g_k(x, it)|^{q_0}\right)^{\frac{1}{q_0}\cdot r}\mu_0(\Theta, t)\,dt\,.$$

Similarly for the second factor on the right-hand side of (8). Hence, (5) and (8) give

$$I\leqq\left[\int_{R_n}\left(\frac{1}{1-\Theta}\int_{R_1}\left(\sum_{k=0}^{\infty}2^{ks_0q_0}|g_k(x, it)|^{q_0}\right)^{\frac{1}{q_0}\cdot r}\mu_0(\Theta, t)\,dt\right)^{(1-\Theta)\frac{p}{r}}\right.$$
$$\left.\times\left(\frac{1}{\Theta}\int_{R_1}\left(\sum_{k=0}^{\infty}2^{ks_1q_1}\,|g_k(x, 1+it)|^{q_1}\right)^{\frac{1}{q_1}\cdot r}\mu_1(\Theta, t)\,dt\right)^{\Theta\frac{p}{r}}dx\right]^{\frac{r}{p}\cdot\frac{1}{r}}.$$

We use (1) and Hölder's inequality for integrals (the counterpart of (8) with p's instead of q's) and obtain that

$$I\leqq\left[\int_{R_n}\left(\frac{1}{1-\Theta}\int_{R_n}\left(\sum_{k=0}^{\infty}2^{ks_0q_0}|g_k(x, it)|^{q_0}\right)^{\frac{1}{q_0}\cdot r}\mu_0(\Theta, t)\,dt\right)^{\frac{p_0}{r}}dx\right]^{\frac{r}{p_0}\cdot\frac{1-\Theta}{r}} \tag{9}$$
$$\times\left[\int_{R_n}\left(\frac{1}{\Theta}\int_{R_1}\left(\sum_{k=0}^{\infty}2^{ks_1q_1}\,|g_k\,(x, 1+it)|^{q_1}\right)^{\frac{1}{q_1}\cdot r}\mu_1(\Theta, t)\,dt\right)^{\frac{p_1}{r}}dx\right]^{\frac{r}{p_1}\cdot\frac{\Theta}{r}}.$$

Recall that $\left[\int_{R_n} (\cdot)^{\frac{p_0}{r}}\, dx\right]^{\frac{r}{p_0}}$ is a norm. Hence, (9) can be estimated from above by

$$I \leqq \left(\frac{1}{1-\Theta} \int_{R_1} \left[\int_{R_n} \left(\sum_{k=0}^{\infty} 2^{ks_0q_0} \, |g_k(x, \mathrm{i}t)|^{q_0}\right)^{\frac{r}{q_0}\cdot\frac{p_0}{r}} dx\right]^{\frac{r}{p_0}} \mu_0(\Theta, t)\right)^{\frac{1-\Theta}{r}}$$

$$\times \left(\frac{1}{\Theta} \int_{R_1} \left[\int_{R_n} \left(\sum_{k=0}^{\infty} 2^{ks_1q_1} \, |g_k(x, 1+\mathrm{i}t)|^{q_1}\right)^{\frac{r}{q_1}\cdot\frac{p_1}{r}} dx\right]^{\frac{r}{p_1}} \mu_1(\Theta, t)\, dt\right)^{\frac{\Theta}{r}}.$$

Now we use (2.4.6/3) to obtain

$$\begin{aligned} I &\leqq \sup_{t\in R_1} \|2^{ks_0}g_k(\cdot\,, \mathrm{i}t) \mid L_{p_0}(R_n, l_{q_0})\|^{1-\Theta} \sup_{t\in R_1} \|2^{ks_1}g_k(\cdot\,, 1+\mathrm{i}t) \mid L_{p_1}(R_n, l_{q_1})\|^{\Theta} \\ &\leqq \|f(z) \mid F(F^{s_0}_{p_0,q_0}, F^{s_1}_{p_1,q_1})\|\,, \end{aligned} \tag{10}$$

cf. (2.4.4/2) and (2.3.1/6). However, (5) and (10) prove (4).

Step 2. We prove that

$$F^s_{p,q}(R_n) \subset (F^{s_0}_{p_0,q_0}(R_n), F^{s_1}_{p_1,q_1}(R_n))_\Theta\,. \tag{11}$$

Let $g \in F^s_{p,q}(R_n)$, $\{\varphi_k(x)\}_{k=0}^{\infty} \in \Phi(R_n)$ and $g_k(x) = F^{-1}\varphi_k Fg$. Let $\psi_k(x) = \sum_{l=-1}^{1} \varphi_{k+l}(x)$ if $k=0, 1, 2, \ldots$ (with $\varphi_{-1}=0$). In particular, $\psi_k(x)=1$ if $x \in \operatorname{supp} \varphi_k$. Finally we recall of the maximal function

$$g_k^*(x) = \sup_{x\in R_n} \frac{|g_k(x-y)|}{1+|2^k y|^a}\,, \quad x\in R_n\,, \tag{12}$$

from Definition 2.3.6/2, where we assume that $a > \frac{n}{\min(p, q)}$. Obviously, $g_k^*(x) > 0$ if $g_k(x)$ does not vanish identically. If $z \in \bar{A}$, we put

$$f(z) = \sum_{k=0}^{\infty} F^{-1}\psi_k F\left[2^{\varrho_1(z)k}\left(\sum_{l=0}^{k} |2^{ls}g_l^*(x)|^q\right)^{\frac{\varrho_2(z)}{q}} \|2^{ls}g_l^* \mid L_p(R_n, l_q)\|^{\varrho_3(z)} g_k^{\varrho_4(z)}(x)\right], \tag{13}$$

where

$$\begin{cases} \varrho_1(z) = sq\left(\dfrac{1-z}{q_0}+\dfrac{z}{q_1}\right) - (1-z)s_0 - zs_1\,, \\ \varrho_2(z) = p\left(\dfrac{1-z}{p_0}+\dfrac{z}{p_1}\right) - q\left(\dfrac{1-z}{q_0}+\dfrac{z}{q_1}\right), \\ \varrho_3(z) = 1-p\left(\dfrac{1-z}{p_0}+\dfrac{z}{p_1}\right), \quad \varrho_4(z) = q\left(\dfrac{1-z}{q_0}+\dfrac{z}{q_1}\right). \end{cases} \tag{14}$$

Theorem 2.3.6. shows that $\|2^{ls}g_l^* \mid L_p(R_n, l_q)\|$ is finite. First we wish to show that $f(z)$ is an $S'(R_n)$-analytic function in A in the sense of 2.4.4. The estimates below will show that the series (13) converges at least in $S'(R_n)$, in particular, $f(z) \in S'(R_n)$ for every $z \in \bar{A}$. Because $g_k(x) \in L_p(R_n)$ it follows from (1.4.1/3) that $g_k(x)$ is a bounded and uniformly continuous function in R_n. Then $g_k^*(x)$ has the same properties. Now it is not complicated to see that the properties (ii) and (iii) from 2.4.4. for an $S'(R_n)$-analytic function are satisfied. Furthermore, (1) and (14) yield

$$f(\Theta) = \sum_{k=0}^{\infty} F^{-1}\psi_k F g_k = \sum_{k=0}^{\infty} F^{-1}\varphi_k F g = g\,. \tag{15}$$

Next we prove that

$$\|f(it) \mid F^{s_0}_{p_0,q_0}(R_n)\| \leqq c\|g \mid F^s_{p,q}(R_n)\|, \tag{16}$$

where c is independent of g and of $t \in R_1$. The proof will also show that (13) converges in $S'(R_n)$ (appropriate modifications). If $[\ldots]_k$ denotes the corresponding brackets in (13), then

$$\begin{aligned} F^{-1}\varphi_k F[\ldots]_k(x, z) &= 2^{\varrho_1(z)k}\|2^{ls}g_l^* \mid L_p(R_n, l_q)\|^{\varrho_3(z)} \\ &\times \int_{R_n} (F^{-1}\varphi_k)(y)\, g_k^{\varrho_4(z)}(x-y) \left(\sum_{l=0}^{k} |2^{ls}g_l^*(x-y)|^q\right)^{\frac{\varrho_2(z)}{q}} dy\,. \end{aligned} \tag{17}$$

By (12) we have

$$\begin{aligned} &|g_k^{\varrho_4(z)}(x-y)| \left|\left(\sum_{l=0}^{k} |2^{ls}g_l^*(x-y)|^q\right)^{\frac{\varrho_2(z)}{q}}\right| \\ &\leqq c g_k^{*\operatorname{Re}\varrho_4(z)}(x) \left(\sum_{l=0}^{k} 2^{lsq}g_l^{*q}(x)\right)^{\frac{\operatorname{Re}\varrho_2(z)}{q}} (1+|2^k y|^a)^{|\operatorname{Re}\varrho_4(x)|+|\operatorname{Re}\varrho_2(z)|}\,. \end{aligned}$$

Now (17) yields

$$\begin{aligned} |F^{-1}\varphi_k F[\ldots]_k(x, z)| &\leqq 2^{\operatorname{Re}\varrho_1(z)k}\|2^{ls}g_l^* \mid L_p(R_n, l_q)\|^{\operatorname{Re}\varrho_3(z)} g_k^{*\operatorname{Re}\varrho_4(z)} \\ &\times \left(\sum_{l=0}^{k} 2^{lsq}g_l^{*q}(x)\right)^{\frac{\operatorname{Re}\varrho_2(z)}{q}} \int_{R_n} |(F^{-1}\varphi_k)(y)|\,(1+|2^k y|^a)^{|\operatorname{Re}\varrho_4(z)|+|\operatorname{Re}\varrho_2(z)|}\, dy\,. \end{aligned} \tag{18}$$

We may assume (without loss of generality) that $\varphi_k(y) = \varphi_1(2^{-k+1}y)$ if $k = 1, 2, \ldots$ Then $(F^{-1}\varphi_k)(y) = 2^{(k-1)n}(F^{-1}\varphi_1)(2^{k-1}y)$. In particular, the integral on the right-hand side of (18) can be estimated from above by a constant which is independent of k and $z \in \bar{A}$. Let $z = it$ with $t \in R_1$. We distinguish two cases, $\operatorname{Re}\varrho_2(it) \geqq 0$ and $\operatorname{Re}\varrho_2(it) < 0$. If $\operatorname{Re}\varrho_2(it) \geqq 0$, i.e. $\frac{p}{p_0} \geqq \frac{q}{q_0}$, then we can replace $\sum_{l=0}^{k}$ in (18) by $\sum_{l=0}^{\infty}$. From this it follows that

$$\|2^{ks_0}F^{-1}\varphi_k F[\ldots]_k(\cdot, it) \mid L_{p_0}(R_n, l_{q_0})\| \leqq c\|2^{ls}g_l^* \mid L_p(R_n, l_q)\|\,. \tag{19}$$

By Theorem 2.3.6, the right-hand side of (19) can be estimated from above by $\|g \mid F^s_{p,q}(R_n)\|$. From Theorem 1.6.3 we obtain that

$$\|f(it) \mid F^{s_0}_{p_0,q_0}(R_n)\| \leqq c\|g \mid F^s_{p,q}(R_n)\|\,. \tag{20}$$

In a similar way one can prove that

$$\|f(1+it) \mid F^{s_1}_{p_1,q_1}(R_n)\| \leqq c\|g \mid F^s_{p,q}(R_n)\|\,. \tag{21}$$

(20), (21) and (15) prove $g \in (F^{s_0}_{p_0,q_0}(R_n), F^{s_1}_{p_1,q_1}(R_n))_\Theta$ and (11). Now we assume that $\operatorname{Re}\varrho_2(it) < 0$, i.e. $\frac{q}{q_0} > \frac{p}{p_0}$. Let $a_l = 2^{lsq}g_l^{*q}$ and $\varkappa = \frac{q_0}{q}\left(\frac{q}{q_0} - \frac{p}{p_0}\right) = -q_0\frac{\operatorname{Re}\varrho_2(it)}{q}$. We apply Lemma 2.4.6/1 to obtain

$$b_k = 2^{ksq}g_k^{*q}\left(\sum_{l=0}^{k} 2^{lsq}g_l^{*q}(x)\right)^{\frac{\operatorname{Re}\varrho_2(it)}{q}q_0} \tag{22}$$

and

$$\sum_{k=0}^{\infty} b_k \leqq c\left(\sum_{k=0}^{\infty} 2^{ksq}g_k^{*q}\right)^{\frac{q_0}{q}\cdot\frac{p}{p_0}}. \tag{23}$$

If one uses (22) und (23), then (19) is a consequence of (18). The rest is the same as in the case $\operatorname{Re} \varrho_2(it) \geqq 0$. This proves (11) in any case. Now (4) and (11) prove (3).

Remark 1. As we remarked in (2.3.5/4), $F^0_{p,2}(R_n) = h_p(R_n)$ are the local Hardy spaces. Restricted to that case, (3) yields

$$(h_{p_0}(R_n), h_{p_1}(R_n))_\Theta = h_p(R_n), \quad 0<\Theta<1, \quad \frac{1}{p} = \frac{1-\Theta}{p_0} + \frac{\Theta}{p_1}, \tag{24}$$

$0<p_0<\infty$ and $0<p_1<\infty$. A corresponding formula holds for the Hardy spaces $H_p(R_n)$. (24) is essentially due to A. P. Calderón, A. Torchinsky [1, II, p. 135]. The proof of (24) in that paper is different, although it is also based on the technique of maximal functions. A proof of the above theorem for the special case $\frac{p_0}{q_0} = \frac{p_1}{q_1}$ (where we have $\varrho_2(z) \equiv 0$) has been given in H. Triebel [15]. In that case the proof is simpler, in particular one does not need Lemma 2.4.6/1. L. Päivärinta [1] combined the methods in A. P. Calderón, A. Torchinsky [1, II] and H. Triebel [15] and proved the above theorem for a wide variety of parameters p_0, q_0, p_1, q_1, but not for all. The full proof was given in H. Triebel [19], cf. also H. Triebel [2]. The above traetment followed those papers.

Remark 2. We return to the problem discussed in Remark 2.4.4/2. If p_0, q_0, p_1 and q_1 are larger than or equal to 1, then both complex methods, the classical method $[\cdot,\cdot]_\Theta$ and the above method $(\cdot,\cdot)_\Theta$ make sense. If $1<p_0<\infty$, $1<q_0<\infty$, $1<p_1<\infty$ and $1<q_1<\infty$ then we proved in [I, pp. 182/185] that

$$[B^{s_0}_{p_0,q_0}(R_n), B^{s_1}_{p_1,q_1}(R_n)]_\Theta = B^s_{p,q}(R_n),$$
$$[F^{s_0}_{p_0,q_0}(R_n), F^{s_1}_{p_1,q_1}(R_n)]_\Theta = F^s_{p,q}(R_n),$$

where the parameters have the same meaning as in the above theorem. In these cases, the two methods yield the same result. On the other hand

$$[B^{s_0}_{p,\infty}(R_n), B^{s_1}_{p,\infty}(R_n)]_\Theta = \mathring{B}^s_{p,\infty}(R_n), \tag{25}$$

where $-\infty<s_0<s_1<\infty$, $1<p<\infty$, $0<\Theta<1$ and $s=(1-\Theta)s_0+\Theta s_1$, cf. [I, p. 182]. Here $\mathring{B}^s_{p,\infty}(R_n)$ is the completion of $S(R_n)$ in $B^s_{p,\infty}(R_n)$ and $\mathring{B}^s_{p,\infty}(R_n) \neq B^s_{p,\infty}(R_n)$. The counterpart of (25) for the above method is

$$(B^{s_0}_{p,\infty}(R_n), B^{s_1}_{p,\infty}(R_\infty))_\Theta = B^s_{p,\infty}(R_n). \tag{26}$$

Hence, in this limiting case, the two methods yield different results.

2.4.8. Fourier Multipliers for the Spaces $F^s_{p,q}(R_n)$

We proved in Proposition 2.4.1 the interpolation property for the real method. The question arises whether the above complex method has a corresponding property. Unfortunately, the situation for this method is more complicated than for the real method, and we have no direct counterpart of Proposition 2.4.1. In the few subsequent applications we shall prove the interpolation property from time to time under additional hypotheses on the underlying operators. One of the most interesting applications will be described in this subsection, namely, an improvement of Theorem 2.3.7, which deals with Fourier multipliers for the spaces $B^s_{p,q}(R_n)$ and $F^s_{p,q}(R_n)$. Fourier multipliers for the spaces $B^s_{p,q}(R_n)$ will be treated extensively in Section 2.6. Thus we restrict ourselves for the moment to the spaces $F^s_{p,q}(R_n)$. If $0<p<\infty$ and $0<q\leqq\infty$, then we put

$$\sigma_{p,q} = n\left(\frac{1}{\min(1,p,q)} - \frac{1}{2}\right), \tag{1}$$

cf. (1.6.4/1). We recall that $H_2^\varkappa(R_n)$ are the special Bessel potential spaces from 1.5.2. If $\varkappa = m = 0, 1, 2, \ldots$ then $H_2^m(R_n) = W_2^m(R_n)$ are the well-known Sobolev spaces. By (2.3.5/2) we have $H_2^\varkappa(R_n) = F_{2,2}^\varkappa(R_n) = B_{2,2}^\varkappa(R_n)$. Let $\psi(x) \in S(R_n)$ and $\varphi(x) \in S(R_n)$ with

$$0 \leqq \psi(x) \leqq 1, \quad \operatorname{supp} \psi \subset \{y \mid |y| \leqq 4\}, \quad \psi(x) = 1 \quad \text{if} \quad |x| \leqq 2, \tag{2}$$

$$0 \leqq \varphi(x) \leqq 1, \quad \operatorname{supp} \varphi \subset \left\{y \,\middle|\, \frac{1}{4} \leqq |y| \leqq 4\right\}, \quad \varphi(x) = 1 \quad \text{if} \quad \frac{1}{2} \leqq |x| \leqq 2. \tag{3}$$

We introduce, for the sake of brevity,

$$\|m \mid h_2^\varkappa(R_n)\| = \|\psi m \mid H_2^\varkappa(R_n)\| + \sup_{k=0,1,2,\ldots} \|\varphi(\cdot)\, m(2^k \cdot) \mid H_2^\varkappa(R_n)\|. \tag{4}$$

Proposition. (i) If $0 < p < \infty$, $0 < q \leqq \infty$, $-\infty < s < \infty$ and $\varkappa > \frac{n}{2} + \frac{n}{\min(p, q)}$, then there exists a constant c such that

$$\|F^{-1} m\, Ff \mid F_{p,q}^s(R_n)\| \leqq c \|m \mid h_2^\varkappa(R_n)\|\, \|f\, F_{p,q}^s(R_n)\| \tag{5}$$

holds for all functions $m(x) \in L_\infty(R_n)$ and all $f \in F_{p,q}^s(R_n)$.

(ii) If $1 < p < \infty$, $1 < q < \infty$, $-\infty < s < \infty$, and $\varkappa > \frac{n}{2}$, then there exists a constant c such that (5) holds for all functions $m(x) \in L_\infty(R_n)$ and all $f \in F_{p,q}^s(R_n)$.

Proof. Step 1. Let $\{\varphi_k(x)\}_{k=0}^\infty \in \Phi(R_n)$, cf. 2.3.1. By (2.3.1/6) we have

$$\|F^{-1} m Ff \mid F_{p,q}^s(R_n)\| = \|2^{sk} F^{-1} m \varphi_k Ff \mid L_p(R_n, l_q)\|.$$

If $k = 1, 2, \ldots$, then

$$2^{sk} F^{-1} m \varphi_k Ff = F^{-1} m(\cdot)\, \varphi(2^{-k} \cdot)\, F[F^{-1} \varphi_k 2^{sk} Ff] \tag{6}$$

(modification if $k = 0$). Application of Theorem 1.6.3 with $f_k = 2^{sk} F^{-1} \varphi_k Ff$ and $M_k(y) = m(y)\, \varphi(2^{-k} y)$ (modification if $k = 0$) yields (5).

Step 2. In order to prove (ii) we recall of the following Fourier multiplier theorem of Michlin-Hörmander type: if $1 < p < \infty$ and $1 < q < \infty$, then there exists a positive constant c such that

$$\|F^{-1} m_k F f_k \mid L_p(R_n, l_q)\| \tag{7}$$
$$\leqq c \sup \left(R^{2|\alpha| - n} \int_{\frac{R}{2} \leqq |x| \leqq 2R} \sum_{k=0}^\infty |D^\alpha m_k(x)|^2 \, dx \right)^{\frac{1}{2}} \|f_k \mid L_p(R_n, l_q)\|$$

holds for all systems $\{m_k(x)\}_{k=0}^\infty \subset L_\infty(R_n)$ and all systems $\{f_k(x)\}_{k=0}^\infty$ of measurable functions in R_n, where the supremum is taken over all $R > 0$ and all multi-indices α with $0 \leqq |\alpha| \leqq 1 + \left[\frac{n}{2}\right]$ (of course, of interest are only the systems $\{m_k\}$ and $\{f_k\}$ for which the right-hand side of (7) makes sense and is finite). A proof of (7) has been given in [I, pp. 161–165] $\left(\text{as usual, } \left[\frac{n}{2}\right] \text{ is the largest natural number which is less than or equal to } \frac{n}{2}\right)$. If φ has the above meaning, cf. (3), then (7) can be reformulated as

$$\|F^{-1} m_k F f_k \mid L_p(R_n, l_q)\| \tag{8}$$
$$\leqq c \sup \left(\sum_{l=0}^\infty \|\varphi(\cdot)\, m_l(2^j \cdot) \mid H_2^\varkappa(R_n)\|^2 \right)^{\frac{1}{2}} \|f_k \mid L_p(R_n, l_q)\|,$$

where the supremum is taken over all integers j and where $\varkappa = 1 + \left[\frac{n}{2}\right]$. But the same proof as in [I, pp. 161–165] shows that (8) is also true if $\varkappa$ is an arbitrary number with $\varkappa > \frac{n}{2}$ (if one starts with such an assumption, then the proof in [I, pp. 161–165] is even more transparent than in the original version). Let $f_k = 2^{sk} F^{-1} \varphi_k F f$ and $m_l(x) = m(x)\,\varphi(2^{-l}x)$ if $k = 0, 1, 2, \ldots$ and $l = 1, 2, \ldots$ Then (8), (6), and (3) yield (ii) (the term with $k = 0$ is covered by Theorem 1.5.2).

Remark 1. The scalar case of (7) is the famous Michlin-Hörmander multiplier theorem, cf. L. Hörmander [1]. Also the scalar case of (8) with $\varkappa > \frac{n}{2}$ is known, cf. J. Peetre [1] and the remarks in A. P. Calderón, A. Torchinsky [1, II, p. 168]. Concerning vector-valued Fourier multiplier theorems for $L_p(R_n, l_q)$ with $1 < p < \infty$ and $1 < q < \infty$ we refer also to the fundamental paper by P. I. Lizorkin [1].

Theorem. If $0 < p < \infty$, $0 < q < \infty$, $-\infty < s < \infty$ and $\varkappa > \sigma_{p,q}$, then there exists a constant c such that

$$\|F^{-1} m F f \mid F^s_{p,q}(R_n)\| \leq c \|m \mid h^{\varkappa}_2(R_n)\| \, \|f \mid F^s_{p,q}(R_n)\| \tag{9}$$

holds for all $m(x) \in L_\infty(R_n)$ and all $f \in F^s_{p,q}(R_n)$.

Proof. *Step 1*. The idea is to interpolate the two parts of the above proposition. We start with a preparation. We assume that (9) is valid for every function $m(x) \in S(R_n)$ with compact support. Then it follows by a limiting argument that (9) is also valid for every $m(x) \in H^{\varkappa}_2(R_n)$ with compact support ($F^{-1} m F f$ is defined via this limiting procedure). Let $m(x)$ be an arbitrary function with $\|m \mid h^{\varkappa}_2(R_n)\| < \infty$. Let $\lambda(x) \in S(R_n)$ with $\operatorname{supp} \lambda \subset \{y \mid |y| \leq 2\}$ and $\lambda(x) = 1$ if $|x| \leq 1$. Then (9), with $\lambda(\varepsilon x)\, m(x)$ instead of $m(x)$, is valid. But

$$\|\lambda(\varepsilon \cdot)\, m(\cdot) \mid h^{\varkappa}_2(R_n)\| \leq c \|m \mid h^{\varkappa}_2(R_n)\| ,$$

where c is independent of $\varepsilon > 0$. Again by a limiting argument it follows that (9) is valid.

Step 2. Let $m(x) \in S(R_n)$ with a compact support. Let $\alpha(x) \in S(R_n)$ and $\beta(x) \in S(R_n)$ be two functions with

$$0 \leq \alpha(x) \leq 1, \quad \operatorname{supp} \alpha \subset \{y \mid |y| \leq 2\},$$

$$0 \leq \beta(x) \leq 1, \quad \operatorname{supp} \beta \subset \left\{y \,\middle|\, \frac{1}{2} \leq |y| \leq 2\right\}, \quad \text{and}$$

$$\sum_{k=1}^{\infty} \beta(2^{-k}x) + \alpha(x) = 1 \quad \text{if} \quad x \in R_n .$$

The function $\psi(x)$ and $\varphi(x)$ have the same meaning as in (2) and (3). Let $\bar{A}$ be the closure of the strip $A = \{z \mid 0 < \operatorname{Re} z < 1\}$ from 2.4.4. If $x \in R_n$ and $z \in \bar{A}$, then we put

$$\begin{aligned} m(x, z) = {} & [F^{-1} (1 + |\xi|^2)^{\frac{\varrho(z)}{2}} F \alpha m](x)\, \psi(x) \\ & + \sum_{k=1}^{\infty} [F^{-1} (1 + |\xi|^2)^{\frac{\varrho(z)}{2}} F \beta(\cdot)\, m(2^k \cdot)]\,(2^{-k}x)\, \varphi(2^{-k}x) , \end{aligned} \tag{10}$$

where $\varrho(z) = \varkappa - \varkappa_0(1 - z) - \varkappa_1 z$ with the positive numbers $\varkappa_0$, $\varkappa_1$ and $\varkappa$. If Θ is a given number with $0 < \Theta < 1$, then we assume that $\varrho(\Theta) = 0$, i.e.

$$\varkappa = (1 - \Theta)\varkappa_0 + \Theta \varkappa_1 . \tag{11}$$

It is clear how to understand (10): the operator

$$F^{-1}(1+|\xi|^2)^{\frac{\varrho(z)}{2}}F=F^{-1}[(1+|\xi|^2)^{\frac{\varrho(z)}{2}}F]$$

acts on $\beta(y)\,m(2^k y)$, and the result is taken at the point $2^{-k}x$, etc. Now it is easy to see that

$$f(z)\to F^{-1}m(\cdot\,,z)\,Ff(z)$$

maps an $S'(R_n)$-analytic function in A into an $S'(R_n)$-analytic function in A, cf. 2.4.4. (by our assumption, the sum in (10) is finite). Furthermore,

$$m(x,\Theta)=m(x)\,. \tag{12}$$

Finally, (4) yields

$$\begin{cases}\|m(\cdot,\text{it})\mid h_2^{\varkappa_0}(R_n)\|\le c\|m\mid h_2^{\varkappa}(R_n)\| \quad\text{and}\\ \|m(\cdot,1+\text{it})\mid h_2^{\varkappa_1}(R_n)\|\le c\|m\mid h_2^{\varkappa}(R_n)\|\,,\end{cases} \tag{13}$$

where c is independent of $t\in R_1$.

Step 3. Now we assume that we have the two multiplier assertions,

$$\|F^{-1}MFf\mid F^{s_l}_{p_l,q_l}(R_n)\|\le c\|M\mid h_2^{\varkappa_l}(R_n)\|\,\|f\mid F^{s_l}_{p_l,q_l}(R_n)\| \tag{14}$$

with $l=0$ or 1. Then (13) yields

$$\|F^{-1}m(\cdot,l+\text{it})\,Ff\mid F^{s_l}_{p_l,q_l}(R_n)\|\le c\|m\mid h_2^{\varkappa}(R_n)\|\,\|f\mid F^{s_l}_{p_l,q_l}(R_n)\| \tag{15}$$

with $l=0$ or 1. Now we can apply Theorem 2.4.7(ii). For that purpose we replace f in (15) by an appropriate $S'(R_n)$-analytic function $f(z)$ in A with $f(\Theta)=f$. If the parameters have the same meaning as in Theorem 2.4.7, then (12) yields

$$\|F^{-1}mFf\mid F^{s}_{p,q}(R_n)\|\le c\|m\mid h_2^{\varkappa}(R_n)\|\,\|f\mid F^{s_l}_{p_l,q_l}(R_n)\|\,. \tag{16}$$

Step 4. Now it is easy to prove the theorem. By part (ii) of the above proposition we may assume that $\min(p,q)\le 1$. Then we choose two couples (p_0,q_0) and (p_1,q_1) as indicated in Fig. 2.4.8. Let $s_0=s_1=s$. Then we apply parts (ii) and (i) of

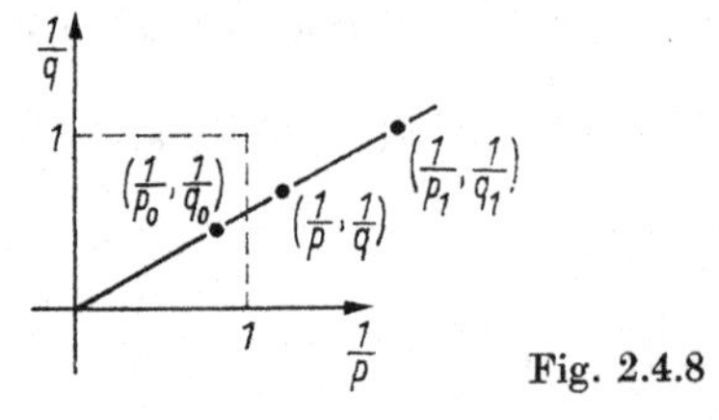

Fig. 2.4.8

the above proposition to $F^{s_0}_{p_0,q_0}(R_n)$ and $F^{s_1}_{p_1,q_1}(R_n)$, respectively, i.e. $\varkappa_0=\frac{n}{2}+\varepsilon$ and $\varkappa_1=\frac{n}{2}+\frac{n}{\min(p,q)}+\varepsilon$ in (14), where $\varepsilon>0$ is an arbitrary number. Then (16) yields (9) with

$$\varkappa=\varepsilon+\frac{n}{2}+\frac{n\Theta}{\min(p_1,q_1)}\,.$$

If $p\le q$ then $p_1\le q_1$. We have $\frac{\Theta}{p_1}=\frac{1}{p}-\frac{1-\Theta}{p_0}$ and

$$\varkappa=\varepsilon+\frac{n}{2}+\frac{n}{p}-n\frac{1-\Theta}{p_0}\,.$$

Let $p_0\downarrow 1$ and then $\Theta\downarrow 0$ (i.e. $p_1\to 0$). Then we see that any number $\varkappa>\sigma_{p,q}$ is admissible. Similarly if $q\le p$.

Remark 2. The theorem is an essential improvement of Theorem 2.3.7. and (2.3.7/6). The question arises whether $\sigma_{p,q}$ is natural. If $1<p<\infty$ and $1<q<\infty$, then $\varkappa>\frac{n}{2}$ in part (ii) of the proposition and in the theorem cannot be improved. The case min $(p, q)\leqq 1$ is more sophisticated. We refer to H. Triebel [19, Remark 11], where we discussed this problem in detail. That paper was also the basis for the above treatment. Fourier multiplier theorems for the spaces $F^s_{p,q}(R_n)$ have also been proved recently by L. Päivärinta [3].

2.4.9. The Spaces $L^\Omega_p(R_n, l_q)$: Complex Interpolation and Fourier Multipliers

The spaces $L^\Omega_p(R_n, l_q)=L^\Omega_p(l_q)$ have been introduced in Definition 1.6.1, where we now prefer to indicate "R_n" in agreement with our notations in this chapter. Our main aim is to improve Theorem 1.6.3 in the sense of 1.6.4. Let (mostly for the sake of simplicity) $\Omega_k=\{y|\ |y|\leqq 2^k\}$ where $k=0, 1, 2, \ldots$ Then one can extend the complex interpolation method from 2.4.4., Theorem 2.4.7 and the considerations in 2.4.8. from the spaces $F^s_{p,q}(R_n)$ to the spaces $L^\Omega_p(R_n, l_q)$. In particular, the proofs of the counterparts of the Theorems 2.4.7 and 2.4.8 go through without any substantial changes. Thus we restrict ourselves to the formulation of the corresponding definitions and results, cf. also H. Triebel [15, 16] and L. Päivärinta [1]. The strip A has the same meaning as in 2.4.4.

Definition 1. If $0<q_0<\infty, 0<q_1<\infty, 0<p_0<\infty$, and $0<p_1<\infty$, then $F(L^\Omega_{p_0}(R_n, l_{q_0}), L^\Omega_{p_1}(R_n, l_{q_1}))$ denotes the space of all systems $f(z)=\{f_k(x, z)\}_{k=0}^\infty\subset S'(R_n)$ of functions with the following properties:
(i) $f_k(x, z)$ with $k=0, 1, 2, \ldots$ is a uniformly continuous and bounded function in $R_n\times\bar{A}$. Furthermore, $f_k(x, z)$ with $k=0, 1, 2, \ldots$ is an analytic function in A for every fixed $x\in R_n$.
(ii) We have

$$\operatorname{supp}\ (Ff_k)\ (\cdot, z)\subset\{y\mid |y|\leqq 2^{k+1}\} \tag{1}$$

for $k=0, 1, 2, \ldots$ and for every fixed $z\in\bar{A}$.
(iii)
$$\begin{aligned}&\|f(z)\mid F(L^\Omega_{p_0}(R_n, l_{q_0}), L^\Omega_{p_1}(R_n, l_{q_1}))\|\\ &=\max_{l=0,1}\sup_{t\in R_1}\|f_k(\cdot, l+it)\mid L_{p_l}(R_n, l_{q_l})\|<\infty\,.\end{aligned} \tag{2}$$

Remark 1. This is the counterpart of an $S'(R_n)$-analytic function in A and of Definition 2.4.4/1 (ii). At first glance one might wish to replace 2^{k+1} in (1) by 2^k. But this is unimportant, and we shall indicate below the reason for this choice. One can prove that

$$F(L^\Omega_{p_0}(R_n, l_{q_0}), L^\Omega_{p_1}(R_n, l_{q_1}))$$

is a quasi-Banach space, cf. Proposition 2.4.5.

Definition 2. Let $0<q_0<\infty$, $0<q_1<\infty$, $0<p_0<\infty$ and $0<p_1<\infty$. If $0<\Theta<1$, then $(L^\Omega_{p_0}(R_n, l_{q_0}), L^\Omega_{p_1}(R_n, l_{q_1}))_\Theta$ is the set of all systems $f=\{f_k(x)\}_{k=0}^\infty$ of functions for which there exists a system

$$g(z)=\{g_k(x, z)\}_{k=0}^\infty\in F(L^\Omega_{p_0}(R_n, l_{q_0}), L^\Omega_{p_1}(R_n, l_{q_1})) \tag{3}$$

with

$$\operatorname{supp}\ (Fg_k)\ (\cdot, \Theta)\subset\{y\mid |y|\leqq 2^k\} \tag{4}$$

and $f_k(x)=g_k(x, \Theta)$ for $k=0, 1, 2, \ldots$ Let

$$\|f\mid (L^\Omega_{p_0}(R_n, l_{q_0}), L^\Omega_{p_1}(R_n, l_{q_1}))_\Theta\|=\inf\|g(z)\mid F(L^\Omega_{p_0}(R_n, l_{p_0}), L^\Omega_{p_1}(R_n, l_{q_1}))\|\,, \tag{5}$$

where the infimum is taken over all admissible systems $g(z)$.

Remark 2. This is the counterpart of 2.4.4/2 (ii). The assumption (4) strengthens (1) (with g_k instead of f_k) if $z=\Theta$.

Theorem 1. Let $0<q_0<\infty$, $0<q_1<\infty$, $0<p_0<\infty$, $0<p_1<\infty$ and $0<\Theta<1$. If

$$\frac{1}{p}=\frac{1-\Theta}{p_0}+\frac{\Theta}{p_1} \quad \text{and} \quad \frac{1}{q}=\frac{1-\Theta}{q_0}+\frac{\Theta}{q_1} \tag{6}$$

then

$$(L_{p_0}^{\Omega}(R_n, l_{q_0}), L_{p_1}^{\Omega}(R_n, l_{q_1}))_{\Theta}=L_p^{\Omega}(R_n, l_q)\,. \tag{7}$$

Remark 3. This is the counterpart of Theorem 2.4.7 (ii). One can use the same proof. The small difficulties with the factor 2 come from the counterpart of (2.4.7/13).

Next we strengthen Theorem 1.6.3 in the sense of 1.6.4. All notations have the same meaning as in Theorem 1.6.3 with the above specialization of Ω_k. In particular $H_2^{\varkappa}=H_2^{\varkappa}(R_n)$ are the Bessel-potential spaces. Furthermore, $\sigma_{p,q}$, has the meaning of (1.6.4/1) or (2.4.8/1).

Theorem 2. If $0<p<\infty$, $0<q<\infty$ and $\varkappa>\sigma_{p,q}$, then there exists a constant c such that

$$\|F^{-1}M_kFf_k \mid L_p(R_n, l_q)\| \leq c \sup_l \|M_l(2^l\cdot) \mid H_2^{\varkappa}(R_n)\| \, \|f_k \mid L_p(R_n, l_q)\| \tag{8}$$

holds for all systems $\{f_k(x)\}_{k=0}^{\infty} \in L_p^{\Omega}(R_n, l_q)$ and all sequences $\{M_k(x)\}_{k=0}^{\infty} \subset H_2^{\varkappa}(R_n)$.

Remark 4. The proof of this theorem is essentially the same as the proof of Theorem 2.4.8. It is based on Theorem 1 above. The restriction $\varkappa>\sigma_{p,q}$ is more natural than the corresponding restriction in Theorem 1.6.3. Later on we shall use this improvement. In contrast to Theorem 1.6.3, the case $q=\infty$ is excluded (but this depends probably only on our method).

2.5. Equivalent Quasi-Norms and Representations

2.5.1. An Orientation

The spaces $B_{p,q}^s(R_n)$ and $F_{p,q}^s(R_n)$ are constructed on the basis of Principle 2.2.4, in particular (2.2.4/13) and (2.2.4/14). We specialized the number a in (2.2.4/13) and (2.2.4/15) to $a=2$, cf. Definition 2.3.1/2. However on the basis of Theorem 1.6.3 (and its scalar counterpart) it is easy to see that any choice of $a>1$ yields the same two scales $B_{p,q}^s(R_n)$ and $F_{p,q}^s(R_n)$ (for all admissible parameters s, p, and q in the sense of Definition 2.3.1/2). Furthermore we proved that all these spaces satisfy Criterion 2.2.3, cf. Corollary 2.3.7. Hence our spaces are meaningful in the sense of the considerations in 2.2.3. and 2.2.4. What about the approximation procedures from 2.2.5., which we described as a further possibility in order to study smoothness properties of functions and distributions? In 2.5.2. and 2.5.3. we shall see that the corresponding spaces, cf. e.g. (2.2.5/7) and (2.2.5/9), coincide with the spaces $B_{p,q}^s(R_n)$ and $F_{p,q}^s(R_n)$ (under some restrictions on the parameters). In 2.2.2. we gave a list of constructive spaces and claimed in Remark 2.2.2/3 and in 2.3.5. that these spaces coincide with special spaces $B_{p,q}^s(R_n)$ and $F_{p,q}^s(R_n)$. This also will be done in this Section 2.5. The Bessel-potential spaces $H_p^s(R_n)$ (which include the classical Sobolev spaces) will be treated in 2.5.6. Subsection 2.5.8. deals with the local Hardy spaces $h_p(R_n)$ (and also with $bmo(R_n)$). Concerning the Besov spaces $\Lambda_{p,q}^s(R_n)$ and the Hölder-Zygmund spaces $\mathcal{C}^s(R_n)$ we describe two rather different approaches. In 2.5.7. we discuss briefly the spaces $\Lambda_{p,q}^s(R_n)$ and $\mathcal{C}^s(R_n)$ from the standpoint of interpolation theory. This approach is very convincing and comparatively easy. The disadvantage is that one needs some statements of the abstract real interpolation method (and we promised in 2.4.1. to use interpolation methods only occasionally). A more direct, but also essentially more complicated approach (which covers also the cases $p<1$ in contrast to the interpolation method in 2.5.7.) is contained in 2.5.12. The Subsections 2.5.9.–2.5.13. are crucial for subsequent considerations.

2.5.2. Nikol'skij Representations

If $0<p\leqq\infty$, then we put

$$\mathfrak{B}_p(R_n)=\{b \mid b=\{b_k(x)\}_{k=0}^{\infty},\ b_k(x)\in S'(R_n)\cap L_p(R_n)\ , \tag{1}$$
$$\operatorname{supp} Fb_0\subset\{y \mid |y|\leqq 2\},\ \operatorname{supp} Fb_j\subset\{y \mid 2^{j-1}\leqq|y|\leqq 2^{j+1}\} \text{ if } j=1,2,3,\ldots\},$$

cf. (2.2.5/8). If $\{\varphi_k(x)\}_{k=0}^{\infty}\in\Phi(R_n)$ and $f\in B^s_{p,q}(R_n)$, then $\{F^{-1}\varphi_k Ff\}_{k=0}^{\infty}\in\mathfrak{B}_p(R_n)$, cf. 2.3.1. This is a typical example which will be useful.

Theorem. Let $0<q\leqq\infty$ and $-\infty<s<\infty$.
(i) If $0<p\leqq\infty$, then

$$B^s_{p,q}(R_n)=\{f \mid f\in S'(R_n),\ \exists\{b_k\}_{k=0}^{\infty}\in\mathfrak{B}_p(R_n) \text{ such that}$$
$$f=\sum_{k=0}^{\infty} b_k(x) \text{ in } S'(R_n),\ \|2^{sk}b_k \mid l_q(L_p(R_n))\|<\infty\}\,. \tag{2}$$

Furthermore,

$$\inf\|2^{sk}b_k \mid l_q(L_p(R_n))\| \tag{3}$$

is an equivalent quasi-norm in $B^s_{p,q}(R_n)$, where the infimum is taken over all admissible representations.
(ii) If $0<p<\infty$, then

$$F^s_{p,q}(R_n)=\{f \mid f\in S'(R_n),\ \exists\{b_k\}_{k=0}^{\infty}\in\mathfrak{B}_p(R_n) \text{ such that}$$
$$f=\sum_{k=0}^{\infty} b_k(x) \text{ in } S'(R_n),\ \|2^{sk}b_k \mid L_p(R_n,l_q)\|<\infty\}\,. \tag{4}$$

Furthermore,

$$\inf\|2^{sk}b_k \mid L_p(R_n,l_q)\| \tag{5}$$

is an equivalent quasi-norm in $F^s_{p,q}(R_n)$, where the infimum is taken over all admissible representations.

Proof. If $\{\varphi_k(x)\}_{k=0}^{\infty}\in\Phi(R_n)$ and $f\in S'(R_n)$, then we have $f=\sum_{k=0}^{\infty} b_k(x)$ in $S'(R_n)$ with $b_k(x)=F^{-1}\varphi_k Ff$. This shows that every $f\in B^s_{p,q}(R_n)$ can be represented by the right-hand side of (2) and that (3) can be estimated from above by $\|f \mid B^s_{p,q}(R_n)\|$. Similarly for $F^s_{p,q}(R_n)$. Conversely, if $f\in S'(R_n)$ is given by

$$f=\sum_{k=0}^{\infty} b_k(x),\ \{b_k\}_{k=0}^{\infty}\in\mathfrak{B}_p(R_n),\ \|2^{sk}b_k \mid l_q(L_p(R_n))\|<\infty$$

then we have

$$F^{-1}\varphi_k Ff=\sum_{r=-1}^{1} F^{-1}\varphi_k Fb_{k+r}$$

(with $b_{-1}=0$). The rest is now the same as in the proof of Proposition 2.3.2/1, including the fact that $\|f \mid B^s_{p,q}(R_n)\|$ can be estimated from above by (3) (multiplied with an appropriate constant). Similarly for $F^s_{p,q}(R_n)$, cf. Step 2 of the proof of Proposition 2.3.2/1.

Remark. Further representations of this type may be found in [S, pp. 43–45]. Representations of the above type have been used extensively by S. M. Nikol'skij and O. V. Besov in connection with the spaces $B^s_{p,q}(R_n)$. A comprehensive treatment and many references may be found in S. M. Nikol'skij [3, 5.5 and 5.6] (cf. also [I, 2.3]). However in comparison with the original

Nikol'skij-Besov approach there is a difference. S. M. Nikol'skij and O. V. Besov use the class $\mathfrak{A}_p(R_n)$ from (2.2.5/6) (with $1 \leq p \leq \infty$) instead of $\mathfrak{B}_p(R_n)$ from (1). The above approach may be characterized as a "decomposition of f by entire analytic functions of exponential type", while the approach which is based on $\mathfrak{A}_p(R_n)$ is an "approximation of f by entire analytic functions of exponential type". It will be studied in detail in the next subsection.

2.5.3. Characterizations by Approximation

We discuss the problem raised in 2.2.5. and start with some more or less classical assertions. Let $f \in L_p(R_n)$ with $1 \leq p \leq \infty$ and let

$$E_p(2^k, f) = \inf \|f - g \mid L_p(R_n)\|, \quad k = 0, 1, 2, \ldots, \tag{1}$$

where the infimum is taken over all $g \in L_p(R_n)$ such that supp $Fg \subset \{y \mid |y| \leq 2^k\}$ (in particular, g is an entire analytic function of exponential type). As we pointed out in 2.2.5., (1) is the counterpart of the best approximation of a periodic function by trigonometric polynomials. Let

$$\|f \mid B^s_{p,q}(R_n)\|^* = \|f \mid L_p(R_n)\| + \left[\sum_{k=0}^{\infty} 2^{skq} E_p(2^k, f)^q\right]^{\frac{1}{q}}, \tag{2}$$

where $1 \leq p \leq \infty$, $1 \leq q \leq \infty$ and $s > 0$ (modification if $q = \infty$). Then it is a well-known fact in the theory of Besov spaces that (under the above restriction on the parameters) (2) is an equivalent norm in $B^s_{p,q}(R_n)$ and that

$$B^s_{p,q}(R_n) = \{f \mid f \in S'(R_n),\ \|f \mid B^s_{p,q}(R_n)\|^* < \infty\}. \tag{3}$$

Cf. S. M. Nikol'skij [3. 5.5 and 5.6] or [**I**, 2.5.4]. The restriction $s > 0$ is natural. The norm in (2) with $s < 0$ is equivalent to $\|f \mid L_p(R_n)\|$. This shows that (3) cannot be true if $s < 0$. Hence, the "approximation approach" (cf. the end of 2.5.2.) is restricted to $s \geq 0$, in contrast to the "decomposition approach" from 2.5.2. In order to extend (2), (3) to the spaces $F^s_{p,q}(R_n)$ we introduce the class $\mathfrak{A}_p(R_n)$, cf. also (2.2.5/6). If $0 < p \leq \infty$, then

$$\begin{aligned} \mathfrak{A}_p(R_n) = \{a \mid a = \{a_j(x)\}_{j=0}^{\infty},\ & a_k(x) \in S'(R_n) \cap L_p(R_n), \\ & \text{supp } Fa_k \subset \{y \mid |y| \leq 2^{k+1}\} \quad \text{for} \quad k = 0, 1, 2, \ldots\}. \end{aligned} \tag{4}$$

Now it is not hard to see that (2), (3) can be reformulated as follows,

$$\begin{aligned} B^s_{p,q}(R_n) = \{f \mid f \in S'(R_n),\ & \exists a = \{a_k\}_{k=0}^{\infty} \in \mathfrak{A}_p(R_n) \\ & \text{such that} \quad f = \lim_{k \to \infty} a_k(x) \quad \text{in} \quad S'(R_n) \quad \text{and} \\ & \|f \mid B^s_{p,q}(R_n)\|^a = \|a_0 \mid L_p(R_n)\| + \|2^{sk}(f - a_k) \mid l_q(L_p(R_n))\| < \infty\}, \end{aligned} \tag{5}$$

where again $s > 0$, $1 \leq p \leq \infty$, $1 \leq q \leq \infty$. Concerning the notations, cf. 1.2.2 (and 2.3.1). Furthermore,

$$\|f \mid B^s_{p,q}(R_n)\|^{\alpha} = \inf \|f \mid B^s_{p,q}(R_n)\|^a, \tag{6}$$

where the infimum is taken over all admissible systems a in the sense of (5), is an equivalent norm in $B^s_{p,q}(R_n)$. We remark that $f = \lim_{k \to \infty} a_k(x)$ in $S'(R_n)$ follows from $\|f \mid B^s_{p,q}(R_n)\|^a < \infty$ under the above restrictions for the parameters. However, we wish to extend (5) to values $p < 1$, and there the situation is different. In that cases there arises also the somewhat delicate question about the nature of $f \in B^s_{p,q}(R_n)$, in particular, how to understand $\|f - a_k \mid L_p(R_n)\|$ etc. We postpone this problem and discuss it in Remark 1 after the proof of the theorem. In the proof itself we temporarily take it for granted that $f(x) \in L_p(R_n)$ makes sense in a natural way for all spaces under consideration. Finally we recall of the important numbers

$$\sigma_p = n\left(\frac{1}{\min(p,1)} - \frac{1}{2}\right) \quad \text{and} \quad \sigma_{p,q} = n\left(\frac{1}{\min(p,q,1)} - \frac{1}{2}\right), \tag{7}$$

where $0 < p \leqq \infty$ and $0 < q \leqq \infty$, cf. 1.5.2., 1.5.4., 1.6.4., 2.4.8., and 2.4.9. Let

$$\tilde{\sigma}_p = n\left(\frac{1}{\min(p,1)} - 1\right) \quad \text{and} \quad \tilde{\sigma}_{p,q} = n\left(\frac{1}{\min(p,q,1)} - 1\right). \tag{8}$$

Theorem. (i) If $0 < p \leqq \infty$, $0 < q \leqq \infty$ and $s > \tilde{\sigma}_p$, then

$$\begin{aligned} B^s_{p,q}(R_n) = \{f \mid f \in S'(R_n),\ \exists a = \{a_k\}_{k=0}^\infty \in \mathfrak{A}_p(R_n) \\ \text{such that} \quad f = \lim_{k\to\infty} a_k \quad \text{in} \quad S'(R_n), \\ \|f \mid B^s_{p,q}(R_n)\|^a = \|a_0 \mid L_p(R_n)\| + \|2^{sk}(f - a_k) \mid l_q(L_p(R_n))\| < \infty\}. \end{aligned} \tag{9}$$

Furthermore,

$$\|f \mid B^s_{p,q}(R_n)\|^\times = \inf \|f \mid B^s_{p,q}(R_n)\|^a, \tag{10}$$

where the infimum is taken over all admissible systems $a \in \mathfrak{A}_p(R_n)$, is an equivalent quasi-norm in $B^s_{p,q}(R_n)$.
(ii) If $0 < p < \infty$, $0 < q < \infty$ and $s > \tilde{\sigma}_{p,q}$, then

$$\begin{aligned} F^s_{p,q}(R_n) = \{f \mid f \in S'(R_n),\ \exists a = \{a_k\}_{k=0}^\infty \in \mathfrak{A}_p(R_n) \\ \text{such that} \quad f = \lim_{k\to\infty} a_k \quad \text{in} \quad S'(R_n), \\ \|f \mid F^s_{p,q}(R_n)\|^a = \|a_0 \mid L_p(R_n)\| + \|2^{sk}(f - a_k) \mid L_p(R_n, l_q)\| < \infty\}. \end{aligned} \tag{11}$$

Furthermore,

$$\|f \mid F^s_{p,q}(R_n)\|^\times = \inf \|f \mid F^s_{p,q}(R_n)\|^a, \tag{12}$$

where the infimum is taken over all admissible systems $a \in \mathfrak{A}_p(R_n)$, is an equivalent quasi-norm in $F^s_{p,q}(R_n)$.

Proof. Step 1. If $0 < p \leqq \infty$, $0 < q \leqq \infty$ and $s > \tilde{\sigma}_p$, then we prove that there exists a constant $c > 0$ such that

$$\|f \mid B^s_{p,q}(R_n)\|^\times \leqq c\|f \mid B^s_{p,q}(R_n)\| \tag{13}$$

holds for all $f \in B^s_{p,q}(R_n)$. If $\{\varphi_k(x)\}_{k=0}^\infty \in \Phi(R_n)$ and $f \in B^s_{p,q}(R_n)$, then

$$a_k(x) = \sum_{l=0}^{k} F^{-1}\varphi_l Ff \to f \quad \text{in} \quad S'(R_n) \quad \text{if} \quad k \to \infty.$$

Furthermore, $a_k(x) \in S'(R_n) \cap L_p(R_n)$ and $\operatorname{supp} Fa_k \subset \{y \mid |y| \leqq 2^{k+1}\}$. Hence, $\{a_k\}_{k=0}^\infty \in \mathfrak{A}_p(R_n)$. If $1 \leqq p \leqq \infty$, then

$$\begin{aligned} 2^{ks}\|f - a_k \mid L_p(R_n)\| &= 2^{ks}\left\| \sum_{l=k+1}^\infty F^{-1}\varphi_l Ff \mid L_p(R_n)\right\| \\ &\leqq \sum_{j=1}^\infty 2^{-js}2^{s(k+j)}\|F^{-1}\varphi_{k+j}Ff \mid L_p(R_n)\|. \end{aligned} \tag{14}$$

If $1 \leqq q \leqq \infty$, the triangle inequality yields

$$\begin{aligned} &\|2^{ks}(f - a_k) \mid l_q(L_p(R_n))\| \\ &\leqq \sum_{j=1}^\infty 2^{-js}\|2^{sk}F^{-1}\varphi_k Ff \mid l_q(L_p(R_n))\| \leqq c\|f \mid B^s_{p,q}(R_n)\|. \end{aligned} \tag{15}$$

If $0<q<1$, then (14) yields

$$2^{ksq}\|f-a_k \mid L_p(R_n)\|^q \leqq \sum_{j=1}^{\infty} 2^{-jsq}2^{s(k+j)q}\|F^{-1}\varphi_{k+j}Ff \mid L_p(R_n)\|^q . \tag{16}$$

Summation over k yields (15). If $0<p<1$, we replace (14) by

$$2^{ksq}\|f-a_k \mid L_p(R_n)\|^p \leqq \sum_{j=1}^{\infty} 2^{-jsp}2^{sp(k+j)}\|F^{-1}\varphi_{k+j}Ff \mid L_p(R_n)\|^p . \tag{17}$$

If $q\geqq p$ or $q<p$, then we use the same technique as in (15) and (16), respectively, and obtain (15).

$$\|a_0 \mid L_p(R_n)\| \leqq c\|f \mid B^s_{p,q}(R_n)\|$$

and (15) prove (13).

Step 2. If $0<p<\infty$, $0<q\leqq\infty$ and $s>\tilde{\sigma}_{p,q}$, then we prove that there exists a constant $c>0$ such that

$$\|f \mid F^s_{p,q}(R_n)\|^\times \leqq c\|f \mid F^s_{p,q}(R_n)\| \tag{18}$$

holds for all $f\in F^s_{p,q}(R_n)$. If $1\leqq q\leqq\infty$, then the counterpart of (14) reads as follows,

$$\begin{aligned}\left(\sum_{k=0}^{\infty} 2^{ksq}|f(x)-a_k(x)|^q\right)^{\frac{1}{q}} &\leqq \sum_{j=1}^{\infty} 2^{-js}\left(\sum_{k=0}^{\infty} 2^{s(k+j)q}|F^{-1}\varphi_{k+j}Ff)(x)|^q\right)^{\frac{1}{q}} \\ &\leqq c\left(\sum_{k=0}^{\infty} 2^{skq}|(F^{-1}\varphi_k Ff)(x)|^q\right)^{\frac{1}{q}},\end{aligned} \tag{19}$$

(modification if $q=\infty$). If $0<q<1$, we have

$$\begin{aligned}\sum_{k=0}^{\infty} 2^{ksq}|f(x)-a_k(x)|^q &\leqq \sum_{j=1}^{\infty} 2^{-jsq}\sum_{k=0}^{\infty} 2^{s(k+j)q}|(F^{-1}\varphi_{k+j}Ff)(x)|^q \\ &\leqq c\sum_{k=0}^{\infty} 2^{skq}|(F^{-1}\varphi_k Ff)(x)|^q .\end{aligned} \tag{20}$$

Taking the $L_p(R_n)$-quasi-norm of (19) and (20), we obtain (18).

Step 3. If $0<p<\infty$, $0<q<\infty$ and $s>\tilde{\sigma}_{p,q}$, we prove that there exists a constant $c>0$ such that

$$\|f \mid F^s_{p,q}(R_n)\| \leqq c\|f \mid F^s_{p,q}(R_n)\|^a \tag{21}$$

holds for any admissible system $a\in\mathfrak{A}_p(R_n)$ in the sense of (11). Let $0<p\leqq q\leqq 1$ (otherwise one has to modify the calculations in the same manner as above; this will not be described here in detail). Let f be given by the right-hand side of (11). If again $\{\varphi_k(x)\}_{k=0}^{\infty}\in\Phi(R_n)$, then we have

$$\begin{aligned}|(F^{-1}\varphi_k Ff)(x)|^q &= \left|\sum_{j=-1}^{\infty} F^{-1}\varphi_k F(a_{k+j}-a_{k+j-1})(x)\right|^q \\ &\leqq \sum_{j=-1}^{\infty} |F^{-1}\varphi_k F(a_{k+j}-a_{k+j-1})(x)|^q\end{aligned}$$

and, as above,

$$\begin{aligned}\|2^{sk}F^{-1}\varphi_k Ff \mid L_p(R_n,l_q)\| &= \left(\int_{R_n}\left[\sum_{k=0}^{\infty} 2^{skq}|F^{-1}\varphi_k Ff(\cdot)|^q\right]^{\frac{p}{q}}\mathrm{d}x\right)^{\frac{1}{p}} \\ &\leqq \left(\sum_{j=-1}^{\infty} 2^{-sjp}\int_{R_n}\left[\sum_{k=0}^{\infty} 2^{s(k+j)q}|F^{-1}\varphi_k F(a_{k+j}-a_{k+j-1})(\cdot)|^q\right]^{\frac{p}{q}}\mathrm{d}x\right)^{\frac{1}{p}} \\ &\leqq \left(\sum_{j=-1}^{\infty} 2^{-sjp}\|2^{s(k+j)}F^{-1}\varphi_k F(a_{k+j}-a_{k+j-1}) \mid L_p(R_n,l_q)\|^p\right)^{\frac{1}{p}}\end{aligned} \tag{22}$$

(of course, $a_k=0$ if $k<0$). We assume, without loss of generality, $\varphi_k(x)=\varphi_1(2^{-k+1}x)$ if $k=1, 2, 3, \ldots$ In order to apply Theorem 2.4.9/2 we remark that $\operatorname{supp}(a_{k+j}-a_{k+j-1})\subset\{y \mid |y|\leqq 2^{k+j+1}\}$ and

$$\|\varphi_k(2^{k+j}\cdot) \mid H_2^{\varkappa}(R_n)\|=\|\varphi_1(2^{j+1}\cdot) \mid H_2^{\varkappa}(R_n)\|\leqq c2^{j\left(\varkappa-\frac{n}{2}\right)}, \tag{23}$$

where c is independent of j. Now Theorem 2.4.9/2 yields

$$\begin{aligned}&\|2^{s(k+j)}F^{-1}\varphi_kF(a_{k+j}-a_{k+j-1}) \mid L_p(R_n, l_q)\|\\ &\leqq c2^{j\left(\varkappa-\frac{n}{2}\right)}\|2^{s(k+j)}(a_{k+j}-a_{k+j-1}) \mid L_p(R_n, l_q)\|,\end{aligned} \tag{24}$$

where $\varkappa>\sigma_{p,q}$ (here we use for the first time that $q<\infty$). Substituting (24) into (22) we obtain

$$\begin{aligned}\|2^{sk}F^{-1}\varphi_kFf \mid L_p(R_n, l_q)\|&\leqq c\left(\sum_{j=-1}^{\infty}2^{j\left(\varkappa-\frac{n}{2}-s\right)p}\right)^{\frac{1}{p}}\|2^{sk}(a_k-a_{k-1}) \mid L_p(R_n, l_q)\|\\ &\leqq c\|2^{sk}(f-a_k) \mid L_p(R_n, l_q)\|+c\|a_0 \mid L_p(R_n)\|,\end{aligned} \tag{25}$$

provided that $s>\varkappa-\frac{n}{2}>\tilde{\sigma}_{p,q}$. This proves (21), where c is independent of the chosen system $a=\{a_k\}_{k=0}^{\infty}$. Similarly for the other pairs p and q. Hence, the proof of part (ii) is complete.

Step 4. If $0<p\leqq\infty$, $0<q\leqq\infty$ and $s>\tilde{\sigma}_p$, then we prove that there exists a constant $c>0$ such that

$$\|f \mid B_{p,q}^s(R_n)\|\leqq c\|f \mid B_{p,q}^s(R_n)\|^{\times} \tag{26}$$

for any admissible system $a\in\mathfrak{A}_p(R_n)$ in the sense of (9). The proof follows the same lines as in Step 3. In the counterpart of (22) we need only the scalar case of Theorem 2.4.9/2, i.e. (1.5.2/13) (including $p=\infty$). Instead of (25) we obtain (for $0<p\leqq 1$)

$$\begin{aligned}&2^{skp}\|F^{-1}\varphi_kFf \mid L_p(R_n)\|^p\\ &\leqq c\sum_{j=-1}^{\infty}2^{j\left(\varkappa-\frac{n}{2}-s\right)p}\|2^{s(k+j)}(a_{k+j}-a_{k+j-1}) \mid L_p(R_n)\|^p,\end{aligned} \tag{27}$$

where $\varkappa>\sigma_p$. If again $0<p\leqq q\leqq 1$ (as in Step 3), then we apply the $l_{\frac{q}{p}}$-norm to (27) to obtain

$$\begin{aligned}&\|2^{sk}F^{-1}\varphi_kFf \mid l_q(L_p(R_n))\|^p\\ &\leqq c\sum_{j=-2}^{\infty}2^{j\left(\varkappa-\frac{n}{2}-s\right)p}\|2^{sk}(a_{k+1}-a_k) \mid l_q(L_p(R_n))\|^p\\ &\leqq c'\|2^{sk}(f-a_k) \mid l_q(L_p(R_n))\|^p+c'\|a_0 \mid L_p(R_n)\|^p,\end{aligned}$$

provided that $s>\varkappa-\frac{n}{2}>\tilde{\sigma}_p$. This proves (26), where c is independent of the chosen system $a=\{a_k\}_{k=0}^{\infty}$. Hence, also the proof of part (i) is complete.

Remark 1. We return to the postponed problem of how to understand $f\in L_p(R_n)$ in the proof (and also in the formulation of the theorem) if $p<1$. Let $f\in B_{p,q}^s(R_n)$. By (1.3.2/5) (and a completion argument) it follows that

$$\|F^{-1}\varphi_kFf \mid L_1(R_n)\|\leqq c2^{k\tilde{\sigma}_p}\|F^{-1}\varphi_kFf \mid L_p(R_n)\|, \tag{28}$$

where c is independent of k. If $s>\tilde{\sigma}_p$, then (28) yields

$$B_{p,q}^s(R_n)\subset B_{1,q}^{s-\tilde{\sigma}_p}(R_n)\subset L_1(R_n). \tag{29}$$

The last inequality follows easily from

$$\|f \mid L_1(R_n)\| \le \sum_{k=0}^{\infty} \|F^{-1}\varphi_k Ff \mid L_1(R_n)\| \le c\|f \mid B^{\tau}_{1,q}(R_n)\| \tag{30}$$

for $\tau>0$ and $0<q\le\infty$. Hence in that cases, i.e. $s>\tilde{\sigma}_p$, every $f\in B^s_{p,q}(R_n)$ is a regular distribution. Of course, this is also valid if $1\le p\le\infty$ and $s>0$, because $B^s_{p,q}(R_n)\subset L_p(R_n)$, cf. (30) with p instead of 1. By (2.3.2/9) and $\tilde{\sigma}_{p,q}\ge\tilde{\sigma}_p$, we have the same argument for the spaces $F^s_{p,q}(R_n)$ with $s>\tilde{\sigma}_{p,q}$. Let again $p<1$ and $f\in B^s_{p,q}(R_n)$ with $s>\tilde{\sigma}_p$. If $\{\varphi_k(x)\}_{k=0}^{\infty}\in\Phi(R_n)$, then it follows by (30) that $a_k(x)$ from Step 1 converges not only in $S'(R_n)$ but also in $L_1(R_n)$ to f if $k\to\infty$. By an argument of type (14) one obtains that $\{a_k(x)\}_{k=0}^{\infty}$ is also a fundamental sequence in $L_p(R_n)$. Let $g(x)\in L_p(R_n)$ be the limit element. However, the limit elements in $L_1(R_n)$ and $L_p(R_n)$ must coincide a. e. Hence $f=g\in L_p(R_n)$. This justifies the calculations in Step 1 and Step 2. What about Step 3 and Step 4? Let $p<1$ and let f be an element of the right-hand side of (9). Then it follows that $f\in L_p(R_n)$ and that a_k converges not only in $S'(R_n)$ but also in $L_p(R_n)$ to f. In order to justify the a. e. pointwise estimate of $F^{-1}\varphi_k Ff$ in Step 3 and in Step 4, we remark that

$$\left\| \sum_{j=-1}^{\infty} F^{-1}\varphi_k F(a_{k+j}-a_{k+j-1}) \mid L_1(R_n) \right\|$$
$$\le c \sum_{j=-1}^{\infty} \|F^{-1}\varphi_k \mid L_1(R_n)\| \, \|a_{k+j}-a_{k+j-1} \mid L_1(R_n)\| .$$

However, $\|F^{-1}\varphi_k \mid L_1(R_n)\|$ is uniformly bounded. Then (28) yields

$$\left\| \sum_{j=-1}^{\infty} F^{-1}\varphi_k F\,(a_{k+j}-a_{k+j-1}) \mid L_1(R_n) \right\| \le c \sum_{j=-1}^{\infty} 2^{(k+j)\tilde{\sigma}_p} \|a_{k+j}-a_{k+j-1} \mid L_p(R_n)\| .$$

Now it follows that $F^{-1}\varphi_k Fa_l$ converges in $L_1(R_n)$ if $l\to\infty$ and the limit element must be $F^{-1}\varphi_k Ff$. This justifies the calculations in Step 4. Similarly in Step 3, because the space on the right-hand side of (11) can be estimated in the same way as in (2.3.2/9) by the space on the right-hand side of (9).

Remark 2. We used Theorem 2.4.9/2 and obtained a sharp and natural result. If one uses the weaker Theorem 1.6.3 then one has for the spaces $F^s_{p,q}(R_n)$ with $\min(p,q)<1$ a weaker result (the case $\infty>p>1$, $\infty>q>1$ can be treated by the known multiplier theorem from Proposition 2.4.8 (ii)). On the other hand, Theorem 1.6.3 covers the case $q=\infty$ (in contrast to Theorem 2.4.9/2). If we use Theorem 1.6.3, then we obtain the assertion of part (ii) of the above theorem under the restriction

$$0<p<\infty, \quad 0<q\le\infty, \quad s>\frac{n}{\min(p,q)}$$

(of course, only $q=\infty$ is of interest). This follows immediately from Step 2 and Step 3 of the above proof. This result has been proved in [S, 2.2.3].

Remark 3. In a formal way (without the justification in Remark 1), the calculation in Step 1 and Step 2 go through if $s>0$. What about the values $0<s\le\tilde{\sigma}_p$, resp. $0<s\le\tilde{\sigma}_{p,q}$? If $0<p<1$ and $s<\tilde{\sigma}_p$, then $\delta\in B^s_{p,q}(R_n)$, where δ is the δ-distribution. The latter assertion follows from $F\delta\equiv c\neq0$ and

$$\|F^{-1}\varphi_k F\delta \mid L_p(R_n)\| = c\left(\int_{R_n} |[F^{-1}\varphi_1(2^{-k+1}\cdot)](x)|^p\,dx\right)^{\frac{1}{p}}$$
$$= c2^{kn}\left(\int_{R_n} |(F^{-1}\varphi_1)(2^{k+1}x)|^p\,dx\right)^{\frac{1}{p}} = c'2^{kn\left(1-\frac{1}{p}\right)} = c'2^{-k\tilde{\sigma}_p},$$

where $k=1,2,3,\ldots$ (again we used $\varphi_k(x)=\varphi_1(2^{-k+1}x)$ for $\{\varphi_k(x)\}_{k=0}^{\infty}\in\Phi(R_n)$). This shows that the calculations in Step 1 do not make sense if $0<p<1$ and $0<s<\tilde{\sigma}_p$, because $f\in B^s_{p,q}(R_n)$ cannot be interpreted in general as a regular distribution which belongs also to $L_p(R_n)$.

2.5.4. Lizorkin Representations

In the definition of the spaces $B^s_{p,q}(R_n)$ and $F^s_{p,q}(R_n)$ in 2.3.1. we used balls and differences of balls as the underlying decomposition of R_n. Now we introduce cubes and corridors. Let

$$K_k=\{x \mid |x_j|<2^k \quad \text{if} \quad j=1,\ldots,n\}-\{x \mid |x_j|\leq 2^{k-1} \quad \text{if} \quad j=1,\ldots,n\} \tag{1}$$

where $k=1, 2, 3, \ldots$ and

$$K_0=\{x \mid |x_j|\leq 1 \quad \text{if} \quad j=1, 2, \ldots\}. \tag{2}$$

We subdivide K_k with $k=1, 2, 3, \ldots$ by the $3n$ hyper-planes $\{x \mid x_m=0\}$ and $\{x \mid x_m=\pm 2^{k-1}\}$, where $m=1, \ldots, n$, in congruent cubes $P_{k,t}$ (cf. Fig. 2.5.4). If k is fixed, we obtain $T=4^n-2^n$ cubes which are numbered by $t=1, \ldots T$ in an arbitrary way. Let $P_{0,t}=K_0$ if

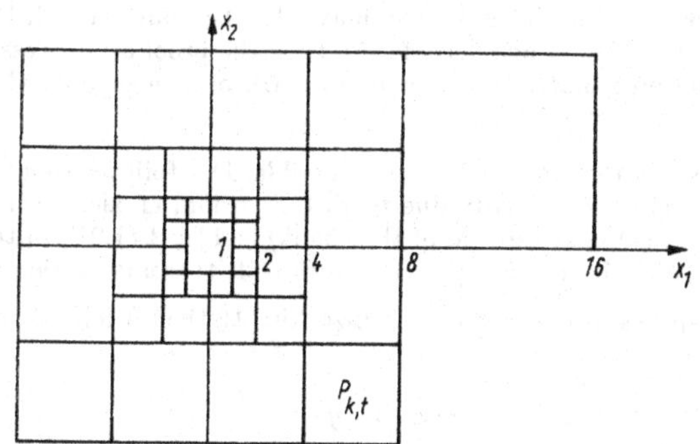

Fig. 2.5.4.

$t=1, \ldots, T$. Of course

$$R_n=\bigcup_{k=0}^{\infty} K_k=\bigcup_{k=0}^{\infty}\bigcup_{t=1}^{T} P_{k,t}\,.$$

Let χ_k and $\chi_{k,t}$ be the characteristic functions of K_k and $P_{k,t}$, respectively, $k=0, 1, \ldots$ and $t=1, \ldots, T$.

Theorem. Let $-\infty<s<\infty$ and $1<p<\infty$.

(i) If $0<q\leq\infty$, then

$$\begin{aligned} B^s_{p,q}(R_n)&=\left\{f \mid f\in S'(R_n),\ \|f \mid B^s_{p,q}(R_n)\|^K=\left(\sum_{k=0}^{\infty} 2^{skq}\|F^{-1}\chi_k Ff \mid L_p(R_n)\|^q\right)^{\frac{1}{q}}<\infty\right\}\\ &=\left\{f \mid f\in S'(R_n),\ \|f \mid B^s_{p,q}(R_n)\|^B=\left(\sum_{k=0}^{\infty}\sum_{t=1}^{T} 2^{skq}\|F^{-1}\chi_{k,t}Ff \mid L_p(R_n)\|^q\right)^{\frac{1}{q}}<\infty\right\} \end{aligned} \tag{3}$$

(modification if $q=\infty$), where $\|f \mid B^s_{p,q}(R_n)\|^K$ and $\|f \mid B^s_{p,q}(R_n)\|^B$ are equivalent quasi-norms in $B^s_{p,q}(R_n)$.

(ii) If $1<q<\infty$, then

$$\begin{aligned} F^s_{p,q}(R_n)=\Bigg\{f \mid f\in S'(R_n),\ &\|f \mid F^s_{p,q}(R_n)\|^K\\ &=\left\|\left(\sum_{k=0}^{\infty} 2^{skq}|(F^{-1}\chi_k Ff)(\cdot)|^q\right)^{\frac{1}{q}} \mid L_p(R_n)\right\|<\infty\Bigg\} \end{aligned} \tag{4}$$

$$=\Big\{f \mid f\in S'(R_n),\ \|f \mid F^s_{p,q}(R_n)\|^B$$

$$=\Big\|\Big(\sum_{k=0}^{\infty}\sum_{t=1}^{T} 2^{skq}|(F^{-1}\chi_{k,t}Ff)(\cdot)|^q\Big)^{\frac{1}{q}} \mid L_p(R_n)\Big\|<\infty\Big\},$$

where $\|f \mid F^s_{p,q}(R_n)\|^K$ and $\|f \mid F^s_{p,q}(R_n)\|^B$ are equivalent norms in $F^s_{p,q}(R_n)$.

Proof. (Outline). The proof is based on the following fact: $\{\chi_k(x)\}_{k=0}^{\infty}$ is a Fourier multiplier for $L_p(R_n, l_q)$ if $1<p<\infty$ and $1<q<\infty$, i. e. there exists a constant $c>0$ such that

$$\|F^{-1}\chi_k Ff_k \mid L_p(R_n, l_q)\| \leqq c\|f_k \mid L_p(R_n, l_q)\| \tag{5}$$

for all systems $\{f_k(x)\}_{k=0}^{\infty}\in L_p(R_n, l_q)$. Similarly for $\{\chi_{k,t}(x)\}_{\substack{k=0,1,2,\dots\\ t=1,\dots,T}}$. This assertion is not covered by Proposition 2.4.8 (ii). It follows from a vector-valued Fourier multiplier theorem in P. I. Lizorkin [1, p. 241] (the one-dimensional case may also be found in N. Dunford, J. T. Schwartz [1, II, 11.11, Lemma 24]). If one accepts (5), then the proof of (ii) is essentially the same as the proof of Proposition 2.3.2/1. In order to prove (i), one needs only the scalar case of (5). In that case, the number q is immaterial.

Remark. An extension of the theorem to values $p\leqq 1$ is not possible. This follows from (1.3.5/3), cf. [S, p. 56]. Part (i) of the theorem is essentially due to P. I. Lizorkin, cf. also S. M. Nikol'skij [3, 8.10.1] (including the comments at the end of that book) and [I, 2.11.2]. On the other hand, representations of type (4) have been used by P. I. Lizorkin [1, 4] as a definition of corresponding spaces. If $\overset{+}{\mathfrak{L}}{}^s_{p,q}$ are the spaces introduced in P. I. Lizorkin [4], then it follows immediately from the above theorem that

$$\overset{+}{\mathfrak{L}}{}^s_{p,q}=F^s_{p,q}(R_n) \quad\text{if}\quad -\infty<s<\infty,\quad 1<p<\infty,\quad 1<q<\infty.$$

2.5.5. Discrete Representations and Schauder Bases for $B^s_{p,q}(R_n)$

Let $P_0=K_0$ and $P_{k,t}$ with $k=1, 2, 3, \dots$ and $t=1, \dots, T$ be the cubes from the previous subsection. Let again χ_0 and $\chi_{k,t}$ be the corresponding characteristic functions of P_0 and $P_{k,t}$, respectively. Let $\sigma_{k,t}$ be the centre of $P_{k,t}$. Let Z_n be the lattice from 1.2.4. with the lattice points $m=(m_1, \dots, m_n)$. Finally, we recall that $xy=\sum_{j=1}^{n} x_j y_j$ denotes the scalar product of $x=(x_1, \dots, x_n)\in R_n$ and $y=(y_1, \dots, y_n)\in R_n$. We combine the representation (2.5.4/3) of $B^s_{p,q}(R_n)$ with Proposition 1.3.5 and Remark 1.3.5/2. The idea is to apply (1.3.5/2) and the equivalence assertion in Remark 1.3.5/2 to $F^{-1}\chi_{k,t}Ff$. We omit details and refer to [S, 2.2.5]. The result is the following.

Theorem. Let $-\infty<s<\infty$, $1<p<\infty$ and $0<q\leqq\infty$.
(i) $f\in S'(R_n)$ belongs to $B^s_{p,q}(R_n)$ if and only if it can be represented as

$$f=\sum_{m\in Z_n}\Big[a_m F^{-1}(\mathrm{e}^{-\mathrm{i}\pi m x}\chi_0)+\sum_{k=1}^{\infty}\sum_{t=1}^{T} a_{m,k,t}F^{-1}(\mathrm{e}^{-\mathrm{i}\pi 2^{-k+2}xm}\chi_{k,t})\Big] \tag{1}$$

(convergence in $S'(R_n)$) with

$$\Big(\sum_{m\in Z_n}|a_m|^p\Big)^{\frac{1}{p}}+\Big[\sum_{k=1}^{\infty}\sum_{t=1}^{T}\Big(\sum_{m\in Z_n}|2^{k\left(s-\frac{n}{p}\right)+kn}a_{m,k,t}|^p\Big)^{\frac{q}{p}}\Big]^{\frac{1}{q}}<\infty \tag{2}$$

(modification if $q=\infty$). Furthermore, (2) is an equivalent quasi-norm in $B^s_{p,q}(R_n)$.

(ii) $f \in S'(R_n)$ belongs to $B^s_{p,q}(R_n)$ if and only if it can be represented as

$$f = \sum_{m \in Z_n} \left[b_m \prod_{j=1}^{n} \frac{\sin x_j}{x_j - \pi m_j} + \sum_{k=1}^{\infty} \sum_{t=1}^{T} b_{m,k,t} e^{-ix\sigma_{k,t}} \prod_{j=1}^{n} \frac{\sin 2^{k-2} x_j}{2^{k-2} x_j - \pi m_j} \right] \tag{3}$$

(convergence in $S'(R_n)$) with

$$\left(\sum_{m \in Z_n} |b_m|^p \right)^{\frac{1}{p}} + \left[\sum_{k=1}^{\infty} \sum_{t=1}^{T} \left(\sum_{m \in Z_n} |2^{k\left(s-\frac{n}{p}\right)} b_{m,k,t}|^p \right)^{\frac{q}{p}} \right]^{\frac{1}{q}} < \infty \tag{4}$$

(modification if $q = \infty$). Furthermore, (4) is an equivalent quasinorm in $B^s_{p,q}(R_n)$.

Remark 1. Let $l_q(l_p)$ be the set of all sequences $\{c_{l,j}\}_{l,j=1}^{\infty}$ of complex numbers $c_{l,j}$ with

$$\left[\sum_{l=1}^{\infty} \left(\sum_{j=1}^{\infty} |c_{l,j}|^p \right)^{\frac{q}{p}} \right]^{\frac{1}{q}} < \infty ,$$

where $0 < p \leqq \infty$ and $0 < q \leqq \infty$ (modification if $p = \infty$ and/or $q = \infty$). It follows immediately from the above theorem that $B^s_{p,q}(R_n)$ with $-\infty < s < \infty$, $1 < p < \infty$ and $0 < q \leqq \infty$ is isomorphic to $l_q(l_p)$. In particular, $B^s_{p,p}(R_n)$ with $-\infty < s < \infty$ and $1 < p < \infty$ is isomorphic to l_p. A first proof of these results has been given in H. Triebel [23], cf. also [I, 2.11.2]. Another proof may be found in J. Peetre [8, p. 180]. In this book (on p. 187), it is also proved that $B^s_{\infty,\infty}(R_n)$ is isomorphic to l_∞. The anisotropic counterpart (in the sense of 10.1.) is due to P. I. Lizorkin [5].

Remark 2. We recall the notion of a Schauder basis. A subset $\{b_j\}_{j=1}^{\infty}$ of a complex separable quasi-Banach space B is said to be a Schauder basis in B if every element $b \in B$ can be represented uniquely as a convergent series

$$b = \sum_{j=1}^{\infty} \beta_j b_j , \quad \beta_j \text{ complex} ,$$

(topological basis). A Schauder basis is said to be unconditional if every rearrangement of the elements in $\{b_j\}_{j=1}^{\infty}$ is also a Schauder basis in B. By Remark 1 it follows immediately that $B^s_{p,q}(R_n)$ with $-\infty < s < \infty$, $1 < p < \infty$ and $0 < q < \infty$ is a separable quasi-Banach space with an unconditional Schauder basis. However, by the above theorem one can strengthen this assertion as follows. If $-\infty < s < \infty$, $1 < p < \infty$ and $0 < q < \infty$, then

$$\{F^{-1}(e^{-i\pi x m}\chi_0),\, F^{-1}(e^{-i\pi 2^{-k+2} x m}\chi_{k,t})\}_{\substack{m \in Z_n \\ k=1,2,\ldots \\ t=1,\ldots,T}}$$

is an unconditional Schauder basis in $B^s_{p,q}(R_n)$ and

$$\left\{ \prod_{j=1}^{n} \frac{\sin x_j}{x_j - \pi m_j},\ e^{-ix\sigma_{k,t}} \prod_{j=1}^{n} \frac{\sin 2^{k-2} x_j}{2^{k-2} x_j - \pi m_j} \right\}_{\substack{m \in Z_n \\ k=1,2,\ldots \\ t=1,\ldots,T}}$$

is an unconditional Schauder basis in $B^s_{p,q}(R_n)$. These are simultaneous unconditional Schauder bases of entire analytic functions in all spaces $B^s_{p,q}(R_n)$ with $-\infty < s < \infty$, $1 < p < \infty$ and $0 < q < \infty$.

2.5.6. The Bessel-Potential Spaces $H^s_p(R_n)$ and the Sobolev Spaces $W^m_p(R_n)$

Let $1 < p < \infty$. It is our aim to prove that $F^s_{p,2}(R_n) = H^s_p(R_n)$ if $-\infty < s < \infty$ and that $H^m_p(R_n) = W^m_p(R_n)$ if $m = 1, 2, 3, \ldots$, cf. (2.2.2/19) and (2.3.5/2). The Sobolev spaces $W^m_p(R_n)$ with $m = 1, 2, 3, \ldots$ and $1 < p < \infty$, and the norm $\|f \mid W^m_p(R_n)\|$

have been defined in (2.2.2/7). Concerning the Bessel-potential spaces $H_p^s(R_n)$ with $-\infty<s<\infty$ and $1<p<\infty$, and the norm $\|f \mid H_p^s(R_n)\|$ we refer to (2.2.2/11). Of course, $H_p^0(R_n)=L_p(R_n)$ if $1<p<\infty$. Then $F_{p,2}^0(R_n)=L_p(R_n)$ is a Littlewood-Paley theorem. It can be proved on the basis of the Hilbert space version of the Michlin-Hörmander multiplier theorem, which reads as follows.

Proposition. If $1<p<\infty$, then there exist a positive constant c such that

$$\left\| F^{-1}\left[\sum_{j=0}^{\infty} m_{k,j}Ff_j\right] \,\middle|\, L_p(R_n, l_2)\right\|$$
$$\leq c \sup \left(R^{2|\alpha|-n} \int_{\frac{R}{2}\leq|x|\leq 2R} \sum_{j,k=0}^{\infty} |D^\alpha m_{k,j}(x)|^2 \,dx\right)^{\frac{1}{2}} \|f_k \mid L_p(R_n, l_2)\| \tag{1}$$

holds for all systems $\{m_{k,j}(x)\}_{k,j=0}^{\infty} \subset L_\infty(R_n)$ and all systems $\{f_k(x)\}_{k=0}^{\infty}$ of measurable functions in R_n, where the supremum is taken over all $R>0$ and all multi-indices α with $0\leq|\alpha|\leq 1+\left[\frac{n}{2}\right]$.

Remark. This is the l_2-version of Hörmander's famous multiplier theorem. A proof of (1) has been given in [I, pp. 161–165]. The l_q-version of (1) (restricted to the "diagonal case", i.e. $m_{k,j}=0$ if $k\neq j$) has been formulated in (2.4.8/7).

Theorem. (i) If $-\infty<s<\infty$ and $1<p<\infty$, then

$$F_{p,2}^s(R_n)=H_p^s(R_n), \tag{2}$$

where $\|f \mid H_p^s(R_n)\|$ is an equivalent norm in $F_{p,2}^s(R_n)$.
(ii). If $m=1, 2, 3, \ldots$ and $1<p<\infty$, then

$$H_p^m(R_n)=W_p^m(R_n)=\{f \mid f\in S'(R_n), \|f \mid W_p^m(R_n)\|^*<\infty\}, \tag{3}$$

where $\|f \mid W_p^m(R_n)\|$ and

$$\|f \mid W_p^m(R_n)\|^*=\|f \mid L_p(R_n)\|+\sum_{j=1}^{n}\left\|\frac{\partial^m f}{\partial x_j^m} \,\middle|\, L_p(R_n)\right\| \tag{4}$$

are equivalent norms in $H_p^m(R_n)$.

Proof. Step 1. Let us assume that $F_{p,2}^0(R_n)=L_p(R_n)$ with $1<p<\infty$ (equivalent norms). Then all the other assertions of the theorem follow from Theorem 2.3.8 (including Step 2 of the proof of that theorem and the Michlin-Hörmander multiplier theorem for $L_p(R_n)$ with $1<p<\infty$).

Step 2. We prove that

$$F_{p,2}^0(R_n)=L_p(R_n) \quad \text{if} \quad 1<p<\infty \tag{5}$$

(equivalent norms). Let $f\in L_p(R_n)$ and $\{\varphi_k(x)\}_{k=0}^{\infty}\in \Phi(R_n)$. Let $m_{k,0}=\varphi_k$ and $m_{k,j}=0$ if $k=0, 1, 2, \ldots$ and $j=1, 2, \ldots$ We apply (1) with $f_0=f$ and $f_j=0$ if $j=1, 2, \ldots$, and obtain

$$\|F^{-1}\varphi_k Ff \mid L_p(R_n, l_2)\|\leq c\|f \mid L_p(R_n)\|.$$

Hence $f\in F_{p,2}^0(R_n)$ and $L_p(R_n)\subset F_{p,2}^0(R_n)$. In order to prove the converse assertion, we assume that $f\in F_{p,2}^0(R_n)$. Let

$$\psi_k(x)=\sum_{r=-1}^{1}\varphi_{k+r}(x) \quad (\text{with } \varphi_{-1}=0),$$

if $k=0, 1, 2, \ldots$ We have $\psi_k(x)=1$ if $x \in \operatorname{supp} \varphi_k$. Let $m_{0,j}=\psi_j$ and $m_{k,j}=0$ if $k=1, 2, 3, \ldots$ and $j=0, 1, 2, \ldots$ We apply (1) with $f_k=F^{-1}\varphi_k Ff$. Then

$$\left\| F^{-1} \sum_{j=0}^{\infty} \psi_j \varphi_j Ff \mid L_p(R_n) \right\| \leq c \| F^{-1}\varphi_k Ff \mid L_p(R_n, l_2)\|$$

and consequently,

$$\|f \mid L_p(R_n)\| \leq c \|f \mid F^0_{p,2}(R_n)\| .$$

This proves $f \in L_p(R_n)$ and $F^0_{p,2}(R_n) \subset L_p(R_n)$. The proof is complete.

2.5.7. The Besov Spaces $\Lambda^s_{p,q}(R_n)$ and the Zygmund Spaces $\mathcal{C}^s(R_n)$

In connection with the Besov spaces $\Lambda^s_{p,q}(R_n)$ from (2.2.2/9), (2.2.2/10), and the Zygmund spaces $\mathcal{C}^s(R_n)$ from (2.2.2/6) we have to prove (2.3.5/3) and (2.3.5/1). Furthermore, we must show that the Slobodeckij spaces $W^s_p(R_n)$ and the Hölder spaces $C^s(R_n)$ are special cases of $\Lambda^s_{p,q}(R_n)$ and $\mathcal{C}^s(R_n)$, respectively, in the sense of (2.2.2/18). All these problems will be solved in the framework of 2.5.12., where we give characterizations of $B^s_{p,q}(R_n)$ in terms of the differences Δ^l_h (and derivatives D^α). This direct approach covers also the values $p<1$, and the above-mentioned results are only special cases. But it is desirable to deal separately with the above problems and to ask for simple proofs. The best way (in the opinion of the author) is to use interpolation theory. This shows also in a very elegant way that the Besov spaces $\Lambda^s_{p,q}(R_n)$ can be obtained by real interpolation from the Sobolev spaces $W^m_p(R_n)$ and that the Zygmund spaces $\mathcal{C}^s(R_n)$ can be obtained by real interpolation from the spaces $C^m(R_n)$, where $m=0, 1, 2, \ldots$, cf. (2.2.2/1). The real interpolation method has been described in 2.4.1. The disadvantage of this approach is that it requires some assertions from abstract interpolation theory in Banach spaces, in particular from the interpolation theory of commutative semigroups of operators in a given Banach space. A reader who is not familiar with interpolation theory (or who is not ready to consult the quoted references) may skip this subsection. As we mentioned previously, the formulas in question, i.e. (2.3.5/3), (2.3.5/1) and (2.2.2/18), are also consequences of the considerations in 2.5.12. This situation allows us to restrict ourselves to an outline of the proof of the theorem.

Let $C^m(R_n)$ with $C(R_n)=C^0(R_n)$ be the spaces from the beginning of 2.2.2. Let $W^m_p(R_n)$ with $L_p(R_n)=W^0_p(R_n)$ be the Sobolev spaces from (2.2.2/7). Here $m=0, 1, 2, \ldots$ and $1<p<\infty$. We extend the definition of $W^m_p(R_n)$ with $m=0, 1, 2, \ldots$ to $p=1$.

Proposition. If $1 \leq p < \infty$, then

$$B^0_{p,1}(R_n) \subset L_p(R_n) \subset B^0_{p,\infty}(R_n) \tag{1}$$

and

$$B^0_{\infty,1}(R_n) \subset C(R_n) \subset B^0_{\infty,\infty}(R_n) . \tag{2}$$

Proof. Step 1. Let $\{\varphi_k(x)\}_{k=0}^\infty \in \Phi(R_n)$, cf. 2.3.1. If $1 \leq p \leq \infty$ and $f \in B^0_{p,1}(R_n)$, then we have

$$\|f \mid L_p(R_n)\| = \left\| \sum_{k=0}^{\infty} F^{-1}\varphi_k Ff \mid L_p(R_n) \right\| \leq \sum_{k=0}^{\infty} \|F^{-1}\varphi_k Ff \mid L_p(R_n)\| = \|f \mid B^0_{p,1}(R_n)\| . \tag{3}$$

This proves the left-hand side of (1). Let $p=\infty$. By (1.4.1/3) (first derivatives), it follows that every $(F^{-1}\varphi_k Ff)(x)$ is bounded and uniformly continuous in R_n. Then (3) yields $f \in C(R_n)$. This proves the left-hand side of (2).

Step 2. Let $f \in L_p(R_n)$ with $1 \leq p \leq \infty$. We have

$$(F^{-1}\varphi_k Ff)(x) = \int_{R_n} (F^{-1}\varphi_k)(y)\, f(x-y)\, dy, \quad x \in R_n . \tag{4}$$

We may assume that $\varphi_k(x)=\varphi_1(2^{-k+1}x)$ if $k=1, 2, 3, \ldots$ Then

$$(F^{-1}\varphi_k)(y) = 2^{(k-1)n}(F^{-1}\varphi_1)(2^{k-1}y) .$$

Now (4) yields

$$\|F^{-1}\varphi_k Ff \mid L_p(R_n)\| \leq \|f \mid L_p(R_n)\| \int_{R_n} |(F^{-1}\varphi_k)(y)|\, dy \leq c\|f \mid L_p(R_n)\| .$$

This proves the right-hand sides of (1) and (2).

Theorem. (i) If $s>0$, $1 \leq p < \infty$ and $1 \leq q \leq \infty$, then

$$\Lambda^s_{p,q}(R_n) = B^s_{p,q}(R_n) . \tag{5}$$

(ii) If $s>0$, then

$$\mathcal{C}^s(R_n) = B^s_{\infty,\infty}(R_n) . \tag{6}$$

Proof. (Outline). Step 1. If $j=1, \ldots, n$ then $\{G_j(t)\}_{0 \leq t < \infty}$ is defined by

$$(G_j(t)\, f)(x) = f(x_1, \ldots, x_{j-1}, x_j + t, x_{j+1}, \ldots, x_n) .$$

It is easy to see that $G_1(t), \ldots, G_n(t)$ are n commutative strongly continuous semi-groups of operators in $L_p(R_n)$ with $1 \leq p < \infty$ and in $C(R_n)$ (here we need that the functions from $C(R_n)$ are uniformly continuous). Concerning strongly continuous semi-groups of operators, and in particular the corresponding interpolation theory, we refer to [I, 1.13]. If $D(\Lambda_j)$ denotes the domain of definition of the infinitesimal generator Λ_j of $\{G_j(t)\}_{0 \leq t < \infty}$, then we have

$$D(\Lambda_j) = \left\{ f \mid f \in L_p(R_n),\ \frac{\partial f}{\partial x_j} \in L_p(R_n) \right\},$$

if $1 \leq p \leq \infty$, and

$$D(\Lambda_j) = \left\{ f \mid f \in C(R_n),\ \frac{\partial f}{\partial x_j} \in C(R_n) \right\},$$

respectively, $j=1, \ldots, n$ (cf. also [I, Lemma 2.5.1]). If m is a natural number, then we obtain by iteration that

$$K^m = \bigcap_{0 \leq j_1 + \cdots + j_n \leq m} D(\Lambda_1^{j_1} \cdots \Lambda_n^{j_n})$$
$$= \{f \mid D^\alpha f \in L_p(R_n) \quad \text{where} \quad |\alpha| \leq m\} = W^m_p(R_n) \quad \text{if} \quad 1 \leq p < \infty$$

and

$$K^m = \{f \mid D^\alpha f \in C(R_n) \quad \text{where} \quad |\alpha| \leq m\} = C^m(R_n) ,$$

respectively. In [I, 1. 13. 6], we developed the real interpolation theory for the spaces K^m in an abstract setting. If we apply the results to the above concrete cases, then we obtain immediately

$$(L_p(R_n), W^m_p(R_n))_{\Theta,q} = \Lambda^{\Theta m}_{p,q}(R_n) \quad \text{if} \quad 1 \leq p < \infty \tag{7}$$

and

$$(C(R_n), C^m(R_n))_{\Theta,\infty} = \mathcal{C}^{\Theta m}(R_n) , \tag{8}$$

respectively (cf. also [I, p. 189]), where $0 < \Theta < 1$ and $1 \leq q \leq \infty$, and also many equivalent norms in the corresponding interpolation spaces. Special cases of these equivalent norms establish (2.2.2/18), i.e.

$$W^s_p(R_n) = \Lambda^s_{p,p}(R_n) \quad \text{and} \quad C^s(R_n) = \mathcal{C}^s(R_n) \tag{9}$$

if $0 < s \neq$ integer and $1 \leq p < \infty$.

Step 2. If $m=1, 2, 3, \ldots$, then (1) and Theorem 2.3.8 yield

$$B^m_{p,1}(R_n) \subset W^m_p(R_n) \subset B^m_{p,\infty}(R_n) \quad \text{if} \quad 1 \leq p < \infty \tag{10}$$

and

$$B^m_{\infty,1}(R_n) \subset C^m(R_n) \subset B^m_{\infty,\infty}(R_n) . \tag{11}$$

Now (10), Theorem 2.4.2 and Remark 2.4.1/4 show that

$$B^{\Theta m}_{p,q}(R_n) = (B^0_{p,1}(R_n), B^m_{p,1}(R_n))_{\Theta,q} \subset (L_p(R_n), W^m_p(R_n))_{\Theta,q}$$
$$\subset (B^0_{p,\infty}(R_n), B^m_{p,\infty}(R_n))_{\Theta,q} = B^{\Theta m}_{p,q}(R_n) ,$$

where $0<\Theta<1$, $1\leqq q\leqq\infty$, $1\leqq p<\infty$ and $m=1, 2, 3, \ldots$ This proves

$$(L_p(R_n), W_p^m(R_n))_{\Theta,q}=B^{\Theta m}_{p,q}(R_n)\,. \tag{12}$$

Similarly,

$$(C(R_n), C^m(R_n))_{\Theta,\infty}=B^{\Theta m}_{\infty,\infty}(R_n)\,. \tag{13}$$

Comparison of (7) and (12) gives (5), and comparison of (8) and (13) gives (6).

Remark. Similar proofs have been given in [I, 2.5.1 and 2.7.2] and in [S, 2.2.9.].

2.5.8. The Local Hardy Spaces $h_p(R_n)$, the Space $bmo(R_n)$

The spaces $h_p(R_n)$ and $bmo(R_n)$ have been defined in (2.2.2/12) and (2.2.2/14), respectively. It is our aim to prove (2.3.5/4) and (2.3.5/5). Then it follows from (2.5.6/2) that (2.2.2/17) also is valid. In other words, after the two theorems of this subsection have been established, all assertions of Remark 2.2.2/3, i.e. (2.2.2/17)–(2.2.2/19)) and (2.3.5/1)–(2.3.5/5) will have been proved.

One can prove the main result of this subsection, i.e. $h_p(R_n)=F^0_{p,2}(R_n)$ if $0<p<\infty$, directly or via a corresponding result for the Hardy spaces $H_p(R_n)$, cf. Remark 2.2.2/2. The spaces $H_p(R_n)$ are more fashionable, but they do not fit in the framework of the non-homogeneous isotropic spaces, so we prefer the direct approach (later on in 5.2.4. we return to $H_p(R_n)$). We use the following Fourier multiplier theorem for $h_p(R_n)$ with $0<p<\infty$, which is due to D. Goldberg [1, 2]: *let $m(x)$ be an infinitely differentiable complex-valued function in R_n such that*

$$\sup_{x\in R_n}\,(1+|x|)^{|\alpha|}\,|D^\alpha m(x)|<\infty \tag{1}$$

for every multi-index α. If $0<p<\infty$, then there exists a constant c such that

$$\|F^{-1}mFf\mid h_p(R_n)\|\leqq c\|f\mid h_p(R_n)\| \tag{2}$$

holds for all $f\in h_p(R_n)$, cf. also Bui Huy Qui [1]. In other words, the space $h_p(R_n)$ has, for every p with $0<p<\infty$, the weak Fourier multiplier property from Definition 2.2.3 and satisfies Criterion 2.2.3. Next we use a well-known rule of thumb, without going into detail. Let H be a separable Hilbert space. The rule is the following: if we have a Fourier multiplier theorem for a space of scalar-valued functions, then there holds also a corresponding Fourier multiplier theorem for the H-valued counterpart of the space under consideration. The idea is that one replaces (in formulations and proofs) pointwise products by scalar products, absolute values by H-norms, etc. A typical example is Proposition 2.5.6, where we formulated a Fourier multiplier assertion for $L_p(R_n, l_2)$. Now we extend this assertion to $h_p(R_n, l_2)$, where $0<p<\infty$. If η has the same properties as the function φ from (2.2.2/12) (including $\eta_t(x)=\eta(tx)$), then the l_2-valued counterpart of $\|f\mid h_p(R_n)\|^\eta$ with $0<p<\infty$ is given by

$$\|f_k\mid h_p(R_n, l_2)\|^\eta=\left\|\sup_{0<t<1}\left(\sum_{k=0}^\infty|(F^{-1}\eta_tFf_k)(\cdot)|^2\right)^{\frac{1}{2}}\,\Big|\,L_p(R_n)\right\|. \tag{3}$$

In the sequel we omit the index η in (3). Then the l_2-valued counterpart of (1), (2) reads as follows.

Proposition. Let $\{m_{k,j}(x)\}_{k,j=0}^\infty$ be a sequence of infinitely differentiable complex-valued functions in R_n such that

$$\sup_{x\in R_n}\,(1+|x|)^{|\alpha|}\left(\sum_{k,j=0}^\infty|D^\alpha m_{k,j}(x)|^2\right)^{\frac{1}{2}}<\infty \tag{4}$$

for every multi-index α. If $0<p<\infty$, then there exists a constant c such that

$$\left\| F^{-1}\left[\sum_{j=0}^{\infty} m_{k,j} F f_j\right] \middle| h_p(R_n, l_2) \right\| \leq c \| f_k \mid h_p(R_n, l_2) \| . \tag{5}$$

Remark 1. This is the somewhat weaker counterpart of Proposition 2.5.6 for the spaces $L_p(R_n, l_2)$ with $1<p<\infty$. We refer also to Bui Huy Qui [1, Lemma 1], where a special case of (5) is formulated. A similar, but slightly more complicated, assertion for the spaces $H_p(R_n, l_2)$ was formulated in [S, p. 165] (where we wrote $H_p(R_n, l_2)$ instead of $H_p(R_n, l_2)$). It is the l_2-version of a Fourier multiplier theorem for the spaces $H_p(R_n)$ due to C. Fefferman, E. M. Stein [2, pp. 188–190]. Cf. also A. P. Calderón, A. Torchinsky [1, II, p. 167], where a more general Fourier multiplier theorem for $H_p(R_n)$ (and generalizations of $H_p(R_n)$) has been proved, and H. Triebel [13] (that paper contains the l_2-version of the Calderón-Torchinsky Fourier multiplier theorem).

Theorem 1. If $0<p<\infty$, then

$$h_p(R_n) = F^0_{p,2}(R_n) \tag{6}$$

(equivalent quasi-norms).

Proof. Step 1. The functions $f(x) \in S(R_n)$ with compact support are dense in $F^0_{p,2}(R_n)$ and dense in $h_p(R_n)$, cf. Theorem 2.3.3 (ii) (including (2.3.3/2)) and D. Goldberg [1] (and Bui Huy Qui [1]). By the above proposition, it follows in exactly the same way as in Step 2 of the proof of Theorem 2.5.6 that

$$\| f \mid h_p(R_n) \| \sim \| F^{-1} \varphi_k F f \mid h_p(R_n, l_2) \| \tag{7}$$

with $\{\varphi_k(x)\}_{k=0}^{\infty} \in \Phi(R_n)$ (equivalent quasi-norms).

Step 2. We recall that $\eta(0)=1$. Then it follows from (3) with $f_k = F^{-1}\varphi_k F f$ and (7) that

$$\| f \mid h_p(R_n) \| \geq c \| F^{-1} \varphi_k F f \mid L_p(R_n, l_2) \| = c \| f \mid F^0_{p,2}(R_n) \| \tag{8}$$

with $c>0$.

Step 3. We prove the converse assertion of (8). By (3) and (7) we have

$$\| f \mid h_p(R_n) \| \leq c \left\| \left(\sum_{k=0}^{\infty} \sup_{0<t<1} |(F^{-1}\eta(t\,\cdot)\, \varphi_k(\cdot)\, Ff)\,(\cdot)|^2 \right)^{\frac{1}{2}} \middle| L_p(R_n) \right\| . \tag{9}$$

Recall that $\eta(x)$ (as the counterpart of the function φ from (2.2.2/12)) has compact support. Hence $\eta(tx)\, \varphi_k(x) \neq 0$ implies $|x| \sim 2^k$, $t|x| \leq c$, and consequently $t \leq c2^{-k}$ if $k=1, 2, \ldots$ Because $t \leq 1$ we obtain for all k (including $k=0$)

$$\begin{aligned} &\sup_{0<t<1} |(F^{-1}\eta(t\,\cdot)\, \varphi_k(\cdot)\, Ff)(x)| \\ &\leq \sup_{t \leq c2^{-k}} \left| \int_{R_n} (F^{-1}\eta(t\,\cdot))(u)\, (F^{-1}\varphi_k Ff)(x-u)\, du \right| \\ &= \sup_{t \leq c2^{-k}} \left| \int_{R_n} (F^{-1}\eta)(v)\, (F^{-1}\varphi_k Ff)(x-tv)\, dv \right| \\ &\leq c \sup_{\substack{v \in R_n \\ t \leq c2^{-k}}} |(F^{-1}\varphi_k Ff)(x-tv)|\, (1+|v|^a)^{-1} , \end{aligned}$$

where a is an arbitrary positive number. We used

$$(F^{-1}\eta(t\,\cdot))\,(u) = t^{-1}(F^{-1}\eta)\,(t^{-1}u)$$

and the transformation of coordinates $v=t^{-1}u$. If $(\varphi_k^* f)(x)$ is the maximal function from (2.3.6/2), then

$$\sup_{0<t<1} |(F^{-1}\eta(t\,\cdot)\,\varphi_k(\cdot)\,Ff)(x)| \leq c \sup_{\substack{v\in R_n \\ t\leq c2^{-k}}} \frac{|(F^{-1}\varphi_k Ff)\,(x-tv)|}{1+2^{ka}|tv|^a}$$
$$\leq c' \sup_{v\in R_n} \frac{|(F^{-1}\varphi_k Ff)\,(x-v)|}{1+|2^k v|^a} = c'(\varphi_k^* f)(x)\,. \tag{10}$$

We substitute (10) in (9) and obtain by Theorem 2.3.6(ii) that

$$\|f \mid h_p(R_n)\| \leq c\|\varphi_k^* f \mid L_p(R_n, l_2)\| \leq c'\|f \mid F^0_{p,2}(R_n)\|\,. \tag{11}$$

Hence, by (8) and (11), the quasi-norms $\|f \mid h_p(R_n)\|$ and $\|f \mid F^0_{p,2}(R_n)\|$ are equivalent on the functions $f\in S(R_n)$ with compact support. (6) follows by completion, cf. Step 1.

Remark 2. Formula (7) is a Littlewood-Paley theorem for $h_p(R_n)$. The counterpart of (6) for the Hardy spaces $H_p(R_n)$ is

$$H_p(R_n) = \dot{F}^0_{p,2}(R_n)\,, \qquad 0<p<\infty\,, \tag{12}$$

where $\dot{F}^0_{p,2}(R_n)$ is the homogeneous space from Chapter 5. This assertion is due to J. Peetre [5, 6]. An extension of this result to anisotropic Hardy spaces in the sense of A P. Calderón, A. Torchinsky [1, II] has been given in H. Triebel [13]. The above proof is a modification of a corresponding proof in [S, pp. 167–169]. However we wish to emphasize that the first proof of (6) is due to Bui Huy Qui [1], cf. also Y. Meyer [1].

Theorem 2. The following equality holds:

$$bmo(R_n) = F^0_{\infty,2}(R_n) \tag{13}$$

(equivalent norms).

Proof. (Outline). The spaces $F^0_{\infty,2}(R_n)$ have been defined in 2.3.4. By (2.3.4/6) and (6) we have

$$(h_1(R_n))' = (F^0_{1,2}(R_n))' = F^0_{\infty,2}(R_n)\,. \tag{14}$$

D. Goldberg [1, p. 36] proved

$$(h_1(R))' = bmo(R_n)\,. \tag{15}$$

Now (13) is a consequence of (14) and (15).

Remark 3. The equality (15) is the non-homogeneous counterpart of the famous assertion

$$(H_1(R_n))' = BMO(R_n)\,. \tag{16}$$

which is due to C. Fefferman [1], cf. also C. Fefferman, E. M. Stein [2]. The space $BMO(R_n)$ has been described in Remark 2.2.2/2. The homogeneous counterpart of (13) (cf. (12) and 5.2.4.) reads as follows:

$$BMO(R_n) = \dot{F}^0_{\infty,2}(R_n)\,. \tag{17}$$

Remark 4. Finally we mention the following assertions, which have been proved in [S, p. 125/126] and in D. Goldberg [1, pp. 33, 36]. Let $\psi(x)$ be an infinitely differentiable function on R_n, such that $\psi(x)=\psi(-x)$,

$$\psi(x)=0 \quad \text{if} \quad |x|\leq 1 \quad \text{and} \quad \psi(y)=1 \quad \text{if} \quad |y|\geq 2\,.$$

Let

$$r_j = F^{-1}\psi\,\frac{x_j}{|x|}\,F, \quad j=1,\ldots,n\,,$$

be the non-homogeneous Riesz transforms. Then

$$h_1(R_n)=\left\{f \mid f\in S'(R_n),\ \|f \mid h_1(R_n)\|^*=\|f \mid L_1(R_n)\|+\sum_{j=1}^{n}\|r_jf \mid L_1(R_n)\|<\infty\right\} \tag{18}$$

and

$$bmo(R_n)=\left\{g \mid g\in S'(R_n),\ \exists\{g_0, g_1, \ldots, g_n\}\subset L_\infty(R_n) \text{ such that } g=g_0+\sum_{j=1}^{n} r_jg_j\right\}, \tag{19}$$

where $\|f \mid h_1(R_n)\|^*$ and

$$\|g \mid bmo(R_n)\|^*=\inf \sum_{k=0}^{n}\|g_k \mid L_\infty(R_n)\|$$

(the infimum is taken over all admissible representations of g in the sense of (19)) are equivalent norms in $h_1(R_n)$ and $bmo(R_n)$, respectively.

2.5.9. Characterizations by Maximal Functions of Differences

The rest of Section 2.5. (including this subsection) is the basis for subsequent considerations. First we introduce some notations. If $f(x)$ is an arbitrary function on R_n and $h\in R_n$, then

$$(\Delta_h^1 f)(x)=f(x+h)-f(x),\quad (\Delta_h^l f)(x)=\Delta_h^1(\Delta_h^{l-1}f)(x), \tag{1}$$

where $l=2, 3, \ldots$ These are the well-known differences of functions which play an important role in the theory of function spaces, cf. the spaces in 2.2.2. and 2.5.7. Let $|\gamma_n|$ be the volume (surface measure) of the unit sphere $\gamma_n=\{x \mid x\in R_n, |x|=1\}$ and let $|\Gamma_n|$ be the volume of $\Gamma_n=\{x \mid x\in R_n, 1\leq|x|\leq 2\}$. If M is a natural number and $r>0$, then

$$(\mathbb{S}_r^M g)(y)=\frac{1}{|\gamma_n|}\int_{\gamma_n}(\Delta_{r\gamma}^M g)(y)\,\mathrm{d}\gamma \tag{2}$$

is the spherical mean value of the differences of g with respect to the sphere of radius r centered at the origin. Here, $\mathrm{d}\gamma$ denotes the surface element of γ_n, and $y\in R_n$. Similarly,

$$(\mathcal{V}_r^M g)(y)=\frac{1}{|\Gamma_n|}\int_{\Gamma_n}(\Delta_{rh}^M g)(y)\,\mathrm{d}h \tag{3}$$

is a truncated volume mean value of the differences of g. Furthermore, we introduce the maximal functions

$$(S_r^M g)(x)=\sup_{y\in R_n}(1+|r^{-1}y|^a)^{-1}\,|(\mathbb{S}_r^M g)(x-y)|, \tag{4}$$

$$(V_r^M g)(x)=\sup_{y\in R_n}(1+|r^{-1}y|^a)^{-1}\,|(\mathcal{V}_r^M g)(x-y)|, \tag{5}$$

and

$$(D_h^M g)(x)=\sup_{y\in R_n}\left(1+\frac{|y|^a}{|h|^a}\right)^{-1}|(\Delta_h^M g)(x-y)|, \tag{6}$$

where $x\in R_n$, $r>0$ and $h\in R_n$ with $h\neq 0$. The other symbols have the above meaning, the positive number a in (4)–(6) will be determined later on. Finally, let

$$G_p=n+3+\frac{3n}{p}\quad\text{and}\quad G_{p,q}=n+3+\frac{3n}{\min(p, q)} \tag{7}$$

where $0<p\leq\infty$ and $0<q\leq\infty$.

Theorem. (i) Let $0<p\leqq\infty$, $0<q\leqq\infty$ and $s\geqq G_p$. If M is an integer with $M>2G_p+s$ and if $a>\frac{n}{p}$ in (4)–(6), then the following five quasi-norms are equivalent quasi-norms in $B^s_{p,q}(R_n)$,

$$\|f \mid L_p(R_n)\| + \|2^{sk} S^M_{2^{-k}} f \mid l_q(L_p(R_n))\| , \tag{8}$$

$$\|f \mid L_p(R_n)\| + \|2^{sk} \sup_{1\leqq\tau\leqq 2} S^M_{\tau 2^{-k}} f \mid l_q(L_p(R_n))\| , \tag{9}$$

$$\|f \mid L_p(R_n)\| + \|2^{sk} V^M_{2^{-k}} f \mid l_q(L_p(R_n))\| , \tag{10}$$

$$\|f \mid L_p(R_n)\| + \|2^{sk} \sup_{1\leqq\tau\leqq 2} V^M_{\tau 2^{-k}} f \mid l_q(L_p(R_n))\| , \tag{11}$$

$$\|f \mid L_p(R_n)\| + \|2^{sk} \sup_{h\in\Gamma_n} D^M_{2^{-k}h} f \mid l_q(L_p(R_n))\| . \tag{12}$$

(ii) Let $0<p<\infty$, $0<q\leqq\infty$ and $s\geqq G_{p,q}$. If M is an integer with $M>2G_{p,q}+s$ and if $a>\frac{n}{\min(p,q)}$ in (4)–(6), then the following five quasi-norms are equivalent quasi-norms in $F^s_{p,q}(R_n)$,

$$\|f \mid L_p(R_n)\| + \|2^{sk} S^M_{2^{-k}} f \mid L_p(R_n, l_q)\| , \tag{13}$$

$$\|f \mid L_p(R_n)\| + \|2^{sk} \sup_{1\leqq\tau\leqq 2} S^M_{\tau 2^{-k}} f \mid L_p(R_n, l_q)\| , \tag{14}$$

$$\|f \mid L_p(R_n)\| + \|2^{sk} V^M_{2^{-k}} f \mid L_p(R_n, l_q)\| , \tag{15}$$

$$\|f \mid L_p(R_n)\| + \|2^{sk} \sup_{1\leqq\tau\leqq 2} V^M_{\tau 2^{-k}} f \mid L_p(R_n, l_q)\| , \tag{16}$$

$$\|f \mid L_p(R_n)\| + \|2^{sk} \sup_{h\in\Gamma_n} D^M_{2^{-k}h} f \mid L_p(R_n, l_q)\| . \tag{17}$$

Proof. We prove in Step 1–4 that (13) and (14) are equivalent quasinorms in $F^s_{p,q}(R_n)$. In Step 5 we prove that (17) is an equivalent quasinorm in $F^s_{p,q}(R_n)$. The proof of the other assertions of the theorem is essentially the same.

Step 1. Let $f\in F^0_{p,q}(R_n)$ and $\operatorname{supp} Ff\subset\{y \mid |y|\geqq 1\}$. Let

$$\varphi(x)=\frac{1}{|x|^m|\gamma_n|}\int_{\gamma_n}(e^{i\gamma x}-1)^M d\gamma \tag{18}$$

and $\varphi_k(x)=\varphi(2^{-k}x)$ if $k=0, 1, 2, \ldots$ Here γx is the scalar product of $\gamma\in\gamma_n$ and $x\in R_n$. Furthermore, m is a positive number and M is a natural number. $\{\varphi_k(x)\}_{k=0}^\infty$ satisfies (2.3.6/1) if $\sigma\geqq 0$ and

$$m=L+\sigma,\quad M\geqq 3L+\sigma . \tag{19}$$

We use the maximal function from (2.3.6/2). Recall that

$$F^{-1}\left((e^{i\gamma x}-1)\, g\right)=\Delta^1_\gamma Fg .$$

Iteration of this formula and (4) yield

$$(\varphi_k^* f)(x)=2^{km}(S^M_{2^{-k}}(F^{-1}|\xi|^{-m}Ff))(x) . \tag{20}$$

Here $k=0, 1, 2, \ldots$ By later considerations and Remark 2.3.6/4 it will be clear that $(\varphi_k^* f)(x)$ makes sense as a function on R_n. If we replace $i\gamma x$ in (18) by $i\tau\gamma x$, where τ is a positive number, then 2^{-k} on the right-hand side of (20) must be replaced by $\tau 2^{-k}$. Now Theorem 2.3.6 and Remark 2.3.6/4 prove that

$$\|2^{km} \sup_{1\leqq\tau\leqq 2} S^M_{\tau 2^{-k}}(F^{-1}|\xi|^{-m}Ff) \mid L_p(R_n, l_q)\| \leqq c\|f \mid F^0_{p,q}(R_n)\| \tag{21}$$

provided that $L=[G_{p,q}]$ (the largest integer which is less than or equal to $G_{p,q}$). By the lifting property from Theorem 2.3.8 and our assumption about the support of Ff it follows that $\|F^{-1}|\xi|^{-m}Ff \mid F^m_{p,q}(R_n)\|$ is equivalent to $\|f \mid F^0_{p,q}(R_n)\|$ (this is essentially an application of Theorem 2.3.7). If $g=F^{-1}|\xi|^{-m}Ff$ and $s=m=$ $=[G_{p,q}]+\sigma$, cf. (19), then we have

$$\|2^{sk} \sup_{1\leq\tau\leq 2} S^M_{\tau 2^{-k}}g \mid L_p(R_n, l_q)\| \leq c\|g \mid F^s_{p,q}(R_n)\| , \tag{22}$$

where c is independent of $g\in F^s_{p,q}(R_n)$ with $\operatorname{supp} Fg\subset\{y \mid |y|\geq 1\}$. By (19) it follows that $M\geq 3[G_{p,q}]+\sigma=2[G_{p,q}]+s$.

Step 2. Let $f\in F^s_{p,q}(R_n)$ and $\operatorname{supp} Ff\subset\{y \mid |y|\geq 1\}$. Let $\{\varphi_k(x)\}^\infty_{k=0}\in\Phi(R_n)$, where we assume that $\varphi_k(x)=\varphi(d2^{-k}x)$ if $k=1, 2, 3, \ldots$ Furthermore we suppose that $\operatorname{supp}\varphi\subset\{y \mid d_1\leq|y|\leq d_2\}$, where d_1 and d_2 are positive numbers, and $d>0$ is an appropriate number (the function φ_0 is uninteresting by our assumption on $\operatorname{supp} Ff$). Let $b(x)$ be a continuous function on R_n such that $b(x)\neq 0$ if $x\in\operatorname{supp}\varphi$. If $b_k(x)=b(d2^{-k}x)$ (with the same d as above), then

$$|(F^{-1}\varphi_kFf)(x)|=\left|\left(F^{-1}\frac{\varphi_k}{b_k}FF^{-1}b_kFf\right)(x)\right| \tag{23}$$
$$=\left|\int_{R_n}\left(F^{-1}\frac{\varphi_k}{b_k}\right)(y)\,(F^{-1}b_kFf)(x-y)\,\mathrm{d}y\right|$$
$$\leq c'\sup_{y\in R_n}\frac{|(F^{-1}b_kFf)\,(x-y)|}{1+|2^ky|^a}\int_{R_n}\left|\left(F^{-1}\frac{\varphi}{b}\right)(d^{-1}2^ky)\right|2^{kn}\,(1+|2^ky|^a)\,\mathrm{d}y\,,$$

where a is a positive number. As we shall see, $(F^{-1}b_kFf)(x)$ makes sense as a function on R_n if we choose

$$b(x)=\frac{1}{|\gamma_n|}\int_{\gamma_n}(\mathrm{e}^{\mathrm{i}x\gamma}-1)^M\,\mathrm{d}y\,. \tag{24}$$

If the above positive numbers d_1 and d_2 are sufficiently small, then $(\mathrm{e}^{\mathrm{i}x\gamma}-1)^M=$ $=(\mathrm{i}x\gamma)^M\,(1+o(1))$ if $x\in\operatorname{supp}\varphi$. If M is an even natural number, then $b(x)\neq 0$ if $x\in\operatorname{supp}\varphi$. If M is an odd natural number, then the integral of $(x\gamma)^M$ over γ_n vanishes. However, the integral over the next term in $(\mathrm{i}x\gamma)^M(1+o(1))$, which is of type $(x\gamma)^{M+1}$, does not vanish. Hence, $b(x)$ satisfies the above hypotheses. Furthermore,

$$(F^{-1}bFf)(x)=\frac{1}{|\gamma_n|}\int_{\gamma_n}(\Delta^M_\gamma f)(x)\,\mathrm{d}\gamma \tag{25}$$

makes sense if $f(x)$ belongs to $L_\varrho(R_n)$ with $1\leq\varrho\leq\infty$. If $1\leq p<\infty$ and $s>0$, then $F^s_{p,q}(R_n)\subset L_p(R_n)$. If $0<p<1$, then (2.5.3/29) and (2.3.2/9) show that $F^s_{p,q}(R_n)\subset$ $\subset L_1(R_n)$ if $s>n\left(\frac{1}{p}-1\right)$. Under the above assumptions for s and p, this is satisfied. Similarly for $(F^{-1}b_kFf)(x)$. This also justifies (23) (the right-hand side of (23) is finite. This follows from $F^s_{p,q}(R_n)\subset C(R_n)$, which holds by the above restrictions for s, cf. 2.7.1.). The last integral in (23) is independent of k. Thus (23) and the counterpart of (25) with b_k instead of b yields

$$|(F^{-1}\varphi_kFf)(x)|\leq c(S^M_{d2^{-k}}f)(x)\,, \tag{26}$$

where c is independent of k. By the above considerations, the numbers d_1 and d_2 are small. Consequently, d is also small. We may assume that $d=2^{-l}$, where l is an appropriate natural number. Because the term with ψ_0 is not of interest, it follows that

$$\|f \mid F^s_{p,q}(R_n)\| \leqq c\|2^{sk} S^M_{2^{-k}} f \mid L_p(R_n, l_q)\| . \tag{27}$$

Now (22) and (27) prove that (13) and (14) without the term $\|f \mid L_p(R_n)\|$ are equivalent quasi-norms on $f \in F^s_{p,q}(R_n)$ with supp $Ff \subset \{y \mid |y| \geqq 1\}$ (under the hypotheses for p, q, s and M).

Step 3. Let $f \in F^s_{p,q}(R_n)$ be an arbitrary element, where $0<p<\infty$, $0<q\leqq\infty$ and $s \geqq G_{p,q}$. First we remark that $f(x) \in L_p(R_n)$ makes sense, cf. Remark 2.5.3/1. If $\psi_0(x) \in S(R_n)$ with

$$\operatorname{supp} \psi_0 \subset \{y \mid |y| \leqq 2\} \quad \text{and} \quad \psi_0(x) = 1 \quad \text{if} \quad |x| \leqq 1 ,$$

then Step 1 and Step 2 yield

$$\|F^{-1}(1-\psi_0)\, Ff \mid F^s_{p,q}(R_n)\| \sim \|2^{sk} S^M_{2^{-k}} F^{-1}(1-\psi_0)\, Ff \mid L_p(R_n, l_q)\| . \tag{28}$$

Consequently,

$$\begin{aligned} \|f \mid F^s_{p,q}(R_n)\| &\leqq c\|F^{-1}\psi_0 Ff \mid L_p(R_n)\| \\ &\quad + c\|2^{sk} S^M_{2^{-k}} F^{-1}\psi_0 Ff \mid L_p(R_n, l_q)\| + c\|2^{sk} S^M_{2^{-k}} f \mid L_p(R_n, l_q)\| . \end{aligned} \tag{29}$$

First we estimate the second term on the right-hand side. By Proposition 2.3.2/2 (and its proof) we have

$$\|2^{sk} S^M_{2^{-k}} F^{-1}\psi_0 Ff \mid L_p(R_n, l_q)\|^p \leqq c \sum_{k=0}^{\infty} 2^{(s+\varepsilon)kp} \|S^M_{2^{-k}} F^{-1}\psi_0 Ff \mid L_p(R_n)\|^p \tag{30}$$

with $\varepsilon>0$. We recall that $a>\frac{n}{p}$, so (1.4.1/2) yields

$$\begin{aligned} \|S^M_{2^{-k}} F^{-1}\psi_0 Ff \mid L_p(R_n)\|^p &\leqq \Big\| \sup_{y \in R_n} \frac{|(\mathcal{S}^M_{2^{-k}} F^{-1}\psi_0 Ff)(\cdot - y)|}{1+|y|^a} \Big| L_p(R_n)\Big\|^p \\ &\leqq c\|\mathcal{S}^M_{2^{-k}} F^{-1}\psi_0 Ff \mid L_p(R_n)\|^p . \end{aligned} \tag{31}$$

Simple calculations give

$$|(\mathcal{S}^M_{2^{-k}} F^{-1}\psi_0 Ff)(x)| \leqq c \sum_{|\alpha|=M} \sup_{|x-y| \leqq 2^{-k}M} |D^\alpha(F^{-1}\psi_0 Ff)(y)|\, 2^{-kM} . \tag{32}$$

So if we put this estimate into (31) and then apply (1.4.1/2) and (1.4.1/3) to $D^\alpha F^{-1}\psi_0 Ff$, we obtain

$$\begin{aligned} \|S^M_{2^{-k}} F^{-1}\psi_0 Ff \mid L_p(R_n)\|^p &\leqq c \sum_{|\alpha|=M} 2^{-kMp} \|D^\alpha |F^{-1}\psi_0 Ff) \mid L_p(R_n)\|^p \\ &\leqq c' 2^{-kMp} \|F^{-1}\psi_0 Ff \mid L_p(R_n)\|^p . \end{aligned}$$

Substituting this estimate into (30) and noting that $s+\varepsilon<M$, we obtain

$$\|2^{sk} S^M_{2^{-k}} F^{-1}\psi_0 Ff \mid L_p(R_n, l_q)\|^p \leqq c\|F^{-1}\psi_0 Ff \mid L_p(R_n)\|^p . \tag{33}$$

By (29) we have

$$\|f \mid F^s_{p,q}(R_n)\| \leqq c\|F^{-1}\psi_0 Ff \mid L_p(R_n)\| + \|2^{sk} S^M_{2^{-k}} f \mid L_p(R_n, l_q)\| . \tag{34}$$

In order to estimate the first term on the right-hand side of (34) we remark that

$$\|F^{-1}\psi_0 Ff \mid L_p(R_n)\| = \|F^{-1}\psi_0 F F^{-1}\psi_0(2^{-l})\, Ff \mid L_p(R_n)\|$$

with $l=2, 3, \ldots$. To apply (1.5.2/13) with ψ_0 as multiplier we calculate

$$\|\psi_0(2^l \cdot) \mid H^\varkappa_2(R_n)\| \leqq c 2^{l\left(\varkappa - \frac{n}{2}\right)} ,$$

where $\varkappa > \sigma_p = n\left(\frac{1}{\min(p,1)} - \frac{1}{2}\right)$. Hence,

$$\|F^{-1}\psi_0 Ff \mid L_p(R_n)\| \leq c 2^{l\sigma} \|F^{-1}\psi_0(2^{-l}\cdot) Ff \mid L_p(R_n)\| . \tag{35}$$

where $\sigma > \tilde{\sigma}_p = n\left(\frac{1}{\min(p,1)} - 1\right)$. On the other hand

$$\begin{aligned} &\|F^{-1}\psi_0(2^{-l}\cdot) Ff \mid L_p(R_n)\| \\ &\leq c\|f \mid L_p(R_n)\| + c\|F^{-1}(1-\psi_0(2^{-l}\cdot)) Ff \mid L_p(R_n)\| \\ &\leq c\|f \mid L_p(R_n)\| + c' 2^{-sl}\|f \mid F^s_{p,q}(R_n)\| . \end{aligned} \tag{36}$$

We put (36) in (35). Because $s > \sigma$ we may choose l in such a way that

$$\|F^{-1}\psi_0 Ff \mid L_p(R_n)\| \leq c_\varepsilon \|f \mid L_p(R_n)\| + \varepsilon \|f \mid F^s_{p,q}(R_n)\|, \tag{37}$$

where ε is a given positive number. If we choose ε sufficiently small then (34) and (37) yield

$$\|f \mid F^s_{p,q}(R_n)\| \leq c\|f \mid L_p(R_n)\| + c\|2^{sk} S^M_{2^{-k}} f \mid L_p(R_n, l_q)\| . \tag{38}$$

Step 4. Again let $f \in F^s_{p,q}(R_n)$ be an arbitrary element, where $0 < p < \infty$, $0 < q \leq \infty$, and $s \geq G_{p,q}$. As we mentioned above, $f(x) \in L_p(R_n)$ makes sense. ψ_0 has the same meaning as above. First we remark that (33) can be strengthened by

$$\|2^{sk} \sup_{1 \leq \tau \leq 2} S^M_{\tau 2^{-k}} F^{-1}\psi_0 Ff \mid L_p(R_n, l_q)\|^p \leq c\|F^{-1}\psi_0 Ff \mid L_p(R_n)\|^p . \tag{39}$$

In order to prove this assertion we replace $\sup\limits_{y \in R_n}$ and $\mathfrak{S}^M_{2^{-k}}$ in the middle term of (31) by $\sup\limits_{1 \leq \tau \leq 2} \sup\limits_{y \in R_n}$ and $\mathfrak{S}^M_{\tau 2^{-k}}$, respectively. Similarly, on the left-hand side of (31). Then we apply (32) with $\mathfrak{S}^M_{\tau 2^{-k}}$ instead of $\mathfrak{S}^M_{2^{-k}}$. The result is independent of τ and can be estimated as above (maximal inequality for $D^\alpha(F^{-1}\psi_0 Ff)$). Now it follows that

$$\begin{aligned} &\|f \mid L_p(R_n)\| + \|2^{sk} \sup_{1 \leq \tau \leq 2} S^M_{\tau 2^{-k}} f \mid L_p(R_n, l_q)\| \\ &\leq \|f \mid L_p(R_n)\| + c\|2^{sk} \sup_{1 \leq \tau \leq 2} S^M_{\tau 2^{-k}} F^{-1}\psi_0 Ff \mid L_p(R_n, l_q)\| \\ &\quad + c\|2^{sk} \sup_{1 \leq \tau \leq 2} S^M_{\tau 2^{-k}} F^{-1}(1-\psi_0) Ff \mid L_p(R_n, l_q)\| . \end{aligned}$$

By (39) and (22) the right-hand side of the last inequality can be estimated from above by

$$\|f \mid L_p(R_n)\| + c\|F^{-1}\psi_0 Ff \mid L_p(R_n)\| + c\|F^{-1}(1-\psi_0) Ff \mid F^s_{p,q}(R_n)\| .$$

This in turn can be estimated from above by $c'\|f \mid F^s_{p,q}(R_n)\|$, cf. Remark 2.5.3/1 as far as $\|f \mid L_p(R_n)\|$ is concerned. Hence,

$$\|f \mid L_p(R_n)\| + \|2^{sk} \sup_{1 \leq \tau \leq 2} S^M_{\tau 2^{-k}} f \mid L_p(R_n, l_q)\| \leq c\|f \mid F^s_{p,q}(R_n)\|. \tag{40}$$

Now (40) and (38) prove that (13) and (14) are equivalent quasi-norms on $F^s_{p,q}(R_n)$ (under the above assumptions for p, q, s and M).

Step 5. Similarly one proves that (15) and (16) are equivalent quasi-norms on $F^s_{p,q}(R_n)$ and that (8)–(11) are equivalent quasi-norms on $B^s_{p,q}(R_n)$ (under the above restrictions for p, q, s and M). We prove that (17) is an equivalent quasi-norm on $F^s_{p,q}(R_n)$. Similarly it follows that (12) is an equivalent quasi-norm on $B^s_{p,q}(R_n)$. Let $f \in F^s_{p,q}(R_n)$ with $\operatorname{supp} Ff \subset \{y \mid |y| \geq 1\}$. We replace $\varphi(x)$ in (18) by $|x|^{-m}(e^{iyx}-1)^M$, where $\gamma \in \Gamma_n$. Then the calculations of Step 1 produce the following counterpart of (22):

$$\|2^{sk} \sup_{h \in \Gamma_n} D^M_{2^{-k}h} f \mid L_p(R_n, l_q)\| \leq c\|f \mid F^s_{p,q}(R_n)\| . \tag{41}$$

On the other hand, the left-hand side of (41) can be estimated from below by $\|2^{sk} V^M_{2^{-k}} f \mid L_p(R_n, l_q)\|$, and this is equivalent to $\|f \mid F^s_{p,q}(R_n)\|$ (we recall that $\operatorname{supp} Ff \subset \{y \mid |y| \geq 1\}$). Hence, the left-hand side of (41) is equivalent to $\|f \mid F^s_{p,q}(R_n)\|$ if $\operatorname{supp} Ff \subset \{y \mid |y| \geq 1\}$. The rest is the same as in Step 3 and Step 4 and the proof is complete.

Corollary 1. (i) Let $0 < p \leq \infty$, $0 < q \leq \infty$ and $s \geq G_p$. If M is an integer with $M > s$ and if $a > \frac{n}{p}$ in (4)–(6), then (8)–(12) are equivalent quasi-norms in $B^s_{p,q}(R_n)$.
(ii) Let $0 < p < \infty$, $0 < q \leq \infty$ and $s > G_{p,q}$. If M is an integer with $M > s$ and if $a > \frac{n}{\min(p, q)}$ in (4)–(6), then (13)–(17) are equivalent quasi-norms in $F^s_{p,q}(R_n)$.

Proof. We have to prove that the unnatural restriction $M > 2G_{p,q} + s$ in part (ii) of the theorem can be replaced by the natural restriction $M > s$. Similarly for part (i) of the theorem. We prove the above claim for the quasi-norm (13). In an analogous way, one proves corresponding assertions for the quasi-norms (8)–(12) and (14)–(17). We needed $M > 2G_{p,q} + s$ only in Step 1 of the proof of the theorem. This shows that it is sufficient to deal with the following inequality: if $M > s$, then there exists a constant c such that

$$\|2^{sk} S^M_{2^{-k}} f \mid L_p(R_n, l_q)\| \leq c\|2^{sk} S^{M+1}_{2^{-k}} f \mid L_p(R_n, l_q)\| + c\|f \mid F^s_{p,q}(R_n)\| \tag{42}$$

for all $f \in F^s_{p,q}(R_n)$ (under the above hypotheses for s, p, and q). If $P(z)$ is an appropriate polynomial then the following identity holds for complex numbers z,

$$(z-1)^M = \frac{1}{2^M} (z^2 - 1)^M + (z-1)^{M+1} P(z) . \tag{43}$$

In particular,

$$(e^{ih\xi} - 1)^M = \frac{1}{2^M} (e^{2ih\xi} - 1)^M + (e^{ih\xi} - 1)^{M+1} P(e^{ih\xi}) , \tag{44}$$

where $\xi \in R_n$ and $h \in R_n$. If we apply (44) to Ff and if we take the inverse Fourier transform of the result, then we obtain

$$(\Delta^M_h f)(x) = \frac{1}{2^M} (\Delta^M_{2h} f)(x) + \Delta^{M+1}_h \left(\sum a_l f(x + lh)\right) , \tag{45}$$

where $\sum$ is a finite sum (the counterpart of $P(z)$). Using (2) and (4) we then have

$$(S^M_{2^{-k}} f)(x) \leq \frac{1}{2^M} (S^M_{2^{-k+1}} f)(x) + c(S^{M+1}_{2^{-k}} f)(x) \tag{46}$$

for all $x \in R_n$ and all $k = 0, 1, 2, \ldots$ If $0 < q \leq 1$, then

$$\sum_{k=1}^{\infty} 2^{skq} (S^M_{2^{-k}} f)^q (x) \leq 2^{(s-M)q} \sum_{k=0}^{\infty} 2^{skq} (S^M_{2^{-k}} f)^q (x) + c \sum_{k=1}^{\infty} 2^{sk} (S^{M+1}_{2^{-k}} f)^q (x)$$

and because $M > s$

$$\left(\sum_{k=0}^{\infty} 2^{skq} (S^M_{2^{-k}} f)^q (x)\right)^{\frac{1}{q}} \leq c \left(\sum_{k=0}^{\infty} 2^{skq} (S^{M+1}_{2^{-k}} f)^q (x)\right)^{\frac{1}{q}} + cR(x) \tag{47}$$

with

$$R(x) = \sup_{y \in R_n} (1 + |x - y|^a)^{-1} |f(y)| .$$

Similarly for $1<q\leq\infty$. We estimate $\|R(x) \mid L_p(R_n)\|$ and for that purpose replace $f(y)$ by $\sum_{k=0}^{\infty} F^{-1}\varphi_k Ff$. Then

$$R(x)\leq c \sum_{k=0}^{\infty} 2^{ka}(\varphi_k^* f)(x)\,.$$

Because $s>a$, it follows that

$$\|R(x) \mid L_p(R_n)\|\leq c\|f \mid F^s_{p,q}(R_n)\|\,.$$

This estimate and (47) establish (42). The proof is complete.

Remark 1. The relation between (43), (44) on the one hand and (45) on the other hand is very useful. It goes back to K. K. Golovkin [1, 3] who used it (and many generalizations) extensively.

Remark 2. The scheme of the proof of the theorem and also of the above corollary can be extended immediately to other mean values of differences. Of interest is the following possibility. Let, e.g.,

$$K_n=\{x \mid x_1^2+\ldots+x_{n-1}^2\leq x_n^2,\ 1<x_n<2\}$$

be a truncated cone. Then one can introduce truncated cone mean values of differences

$$(\mathcal{K}_r^M g)(y)=\frac{1}{|K_n|}\int_{K_n}(\Delta_{rh}^M g)(y)\,dh$$

and the corresponding maximal functions

$$(K_r^M g)(x)=\sup_{y\in R_n}\,(1+|r^{-1}y|^a)^{-1}\,|\,(\mathcal{K}_r^M g)\,(x-y)|\,.$$

The numbers r, M and a have the same meaning as above, cf. e.g. (2) and (4). The above theorem and its corollary remain valid (as do the corresponding proofs) if one replaces $S^M_{2^{-k}}$ and $S^M_{\tau 2^{-k}}$ by $K^M_{2^{-k}}$ and $K^M_{\tau 2^{-k}}$, respectively. Furthermore, one can replace Γ_n in (12) and (17) by K_n. This possibility may be of interest in connection with "inner" characterizations of the spaces $B^s_{p,q}$ and $F^s_{p,q}$ on domains.

Remark 3. On the basis of Theorem 2.3.8 one can replace some differences Δ_h^M in the above theorem and corollary by derivatives. We do not go into detail. In connection with other equivalent quasi-norms we describe this possibility more explicitely in the following subsection. There we shall also give some references.

Corollary 2. (i) Let $0<p\leq\infty$, $0<q\leq\infty$ and $a>\frac{n}{p}$ in (6). Then there exists a positive number $s_B(p,q)$ with the following property. If $s>s_B(p,q)$ and if M is an integer with $M>s$, then

$$\|f \mid L_p(R_n)\|+\|2^{sk}\sup_{|h|\leq 2} D^M_{2^{-k}h}f \mid l_q(L_p(R_n))\| \tag{48}$$

is an equivalent quasi-norm in $B^s_{p,q}(R_n)$.

(ii) Let $0<p<\infty$, $0<q\leq\infty$ and $a>\frac{n}{\min(p,q)}$ in (6). Then there exists a positive number $s_F(p,q)$ with the following property. If $s>s_F(p,q)$ and if M is an integer with $M>s$, then

$$\|f \mid L_p(R_n)\|+\|2^{sk}\sup_{|h|\leq 2} D^M_{2^{-k}h}f \mid L_p(R_n,l_q)\| \tag{49}$$

is an equivalent quasi-norm in $F^s_{p,q}(R_n)$.

Proof. We prove (ii) (the proof of (i) being the same). By the above theorem and Corollary 1, it is sufficient to prove that there exists a constant c such that

$$\|2^{sk} \sup_{|h|\leq 2} D^M_{2^{-k}h} f \mid L_p(R_n, l_q)\| \leq c \|2^{sk} \sup_{h\in\Gamma_n} D^M_{2^{-k}h} f \mid L_p(R_n, l_q)\| \tag{50}$$

for all $f \in F^s_{p,q}(R_n)$. Let $s_F(p, q) \geq G_{p,q}$. Then

$$\begin{aligned}(2^{sk} \sup_{|h|\leq 2} D^M_{2^{-k}h} f)(x) &\leq 2^{sk} \sum_{l=0}^{\infty} (\sup_{2^{-l}\leq|h|\leq 2^{-l+1}} D^M_{2^{-k}h} f)(x) \\ &\leq \sum_{l=0}^{\infty} 2^{-sl} 2^{s(k+l)} (\sup_{h\in\Gamma_n} D^M_{2^{-k-l}h} f)(x) .\end{aligned} \tag{51}$$

If C is the constant in the generalized triangle inequality for the $L_p(R_n, l_q)$-quasi-norm, it follows that

$$\|2^{sk} \sup_{|h|\leq 2} D^M_{2^{-k}h} f \mid L_p(R_n, l_q)\| \leq C \sum_{l=0}^{\infty} (2^{-s}C)^l \|2^{sk} \sup_{h\in\Gamma_n} D^M_{2^{-k}h} f \mid L_p(R_n, l_q)\| .$$

If $2^{-s}C < 1$, we obtain (50). The proof is complete.

2.5.10. Characterizations of the Spaces $F^s_{p,q}(R_n)$ by Differences

The theorem of the last subsection is very important for later considerations. On the other hand there also arise many questions. One may ask whether one can replace the maximal functions in (2.5.9/13), (2.5.9/15) and (2.5.9/17) by $\mathcal{S}^M_r f$, $\mathcal{V}^M_r f$, and $\Delta^M_h f$, respectively (cf. (2.5.9/2) and (2.5.9/3)). Furthermore, one would try to replace the somewhat artificial restriction $s \geq G_{p,q}$ in Theorem 2.5.9(ii) by a more natural one (the corresponding problem for the number M has been solved in Corollary 2.5.9/1 in a satisfactory way). This subsection deals with equivalent quasi-norms for $F^s_{p,q}(R_n)$, where only the differences Δ^M_h (and not their maximal functions) are involved. Similar problems for $F^s_{p,q}(R_n)$, with mean values of Δ^M_h in the sense of (2.5.9/2) and (2.5.9/3), are treated in the next subsection. Corresponding considerations for the spaces $B^s_{p,q}(R_n)$ are given in 2.5.12. The differences $\Delta^M_h f$ have the same meaning as in (2.5.9/1).

Theorem. Let $0<p<\infty$, $0<q\leq\infty$ and $s>\frac{n}{\min(p, q)}$. If M is an integer such that $M>s$, then

$$\|f \mid F^s_{p,q}(R_n)\|^{(1)}_M = \|f \mid L_p(R_n)\| + \left\| \left(\int_{R_n} |h|^{-sq} \sup_{\substack{|\varrho|\leq|h| \\ \varrho\in R_n}} |(\Delta^M_\varrho f)(\cdot)|^q \frac{dh}{|h|^n} \right)^{\frac{1}{q}} \middle| L_p(R_n) \right\| \tag{1}$$

and

$$\|f \mid F^s_{p,q}(R_n)\|^{(2)}_M = \|f \mid L_p(R_n)\| + \left\| \left(\int_{R_n} |h|^{-sq} |(\Delta^M_h f)(\cdot)|^q \frac{dh}{|h|^n} \right)^{\frac{1}{q}} \middle| L_p(R_n) \right\| \tag{2}$$

are equivalent quasi-norms in $F^s_{p,q}(R_n)$ (modification if $q=\infty$).

Proof. Step 1. First we prove that there exists a constant c such that

$$\|f \mid F^s_{p,q}(R_n)\|^{(1)}_M \leq c \|f \mid F^s_{p,q}(R_n)\| \tag{3}$$

holds for all $f \in F^s_{p,q}(R_n)$. Let $f \in F^s_{p,q}(R_n)$. By Theorem 2.5.3 (and Remark 2.5.3/1) the term $\|f \mid L_p(R_n)\|$ in (1) can be estimated from above by $\|f \mid F^s_{p,q}(R_n)\|$. Furthermore, under the above restrictions for s, any $f \in F^s_{p,q}(R_n)$ is bounded and continuous on R_n, cf. 2.7.1. (this remark is useful, but not absolutely necessary for the calculations below). We have

$$\left\| \left(\int_{R_n} |h|^{-sq} \sup_{\substack{|\varrho| \leq |h| \\ \varrho \in R_n}} |(\Delta^M_\varrho f)(\cdot)|^q \frac{dh}{|h|^n} \right)^{\frac{1}{q}} \middle| L_p(R_n) \right\| \tag{4}$$
$$\leq c \left\| \left(\sum_{k=-\infty}^{\infty} 2^{ksq} \sup_{0<|h|\leq 2^{-k}} |(\Delta^M_h f)(\cdot)|^q \right)^{\frac{1}{q}} \middle| L_p(R_n) \right\| .$$

Let $\{\varphi_k(x)\}_{k=0}^\infty \in \Phi(R_n)$, cf. 2.3.1. Let $\varphi_k(x)=0$ if $k=-1, -2, \ldots$ Then, if $0<q\leq 1$,

$$\sum_{k=-\infty}^{\infty} 2^{ksq} \sup_{0<|h|\leq 2^{-k}} |(\Delta^M_h f)(x)|^q \tag{5}$$
$$\leq \sum_{m=-\infty}^{\infty} \sum_{k=-\infty}^{\infty} 2^{ksq} \sup_{0<|h|\leq 2^{-k}} |(\Delta^M_h F^{-1}\varphi_{k+m} Ff)(x)|^q .$$

We used the fact that

$$\left(\sum b_j\right)^q \leq \sum b_j^q \quad \text{if} \quad 0<q\leq 1 \quad \text{and} \quad b_j>0 .$$

Split $\sum_{m=-\infty}^{\infty}$ into $\sum_{m=0}^{\infty}$ and $\sum_{m=-\infty}^{-1}$. First let $m<0$ and $|h|\leq 2^{-k}$. Then

$$|(\Delta^M_h F^{-1}\varphi_{k+m} Ff)(x)| \leq 2^{-kM} \sup_{|x-y|\leq c2^{-k}} \sum_{|\alpha|=M} |(D^\alpha F^{-1}\varphi_{k+m} Ff)(y)| , \tag{6}$$

where c is an appropriate positive constant. Let $(\varphi_k^* f)(x)$ be the maximal function from (2.3.6/2). By (1.3.1/2), a homogeneity consideration similar to the one in Remark 1.3.2/1, and a completion argument, it follows that

$$|D^\alpha (F^{-1}\varphi_{k+m} Ff)(y)| \leq c 2^{(k+m)M} (\varphi^*_{k+m} f)(y), \quad |\alpha| = M , \tag{7}$$

where c is independent of $k+m$ and $y \in R_n$. Putting this estimate into (6), we have

$$\sum_{m=-\infty}^{-1} \sum_{k=-\infty}^{\infty} 2^{ksq} \sup_{0<|h|\leq 2^{-k}} |(\Delta^M_h F^{-1}\varphi_{k+m} Ff)(x)|^q \tag{8}$$
$$\leq c \sum_{m=-\infty}^{-1} \sum_{k=-\infty}^{\infty} 2^{ksq} 2^{-kMq} 2^{(k+m)Mq} (\varphi^*_{k+m} f)^q (x) \leq c' \sum_{j=0}^{\infty} 2^{sjq} (\varphi_j^* f)^q (x) .$$

In the last estimate we used the fact that $M>s$ and we introduced $j=k+m$. Let $m\geq 0$ and $|h|\leq 2^{-k}$. Then

$$|(\Delta^M_h F^{-1}\varphi_{k+m} Ff)(x)| \leq c 2^{ma} (\varphi^*_{k+m} f)(x) ,$$

where a has the same meaning as in (2.3.6/2). Let $s>a>\frac{n}{\min(p,q)}$. Then

$$\sum_{m=0}^{\infty} \sum_{k=-\infty}^{\infty} 2^{ksq} \sup_{0<|h|<2^{-k}} |(\Delta^M_h F^{-1}\varphi_{k+m} Ff)(x)|^q \tag{9}$$
$$\leq c \sum_{m=0}^{\infty} \sum_{k=-\infty}^{\infty} 2^{s(k+m)q} (\varphi^*_{k+m} f)^q (x)\, 2^{-m(s-a)q} \leq c' \sum_{j=0}^{\infty} 2^{jsq} (\varphi_j^* f)^q (x) .$$

The sum of the left-hand sides of (8) and (9) yields the righthand side of (5). Consequently, we have by (4) and (5) (and the estimates (8) and (9))

$$\left\| \left(\int_{R_n} |h|^{-sq} \sup_{\substack{|\varrho|\leq|h| \\ \varrho\in R_n}} |(\Delta_\varrho^M f)(\cdot)|^q \frac{dh}{|h|^n} \right)^{\frac{1}{q}} \Big| L_p(R_n) \right\| \leq c \|2^{sj} \varphi_j^* f \mid L_p(R_n, l_q)\| . \quad (10)$$

If $1<q\leq\infty$, we use in (5) the triangle inequality for l_q-spaces and obtain (10) once more. Because $a > \frac{n}{\min(p, q)}$ the right-hand side of (10) can be estimated from above by $c\|f \mid F^s_{p,q}(R_n)\|$. This follows from Theorem 2.3.6(ii), as proves (3).

Step 2. We prove that there exists a constant c such that

$$\|f \mid F^s_{p,q}(R_n)\| \leq c\|f \mid F^s_{p,q}(R_n)\|^{(2)}_M \quad (11)$$

holds for all $f\in F^s_{p,q}(R_n)$. For this purpose we recall Theorem 2.5.3(ii) and Remark 2.5.3/2 (as far as the case $q=\infty$ is concerned). In other words, it is sufficient to find a sequence $\{a_k\}_{k=0}^\infty \in \mathfrak{A}_p(R_n)$ such that $f=\lim_{k\to\infty} a_k$ in $S'(R_n)$ and

$$\|2^{sk}(f-a_k) \mid L_p(R_n, l_q)\| \leq c\|f \mid F^s_{p,q}(R_n)\|^{(2)}_M \quad (12)$$

If $f\in F^s_{p,q}(R_n)$ (with the above restrictions on the parameters), then f is in any case a regular distribution (either $f\in L_p(R_n)$ if $1\leq p<\infty$ or $f\in L_1(R_n)$ if $p<1$, cf. Remark 2.5.3/1). Now we use a classical construction from approximation theory (and in particular in connection with the classical Besov spaces $\Lambda^s_{p,q}(R_n)$), which can be found in S. M. Nikol'skij [3, 5.2.1]: let $\gamma_n=\{x \mid x\in R_n, |x|=1\}$ be the unit sphere. One finds for any $f\in F^s_{p,q}(R_n)$ a system $\{a_k(x)\}_{k=0}^\infty$ with supp $Fa_k\subset\{y \mid |y|\leq$ $\leq 2^{k+1}\}$ and

$$(-1)^{M-1}(a_k(x)-f(x)) = \int_0^\infty g(r) \int_{\gamma_n} (\Delta^M_{2^{-k}r\gamma} f)(x)\, d\gamma r^{n-1}\, dr , \quad (13)$$

where $g(r)$ is a non-negative rapidly decreasing function on $[0, \infty)$, which is independent of $f(x)$ and k. Furthermore, M is an arbitrary natural number. Although the proof is comparatively simple, we shall not go into detail and refer instead to S. M. Nikol'skij [3, 5.2.1]. In any case, $a_k\to f$ in $S'(R_n)$ is ensured under the above hypotheses. If Γ_n has the same meaning as in 2.5.9., then

$$\begin{aligned} |a_k(x)-f(x)| &\leq c \sum_{\varkappa=-\infty}^{0} \int_{\Gamma_n} |(\Delta^M_{2^{-k+\varkappa}\gamma} f)(x)|\, d\gamma \\ &+ c\sum_{\varkappa=1}^{\infty} 2^{-\varkappa\sigma} \int_{\Gamma_n} |(\Delta^M_{2^{-k+\varkappa}\gamma} f)(x)|\, d\gamma . \end{aligned} \quad (14)$$

where σ is a positive number which may be chosen arbitrarily large. If $0<q\leq 1$ and $\sigma>s$, then

$$\begin{aligned} \sum_{k=0}^\infty 2^{skq}|a_k(x)-f(x)|^q &\leq c \sum_{k=0}^\infty \sum_{\varkappa=-\infty}^{0} 2^{skq}\Big(\int_{\Gamma_n} |(\Delta^M_{2^{-k+\varkappa}\gamma} f)(x)|\, d\gamma\Big)^q \\ &+ c\sum_{k=0}^\infty \sum_{\varkappa=1}^\infty 2^{skq-\varkappa\sigma q}\Big(\int_{\Gamma_n} |(\Delta^M_{2^{-k+\varkappa}\gamma} f)(x)|\, d\gamma\Big)^q \\ &\leq c \sum_{k=-\infty}^{\infty} 2^{skq}\Big(\int_{\Gamma_n} |(\Delta^M_{2^{-k}\gamma} f)(x)|\, d\gamma\Big)^q . \end{aligned} \quad (15)$$

If $1<q\leq\infty$, we use the triangle inequality for l_q-spaces and obtain a corresponding inequality. Hence, the left-hand side of (15) can be estimated from above by the right-hand side of (15) for all $0<q\leq\infty$ (modification if $q=\infty$). If $1\leq q\leq\infty$, we have

$$\Big(\int_{\Gamma_n}|(\Delta^M_{2^{-k},\gamma}f)(x)|\,\mathrm{d}\gamma\Big)^q\leq c\int_{\Gamma_n}|(\Delta^M_{2^{-k},\gamma}f)(x)|^q\,\mathrm{d}\gamma\,. \tag{16}$$

We put (16) into (15), take the $\frac{1}{q}$-power and apply the L_p-quasi-norm. The result is (12), which establishes (11). If $0<q<1$, then (15) yields

$$\begin{aligned}&\sum_{k=0}^{\infty}2^{skq}\,|a_k(x)-f(x)|^q\\ &\leq c\sum_{k=-\infty}^{\infty}2^{skq}\Big(\int_{\Gamma_n}|(\Delta^M_{2^{-k},\gamma}f)(x)|^q\,\mathrm{d}\gamma\Big)^q\sup_{\gamma\in\Gamma_n}|(\Delta^M_{2^{-k},\gamma}f)(x)|^{q(1-q)}\,.\end{aligned} \tag{17}$$

By Hölder's inequality, we have

$$\begin{aligned}&\sum_{k=0}^{\infty}2^{skq}\,|a_k(x)-f(x)|^q\\ &\leq c\left(\sum_{k=-\infty}^{\infty}2^{skq}\int_{\Gamma_n}|(\Delta^M_{2^{-k},\gamma}f)(x)|^q\,\mathrm{d}\gamma\right)^q\left(\sum_{k=-\infty}^{\infty}2^{skq}\sup_{\gamma\in\Gamma_n}|(\Delta^M_{2^{-k},\gamma}f)(x)|^q\right)^{1-q}.\end{aligned} \tag{18}$$

Taking the $\frac{1}{q}$-power, applying the L_p-quasi-norm, and using Hölder's inequality again we obtain

$$\|2^{sk}\,(f-a_k)\mid L_p(R_n,\,l_q)\|\leq c(\|f\mid F^s_{p,q}(R_n)\|^{(2)}_M)^q\,(\|f\mid F^s_{p,q}(R_n)\|^{(1)}_M)^{1-q}\,, \tag{19}$$

cf. (1) and (4). We can add the term $\|f\mid L_p(R_n)\|$ on the left-hand side (because both terms on the right-hand side contain that term). Then it follows from Theorem 2.5.3(ii) and (3) that

$$\|f\mid F^s_{p,q}(R_n)\|\leq c(\|f\mid F^s_{p,q}(R_n)\|^{(2)}_M)^q\,\|f\mid F^s_{p,q}(R_n)\|^{1-q}.$$

This proves (11). The theorem is now an easy consequence of (3) and (11).

Remark 1. If one compares (1) and (2) with the classical norms (2.2.2/9) and (2.2.2/10) for the Besov spaces $\Lambda^s_{p,q}(R_n)$, the question arises whether one can replace some differences in (1) and (2) by derivatives. The basis for such a possibility is Theorem 2.3.8. We describe an example. Let $s>\frac{n}{\min(p,q)}$ and

$$s=\sigma+m\quad\text{with}\quad\frac{n}{\min(p,q)}<\sigma\leq\frac{n}{\min(p,q)}+1\quad\text{and}\quad m=0,1,2,\ldots \tag{20}$$

If M is an integer such that $M>\sigma$, then (2) and (2.3.8/3) show that

$$\sum_{|\alpha|\leq m}\|D^\alpha f\mid L_p(R_n)\|+\sum_{|\alpha|\leq m}\left\|\left(\int_{R_n}|h|^{-\sigma q}|(\Delta^M_h D^\alpha f)(\cdot)|^q\frac{\mathrm{d}h}{|h|^n}\right)^{\frac{1}{q}}\mid L_p(R_n)\right\| \tag{21}$$

is an equivalent quasi-norm in $F^s_{p,q}(R_n)$. The smallest number M is either $1+\left[\frac{n}{\min(p,q)}\right]$ or $2+\left[\frac{n}{\min(p,q)}\right]$.

Remark 2. We mention a simple consequence of (2) (and also of (1)) which seems to be of some interest. If $1<p<\infty$, then Proposition 2.3.2/2 and (2.5.6/2) show that

$$H_p^s(R_n)=F_{p,2}^s(R_n)\subset F_{p,\infty}^s(R_n), \quad (s \text{ real}), \tag{22}$$

where $H_p^s(R_n)$ are the Bessel-potential spaces. Hence, it follows from (2) (with $q=\infty$) that

$$\| \sup_{h\in R_n} |h|^{-s}\, |(\Delta_h^M f)(\cdot)|\; |L_p(R_n)\| \leq c\|f|\, H_p^s(R_n)\| \tag{23}$$

holds for all $f\in H_p^s(R_n)$ if $1<p<\infty$, $s>\frac{n}{p}$ and $M>s$ (of course, c in (23) is independent of f).

Remark 3. As we mentioned in (22), $H_p^s(R_n)=F_{p,2}^s(R_n)$ with $1<p<\infty$ and $-\infty<s<\infty$ are the usual Bessel-potential spaces. If $1<p<\infty$ and $s>0$, then equivalent norms in $H_p^s(R_n)$ are known, which are constructed with the help of truncated volume means of differences and derivatives, and corresponding spherical means (cf. 2.5.9. and the following subsection 2.5.11.). We refer to R. S. Strichartz [1] and P. I. Lizorkin [2,3], cf. also E. M. Stein [1, pp. 160–163]. A formula similar to (2) $\left(\text{with } 1<p<\infty,\ q=2,\ s>\frac{n}{2} \text{ and } M>s\right)$ may be found in P. I. Lizorkin [2, p. 240]. Equivalent norms in $F_{p,q}^s(R_n)$ with $1<p<\infty$ and $1<q<\infty$, which are related to the above considerations (including the one in 2.5.11.) and to P. I. Lizorkin [2] have been studied by G. A. Kaljabin [2, 5]. We refer also to H. Triebel [12, 14] (the above theorem is an improvement of H. Triebel [14, Theorem 1]). Finally we mention that inequalities of type (23) have been studied by C. P. Calderón, E. B. Fabes, N. M. Riviere [1] (cf. also C. J. Neugebauer [1], where corresponding inequalities for Besov spaces with fractional differences are considered).

2.5.11. Characterizations of the Spaces $F_{p,q}^s(R_n)$ by Ball Means of Differences

Theorem 2.5.10 is useful and sheds light upon the structure of the spaces $F_{p,q}^s(R_n)$ if $s>\frac{n}{\min(p,q)}$. On the one hand, this restriction for s is too strong for our later purposes while on the other, it cannot be essentially relaxed. Another question of the same type is connected with Theorem 2.5.9: is it possible to replace $V_{2^{-k}}^M f$ in (2.5.9/15) by the truncated ball means of differences $\mathcal{V}_{2^{-k}}^M f$ from (2.5.9/3) and what are the natural restrictions for s? We study these questions in this subsection and introduce (beside the truncated ball means of differences $\mathcal{V}_r^M f$ from (2.5. 9/3)) the ball means of differences

$$(\mathcal{B}_r^M f)(y)=\frac{1}{|B_n|}\int_{B_n}(\Delta_{rh}^M f)(y)\,dh\,, \tag{1}$$

where $B_n=\{y \mid y\in R_n,\ |y|\leq 1\}$ is the unit ball, and $|B_n|$ is its volume. Let again

$$\tilde{\sigma}_{p,q}=n\left(\frac{1}{\min(p,q,1)}-1\right), \tag{2}$$

where $0<p<\infty$ and $0<q\leq\infty$, cf. (2.5.3/8).

Theorem. Let $0<p<\infty$, $0<q<\infty$ and $s>\tilde{\sigma}_{p,q}$. If M is an integer such that $M>s$, then

$$\|f \mid F_{p,q}^s(R_n)\|_M^{(3)}=\|f \mid L_p(R_n)\|+\left\|\left(\sum_{k=-\infty}^{\infty}2^{ksq}\,|(\mathcal{B}_{2^{-k}}^M f)(\cdot)|^q\right)^{\frac{1}{q}} \Big| L_p(R_n)\right\| \tag{3}$$

and

$$\|f \mid F^s_{p,q}(R_n)\|^{(4)}_M = \|f \mid L_p(R_n)\| + \left\| \left(\sum_{k=-\infty}^{\infty} 2^{ksq} |(\mathcal{V}^M_{2^{-k}} f)(\cdot)|^q \right)^{\frac{1}{q}} \Big| L_p(R_n) \right\| \tag{4}$$

are equivalent quasi-norms in $F^s_{p,q}(R_n)$.

Proof. Step 1. First we remark that $\|f \mid L_p(R_n)\|$ makes sense if $f \in F^s_{p,q}(R_n)$ with $s > \tilde{\sigma}_{p,q}$, cf. Remark 2.5.3/1. Furthermore, $\|f \mid L_p(R_n)\|$ can be estimated from above by $\|f \mid F^s_{p,q}(R_n)\|$, cf. Theorem 2.5.3 (ii). In this step we prove that the two quasi-norms in (3) and (4) can be estimated from above by $c\|f \mid F^s_{p,q}(R_n)\|$. We actually prove a bit more (which covers both cases, cf. formula (6) below): there exists a constant c such that

$$\left\| \left(\int_0^\infty r^{-sq} \left(\int_{B_n} |(\Delta^M_{rh} f)(\cdot)| \, dh \right)^q \frac{dr}{r} \right)^{\frac{1}{q}} \Big| L_p(R_n) \right\| \leqq c \|f \mid F^s_{p,q}(R_n)\| \tag{5}$$

holds for all $f \in F^s_{p,q}(R_n)$. We begin in the same way as in Step 1 of the proof of Theorem 2.5.10:

$$\begin{aligned} &\left\| \left(\int_0^\infty r^{-sq} \left(\int_{B_n} |(\Delta^M_{rh} f)(\cdot)| \, dh \right)^q \frac{dr}{r} \right)^{\frac{1}{q}} \Big| L_p(R_n) \right\| \\ &\leqq c \left\| \left(\sum_{k=-\infty}^{\infty} 2^{ksq} \left(\int_{B_n} |(\Delta^M_{2^{-k}h} f)(\cdot)| \, dh \right)^q \right)^{\frac{1}{q}} \Big| L_p(R_n) \right\| \end{aligned} \tag{6}$$

is the counterpart of (2.5.10/5). Instead of (2.5.10/5) we have

$$\begin{aligned} &\sum_{k=-\infty}^{\infty} 2^{ksq} \left(\int_{B_n} |(\Delta^M_{2^{-k}h} f)(x)| \, dh \right)^q \\ &\leqq \sum_{m=-\infty}^{\infty} \sum_{k=-\infty}^{\infty} 2^{ksq} \left(\int_{B_n} |(\Delta^M_{2^{-k}h} F^{-1} \varphi_{k+m} Ff)(x)| \, dh \right)^q . \end{aligned} \tag{7}$$

where again $0 < q \leqq 1$ and $\varphi_j(x) = 0$ if $j < 0$. We split $\sum_{m=-\infty}^{\infty}$ in $\sum_{m=0}^{\infty}$ and $\sum_{m=-\infty}^{-1}$. The terms with $m < 0$ can be estimated from above by the corresponding terms in (2.5.10/5). Then (2.5.10/8) yields

$$\sum_{m=-\infty}^{-1} \sum_{k=-\infty}^{\infty} 2^{ksq} \left(\int_{B_n} |(\Delta^M_{2^{-k}h} F^{-1} \varphi_{k+m} Ff)(x)| \, dh \right)^q \leqq c \sum_{j=0}^{\infty} 2^{sjq} (\varphi_j^* f)^q(x) . \tag{8}$$

Here, one needs only that $M > s$. If $1 < q < \infty$, one has to modify in the usual way. Hence, (8) holds for all $0 < q < \infty$. Now we estimate the terms with $m \geqq 0$. If $\min(p, q) \leqq 1$, then

$$s > n \left(\frac{1}{\min(p, q)} - 1 \right) = \frac{n}{\min(p, q)} (1 - \min(p, q)) ,$$

and we can choose two numbers λ and a such that

$$0 < \lambda < \min(p, q), \quad a > \frac{n}{\min(p, q)}, \quad s > a(1 - \lambda) . \tag{9}$$

Let $\lambda=1$ if $\min(p,q)>1$ (in that case we have $s>0$). Temporarily, we denote the Hardy-Littlewood maximal function from (1.2.3/1) by $(\mathfrak{M}f)(x)$ (in order to avoid confusions with the above number M). Also, we use the estimate preceding (2.5.10/9). Then

$$\begin{aligned}
&\int_{B_n} |(\Delta^M_{2^{-k}h} F^{-1}\varphi_{k+m}Ff)(x)|\,dh \qquad (10)\\
&\leq \int_{B_n} |(\Delta^M_{2^{-k}h} F^{-1}\varphi_{k+m}Ff)(x)|^{\lambda}\,dh \sup_{|h|\leq 2^{-k}} |(\Delta^M_h F^{-1}\varphi_{k+m}Ff)(x)|^{1-\lambda}\\
&\leq c(\mathfrak{M}\,|F^{-1}\varphi_{k+m}Ff|^{\lambda})(x)\, 2^{ma(1-\lambda)}(\varphi^*_{k+m}f)^{1-\lambda}(x)\,.
\end{aligned}$$

Because $s>a(1-\lambda)$, Hölder's inequality gives

$$\begin{aligned}
&\sum_{m=0}^{\infty}\sum_{k=-\infty}^{\infty} 2^{ksq}\Big(\int_{B_n} |(\Delta^M_{2^{-k}h} F^{-1}\varphi_{k+m}Ff)(x)|\,dh\Big)^q \qquad (11)\\
&\leq c\sum_{j=0}^{\infty} 2^{sjq}(\mathfrak{M}\,|F^{-1}\varphi_jFf|^{\lambda})^q(x)\,(\varphi^*_jf)^{(1-\lambda)q}(x)\\
&\leq c\left(\sum_{j=0}^{\infty} 2^{sjq}(\mathfrak{M}\,|F^{-1}\varphi_jFf|^{\lambda})^{\frac{q}{\lambda}}(x)\right)^{\lambda}\left(\sum_{j=0}^{\infty} 2^{sjq}(\varphi^*_jf)^q(x)\right)^{1-\lambda}.
\end{aligned}$$

Taking the $\frac{1}{q}$-power of (11), applying the L_p-quasi-norm, and using Hölder's inequality again (if $\lambda<1$), we obtain

$$\begin{aligned}
&\left\|\left[\sum_{m=0}^{\infty}\sum_{k=-\infty}^{\infty} 2^{ksq}\Big(\int_{B_n} |(\Delta^M_{2^{-k}h} F^{-1}\varphi_{k+m}Ff)(x)|\,dh\Big)^q\right]^{\frac{1}{q}} \,\Big|\, L_p(R_n)\right\|\\
&\leq \|2^{sj}\varphi^*_jf \mid L_p(R_n,l_q)\|^{1-\lambda}\,\|2^{sj}(\mathfrak{M}|F^{-1}\varphi_jFf|^{\lambda})^{\frac{1}{\lambda}} \mid L_p(R_n,l_q)\|^{\lambda} \qquad (12)\\
&= \|2^{sj}\varphi^*_jf \mid L_p(R_n,l_q)\|^{1-\lambda}\,\|\mathfrak{M}\,|2^{sj}F^{-1}\varphi_jFf|^{\lambda} \mid L_{\frac{p}{\lambda}}(R_n,l_{\frac{q}{\lambda}})\|\,.
\end{aligned}$$

However, the first term on the right-hand side of (12) can be estimated from above by $c\|f \mid F^s_{p,q}(R_n)\|^{1-\lambda}$, cf. Theorem 2.3.6(ii) $\left(\text{here we need } a>\frac{n}{\min(p,q)}\right)$. Furthermore, because $\frac{q}{\lambda}>1$ and $\frac{p}{\lambda}>1$, it follows from Theorem 1.2.3, that the second term on the right-hand side of (12) can be estimated from above by

$$\begin{aligned}
&c\||2^{sj}F^{-1}\varphi_jFf|^{\lambda} \mid L_{\frac{p}{\lambda}}(R_n,l_{\frac{q}{\lambda}})\| \qquad (13)\\
&= c\|2^{sj}F^{-1}\varphi_jFf \mid L_p(R_n,l_q)\|^{\lambda} = c\|f \mid F^s_{p,q}(R_n)\|^{\lambda}\,.
\end{aligned}$$

In a similar way one deals with (8). Now (5) follows from (6), (7) and the indicated estimates.

Step 2. We prove that there exists a constant c such that

$$\|f \mid F^s_{p,q}(R_n)\| \leq c\|f \mid F^s_{p,q}(R_n)\|^{(4)}_M \qquad (14)$$

for all $f\in F^s_{p,q}(R_n)$. We use the same method as in Step 2 of the proof of Theorem 2.5.10 (by our hypotheses, Theorem 2.5.3(ii) is applicable). Let $\{a_k(x)\}_{k=0}^{\infty}$ be the same system as in (2.5.10/13). It is easy to see that (2.5.10/14) and (2.5.10/15) can be modified by

$$\sum_{k=0}^{\infty} 2^{skq} |a_k(x)-f(x)|^q \leq c \sum_{k=-\infty}^{\infty} 2^{skq} |(\mathcal{V}^M_{2^{-k}}f)(x)|^q \tag{15}$$

cf. the definition of $\mathcal{V}^M_r$ in (2.5.9/3). Formula (15) holds for $0<q<\infty$. We raise (15) to the $\frac{1}{q}$-power and take the L_p-quasi-norm. Then Theorem 2.5.3(ii) produces (14) (we remark that $\|a_0 \mid L_p(R_n)\|$ can be estimated from above by $\|f \mid L_p(R_n)\|$ and $\|f-a_0 \mid L_p(R_n)\|$, where the last term is covered by (15)). Now (5) (cf. also (6)) and (14) prove that $\|f \mid F^s_{p,q}(R_n)\|^{(4)}$ is an equivalent quasi-norm in $F^s_{p,q}(R_n)$.

Step 3. We prove (14) with $\|f \mid F^s_{p,q}(R_n)\|^{(3)}$ instead of $\|f \mid F^s_{p,q}(R_n)\|^{(4)}$. For that purpose we remark that

$$|(\mathcal{V}^M_{2^{-k}}f)(x)| \leq c|(\mathcal{B}^M_{2^{-k+1}}f)(x)| + c|(\mathcal{B}^M_{2^{-k}}f)(x)| \,.$$

Substituting this into (15), we obtain the desired estimate in the same way as above. Hence, $\|f \mid F^s_{p,q}(R_n)\|^{(3)}$ is also an equivalent quasi-norm in $F^s_{p,q}(R_n)$.

Corollary. Let $0<p<\infty$, $0<q<\infty$ and $s>\tilde{\sigma}_{p,q}$. If M is an integer such that $M>s$, then

$$\|f \mid F^s_{p,q}(R_n)\|^{(5)}_M = \|f \mid L_p(R_n)\| + \left\| \left[\int_0^\infty r^{-sq} \left(\int_{B_n} |(\Delta^M_{rh}f)(\cdot)| \,\mathrm{d}h \right)^q \frac{\mathrm{d}r}{r} \right]^{\frac{1}{q}} \middle| L_p(R_n) \right\| \tag{16}$$

and

$$\|f \mid F^s_{p,q}(R_n)\|^{(6)}_M = \|f \mid L_p(R_n)\| + \left\| \left[\int_0^\infty r^{-sq} \left(\int_{\Gamma_n} |(\Delta^M_{rh}f)(\cdot)| \,\mathrm{d}h \right)^q \frac{\mathrm{d}r}{r} \right]^{\frac{1}{q}} \middle| L_p(R_n) \right\| \tag{17}$$

are equivalent quasi-norms in $F^s_{p,q}(R_n)$. Replacing $\int_0^\infty$ in (16) and (17) by $\int_0^1$, we obtain further equivalent quasi-norms in $F^s_{p,q}(R_n)$.

Proof. Step 1. The equivalence of $\|f \mid F^s_{p,q}(R_n)\|^{(5)}_M$ and $\|f \mid F^s_{p,q}(R_n)\|$ follows from (5) and the equivalent quasi-norm $\|f \mid F^s_{p,q}(R_n)\|^{(3)}_M$ in (3). The corresponding assertion for $\|f \mid F^s_{p,q}(R_n)\|^{(6)}_M$ is a consequence of

$$\int_{B_n} |(\Delta^M_{rh}f)(x)| \,\mathrm{d}h \leq \sum_{j=1}^{\infty} \int_{\Gamma_n} |(\Delta^M_{r2^{-j}h}f)(x)| \,\mathrm{d}h$$

and

$$\int_0^\infty r^{-sq} \left(\int_{B_n} |(\Delta^M_{rh}f)(x)| \,\mathrm{d}h \right)^q \frac{\mathrm{d}r}{r} \leq \sum_{j=1}^{\infty} \int_0^\infty r^{-sq} \left(\int_{\Gamma_n} |(\Delta^M_{r2^{-j}h}f)(x)| \,\mathrm{d}h \right)^q \frac{\mathrm{d}r}{r}$$

$$\leq c \int_0^\infty r^{-sq} \left(\int_{\Gamma_n} |(\Delta^M_{rh}f)(x)| \,\mathrm{d}h \right)^q \frac{\mathrm{d}r}{r}$$

if $0<q\leq 1$ (similarly if $1<q<\infty$).

Step 2. We prove that $\int_0^\infty$ in (16) can be replaced by $\int_0^1$. If $\varepsilon>0$ is sufficiently small, we have in the same way as in (6) that

$$\left\| \left(\int_1^\infty r^{-sq} \left(\int_{B_n} |(\Delta_{rh}^M f)(\cdot)| \, dh \right)^q \frac{dr}{r} \right)^{\frac{1}{q}} \Big| L_p(R_n) \right\|$$

$$\leq c \left\| \left(\sum_{k=-\infty}^{0} 2^{k(s-\varepsilon)p} \left(\int_{B_n} |(\Delta_{2^{-k}h}^M f)(\cdot)| \, dh \right)^p \right)^{\frac{1}{p}} \Big| L_p(R_n) \right\| \leq c' \|f \mid B_{p,p}^{s-\varepsilon}(R_n)\| .$$

If $0<\varepsilon'<\varepsilon$, then it follows easily from (2.5.12/4) that

$$\begin{aligned} \|f \mid B_{p,p}^{s-\varepsilon}(R_n)\| &\leq c_\delta \|f \mid L_p(R_n)\| + \delta \|f \mid B_{p,p}^{s-\varepsilon'}(R_n)\| \\ &\leq c_\delta \|f \mid L_p(R_n)\| + \delta \|f \mid F_{p,q}^s(R_n)\| , \end{aligned}$$

where $\delta>0$ is a given positive number. If we put the last estimate in the last but one, then we obtain the desired estimate by standard arguments. We obtain the corresponding assertion for (17) in the same way as in the first step.

Remark 1. References were given in Remark 2.5.10/3. The last part of the corollary was inspired by P. Nilsson [2].

Remark 2. As in Remark 2.5.10/1, one can replace some differences in the above quasi-norms by derivatives. The number σ from (2.5.10/20) can be chosen in this case as

$$s=\sigma+m, \quad \tilde{\sigma}_{p,q}<\sigma\leq\tilde{\sigma}_{p,q}+1, \quad \text{and} \quad m=0, 1, 2, \ldots \tag{18}$$

2.5.12. Characterizations of the Spaces $B_{p,q}^s(R_n)$ by Differences; the Spaces $\Lambda_{p,q}^s(R_n)$ and $\mathcal{C}^s(R_n)$

It is well-known that the norms of the classical Besov spaces $B_{p,q}^s(R_n)=\Lambda_{p,q}^s(R_n)$ with $s>0$, $1\leq p<\infty$ and $1\leq q\leq\infty$ can be characterized by differences and derivatives of the functions involved, cf. (2.5.7/5). For a wider range of the parameters p and q, we have given in Theorem 2.5.9 (and in the Corollaries 2.5.9/1 and 2.5.9/2) a description of the quasi-norms of $B_{p,q}^s(R_n)$ and $F_{p,q}^s(R_n)$ by maximal functions of differences. Subsections 2.5.10. and 2.5.11. dealt with the question of whether the maximal function of differences in the quasi-norms of the spaces $F_{p,q}^s(R_n)$ can be replaced by the differences themselves. What about the corresponding problem for the spaces $B_{p,q}^s(R_n)$? There is no doubt that the corresponding considerations in 2.5.10. and 2.5.11. have immediate counterparts for the spaces $B_{p,q}^s(R_n)$ (by the rule of thumb that all statements for $F_{p,q}^s(R_n)$ can be carried over to $B_{p,q}^s(R_n)$ and that the corresponding proofs are easier than for $F_{p,q}^s(R_n)$). However, we did not formulate the corresponding assertions, the main reason being that we wish to use a fact from approximation theory which yields a better result for the spaces $B_{p,q}^s(R_n)$ than the immediate counterparts of the corresponding assertions for the spaces $F_{p,q}^s(R_n)$ from 2.5.10. and 2.5.11. For this purpose we extend the definition of $E_p(2^k, f)$ from (2.5.3/1). Let $0<p\leq\infty$ and $b>0$. If $f\in L_p(R_n)$, then

$$E_p(b, f)=\inf \|f-g \mid L_p(R_n)\| , \tag{1}$$

where the infimum is taken over all $g\in S'(R_n)\cap L_p(R_n)$ such that $\operatorname{supp} Fg\subset \subset\{y \mid |y|\leq b\}$. This is an approximation of $f(x)$ by entire analytic functions of exponential type, cf. also 2.5.3.

Proposition. Let M be a natural number and $0<p\leq\infty$. Then there exists a positive constant c (which depends on M and p) such that

$$E_p(b,f)\leq c \sup_{|h|\leq b^{-1}} \|\Delta_h^M f \mid L_p(R_n)\| \tag{2}$$

for all $b\geq 1$ and all $f\in L_p(R_n)$.

Remark 1. Of course, the supremum in (2) is taken over all $h\in R_n$ with $|h|\leq b^{-1}$. Also, the differences Δ_h^M have the same meaning as in (2.5.9/1). We shall not prove (2), but we add some comments. If $1\leq p\leq\infty$, then (2) is a well-known famous classical assertion. It follows easily from (2.5.10/13), cf. S. M. Nikol'skij [3, 5.2.1] for details. However, if $0<p<1$, then the method from S. M. Nikol'skij [3, 5.2.1] cannot be applied. In that case, (2) is due to É. A. Storoženko, P. Oswald [3, Theorem 3], cf. also É. A. Storoženko, P. Oswald [1] (these two authors proved a corresponding formula for the periodic case, the n-torus; however, by standard arguments one can carry over the periodic case to the above non-periodic case). Historical remarks for the classical case, i.e. $1\leq p\leq\infty$, may be found in the comments of Subsection 5.2.1. in S. M. Nikol'skij [3]. Concerning approximation problems (also via algebraic polynomials and spline functions) in the case $0<p<1$ we refer also (beside the above papers) to V. I. Ivanov [1], É. A. Storoženko, V. G. Krotov, P. Oswald [1], É. A. Storoženko, P. Oswald [2], P. Oswald [1, 2, 3], É. A. Storoženko [1], V. G. Krotov [1] and L. B. Chodak [1],

Let again $\tilde{\sigma}_p=n\left(\dfrac{1}{\min(p,1)}-1\right)$ if $0<p\leq\infty$, cf. (2.5.3/8).

Theorem. Let $0<p\leq\infty$, $0<q\leq\infty$ and $s>\tilde{\sigma}_p$. If M is an integer such that $M>s$, then

$$\|f \mid B^s_{p,q}(R_n)\|^{(1)}_M=\|f \mid L_p(R_n)\|+\left(\int_{R_n}|h|^{-sq}\sup_{\substack{|\varrho|\leq|h|\\ \varrho\in R_n}}\|\Delta_\varrho^M f \mid L_p(R_n)\|^q\frac{dh}{|h|^n}\right)^{\frac{1}{q}} \tag{3}$$

and

$$\|f \mid B^s_{p,q}(R_n)\|^{(2)}_M=\|f \mid L_p(R_n)\|+\left(\int_{R_n}|h|^{-sq}\|\Delta_h^M f \mid L_p(R_n)\|^q\frac{dh}{|h|^n}\right)^{\frac{1}{q}} \tag{4}$$

are equivalent quasi-norms in $B^s_{p,q}(R_n)$ (modification if $q=\infty$). One can replace $\int_{R_n}$ in (3) and (4) by $\int_{|h|\leq 1}$.

Proof. Step 1. First we remark that $\|f \mid L_p(R_n)\|$ makes sense if $f\in B^s_{p,q}(R_n)$ with $s>\tilde{\sigma}_p$, cf. Remark 2.5.3/1, and that $\|f \mid L_p(R_n)\|$ can be estimated from above by $\|f \mid B^s_{p,q}(R_n)\|$, cf. Theorem 2.5.3 (i). Furthermore, the last assertion of the theorem is obvious. In this step we prove that there exists a positive constant c such that

$$\|f \mid B^s_{p,q}(R_n)\|^{(1)}_M\leq c\|f \mid B^s_{p,q}(R_n)\| \tag{5}$$

holds for all $f\in B^s_{p,q}(R_n)$. Let $f\in B^s_{p,q}(R_n)$ and $\{\varphi_j(x)\}_{j=0}^\infty\in\Phi(R_n)$, cf. Definition 2.3.1/1. Then we have

$$f=\sum_{j=0}^\infty F^{-1}\varphi_j Ff, \tag{6}$$

where (6) converges not only in $S'(R_n)$ but also in $L_p(R_n)$, cf. Remark 2.5.3/1 and Step 1 of the proof of Theorem 2.5.3. By (2.5.10/6) and (2.5.10/7) (and the definition of $(\varphi_j^*f)(x)$ in (2.3.6/2)) we have

$$|(\Delta_\varrho^M F^{-1}\varphi_j Ff)(x)|\leq c2^{(j-k)M}(\varphi_j^*f)(x), \quad x\in R_n, \tag{7}$$

if $|\varrho|\leq 2^{-k}$ and $j=0,\ldots,k$. Hence, we obtain in any case, i.e. for $j=0,1,2,\ldots$ that

$$\sup_{|\varrho|\leq 2^{-k}} \|\Delta_\varrho^M F^{-1}\varphi_j F f \mid L_p(R_n)\| \leq c \min\left(1, 2^{(j-k)M}\right) \|\varphi_j^* f \mid L_p(R_n)\|, \tag{8}$$

where c is independent of f. Let $0<p\leq 1$ (otherwise one has to modify the calculations below in an obvious way). Then

$$\begin{aligned} \sup_{|\varrho|\leq 2^{-k}} \|\Delta_\varrho^M f \mid L_p(R_n)\|^p &\leq \sum_{j=0}^{\infty} \sup_{|\varrho|\leq 2^{-k}} \|\Delta_\varrho^M F^{-1}\varphi_j F f \mid L_p(R_n)\|^p \\ &\leq c \sum_{j=0}^{k} 2^{(j-k)Mp} \|\varphi_j^* f \mid L_p(R_n)\|^p + c \sum_{j=k+1}^{\infty} \|\varphi_j^* f \mid L_p(R_n)\|^p . \end{aligned} \tag{9}$$

It follows (with the usual modification if $q=\infty$) that

$$\begin{aligned} \|f \mid B_{p,q}^s(R_n)\|^{(1)q} &\leq c\|f \mid L_p(R_n)\|^q + c \sum_{k=0}^{\infty} 2^{skq} \sup_{|\varrho|\leq 2^{-k}} \|\Delta_\varrho^M f \mid L_p(R_n)\|^q \\ &\leq c\|f \mid L_p(R_n)\|^q + c' \sum_{k=0}^{\infty} 2^{skq} \left(\sum_{j=0}^{k} 2^{(j-k)Mp} \|\varphi_j^* f \mid L_p(R_n)\|^p \right)^{\frac{q}{p}} \\ &\quad + c' \sum_{k=0}^{\infty} 2^{skq} \left(\sum_{j=k+1}^{\infty} \|\varphi_j^* f \mid L_p(R_n)\|^p \right)^{\frac{q}{p}} . \end{aligned} \tag{10}$$

Let ε be a positive number such that $0<\varepsilon<s<s+\varepsilon<M$. Then

$$\begin{aligned} &\sum_{k=0}^{\infty} 2^{skq} \left(\sum_{j=0}^{k} 2^{(j-k)Mp} \|\varphi_j^* f \mid L_p(R_n)\|^p \right)^{\frac{q}{p}} \\ &\leq c \sum_{k=0}^{\infty} \sum_{j=0}^{k} 2^{(j-k)(M-\varepsilon-s)q} 2^{jsq} \|\varphi_j^* f \mid L_p(R_n)\|^q \leq c' \sum_{j=0}^{\infty} 2^{jsq} \|\varphi_j^* f \mid L_p(R_n)\|^q \end{aligned} \tag{11}$$

and

$$\begin{aligned} &\sum_{k=0}^{\infty} 2^{skq} \left(\sum_{j=k+1}^{\infty} \|\varphi_j^* f \mid L_p(R_n)\|^p \right)^{\frac{q}{p}} \\ &\leq c \sum_{k=0}^{\infty} \sum_{j=k+1}^{\infty} 2^{(k-j)(s-\varepsilon)q} 2^{jsq} \|\varphi_j^* f \mid L_p(R_n)\|^q \leq c' \sum_{j=0}^{\infty} 2^{jsq} \|\varphi_j^* f \mid L_p(R_n)\|^q . \end{aligned} \tag{12}$$

We put (11) and (12) in (10). Then (5) is a consequence of (2.3.6/19) and the above remarks about $\|f \mid L_p(R_n)\|$.

Step 2. Let $f\in B_{p,q}^s(R_n)$. By Theorem 2.5.3(i) and (2) we have

$$\begin{aligned} \|f \mid B_{p,q}^s(R_n)\| &\leq c\|f \mid L_p(R_n)\| + c\left[\sum_{k=0}^{\infty} 2^{skq} E_p(2^k, f)^q \right]^{\frac{1}{q}} \\ &\leq c\|f \mid L_p(R_n)\| + c' \left[\sum_{k=0}^{\infty} 2^{skq} \sup_{|\varrho|\leq 2^{-k}} \|\Delta_\varrho^M f \mid L_p(R_n)\|^q \right]^{\frac{1}{q}} \\ &\leq c''\|f \mid B_{p,q}^s(R_n)\|_M^{(1)} . \end{aligned} \tag{13}$$

(5) and (13) prove that $\|f \mid B_{p,q}^s(R_n)\|_M^{(1)}$ is an equivalent quasi-norm on $B_{p,q}^s(R_n)$.

Step 3. In order to prove that $\|f \mid B^s_{p,q}(R_n)\|^{(2)}_M$ is also an equivalent quasi-norm in $B^s_{p,q}(R_n)$, we begin with

$$\begin{aligned} I &= \int_{R_n} |h|^{-sq} \sup_{|\varrho|\leq|h|} \|\Delta^M_\varrho f \mid L_p(R_n)\|^q \frac{dh}{|h|^n} \\ &\leq \int_{R_n} |h|^{-sq} \sup_{\frac{|h|}{2}\leq|\varrho|\leq|h|} \|\Delta^M_\varrho f \mid L_p(R_n)\|^q \frac{dh}{|h|^n} \\ &\quad + \int_{R_n} |h|^{-sq} \sup_{|\varrho|\leq\frac{|h|}{2}} \|\Delta^M_\varrho f \mid L_p(R_n)\|^q \frac{dh}{|h|^n} \\ &\leq \int_{R_n} |h|^{-sq} \sup_{\frac{|h|}{2}\leq|\varrho|\leq|h|} \|\Delta^M_\varrho f \mid L_p(R_n)\|^q \frac{dh}{|h|^n} + 2^{-sq} I\,. \end{aligned} \tag{14}$$

This shows that I can be estimated from above by the first term on the right-hand side of (14) (modification if $q=\infty$). Hence, the right-hand side of (3) is also an equivalent quasi-norm in $B^s_{p,q}(R_n)$ if one replaces $\sup\limits_{|\varrho|\leq|h|}$ by $\sup\limits_{\frac{|h|}{2}\leq|\varrho|\leq|h|}$. Let $h\in R_n$ and $\varrho\in R_n$ be given with $\frac{|h|}{2}\leq|\varrho|\leq|h|$. Let K be the indicated ball (cf. the figure).

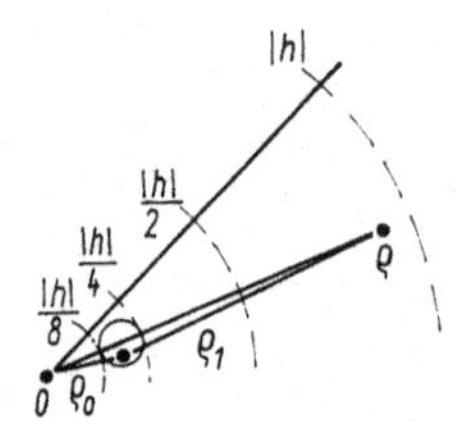

Fig. 2.5.12.

Let $\varrho=\varrho_0+\varrho_1$ with $\varrho_0\in K$. Then

$$e^{i\varrho\xi}-1=e^{i\varrho_0\xi}\,(e^{i\varrho_1\xi}-1)+e^{i\varrho_0\xi}-1\,. \tag{15}$$

We raise (15) to the power $2M$, where M is a natural number, and use the corresponding counterpart of (2.5.9/44) and (2.5.9/45) to obtain

$$\|\Delta^{2M}_\varrho f \mid L_p(R_n)\|^q \leq c\|\Delta^M_{\varrho_0} f \mid L_p(R_n)\|^q + c\|\Delta^M_{\varrho_1} f \mid L_p(R_n)\|^q \tag{16}$$

(modification if $q=\infty$). Integration over K yields

$$\begin{aligned} \|\Delta^{2M}_\varrho f \mid L_p(R_n)\|^q &\leq c \int_{\frac{|h|}{8}\leq|\lambda|\leq|h|} \|\Delta^M_\lambda f \mid L_p(R_n)\|^q \, d\lambda \cdot |h|^{-n} \\ &\leq c' \int_{\frac{1}{8}\leq|\lambda|\leq 1} \|\Delta^M_{\lambda|h|} f \mid L_p(R_n)\|^q d\lambda\,, \end{aligned} \tag{17}$$

where c and c' are independent of ϱ and h. We take the supremum with respect to ϱ, where $\frac{|h|}{2}\leq|\varrho|\leq|h|$, multiply by $|h|^{-sq-n}$, and integrate over $h\in R_n$. Then it follows by standard arguments that

$$\begin{aligned} &\int_{R_n} |h|^{-sq} \sup_{\frac{|h|}{2}\leq|\varrho|\leq|h|} \|\Delta^{2M}_\varrho f \mid L_p(R_n)\|^q \frac{dh}{|h|^n} \\ &\leq c \int_{R_n} |h|^{-sq} \|\Delta^M_h f \mid L_p(R_n)\|^q \frac{dh}{|h|^n}\,. \end{aligned} \tag{18}$$

Together with the above remarks this proves that $\|f \mid B^s_{p,q}(R_n)\|^{(1)}_{2M}$ can be estimated from above by $c\|f \mid B^s_{p,q}(R_n)\|^{(2)}_M$. Furthermore.

$$\|f \mid B^s_{p,q}(R_n)\|^{(2)}_M \leqq \|f \mid B^s_{p,q}(R_n)\|^{(1)}_M$$

is obvious. Now Step 2 proves that $\|f \mid B^s_{p,q}(R_n)\|^{(2)}_M$ is an equivalent quasi-norm on $B^s_{p,q}(R_n)$.

Remark 2. As in Remark 2.5.10/1 and Remark 2.5.11/2 one can try to replace differences in (3) and (4) by derivatives. Let $0<p\leqq\infty$, $0<q\leqq\infty$ and

$$s=\sigma+m, \quad \tilde{\sigma}_p<\sigma\leqq\tilde{\sigma}_p+1 \quad \text{and} \quad m=0, 1, 2, \ldots \tag{19}$$

Then (4) and (2.3.8/3) show that

$$\sum_{|\alpha|\leqq m} \|D^\alpha f \mid L_p(R_n)\| + \sum_{|\alpha|\leqq m} \left(\int_{R_n} |h|^{-\sigma q} \|\Delta^M_h D^\alpha f \mid L_p(R_n)\|^q \frac{dh}{|h|^n} \right)^{\frac{1}{q}} \tag{20}$$

(modification if $q=\infty$) is an equivalent quasi-norm in $B^s_{p,q}(R_n)$. The smallest number M is either $[\tilde{\sigma}_p]+1$ or $[\tilde{\sigma}_p]+2$.

Remark 3. As a matter of fact, we proved a bit more than was stated in the theorem. We have by Step 1 that both $\|f \mid B^s_{p,q}(R_n)\|^{(1)}_M$ and $\|f \mid B^s_{p,q}(R_n)\|^{(2)}_M$ are finite if $f\in B^s_{p,q}(R_n)$. Conversely, let $f\in L_p(R_n)\cap S'(R_n)$ such that $\|f \mid B^s_{p,q}(R_n)\|^{(2)}_M<\infty$. On the basis of Step 3 one can prove that $\|f \mid B^s_{p,q}(R_n)\|^{(1)}_{2M}<\infty$ and by Step 2 it follows $f\in B^s_{p,q}(R_n)$ (under the above restrictions for p, q, s and M), cf. Theorem 2.5.3 (i). Similarly for the quasi-norm (20), cf. Theorem 2.3.8 (i). We claim that $f\in L_p(R_n)\cap S'(R_n)$ in the above hypotheses can be replaced by $f\in L_p(R_n)$. This is trivial if $1\leqq p\leqq\infty$. Let $0<p<1$, $f\in L_p(R_n)$ and $\|f \mid B^s_{p,q}(R_n)\|^{(2)}_M<\infty$. By Step 3 of the proof and the Proposition we have

$$A=\left(\sum_{k=0}^{\infty} 2^{skq} \|f-a_k \mid L_p(R_n)\|^q\right)^{\frac{1}{q}} \leqq c\|f \mid B^s_{p,q}(R_n)\|^{(2)}_M<\infty ,$$

where $\{a_k(x)\}_{k=0}^\infty \in \mathfrak{A}_p(R_n)$, cf. (2.5.3/4), is an appropriate sequence. Then it follows from (1.3.2/5) (and a completion argument) that

$$\|f \mid L_1(R_n)\| \leqq \|a_0 \mid L_1(R_n)\| + \sum_{k=0}^{\infty} \|a_{k+1}-a_k \mid L_1(R_n)\|$$

$$\leqq c\|a_0 \mid L_p(R_n)\| + c \sum_{k=0}^{\infty} 2^{k\tilde{\sigma}_p} \|a_{k+1}-a_k \mid L_p(R_n)\| \leqq c'\|f \mid L_p(R_n)\| + c'A .$$

This proves $f\in L_1(R_n)$, and consequently $f\in S'(R_n)$. Hence, the theorem (and, in a similar way, Remark 2 also) can be strengthened in the following way: *let $0<p\leqq\infty$, $0<q\leqq\infty$ and $s>\tilde{\sigma}_p$. Then $f\in B^s_{p,q}(R_n)$ if and only if $f\in L_p(R_n)$ and $\|f \mid B^s_{p,q}(R_n)\|^{(2)}_M<\infty$ with $M>s$.* Similarly for $\|f \mid B^s_{p,q}(R_n)\|^{(1)}_M$.

Corollary. (i) If $s>0$, $1\leqq p<\infty$ and $1\leqq q\leqq\infty$, then

$$\Lambda^s_{p,q}(R_n)=B^s_{p,q}(R_n) . \tag{21}$$

(ii) If $s>0$, then

$$\mathcal{C}^s(R_n)=B^s_{\infty,\infty}(R_n) . \tag{22}$$

Proof. Concerning the definitions of $\Lambda^s_{p,q}(R_n)$ and $\mathcal{C}^s(R_n)$ we refer to (2.2.2/9), (2.2.2/10), and (2.2.2/6). In these cases we have $\tilde{\sigma}_p=0$ and (19) coincides essentially with (2.2.2/4). Then $M=2$ is an admissible number in (20). The rest follows from (20) and the considerations in Remark 3.

Remark 4. The above corollary coincides with Theorem 2.5.7, where we proved (21) and (22) via interpolation theory. In 2.5.7. we made some additional comments. In particular, we mention that (2.2.2/18) is also an immediate consequence of (20) and Remark 3.

2.5.13. Fubini Type Theorems

If $1<p<\infty$ and $m=0, 1, 2, \ldots$, then $W_p^m(R_n)=F_{p,2}^m(R_n)$ are the usual Sobolev spaces, cf. Theorem 2.5.6. Furthermore, the equivalent norm in (2.5.6/4) shows that

$$\sum_{j=1}^{n} \| \|f(x_1, \ldots, x_{j-1}, \cdot, x_{j+1}, \ldots, x_n) \mid W_p^m(R_1)\| \, L_p(R_{n-1}) \| \tag{1}$$

is also an equivalent norm in $W_p^m(R_n)$. The inner norm

$$\|f(x_1, \ldots, x_{j-1}, \cdot, x_{j+1}, \ldots, x_n) \mid W_p^m(R_1)\|$$

means that $f(x_1, \ldots, x_j, \ldots, x_n)$ is considered as a function of x_j, where $x_1, \ldots, x_{j-1}$, $x_{j+1}, \ldots, x_n$ are fixed. The $W_p^m(R_1)$-norm (2.5.6/4) (with $n=1$) of this function is taken with respect to x_j, the result being a function of $x_1, \ldots, x_{j-1}, x_{j+1}, \ldots, x_n$, and one has to take the $L_p(R_{n-1})$-norm with respect to these $n-1$ variables. (1) says that the $W_p^m(R_n)$-norm can be reduced to $W_p^m(R_1)$-norms. This is reminiscent of Fubini's theorem for integrable functions of several variables with a similar interpretation: if $f \in W_p^m(R_n)$, then

$$\|f(x_1, \ldots, x_{j-1}, \cdot, x_{j+1}, \ldots, x_n) \mid W_p^m(R_1)\|$$

is finite a. e. in R_{n-1} and p-integrable $\left(\text{the derivatives } \frac{\partial^m f}{\partial x_j^m} \text{ must always be understood as elements of } S'(R_n)\right)$. What about corresponding assertions for $B_{p,q}^s(R_n)$ and $F_{p,q}^s(R_n)$? If $s>\tilde{\sigma}_p=n\left(\frac{1}{\min(p, 1)}-1\right)$, then $f \in B_{p,q}^s(R_n)$ or $f \in F_{p,q}^s(R_n)$ is a regular distribution, cf. Remark 2.5.3/1. In particular, the above question makes sense. On the other hand, Theorem 2.5.12 shows that one cannot expect an assertion of type (1) (with $B_{p,q}^s$ instead of W_p^m) for $B_{p,q}^s(R_n)$ if $p \neq q$. So we restrict our attention to $B_{p,p}^s(R_n)$ with $0<p\leqq\infty$ and $F_{p,q}^s(R_n)$ with $0<p<\infty$ and $0<q\leqq\infty$ (recall that $B_{p,p}^s(R_n)=F_{p,p}^s(R_n)$ if $0<p<\infty$).

Definition. (i) Let $0<p\leqq\infty$ and $s>\tilde{\sigma}_p$. Then $B_{p,p}^s(R_n)$ has the Fubini property if

$$\sum_{j=1}^{n} \| \|f(x_1, \ldots, x_{j-1}, \cdot, x_{j+1}, \ldots, x_n) \mid B_{p,p}^s(R_1)\| \, L_p(R_{n-1}) \| \tag{2}$$

is an equivalent quasi-norm in $B_{p,p}^s(R_n)$.

(ii) Let $0<p<\infty$, $0<q\leqq\infty$ and $s>\frac{n}{\min(p, q)}$. Then $F_{p,q}^s(R_n)$ has the Fubini property if

$$\sum_{j=1}^{n} \| \|f(x_1, \ldots, x_{j-1}, \cdot, x_{j+1}, \ldots, x_n) \mid F_{p,q}^s(R_1)\| \, L_p(R_{n-1}) \| \tag{3}$$

is an equivalent quasi-norm in $F_{p,q}^s(R_n)$.

Remark 1. In order to avoid difficulties about the interpretation of (2) and (3) we assume that $\|f(\ldots) \mid B_{p,p}^s(R_1)\|$ and $\|f(\ldots) \mid F_{p,q}^s(R_1)\|$ have the meaning of (2.5.12/4) and (2.5.10/2), respectively (with $n=1$). This makes sense. Of course, there is no difficulty in extending part (ii) of the definition to $s>\tilde{\sigma}_{p,q}$ if one interprets $\|f(\ldots) \mid F^s_{p,q}(R_1)\|$ via Theorem 2.5.11. Finally, in order to have explicit descriptions of (2) and (3) we introduce the differences

$$(\Delta^1_{h,j}f)(x)=f(x_1,\ldots,x_{j-1},x_j+h,x_{j+1},\ldots,x_n)-f(x),\ (\Delta^l_{h,j}f)(x)=\Delta^1_{h,j}(\Delta^{l-1}_{h,j}f)(x) \tag{4}$$

where $h\in R_1$, $x\in R_n$ and $l=2, 3, \ldots$, cf. (2.2.2/5). Then (2) means essentially that

$$\|f \mid L_p(R_n)\| + \sum_{j=1}^n \left\| \int_{R_1} |h|^{-sp} \|(\Delta^M_{h,j}f)(x_1,\ldots,x_{j-1},\cdot,x_{j+1},\ldots,x_n) \mid L_p(R_1)\|^p \frac{dh}{|h|} \mid L_p(R_{n-1}) \right\| \tag{5}$$

is an equivalent quasi-norm in $B^s_{p,p}(R_n)$, provided that $M>s$. Similarly for (3).

Theorem. (i) Let $0<p\leqq\infty$, $0<q\leqq\infty$ and $s>\tilde{\sigma}_p=n\left(\frac{1}{\min(p,1)}-1\right)$. If M is an integer such that $M>s$, then

$$\|f \mid L_p(R_n)\| + \sum_{j=1}^n \left(\int_{R_1} |h|^{-sq} \|\Delta^M_{h,j}f \mid L_p(R_n)\|^q \frac{dh}{|h|} \right)^{\frac{1}{q}} \tag{6}$$

is an equivalent quasi-norm on $B^s_{p,q}(R_n)$ (modifcation if $q=\infty$). Furthermore, $B^s_{p,p}(R_n)$ has the Fubini property.
(ii) Let $0<p<\infty$, $0<q<\infty$ and $s>\max\left(\frac{n}{\min(p,q)}, 4+\frac{3}{\min(p,q)}\right)$. If M is an integer such that $M>s$, then

$$\|f \mid L_p(R_n)\| + \sum_{j=1}^n \left\| \left(\int_{R_1} |h|^{-sq} |(\Delta^M_{h,j}f)(\cdot)|^q \frac{dh}{|h|} \right)^{\frac{1}{q}} \mid L_p(R_n) \right\| \tag{7}$$

is an equivalent quasi-norm on $F^s_{p,q}(R_n)$, and $F^s_{p,q}(R_n)$ has the Fubini property.

Proof. Step 1. The second term on the right-hand side of (2.5.12/3) can be written as

$$\left(\int_0^\infty r^{-sq} \sup_{|\varrho|\leqq r} \|\Delta^M_\varrho f \mid L_p(R_n)\|^q \frac{dr}{r} \right)^{\frac{1}{q}}. \tag{8}$$

Then it follows immediately that the quasi-norm in (6) can be estimated from above by $c\|f \mid B^s_{p,q}(R_n)\|^{(1)}_M$ from Theorem 2.5.12, which is an equivalent quasi-norm in $B^s_{p,q}(R_n)$. In a similar way, the quasi-norm in (7) can be estimated from above by $c\|f \mid F^s_{p,q}(R_n)\|^{(1)}_M$ from Theorem 2.5.10.

Step 2. We prove that there exists a constant c such that

$$\left(\int_{R_n} |h|^{-sq} \|\Delta^{Mn}_h f \mid L_p(R_n)\|^q \frac{dh}{|h|^n} \right)^{\frac{1}{q}} \leqq c \sum_{j=1}^n \left(\int_{R_1} |h|^{-sq} \|\Delta^M_{h,j}f \mid L_p(R_n)\|^q \frac{dh}{|h|} \right)^{\frac{1}{q}} \tag{9}$$

holds for all $f\in B^s_{p,q}(R_n)$. Of course, $0<p\leqq\infty$, $0<q\leqq\infty$, $s>\tilde{\sigma}_p$ and $M>s$. We use the same technique as in (2.5.9/44) and (2.5.9/45). Let $h=(h_1,\ldots,h_n)$ and $\xi=(\xi_1,\ldots,\xi_n)$. Then

$$e^{ih\xi}-1=e^{ih_1\xi_1}-1+e^{ih_1\xi_1}(e^{ih_2\xi_2}-1)+\ldots + e^{i(h_1\xi_1+\ldots+h_{n-1}\xi_{n-1})}(e^{ih_n\xi_n}-1). \tag{10}$$

We raise (10) to the power nM and apply the counterpart of (2.5.9/45). Then

$$\|\Delta_h^{Mn} f \mid L_p(R_n)\| \leqq c \sum_{j=1}^{n} \|\Delta_{h_j,j}^M f \mid L_p R_n)\| . \tag{11}$$

Furthermore, we have

$$\begin{aligned} &\int_{R_n} |h|^{-sq} \|\Delta_{h_1,1}^M f \mid L_p(R_n)\|^q \frac{dh}{|h|^n} \\ &= \int_{R_1} \|\Delta_{h_1,1}^M f \mid L_p(R_n)\|^q \left(\int_{R_{n-1}} (h_1^2 + h_2^2 + \ldots + h_n^2)^{-\frac{1}{2}(sq+n)} dh_2 \ldots dh_n \right) dh_1 \\ &= c \int_{R_1} |h_1|^{-sq} \|\Delta_{h_1,1}^M f \mid L_p(R_n)\|^q \frac{dh_1}{|h_1|} . \end{aligned} \tag{12}$$

Now (11) and (12) prove (9). By (9) and Theorem 2.5.12 we have that $\|f \mid B_{p,q}^s(R_n)\|$ can be estimated from above by the quasi-norm in (6). This statement and Step 1 prove that (6) is an equivalent quasi-norm in $B_{p,q}^s(R_n)$. If $p=q$, one can change the order of integration in (6) and obtain (5). This proves that $B_{p,p}^s(R_n)$ has the Fubini property. The proof of (i) is complete.

Step 3. We prove that $\|f \mid F_{p,q}^s(R_n)\|$ can be estimated from above by (7) (multiplied by a positive constant). For that purpose we modify the calculations in Step 2 of the proof of Theorem 2.5.9. We use the decomposition of R_n in the cubes $P_{k,t}$ from 2.5.4. Let $\{\varphi_{j,t}(x)\}_{\substack{j=0,1,\ldots \\ t=1,\ldots,T}} \subset S(R_n)$ be the counterpart of the system from Definition 2.3.1/1:

$$\operatorname{supp} \varphi_{k,t} \subset \cup \bar{P}_{l,\tau} ,$$

where the union is taken over all pairs (l, τ) such that $\bar{P}_{l,\tau} \cap \bar{P}_{k,t} \neq \emptyset$; for every multi-index α there exists a positive number c_α such that

$$2^{j|\alpha|} |D^\alpha \varphi_{j,t}(x)| \leqq c_\alpha \quad \text{for any} \quad j=0, 1, 2, \ldots; \quad \text{any} \quad t=1, \ldots, T; \\ \text{any} \quad x \in R_n;$$

and

$$\sum_{k=0}^{\infty} \sum_{t=1}^{T} \varphi_{k,t}(x) \equiv 1 \quad \text{for any} \quad x \in R_n .$$

By the same proof as in Proposition 2.3.2/1 it follows that

$$\left\| \left(\sum_{k=0}^{\infty} \sum_{t=1}^{T} 2^{skq} |F^{-1} \varphi_{k,t} F f|^q \right)^{\frac{1}{q}} \mid L_p(R_n) \right\| \tag{13}$$

is an equivalent quasi-norm in $F_{p,q}^s(R_n)$ for all values s, p, q with $-\infty < s < \infty$, $0 < p < \infty$ and $0 < q \leqq \infty$. Now we return to the original problem. If $j=1, \ldots, n$, then

$$b^j(x) = \int_1^2 (e^{i\tau x_j} - 1)^M d\tau, \quad x = (x_1, \ldots, x_n) \in R_n$$

are the counterparts of the function $b(x)$ from (2.5.9/24). We have

$$(F^{-1} b^j F f)(x) = \int_1^2 (\Delta_{\tau,j}^M f)(x) \, d\tau . \tag{14}$$

If t with $t=1, \ldots, T$ is fixed, then we find a number $j\,(=j(t))$ such that b^j has the same properties with respect to $\varphi_{k,t}$ as b with respect to ψ_k in (2.5.9/23). We thus obtain the counterparts of (2.5.9/26) and (2.5.9/27). The restrictions for p, q, and s in these calculations are given by $0<p<\infty$, $0<q\leqq\infty$ and $s>\tilde{\sigma}_p$. Under the same restrictions on the parameters, we carry over the arguments from Step 3 of the proof of Theorem 2.5.9 (with obvious modifications) to obtain the counterpart of (2.5.9/38). By the one-dimensional case of Theorem 2.5.9. $\left(\text{here we use } s>G_{p,q}=4+\frac{3}{\min(p,q)}\right)$ we have

$$\|f \mid F^s_{p,q}(R_n)\| \leqq c \sum_{j=1}^{n} \big\| \|f(x_1, \ldots, x_{j-1}, \cdot, x_{j+1}, \ldots, x_n) \mid F^s_{p,q}(R_1)\| L_p(R_{n-1})\big\|. \quad (15)$$

Because $s>\frac{n}{\min(p,q)}\geqq\frac{1}{\min(p,q)}$, Theorem 2.5.10 may be applied with $n=1$. This shows that the right-hand side of (15) can be estimated from above by the quasi-norm (7) (multiplied by a positive constant). This proves that (7) is an equivalent quasi-norm in $F^s_{p,q}(R_n)$. The Fubini property of $F^s_{p,q}(R_n)$ is an obvious consequence of (7).

Remark 2. The last arguments of the proof are valid for $f\in S(R_n)$. On the other hand, $S(R_n)$ is dense in $F^s_{p,q}(R_n)$, because $q<\infty$, cf. Theorem 2.3.3 (ii). The rest is a matter of completion. Only for that reason did we exclude $q=\infty$. However, there is little doubt that part (ii) of the theorem is valid for $q=\infty$, too. Furthermore, it should be possible to improve the unnatural restrictions for s. However, the main problem is to show that $F^s_{p,q}(R_n)$ has the Fubini property if $0<p<\infty$, $0<q\leqq\infty$ and $s>\tilde{\sigma}_{p,q}=n\left(\frac{1}{\min(p,q,1)}-1\right)$. It is not complicated to see that this conjecture is valid for the Bessel-potential spaces $H^s_p(R_n)=F^s_{p,2}(R_n)$ with $1<p<\infty$ and $s>0$, cf. also (1). The above theorem is an improvement of a corresponding statement in H. Triebel [10].

2.6. Fourier Multipliers

2.6.1. Definitions and Preliminaries

We have considered several times Fourier multipliers for the spaces $B^s_{p,q}(R_n)$ and $F^s_{p,q}(R_n)$ and obtained fairly definitive final theorems of Michlin-Hörmander type: let again

$$\sigma_p=n\left(\frac{1}{\min(p,1)}-\frac{1}{2}\right) \quad \text{and} \quad \sigma_{p,q}=n\left(\frac{1}{\min(p,q,1)}-\frac{1}{2}\right), \quad (1)$$

where $0<p\leqq\infty$ and $0<q\leqq\infty$, cf. Theorem 1.5.2 and (1.6.4/1). Let $h^\varkappa_2(R_n)$ be the space from (2.4.8/4). Let $-\infty<s<\infty$. If $0<p\leqq\infty$, $0<q\leqq\infty$ and $\varkappa>\sigma_p$, then

$$\|F^{-1}mFf \mid B^s_{p,q}(R_n)\| \leqq c\|m \mid h^\varkappa_2(R_n)\|\, \|f \mid B^s_{p,q}(R_n)\| \quad (2)$$

for all $m\in h^\varkappa_2(R_n)$ and all $f\in B^s_{p,q}(R_n)$; if $0<p<\infty$, $0<q<\infty$ and $\varkappa>\sigma_{p,q}$, then

$$\|F^{-1}mFf \mid F^s_{p,q}(R_n)\| \leqq c\|m \mid h^\varkappa_2(R_n)\|\, \|f \mid F^s_{p,q}(R_n)\| \quad (3)$$

for all $m\in h^\varkappa_2(R_n)$ and all $f\in F^s_{p,q}(R_n)$. Of course, c in (2) and (3) is independent of m and f. (3) coincides with Theorem 2.4.8, and (2) is an immediate consequence of Remark 1.5.2/3 (and the definition of $B^s_{p,q}(R_n)$). These are quite natural statements which are completely sufficient for our purposes (even the weaker version in Theorem 2.3.7 was of great service). On the other hand, since the days of the milestone papers of S. G. Michlin [1, 2] and L. Hörmander [1], Fourier multipliers for function spaces have attracted much attention for their own sake. Hence it is quite natural to deal with Fourier multipliers for $B^s_{p,q}(R_n)$ and for $F^s_{p,q}(R_n)$ in some detail. However a reader who is not interested in such questions may skip Section 2.6. (the statements in this section will not be needed later).

Definition 2.3.7 of Fourier multipliers for $B^s_{p,q}(R_n)$ and $F^s_{p,q}(R_n)$, respectively, and also Remark 2.3.7/1, are not satisfactory for a detailed study of Fourier multipliers, because the a priori restrictions for Fourier multipliers $m(x)$ are too strong. Also, the assumption $m\in FL_1$ in Definition 1.5.1 is convenient but inadequate. For instance, the characteristic function of a cube in R_n is a Fourier multiplier in $L_p(R_n)=F^0_{p,2}(R_n)$ with $1<p<\infty$. However this important fact is covered neither by Definition 2.3.7 (and Remark 2.3.7/1) nor by Definition 1.5.1. Thus our first task will be to find more natural definitions for Fourier multipliers. With the exception of a few remarks in 2.6.6., we deal with Fourier multipliers for $L_p(R_n)$ with $1\leqq p\leqq\infty$ and for $B^s_{p,q}(R_n)$ with $-\infty<s<\infty$, $0<p\leqq\infty$ and $0<q\leqq\infty$. The reason is that we have some fairly definitive statements for these classes of Fourier multipliers, in contrast to the corresponding assertions for Fourier multipliers for $F^s_{p,q}(R_n)$.

Definition. (i) Let $-\infty<s<\infty$, $0<p\leqq\infty$, and $0<q\leqq\infty$. Then $m\in S'(R_n)$ is said to be a Fourier multiplier for $B^s_{p,q}(R_n)$ if there exists a positive number c such that

$$\|F^{-1}mFf \mid B^s_{p,q}(R_n)\|\leqq c\|f \mid B^s_{p,q}(R_n)\| \tag{4}$$

holds for all $f\in S(R_n)$.
(ii) Let $-\infty<s<\infty$, $0<p\leqq\infty$, and $0<q\leqq\infty$. Then $m\in S'(R_n)$ is said to be a homogeneous Fourier multiplier for $B^s_{p,q}(R_n)$ if there exists a positive number c such that

$$\|F^{-1}m(a\,\cdot)\,Ff \mid B^s_{p,q}(R_n)\|\leqq c\|f \mid B^s_{p,q}(R_n)\| \tag{5}$$

holds for all $a>0$ and all $f\in S(R_n)$.
(iii) Let $1\leqq p\leqq\infty$. Then $m\in S'(R_n)$ is said to be a Fourier multiplier for $L_p(R_n)$ if there exists a positive number c such that

$$\|F^{-1}mFf \mid L_p(R_n)\|\leqq c\|f \mid L_p(R_n)\| \tag{6}$$

holds for all $f\in S(R_n)$.

Remark 1. Because we assumed that $f\in S(R_n)$ in the above definition we have $F^{-1}mFf\in S'(R_n)$ in any case. Consequently, the definition makes sense. Furthermore, if $m\in S'(R_n)$, then $m(a\,\cdot)$ with $a>0$ has the usual meaning, i.e.

$$m(a\,\cdot)\,[\varphi]=a^{-n}m[\varphi(a^{-1}\cdot)] \quad \text{for all} \quad \varphi\in S(R_n)\,.$$

Finally, we recall the well-known fact that for $m\in S'(R_n)$ and $f\in S(R_n)$

$$F^{-1}mFf=c(F^{-1}m)*f=c(F^{-1}m)\,[f\,(x-\,\cdot)]\,,\quad x\in R_n\,, \tag{7}$$

is an infinitely differentiable function in R_n of at most polynomial growth, cf. E. M. Stein, G. Weiss [2, p. 23/24] and [I, p. 152].

Remark 2. Of course, there is no difficulty in defining (homogeneous) Fourier multipliers for the spaces $F^s_{p,q}(R_n)$, cf. 2.6.6.

Remark 3. It is clear what is meant by a homogeneous Fourier multiplier for $L_p(R_n)$ with $1\leqq p\leqq\infty$: one must replace $B^s_{p,q}(R_n)$ in part (ii) of the above definition by $L_p(R_n)$. But it is easy to see that every Fourier multiplier for $L_p(R_n)$ is automatically a homogeneous Fourier multiplier for $L_p(R_n)$, cf. (1.3.2/6), which is the basis of this statement. On the other hand, as we shall see later on, there exist nonhomogeneous Fourier multipliers for $B^s_{p,q}(R_n)$ with $p \neq 2$. However the notion of a homogeneous Fourier multiplier for $B^s_{p,q}(R_n)$ is only of interest for the values $1 \leq p \leq \infty$. Let $0<p<1$, $0<q\leq\infty$ and $-\infty<s<\infty$, and let m be a homogeneous Fourier multiplier for $B^s_{p,q}(R_n)$ with compact support, e.g. supp $m \subset \left\{y \mid |y| \leq \frac{1}{2}\right\}$. If $f \in S(R_n)$ with $(Ff)(x)=1$ when $|x|\leq 1$, then (5) yields

$$\sup_{a>1} \|F^{-1}m(a\,\cdot) \mid L_p(R_n)\| \leq c < \infty .$$

By (1.3.2/6) we have

$$c \geq \|F^{-1}m(a\,\cdot) \mid L_p(R_n)\| = a^{-n} a^{\frac{n}{p}} \|F^{-1}m \mid L_p(R_n)\|$$

for $a \geq 1$. If $a \to \infty$, it follows that $F^{-1}m=0$ and $m=0$. On the other hand, it is easy to see that $m(x)=e^{ibx}$ with $b \in R_n$ and $x \in R_n$ is a homogeneous Fourier multiplier for all spaces $B^s_{p,q}(R_n)$.

Remark 4. We restrict our attention to Fourier multipliers which map a space $B^s_{p,q}(R_n)$ or $L_p(R_n)$ into itself. Of course, one can generalize this problem and study Fourier multipliers which map one space of the above type into another space of the same type. Fourier multipliers for $L_p(R_n)$-spaces have a long history, cf. E. M. Stein [1], E. M. Stein, G. Weiss [2] or [I, 2.2]. Concerning Fourier multipliers for Besov spaces (and spaces of Hardy type) we refer to J. Peetre [8, Chapter 7], R. Johnson [1–3; 5] and T. Mizuhara [1–5]. It would be of interest to develop a more general theory of Fourier multipliers between (different) spaces $B^s_{p,q}(R_n)$ and $F^s_{p,q}(R_n)$ in the sense of the theory of this section and the above-cited papers.

Proposition. (i) Let $-\infty<s<\infty$, $0<p\leq\infty$ and $0<q\leq\infty$. If $m \in S'(R_n)$ is a Fourier multiplier for $B^s_{p,q}(R_n)$ then there exists a translation-invariant bounded linear operator from $B^s_{p,q}(R_n)$ into itself whose restriction to $S(R_n)$ coincides with $F^{-1}mF$. Conversely, for every translation-invariant bounded linear operator T from $B^s_{p,q}(R_n)$ into itself there exists a uniquely determined $m \in S'(R_n)$ such that the restriction of T to $S(R_n)$ coincides with $F^{-1}mF$. Furthermore, m is a Fourier multiplier for $B^s_{p,q}(R_n)$.
(ii) Let $1\leq p\leq\infty$. The statement of part (i) remains valid if $B^s_{p,q}(R_n)$ is replaced by $L_p(R_n)$.

Proof. (Outline). *Step 1.* It is clear what is meant by a translation-invariant operator in $S'(R_n)$. Let $(\tau_h f)(\cdot)=f(h+\cdot)$ (in the sense of distributions). Then a linear operator T is said to be translation-invariant if $T\tau_h=\tau_h T$ for every $h \in R_n$. We prove (ii). Let m be a Fourier multiplier for $L_p(R_n)$. Then $F^{-1}m\,Ff$ with $f \in S(R_n)$ is translation-invariant. If $1 \leq p<\infty$, then the completion of this operator has the desired properties. If $p=\infty$, then m is also a Fourier multiplier for $L_1(R_n)$ (this follows from a duality argument). Let T be the completion of $F^{-1}mFf$ in $L_1(R_n)$. Then the dual operator of T has the desired properties. Conversely, if T is a bounded linear translation-invariant operator in $L_p(R_n)$, it follows from L. Hörmander [1] or E. M. Stein, G. Weiss [2, pp. 26–28] that there exists a uniquely determined $m \in S'(R_n)$ such that $Tf=F^{-1}mFf$ for $f \in S(R_n)$. Then m is a Fourier multiplier for $L_p(R_n)$.

Step 2. We prove (i). Let m be a Fourier multiplier for $B^s_{p,q}(R_n)$. If $0<p<\infty$ and $0<q<\infty$, then $S(R_n)$ is dense in $B^s_{p,q}(R_n)$, cf. Theorem 2.3.3. The completion of $F^{-1}mF$ is therefore translation-invariant. We consider the limiting cases. Let m be a Fourier multiplier for $B^s_{p,\infty}(R_n)$ with $0<p<\infty$. Then it follows from Proposition 2.6.2 that m is a Fourier multiplier for all spaces $B^\sigma_{p,q}(R_n)$ with $-\infty<\sigma<\infty$ and $0<q\leq\infty$. The above completion argument (with $q<\infty$) and the interpolation formula (2.4.2/1) prove that there exists a translation-invariant bounded linear operator from $B^s_{p,\infty}(R_n)$ into itself whose restriction to $S(R_n)$ coincides with $F^{-1}mF$. Finally, let m be a Fourier multiplier for $B^s_{\infty,q}(R_n)$ with $0<q\leq\infty$. Then it follows from Proposition 2.6.2 and Theorem 2.6.3 that m is also a Fourier multiplier for $B^\sigma_{1,r}(R_n)$ with $-\infty<\sigma<\infty$

and $0<r\leqq\infty$. Using the duality formula (2.11.2/1), we see that the dual operator of the completion of $F^{-1}mF$ in $B^{\sigma}_{1,r}(R_n)$ has the desired properties, provided $1<q\leqq\infty$ in $B^{s}_{\infty,q}(R_n)$. If $0<q\leqq 1$, we use also (2.4.2/1). Conversely, if T is a translation-invariant bounded linear operator in $B^{s}_{p,q}(R_n)$ with $1\leqq p\leqq\infty$ and $1\leqq q\leqq\infty$, then it has been proved by M. H. Taibleson [1, II, pp. 824–825] that there exists a uniquely determined $m\in S'(R_n)$ such that the restriction of T to $S(R_n)$ coincides with $F^{-1}mF$. Of course, m is a Fourier multiplier for $B^{s}_{p,q}(R_n)$. It is not hard to see that the proof in E. M. Stein, G. Weiss [2, pp. 26–28] and M. H. Taibleson [1, II] can be extended to $0<p\leqq\infty$ and $0<q\leqq\infty$. The proof is therefore complete.

Remark 5. The last proposition shows that our definition of a Fourier multiplier is very natural.

2.6.2. The Classes $\mathfrak{M}_p$ and $\mathfrak{M}_p^H$

Proposition. Let $0<p\leqq\infty$, $0<q_0\leqq\infty$, $0<q_1\leqq\infty$, $-\infty<s_0<\infty$ and $-\infty<s_1<\infty$.
(i) m is a Fourier multiplier for $B^{s_0}_{p,q_0}(R_n)$ if and only if m is a Fourier multiplier for $B^{s_1}_{p,q_1}(R_n)$.
(ii) m is a homogeneous Fourier multiplier for $B^{s_0}_{p,q_0}(R_n)$ if and only if m is a homogeneous Fourier multiplier for $B^{s_1}_{p,q_1}(R_n)$.

Proof. Let m be a Fourier multiplier for $B^{s_0}_{p,q_0}(R_n)$. Let $\{\varphi_j(x)\}_{j=0}^{\infty}\in\Phi(R_n)$, cf. 2.3.1. Let $\psi_j(x)=\sum_{r=-1}^{1}\varphi_{j+r}(x)$ with $\varphi_{-1}=0$ Then $\psi_j(x)=1$ if $x\in\operatorname{supp}\varphi_j$. If $f\in S(R_n)$, we obtain

$$\begin{aligned}\|F^{-1}m\varphi_jFf\mid L_p(R_n)\|&=\|F^{-1}m\varphi_jFF^{-1}\psi_jFf\mid L_p(R_n)\|\\&\leqq 2^{-s_0j}\|F^{-1}mFF^{-1}\psi_jFf\mid B^{s_0}_{p,q_0}(R_n)\|\\&\leqq c2^{-s_0j}\|F^{-1}\psi_jFf\mid B^{s_0}_{p,q_0}(R_n)\|\leqq c'\sum_{r=-1}^{1}\|F^{-1}\varphi_{j+r}Ff\mid L_p(R_n)\|\,,\end{aligned}\tag{1}$$

where we have used (1.5.2/13). It follows immediately that m is also a Fourier multiplier for $B^{s_1}_{p,q_1}(R_n)$. This proves part (i). The proof of (ii) is the same.

Definition. (i) If $0<p\leqq\infty$, the set of all Fourier multipliers for $B^{s}_{p,q}(R_n)$ (with arbitrary real s and $0<q\leqq\infty$) is denoted by $\mathfrak{M}_p$.
(ii) If $0<p\leqq\infty$, the set of all homogeneous Fourier multipliers for $B^{s}_{p,q}(R_n)$ (with arbitrary real s and $0<q\leqq\infty$) is denoted by $\mathfrak{M}_p^H$.
(iii) If $1\leqq p\leqq\infty$, the set of all Fourier multipliers for $L_p(R_n)$ is denoted by M_p.

Remark. By the above proposition, the set of all (homogeneous) Fourier multipliers for $B^{s}_{p,q}(R_n)$ is independent of s and q. This justifies the above definition. A corresponding assertion for Fourier multipliers for $F^{s}_{p,q}(R_w)$ is not valid: the set of all Fourier multipliers for $F^{s}_{p,q}(R_n)$ is independent of s, but depends on q (and p), cf. 2.6.6.

2.6.3. Properties of the Classes $\mathfrak{M}_p$

We recall some well-known properties of the classes M_p, cf. e.g. E. M. Stein, G. Weiss [2, pp. 28–33] or E. M. Stein [1, pp. 94–95]:

$$M_p=M_{p'}\quad\text{if}\quad 1\leqq p\leqq\infty\quad\text{and}\quad\frac{1}{p}+\frac{1}{p'}=1\,,\tag{1}$$

$$M_p\subset M_q\subset M_2=L_\infty(R_n)\quad\text{if}\quad 1\leqq p\leqq q\leqq 2\,,\tag{2}$$

$$M_1=M_\infty=\{m\mid m=F\mu,\ \mu\ \text{finite complex Borel measure in}\ R_n\}\,.\tag{3}$$

The question arises whether $\mathfrak{M}_p$ has similar properties. If A is a subset of $S'(R_n)$, we put

$$F[A]=\{f \mid \exists g\in A \quad \text{such that} \quad f=Fg\}$$

(Fourier image of A). Furthermore, let $\mathfrak{M}_p^{\mathrm{com}}$ be the set of all $m\in\mathfrak{M}_p$ with compact support. Similarly M_p^{com}.

Theorem. (i) If $1\leqq p\leqq\infty$ and $\frac{1}{p}+\frac{1}{p'}=1$ then

$$\mathfrak{M}_p=\mathfrak{M}_{p'}. \tag{4}$$

(ii) If $0<p\leqq q\leqq 2$, then

$$\mathfrak{M}_p\subset\mathfrak{M}_q\subset\mathfrak{M}_2=M_2=L_\infty(R_n)\,. \tag{5}$$

(iii) The following statements hold:

$$F[B_{p,\infty}^{n(\frac{1}{p}-1)}(R_n)]=\mathfrak{M}_p \quad \text{if} \quad 0<p\leqq 1 \tag{6}$$

and

$$F[B_{1,\infty}^{0}(R_n)]\subset\mathfrak{M}_p\subset F[B_{p,\infty}^{n(\frac{1}{p}-1)}(R_n)] \quad \text{if} \quad 1\leqq p\leqq\infty\,. \tag{7}$$

(iv) If $1\leqq p\leqq\infty$, then

$$\mathfrak{M}_p^{\mathrm{com}}=M_p^{\mathrm{com}}\,. \tag{8}$$

Proof. Step 1. We prove that

$$\mathfrak{M}_p\subset F[B_{p,\infty}^{n(\frac{1}{p}-1)}(R_n)] \quad \text{if} \quad 0<p\leqq\infty\,. \tag{9}$$

Let $m\in\mathfrak{M}_p$ and $\{\varphi_j(x)\}_{j=0}^{\infty}\in\Phi(R_n)$, cf. 2.3.1. We may assume without loss of generality that $\varphi_j(x)=\varphi_1(2^{-j+1}x)$ if $j=1, 2, 3, \ldots$ Furthermore, $\psi_j(x)$ has the same meaning as in the proof of Proposition 2.6.2. We have

$$\begin{aligned}\|F^{-1}\varphi_jFF^{-1}m \mid L_p(R_n)\| &= \|F^{-1}\varphi_j mFF^{-1}\psi_j \mid L_p(R_n)\| \\ &\leqq \|F^{-1}mFF^{-1}\psi_j \mid B_{p,\infty}^{0}(R_n)\| \leqq c\|F^{-1}\psi_j \mid B_{p,\infty}^{0}(R_n)\| \\ &\leqq c'\sum_{r=-2}^{2}\|F^{-1}\psi_j\varphi_{j+r} \mid L_p(R_n)\| \leqq c''2^{jn}2^{-j\frac{n}{p}},\end{aligned} \tag{10}$$

where we have used

$$(F^{-1}\varphi_j)(x)=(F^{-1}\varphi_1(2^{-j}\cdot))(x)=2^{jn}(F^{-1}\varphi_1)(2^jx) \quad \text{if} \quad j=1, 2, \ldots$$

and a similar formula for $F^{-1}\psi_j\varphi_{j+r}$ if j is large. Now (10) proves (9).

Step 2. We prove that

$$F[B_{p,\infty}^{n(\frac{1}{p}-1)}(R_n)]\subset\mathfrak{M}_p \quad \text{if} \quad 0<p\leqq 1\,. \tag{11}$$

Let $F^{-1}m\in B_{p,\infty}^{n(\frac{1}{p}-1)}(R_n)$. Let $f\in S(R_n)$ and let $\{\varphi_j\}$ and $\{\psi_j\}$ be the systems from the first step. If $*$ stands for the convolution in the sense of 1.5.3., then

$$F^{-1}m\varphi_jFf=F^{-1}m\varphi_j\psi_jFf=F^{-1}m\varphi_j * F^{-1}\psi_jFf\,.$$

Remark 1.5.3/2 yields

$$\begin{aligned}&\|F^{-1}m\varphi_jFf \mid L_p(R_n)\| \\ &\leqq c2^{jn(\frac{1}{p}-1)}\|F^{-1}m\varphi_j \mid L_p(R_n)\|\,\|F^{-1}\psi_jFf \mid L_p(R_n)\| \\ &\leqq c'\|F^{-1}m \mid B_{p,\infty}^{n(\frac{1}{p}-1)}(R_n)\|\,\|F^{-1}\psi_jFf \mid L_p(R_n)\|\,,\end{aligned} \tag{12}$$

where c and c' are independent of j. But (12) shows that m belongs to $\mathfrak{M}_p$. This proves (11). Now (9) and (11) yield (6) and the right-hand side of (7).

Step 3. We prove that

$$F[B^0_{1,\infty}(R_n)]=\mathfrak{M}_\infty\,. \tag{13}$$

If $m\in F[B^0_{1,\infty}(R_n)]$, then (6) yields $m\in\mathfrak{M}_1$. We recall that $S(R_n)$ is dense in $B^0_{1,1}(R_n)$, cf. Theorem 2.3.3. Hence the completion of the operator $F^{-1}mF$, which is defined on $S(R_n)$, yields a linear and bounded mapping T from $B^0_{1,1}(R_n)$ into itself. We now use the duality formula

$$(B^0_{1,1}(R_n))'=B^0_{\infty,\infty}(R_n)\,,$$

cf. (2.11.2/1). The restriction of the dual operator T' of T to $S(R_n)$ coincides with FmF^{-1}. Hence, $m\in\mathfrak{M}_\infty$. This proves $F[B^0_{1,\infty}(R_n)]\subset\mathfrak{M}_\infty$. In order to establish the converse assertion, we assume that $m\in\mathfrak{M}_\infty$. Then m is a Fourier multiplier for $B^0_{\infty,\infty}(R_n)$. Let $f\in S(R_n)$ and let $\{\varphi_j\}$ and $\{\psi_j\}$ be the systems from the first step. We obtain

$$\begin{aligned}
&\Big|\int_{R_n}(F^{-1}m\varphi_j)(-y)\,f(y)\,dy\Big|=\Big|\int_{R_n}(m\varphi_j)(y)\,\psi_j(y)\,(Ff)(y)\,dy\Big|\\
&=\Big|\int_{R_n}(F^{-1}m\varphi_j)(-y)\,(F^{-1}\psi_jFf)(y)\,dy\Big|\\
&\le\Big\|\int_{R_n}(F^{-1}m\varphi_j)(x-y)\,(F^{-1}\psi_jFf)(y)\,dy\mid L_\infty(R_n)\Big\|\\
&=c\|F^{-1}m\varphi_jFF^{-1}\psi_jFf\mid L_\infty(R_n)\|\le c'\|F^{-1}mFF^{-1}\psi_jFf\mid B^0_{\infty,\infty}(R_n)\|\\
&\le c''\|F^{-1}\psi_jFf\mid B^0_{\infty,\infty}(R_n)\|\le C\|f\mid L_\infty(R_n)\|\,,
\end{aligned}\tag{14}$$

where we used the special structure of φ_j and ψ_j in the last estimate. Of course, c, c', c'' and C are independent of j (and f). Let $C_0(R_n)$ be the completion of $S(R_n)$ in the L_∞-norm. (14) shows that

$$f\to\int_{R_n}(F^{-1}m\varphi_j)(-y)\,f(y)\,dy$$

is a continuous linear functional on $C_0(R_n)$. By Riesz's famous representation theorem, cf. e.g. Z. Semadeni [1, p. 312], we have

$$\int_{R_n}(F^{-1}m\varphi_j)(-y)\,f(y)\,dy=\int_{R_n}f(y)\,d\mu_j\,,$$

where μ_j is a finite complex Radon measure on R_n such that $\|\mu_j\|\le C$, cf. (14). This proves that

$$\int_{R_n}|(F^{-1}m\varphi_j)(y)\mid dy=\|\mu_j\|\le C$$

and consequently $F^{-1}m\in B^0_{1,\infty}(R_n)$. The proof of (13) is therefore complete.

Step 4. We prove (4). $\mathfrak{M}_1=\mathfrak{M}_\infty$ follows from (13) and (6). Let $1<p<\infty$. Then

$$(B^0_{p,p}(R_n))'=B^0_{p',p'}(R_n)\quad\text{with}\quad\frac{1}{p}+\frac{1}{p'}=1\,, \tag{15}$$

cf. (2.11.2/1). The same duality argument as in the third step yields $\mathfrak{M}_p=\mathfrak{M}_{p'}$.

Step 5. We prove that

$$F[B^0_{1,\infty}(R_n)]\subset L_\infty(R_n)\,. \tag{16}$$

Let $F^{-1}m\in B^0_{1,\infty}(R_n)$ and $\{\varphi_j(x)\}_{j=0}^{\infty}\in\Phi(R_n)$. We have

$$\|\varphi_j m \mid L_\infty(R_n)\|\leqq c\|F^{-1}\varphi_j m \mid L_1(R_n)\|\leqq c\|F^{-1}m \mid B^0_{1,\infty}(R_n)\| .$$

This proves (16). Furthermore, we remark that

$$B^{n\left(\frac{1}{p_0}-1\right)}_{p_0,r}(R_n)\subset B^{n\left(\frac{1}{p_1}-1\right)}_{p_1,r}(R_n) \quad \text{if} \quad 0<r\leqq\infty \quad \text{and} \quad 0<p_0\leqq p_1\leqq\infty. \tag{17}$$

This follows immediately from

$$\|F^{-1}\varphi_j Ff \mid L_{p_1}(R_n)\|\leqq c2^{jn\left(\frac{1}{p_0}-\frac{1}{p_1}\right)}\|F^{-1}\varphi_j Ff \mid L_{p_0}(R_n)\| \tag{18}$$

with $\{\varphi_j\}\in\Phi(R_n)$, cf. (1.3.2/5) (which is applicable after a completion argument in the sense of Theorem 1.4.1).

Step 6. We prove (5). If $0<p\leqq q\leqq 1$, then (5) is an obvious consequence of (6) and (17). Furthermore, because $B^0_{2,2}(R_n)=L_2(R_n)$ we have $\mathfrak{M}_2=M_2=L_\infty(R_n)$, cf. (2). Now $\mathfrak{M}_1\subset\mathfrak{M}_2$ is an immediate consequence of (6) and (16). Let $m\in\mathfrak{M}_1$. Then $m\in\mathfrak{M}_2$ and by the complex interpolation formula (2.4.7/2) or the real interpolation formula (2.4.3/2) (with $p_0=1$ and $p_1=2$) we obtain $m\in\mathfrak{M}_p$ with $1<p<2$ (cf. the following Remark 1 for some comments). Hence, $\mathfrak{M}_1\subset\mathfrak{M}_p$ if $1\leqq p\leqq 2$. Let $m\in\mathfrak{M}_p$ with $1<p<2$. By (4) and the complex interpolation formula

$$(B^0_{p,p}(R_n),\ B^0_{p',p'}(R_n))_{\frac{1}{2}}=B^0_{2,2}(R_n)=L_2(R_n)$$

(or its real counterpart (2.4.3/2)) we have $m\in\mathfrak{M}_2$. Hence, $\mathfrak{M}_p\subset\mathfrak{M}_2$. Again by interpolation it follows that $\mathfrak{M}_p\subset\mathfrak{M}_q\subset\mathfrak{M}_2$ if $p<q<2$. This proves (5). Formula (7) follows immediately from $\mathfrak{M}_1\subset\mathfrak{M}_p$ if $p\geqq 1$, (6) and (9).

Step 7. We prove (8). Let $\{\varphi_j\}_{j=0}^{\infty}\in\Phi(R_n)$. Then $\varphi_0(x)=1$ if $|x|\leqq 1$. Let $m\in S'(R_n)$ with $\operatorname{supp} m\subset\{y \mid |y|\leqq 1\}$. If $m\in M_p$ with $1\leqq p\leqq\infty$ and $f\in S(R_n)$, then

$$\begin{aligned}\|F^{-1}mFf \mid B^0_{p,p}(R_n)&=\|F^{-1}mFF^{-1}\varphi_0 Ff \mid L_p(R_n)\|\\ &\leqq c\|F^{-1}\varphi_0 Ff \mid L_p(R_n)\|\leqq c'\|f \mid B^0_{p,p}(R_n)\| .\end{aligned}$$

Hence, $m\in\mathfrak{M}_p$. Conversely, if $m\in\mathfrak{M}_p$ with $1\leqq p\leqq\infty$, then

$$\begin{aligned}\|F^{-1}mFf \mid L_p(R_n)\|&=\|F^{-1}mFF^{-1}\varphi_0 Ff \mid B^0_{p,p}(R_n)\|\\ &\leqq c\|F^{-1}\varphi_0 Ff \mid B^0_{p,p}(R_n)\|\\ &\leqq c'\|F^{-1}\varphi_0^2 Ff \mid L_p(R_n)\|+c'\|F^{-1}\varphi_0\varphi_1 Ff \mid L_p(R_n)\|\\ &\leqq c''\|f \mid L_p(R_n)\| .\end{aligned}$$

Hence, $m\in M_p$. If m is an arbitrary Fourier multiplier with compact support, then we apply the above considerations to $m(ax)$, where $a>0$ is an appropriate number. This proves (8).

Remark 1. We add some comments about the application of interpolation formulas in Step 6. For the real interpolation method we have the interpolation property from Proposition 2.4.1. If e.g. $m\in\mathfrak{M}_1$ then it follows from Proposition 2.6.1 that Proposition 2.4.1 is applicable to the completion of $F^{-1}mF$ as a bounded mapping from $B^0_{1,1}(R_n)$ into itself and from $B^0_{2,2}(R_n)=L_2(R_n)$ into itself, respectively. As a result, one obtains by real interpolation that the completion of $F^{-1}mF$ yields a bounded mapping from $B^0_{p,p}(R_n)$ into itself, $1<p<2$. Hence, $m\in\mathfrak{M}_p$. In Remark 2.4.4/2 and Remark 2.4.7/2 we mentioned briefly the classical complex interpolation method $[\cdot,\cdot]_\Theta$. This method has the interpolation property from Proposition 2.4.1. For the complex method from 2.4.4., the situation is a bit more complicated. If the involved operators are of the form $F^{-1}mF$ with an infinitely differentiable function m, then one again has the interpolation property. This follows immediately from the definitions in 2.4.4, cf. also Step 2 of the proof of Theorem 2.4.8. If m is not smooth (as may happen in the above application), we use the construction from Step 2 of the proof of Theorem 2.4.7 with $B^s_{p,q}(R_n)$ instead of $F^s_{p,q}(R_n)$. If $f\in S(R_n)$ (and this is sufficient), then the counterpart of $f(z)$ from (2.4.7/13) is simpler and belongs to $S(R_n)$. In any case $F^{-1}mFf(z)$ makes sense for $m\in S'(R_n)$, and it is an $S'(R_n)$-analytic function in the sense of 2.4.4. By Step 2 of Theorem 2.4.7 (and its modification with $B^s_{p,q}(R_n)$ instead of $F^s_{p,q}(R_n)$) this is sufficient to prove the interpolation property for $F^{-1}mF$.

Remark 2. Let $2\leqq p\leqq\infty$ and $\frac{1}{p}+\frac{1}{p'}=1$. Then (4), (7) and (17) prove that

$$\mathfrak{M}_p=\mathfrak{M}_{p'}\subset F[B^{n\left(\frac{1}{p'}-1\right)}_{p',\infty}(R_n)]\subset F[B^{n\left(\frac{1}{p}-1\right)}_{p,\infty}(R_n)]\,, \tag{19}$$

which strengthens (7).

Remark 3. $\mathfrak{M}_1=F[B^0_{1,\infty}(R_n)]$ is known. The periodic counterpart of this statement is due to A. Zygmund [2]. Further results for the periodic case may be found in T. Mizuhara [1–4]. The above non-periodic case has been proved by M. H. Taibleson [1, II, p. 827], cf. also J. Peetre [8, p. 249]. For the anisotropic counterpart we refer to W. R. Madych, N. M. Riviere [1, p. 372]. Formula (6) for all $0<p\leqq 1$ has also been proved by J. Peetre [8, p. 249]. Similar results for Fourier multipliers between different Besov spaces may be found in R. Johnson [5, Theorem 6] and D. G. Orlovskij [1]. This subsection is a modification of [S, 2.3.3].

2.6.4. Properties of the Classes $\mathfrak{M}^H_p$

The compact Fourier multipliers for $L_p(R_n)$ with $1\leqq p\leqq\infty$ coincide with the ones for $B^0_{p,q}(R_n)$, cf. (2.6.3/8). What about for arbitrary Fourier multipliers? It is the main goal of this subsection to prove that M_p coincides with $\mathfrak{M}^H_p$ (cf. Definition 2.6.2(ii)) and that there exists a function $m\in\mathfrak{M}_p$ which is not a Fourier multiplier for $L_p(R_n)$ (with the exception of the trivial case $p=2$). We always assume that $1\leqq p\leqq\infty$. This is natural, because the definition of M_p is restricted to these values and $\mathfrak{M}^H_p$ is not interesting if $p<1$, cf. Remark 2.6.1/3.

Proposition. Let $\psi(x)\in S(R_n)$ be a not identically vanishing function with $\operatorname{supp}\psi\subset\left\{y\mid |y|\leqq\frac{1}{2}\right\}$. Let $\sigma_k\in R_n$ with $|\sigma_k|=2^k$ if $k=0, 1, 2, \ldots$ and let

$$m(x)=\sum_{k=0}^{\infty} e^{i\sigma_k x}\psi(x-\sigma_k)\,. \tag{1}$$

Then

$$m\in\bigcap_{1\leqq p\leqq\infty}\mathfrak{M}_p \quad\text{and}\quad m\notin\bigcup_{\substack{1\leqq p\leqq\infty\\ p\neq 2}} M_p\,. \tag{2}$$

Proof. Step 1. We prove that $m\in\mathfrak{M}_p$ if $1\leqq p\leqq\infty$. Let $\{\varphi_j\}_{j=0}^{\infty}\in\Phi(R_n)$ be an appropriate system, cf. Definition 2.3.1/1. If $f\in S(R_n)$, then

$$\begin{aligned}\|F^{-1}\varphi_k mFf\mid L_p(R_n)\|&=\|F^{-1}e^{i\sigma_k x}\psi(x-\sigma_k)Ff\mid L_p(R_n)\|\\ &=\|F^{-1}\psi(\cdot-\sigma_k)FF^{-1}\varphi_kFf\mid L_p(R_n)\|\leqq c\|F^{-1}\varphi_kFf\mid L_p(R_n)\|\end{aligned} \tag{3}$$

if $1\leqq p\leqq\infty$ and $k=0, 1, 2, \ldots$ Of course, c in (3) is independent of k. This proves the first statement in (2).

Step 2. In order to prove the second statement in (2) it is sufficient to show that

$$m \notin M_p \quad \text{if} \quad 2 < p < \infty , \tag{4}$$

cf. (2.6.3/1) and (2.6.3/2). Let $N = 1, 2, \ldots$ We consider the function f_N which is given by

$$F f_N = \sum_{k=0}^{N} e^{-i\sigma_k x} \psi(x - \sigma_k) . \tag{5}$$

where ψ has the same meaning as in the first step. Then

$$|f_N(x)| \leq \sum_{k=0}^{N} |(F^{-1}\psi)(x - \sigma_k)| \leq c < \infty , \tag{6}$$

where c is independent of x and N. On the other hand,

$$\|f_N \mid L_2(R_n)\| = \|F f_N \mid L_2(R_n)\| = (N+1)^{\frac{1}{2}} \| \psi \mid L_2(R_n)\| . \tag{7}$$

If $2 < p < \infty$, then (6) and (7) yield

$$\|f_N \mid L_p(R_n)\| \leq \|f_N \mid L_\infty(R_n)\|^{1-\frac{2}{p}} \|f_N \mid L_2(R_n)\|^{\frac{2}{p}} \leq c N^{\frac{1}{p}} , \tag{8}$$

where c is independent of N. We recall that $L_p(R_n) = F^0_{p,2}(R_n)$ if $1 < p < \infty$, cf. (2.5.6/2). Then (2.3.1/6) with an appropriate system $\{\varphi_j(x)\}_{j=0}^{\infty} \in \Phi(R_n)$ yields

$$\begin{aligned} \|F^{-1} m F f_N \mid L_p(R_n)\| &= \left\| F^{-1}\left[\sum_{k=0}^{N} \psi^2(\cdot - \sigma_k)\right](\cdot) \mid F^0_{p,2}(R_n) \right\| \\ &= \left\| \left(\sum_{k=0}^{N} |F^{-1}\psi^2(\cdot - \sigma_k)|^2\right)^{\frac{1}{2}} \middle| L_p(R_n) \right\| = N^{\frac{1}{2}} \|F^{-1}\psi^2 \mid L_p(R_n)\| . \end{aligned} \tag{9}$$

Now we assume that m is a Fourier multiplier for $L_p(R_n)$ with $2 < p < \infty$. Then (8) and (9) give

$$N^{\frac{1}{2}} \leq c \|F^{-1} m F f_N \mid L_p(R_n)\| \leq c' \|f_N \mid L_p(R_n)\| \leq c'' N^{\frac{1}{p}} , \tag{10}$$

where c, c' and c'' are independent of N. This is a contradiction. The proof is therefore complete.

Remark 1. In particular, m is a Fourier multiplier for the Hölder-Zygmund spaces $\mathcal{C}^s(R_n)$ with $s > 0$, but it is not a Fourier multiplier for $L_p(R_n)$ with $1 \leq p \leq \infty$ and $p \neq 2$ (we recall that $\mathcal{C}^s(R_n) = B^s_{\infty,\infty}(R_n)$, cf. (2.5.7/6) or (2.5.12/22)). The first examples of this type were constructed by E. M. Stein, A. Zygmund [1]. Functions of type (1) have also been used in other contexts, cf. S. M. Nikol'skij [3, 9.7] and W. Littman, C. McCarthy, N. Riviere [1, p. 206].

Theorem. (i) If $1 \leq p \leq \infty$, then $M_p = \mathfrak{M}^H_p$.
(ii) If $1 \leq p \leq \infty$ and $p \neq 2$, then $\mathfrak{M}^H_p \neq \mathfrak{M}_p$.

Proof. Step 1. We prove that $M_p \subset \mathfrak{M}^H_p$ if $1 < p < \infty$. It follows immediately from (2.2.2/11) that m is a Fourier multiplier for $L_p(R_n)$ if and only if it is a Fourier multiplier for $H^s_p(R_n)$, where s is an arbitrary real number. Recall that $H^s_p(R_n) = F^s_{p,2}(R_n)$, cf. (2.5.6/2). In other words, if we assume that $m \in M_p$, then m is a Fourier multiplier for all spaces $F^s_{p,2}(R_n)$. The interpolation formula (2.4.2/2) shows that m also is a Fourier multiplier for $B^s_{p,q}(R_n)$, cf. Remark 2.6.3/1. Consequently, $m \in \mathfrak{M}_p$. If $m \in M_p$, then m is a homogeneous Fourier multiplier for $L_p(R_n)$, cf. Remark 2.6.1/3. Hence it is also a homogeneous Fourier multiplier for $H^s_p(R_n)$. Interpolation preserves this property, cf. (2.4.1/7). Consequently, $m \in \mathfrak{M}^H_p$. This proves $M_p \subset \mathfrak{M}^H_p$ if $1<p<\infty$. Me extend this inclusion to $p=1$ and $p=\infty$. If I_σ is the lifting operator from (2.3.8/1), we put

$$I_{-s}L_1(R_n)=\{f \mid f \in S'(R_n),\ \exists g \in L_1(R_n) \quad \text{with} \quad f=I_{-s}g\}\,.$$

Of course, $\|f \mid I_{-s}L_1(R_n)\|=\|g \mid L_1(R_n)\|$. If $m \in M_1$, then m is also a Fourier multiplier for $I_{-s}L_1(R_n)$. By (2.5.7/1), the lifting property from Theorem 2.3.8, and (2.4.2/1) (and obvious monotonicity properties of the real interpolation method), we have for $s \neq 0$

$$(I_{-s}L_1(R_n),\, L_1(R_n))_{\Theta,q}=B^{s(1-\Theta)}_{1,q}(R_n)\,.$$

Now the rest is the same as above. If $p=\infty$, we use (2.5.7/2) with $L_\infty(R_n)$ instead of $C(R_n)$ and argue in the same way as for $p=1$. This proves $M_p \subset \mathfrak{M}^H_p$ for all values $1 \leqq p \leqq \infty$.

Step 2. We prove $\mathfrak{M}^H_p \subset M_p$ if $1 \leqq p \leqq \infty$. Let $m \in \mathfrak{M}^H_p$. In order to prove $m \in M_p$ it will suffice to show that there exists a positive constant c such that

$$\|F^{-1}mFf \mid L_p(R_n)\| \leqq c\|f \mid L_p(R_n)\| \tag{11}$$

for all $f \in S(R_n)$ for which Ff has compact support. This follows from Definition 2.6.1 (iii) and the fact that

$$S_0(R_n)=\{f \mid f \in S(R_n),\ \text{supp}\ Ff\ \text{compact}\}$$

is dense in $S(R_n)$. Let $f \in S_0(R_n)$ and $a>0$. Then

$$\begin{aligned}\|F^{-1}mFf \mid L_p(R_n)\| &= a^{\frac{n}{p}}\,\|(F^{-1}mFf)\,(a\,\cdot) \mid L_p(R_n)\| \\ &= a^{\frac{n}{p}}\,\|F^{-1}m\,(a^{-1}\,\cdot)\,F[f\,(a\,\cdot)] \mid L_p(R_n)\|\,.\end{aligned} \tag{12}$$

If a is small enough, then the support of

$$F[f\,(a\,\cdot)](\xi)=a^{-n}(Ff)(a^{-1}\xi)$$

is contained in $\{y \mid |y| \leqq 1\}$. Hence it follows from (11) and the definition of $\mathfrak{M}^H_p$ that

$$\begin{aligned}&\|F^{-1}mFf \mid L_p(R_n)\| = a^{\frac{n}{p}}\,\|\,F^{-1}m\,(a^{-1}\,\cdot)\,F[f\,(a\,\cdot)] \mid B^0_{p,q}(R_n)\| \\ &\leqq c a^{\frac{n}{p}}\,\|f\,(a\,\cdot) \mid B^0_{p,q}(R_n)\| = c a^{\frac{n}{p}}\,\|f\,(a\,\cdot) \mid L_p(R_n)\| = c\|f \mid L_p(R_n)\|\,,\end{aligned}$$

where c is independent of a and $f \in S_0(R_n)$. This proves (11) and $m \in M_p$. Together with Step 1, we obtain part (i) of the theorem.

Step 3. Part (ii) follows from part (i) and the above proposition.

Remark 2. We again followed essentially the treatment in [S, 2.3.4 and 2.3.5]. Another characterization of M_p via $\mathfrak{M}_p$ has been given by R. Johnson [4]. Instead of multiplications $m(x) \to m(ax)$ with $a>0$ in the definition of homogeneous Fourier multipliers, he considered translations $m(x) \to m(x+b)$. For details we refer to the cited paper.

Remark 3. By Remark 2.6.1/3 and the above theorem there exist non-homogeneous Fourier multipliers for $B^s_{p,q}(R_n)$ if $0<p \leqq \infty$ and $p \neq 2$. For the spaces $L_p(R_n)$ with $1 \leqq p \leqq \infty$ the situation is different: every Fourier multiplier for $L_p(R_n)$ is a homogeneous Fourier multiplier.

2.6.5. Convolution Algebras

If $f \in L_1(R_n)$ and $g \in L_1(R_n)$ then the convolution

$$(f * g)(x) = \int_{R_n} f(x-y)\, g(y)\, dy \tag{1}$$

makes sense and belongs to $L_1(R_n)$. It is well-known that $L_1(R_n)$ is a Banach algebra if one takes the convolution as the multiplication. Recall that

$$f * g = c\, F^{-1}(Ff \cdot Fg), \quad f \in S(R_n) \quad \text{and} \quad g \in S(R_n) \tag{2}$$

where c is an immaterial positive constant. This shows that convolutions are closely related to Fourier multipliers. We ask the question whether some spaces $B^s_{p,q}(R_n)$ or $F^s_{p,q}(R_n)$ have properties similar to the Banach algebra $L_1(R_n)$.

Definition. (i) $B^s_{p,q}(R_n)$ with $-\infty < s < \infty$, $0 < p \leqq \infty$ and $0 < q \leqq \infty$ is said to be a convolution algebra if there exists a positive number c such that

$$\|f * g \mid B^s_{p,q}(R_n)\| \leqq c \|f \mid B^s_{p,q}(R_n)\| \cdot \|g \mid B^s_{p,q}(R_n)\| \tag{3}$$

holds for all $f \in S(R_n)$ and all $g \in S(R_n)$.
(ii) $F^s_{p,q}(R_n)$ with $-\infty < s < \infty$, $0 < p < \infty$ and $0 < q \leqq \infty$ is said to be a convolution algebra if there exists a positive number c such that

$$\|f * g \mid F^s_{p,q}(R_n)\| \leqq c \|f \mid F^s_{p,q}(R_n)\| \cdot \|g \mid F^s_{p,q}(R_n)\| \tag{4}$$

holds for all $f \in S(R_n)$ and all $g \in S(R_n)$.

Remark 1. The situation is similar to that for the Fourier multipliers, cf. Definition 2.6.1 and Proposition 2.6.1. If $p < \infty$ and $q < \infty$, then $S(R_n)$ is dense in $B^s_{p,q}(R_n)$ and in $F^s_{p,q}(R_n)$, cf. Theorem 2.3.3. If we assume that $B^s_{p,q}(R_n)$ or $F^s_{p,q}(R_n)$ with $p < \infty$ and $q < \infty$ is a convolution algebra in the above sense, then (3) or (4), respectively, can be extended by completion to the whole space. There is a similar situation in the limiting cases, too, cf. Remark 3. In other words, if $B^s_{p,q}(R_n)$ or $F^s_{p,q}(R_n)$ is such a convolution algebra which is covered by the following theorem, then (3) or (4), respectively, make sense for all f and g in the corresponding spaces. This shows that the above definition is reasonable.

Theorem. (i) Let $-\infty < s < \infty$, $0 < p \leqq \infty$ and $0 < q \leqq \infty$. Then $B^s_{p,q}(R_n)$ is a convolution algebra if and only if $0 < p \leqq 1$ and $s \geqq n\left(\frac{1}{p} - 1\right)$.
(ii) Let $-\infty < s < \infty$, $0 < p < \infty$ and $0 < q \leqq \infty$. If $F^s_{p,q}(R_n)$ is a convolution algebra, then $0 < p \leqq 1$ and $s \geqq n\left(\frac{1}{p} - 1\right)$. Conversely, if $0 < p \leqq 1$, $p \leqq q$ and $s \geqq n\left(\frac{1}{p} - 1\right)$, then $F^s_{p,q}(R_n)$ is a convolution algebra.

Remark 2. A proof of the theorem may be found in [S, 2.3.8]. For the spaces $B^s_{p,q}(R_n)$ one has a final answer. Similarly for the spaces $F^s_{p,q}(R_n)$ if one assumes additionally that $p \leqq q$. If $q < p$, there remains a gap.

Remark 3. The "if"-part of the theorem can be strengthened in the following way. If $-\infty < s < \infty$, $0 < p \leqq 1$ and $0 < q \leqq \infty$, then there exists a positive number c such that

$$\|f * g \mid B^s_{p,q}(R_n)\| \leqq c \|f \mid B^s_{p,q}(R_n)\| \, \|g \mid B^{n\left(\frac{1}{p}-1\right)}_{p,\infty}(R_n)\| \tag{5}$$

holds for all $f \in S(R_n)$ and all $g \in S(R_n)$. By a limiting procedure, (5) can be extended to arbitrary $f \in B^s_{p,q}(R_n)$ and $g \in B^{n\left(\frac{1}{p}-1\right)}_{p,\infty}(R_n)$, cf. [S, p. 99]. Similarly, there is the following statement.

If $-\infty<s<\infty$, $0<p\leq 1$ and $p\leq q\leq\infty$, then there exists a positive number c such that

$$\|f*g \mid F^s_{p,q}(R_n)\|\leq c\|f \mid F^s_{p,q}(R_n)\|\;\|g \mid F^{n(\frac{1}{p}-1)}_{p,\infty}(R_n)\| \tag{6}$$

holds for all $f\in F^s_{p,q}(R_n)$ and all $g\in F^{n(\frac{1}{p}-1)}_{p,\infty}(R_n)$.

Remark 4. Mapping properties of operators of convolution type acting between spaces of Hardy-Besov type have been treated in several papers. We refer to R. Johnson [1–3], T. Mizuhara [1–4] and Y. Uno [1]. The theorem shows that the local Hardy space $h_1(R_n)=F^0_{1,2}(R_n)$, cf. Theorem 2.5.8, is a convolution algebra. It is not complicated to extend this statement to the Hardy spaces $H_1(R_n)$ in the sense of Remark 2.2.2/2. This is a known fact, cf. e.g. U. Neri [1, Lemma 2.0] (concerning corresponding assertions for the classical Hardy spaces of analytic functions in the unit disc in the complex plane we refer to N. M. Wigley [1]).

2.6.6. The Classes $\mathfrak{M}_{p,q}$

The counterpart of Definition 2.6.1 (i) for the spaces $F^s_{p,q}(R_n)$ reads as follows: *let $-\infty<s<\infty$, $0<p<\infty$ and $0<q\leq\infty$. Then $m\in S'(R_n)$ is said to be a Fourier multiplier for $F^s_{p,q}(R_n)$ if there exists a positive number c such that*

$$\|F^{-1}mFf \mid F^s_{p,q}(R_n)\|\leq c\|f \mid F^s_{p,q}(R_n)\| \tag{1}$$

holds for all $f\in S(R_n)$. One obtains, as an immediate consequence of the lifting property from Theorem 2.3.8, that the set of all Fourier multipliers for $F^s_{p,q}(R_n)$ is independent of s. *Let $\mathfrak{M}_{p,q}$ be the set of all Fourier multipliers for $F^s_{p,q}(R_n)$.* We recall that $B^s_{p,p}(R_n)=F^s_{p,p}(R_n)$ if $0<p<\infty$ and that $L_p(R_n)=F^0_{p,2}(R_n)$ if $1<p<\infty$. Then it follows from Definition 2.6.2 that

$$\mathfrak{M}_{p,p}=\mathfrak{M}_p \quad\text{if}\quad 0<p<\infty \quad\text{and}\quad \mathfrak{M}_{p,2}=M_p \quad\text{if}\quad 1<p<\infty\,. \tag{2}$$

Because $\mathfrak{M}_p\neq M_p$ if $p\neq 2$ and $1\leq p\leq\infty$, cf. Proposition 2.6.4 and Theorem 2.6.4, one cannot in general expect $\mathfrak{M}_{p,q}$ to be independent of q (in contrast to the corresponding situation for the spaces $B^s_{p,q}(R_n)$). A detailed study of $\mathfrak{M}_{p,q}$ would be desirable. We list some results which can be obtained comparatively easily on the basis of duality and interpolation. (2.4.2/2) yields

$$\mathfrak{M}_{p,q}\subset\mathfrak{M}_p \quad\text{if}\quad 0<p<\infty \quad\text{and}\quad 0<q\leq\infty\,. \tag{3}$$

Now it follows from (2.4.7/3) that

$$\mathfrak{M}_{p,q_0}\subset\mathfrak{M}_{p,q_1}\subset\mathfrak{M}_p \quad\text{if either}\quad 0<q_0\leq q_1\leq p \quad\text{or}\quad p\leq q_1\leq q_0\leq\infty \tag{4}$$

(and $0<p<\infty$). By (3), $\mathfrak{M}_p\subset\mathfrak{M}_2=M_2$, and (2.4.7/3) we have

$$\mathfrak{M}_{p_0,q_0}\subset\mathfrak{M}_{p_1,q_1} \quad\text{if}\quad \frac{1}{p_1}=\frac{1-\Theta}{p_0}+\frac{\Theta}{2}\,,\quad \frac{1}{q_1}=\frac{1-\Theta}{q_0}+\frac{\Theta}{2}\,, \tag{5}$$

where $0<\Theta<1$, $0<p_0<\infty$ and $0<q_0\leq\infty$. By duality, cf. (2.11.2/2) it follows that

$$\mathfrak{M}_{p,q}=\mathfrak{M}_{p',q'} \quad\text{if}\quad 1<p<\infty,\quad 1<q<\infty,\quad\text{and}\quad \frac{1}{p}+\frac{1}{p'}=\frac{1}{q}+\frac{1}{q'}=1\,. \tag{6}$$

Furthermore, (5), (6), and a modification of the proof of Proposition 2.6.4 yield the following result. If m is the function from (2.6.4/1), then

$$m\in\bigcap\mathfrak{M}_p \quad\text{and}\quad m\notin\bigcup\mathfrak{M}_{p,q}\,; \tag{7}$$

where the intersection is taken over all p with $1\leq p\leq\infty$ and the union is taken over all p and q with $0<p<\infty$, $0<q\leq\infty$ and $p\neq q$. Finally, if $0<p\leq 1$ and $p\leq q\leq\infty$, then

$$F[F^{n(\frac{1}{p}-1)}_{p,\infty}(R_n)]\subset\mathfrak{M}_{p,q}\subset F[B^{n(\frac{1}{p}-1)}_{p,\infty}(R_n)]\,. \tag{8}$$

The right-hand side follows from (3) and (2.6.3/6). The left-hand side is a consequence of (2.6.5/6) and $F(f*g)=cFf\cdot Fg$.

2.7. Embedding Theorems

2.7.1 Embedding Theorems for Different Metrics

Section 2.7. deals with two types of embedding theorems: embedding theorems for different metrics and traces on hyperplanes. For the classical spaces of Sobolev-Besov type (including Bessel-potential spaces) such problems have been treated extensively, cf. S. M. Nikol'skij [3] and [I, 2.8 and 2.9]. They play an important role in boundary value problems for elliptic differential operators. Traces will be treated in the following subsection, this subsection dealing with embedding theorems for different metrics.

We proved in Proposition 2.3.2/2 some elementary embeddings. Recall that $\subset$ always stands for a continuous embedding.

Theorem. (i) Let $0<p_0\leqq p_1\leqq\infty$, $0<q\leqq\infty$, and $-\infty<s_1\leqq s_0<\infty$. Then

$$B^{s_0}_{p_0,q}(R_n)\subset B^{s_1}_{p_1,q}(R_n) \quad \text{if} \quad s_0-\frac{n}{p_0}=s_1-\frac{n}{p_1}. \tag{1}$$

(ii) Let $0<p_0<p_1<\infty$, $0<q\leqq\infty$, $0<r\leqq\infty$, and $-\infty<s_1<s_0<\infty$. Then

$$F^{s_0}_{p_0,q}(R_n)\subset F^{s_1}_{p_1,r}(R_n) \quad \text{if} \quad s_0-\frac{n}{p_0}=s_1-\frac{n}{p_1}. \tag{2}$$

Proof. Step 1. In order to prove (1), we assume that $\{\varphi_j(x)\}_{j=0}^{\infty}\in\Phi(R_n)$, cf. Definition 2.3.1/1. Let $f\in B^{s_0}_{p_0,q}(R_n)$. By (1.3.2/5) and Remark 1.4.1/4 we have

$$\|F^{-1}\varphi_jFf \mid L_{p_1}(R_n)\|\leqq c2^{jn\left(\frac{1}{p_0}-\frac{1}{p_1}\right)}\|F^{-1}\varphi_jFf \mid L_{p_0}(R_n)\|, \tag{3}$$

where c is independent of $j=0, 1, 2, \ldots$ However, (1) is an immediate consequence of (3).

Step 2. We prove (2). We may assume that $s_0=0$. This follows from the lifting property from Theorem 2.3.8. Then we have $s_1<0$. Furthermore, we may assume that $q=\infty$ and $0<r<1$. Let $f\in F^0_{p_0,\infty}(R_n)$ with $\|f \mid F^0_{p_0,\infty}(R_n)\|=1$. Then (3) yields

$$\|F^{-1}\varphi_kFf \mid L_\infty(R_n)\|\leqq c2^{k\frac{n}{p_0}}\|F^{-1}\varphi_kFf \mid L_{p_0}(R_n)\|\leqq c2^{k\frac{n}{p_0}}\|f \mid F^0_{p_0,\infty}(R_n)\|=c2^{k\frac{n}{p_0}}.$$

If $K=0, 1, 2, \ldots$, then

$$\sum_{k=0}^{K}|2^{ks_1}(F^{-1}\varphi_kFf)(x)|^r\leqq c\sum_{k=0}^{K}2^{k\frac{n}{p_1}r}\leqq C2^{K\frac{n}{p_1}r}, \tag{4}$$

where C is independent of K. On the other hand, because $s_1<0$ it follows that

$$\sum_{k=K+1}^{\infty}|2^{ks_1}(F^{-1}\varphi_kFf)(x)|^r\leqq c2^{Ks_1r}\sup_{k=0,1,2,\ldots}|(F^{-1}\varphi_kFf)(x)|^r. \tag{5}$$

We recall that

$$\|g \mid L_p(R_n)\|^p=\int_{R_n}|g(x)|^p\,dx=p\int_0^\infty t^{p-1}\,|\{x \mid |g(x)|>t\}|\,dt, \tag{6}$$

where $|\{x \mid \ldots\}|$ is the Lebesgue measure of $\{x \mid \ldots\}$ and $0<p<\infty$. Consequently,

$$\|f \mid F^{s_1}_{p_1,r}(R_n)\|^{p_1}=p_1\int_0^\infty t^{p_1-1}\left|\left\{x \,\middle|\, \left(\sum_{k=0}^{\infty}2^{krs_1}|(F^{-1}\varphi_kFf)(x)|^r\right)^{\frac{1}{r}}>t\right\}\right|dt. \tag{7}$$

We split the integral in (7) in an integral over $(0, (2C)^{\frac{1}{r}})$ and an integral over $((2C)^{\frac{1}{r}}, \infty)$, where C has the same meaning as in (4). By (5) with $K=-1$, it follows that

$$\int_0^{(2C)^{\frac{1}{r}}} t^{p_1-1}\left|\left\{x\,\Big|\left(\sum_{k=0}^{\infty} 2^{krs_1}|(F^{-1}\varphi_k Ff)(x)|^r\right)^{\frac{1}{r}}>t\right\}\right|\,dt$$
$$\leqq c\int_0^{c'(2C)^{\frac{1}{r}}} t^{p_0-1}\,|\{x\mid \sup_{k=0,1,2,\ldots}|(F^{-1}\varphi_k Ff)(x)|>t\}|\,dt\leqq c''\|f\mid F^0_{p_0,\infty}(R_n)\|=c''\,. \tag{8}$$

If $t>(2C)^{\frac{1}{r}}$, we determine K in (4) as the largest non-negative integer such that $C2^{K\frac{n}{p_1}r}\leqq\frac{t^r}{2}$. This is possible (and this is also the reason for the above splitting number $(2C)^{\frac{1}{r}}$). Now it follows from (4) and (5) that

$$\left|\left\{x\,\Big|\sum_{k=0}^{\infty} 2^{krs_1}|(F^{-1}\varphi_k Ff)(x)|^r>t^r\right\}\right|\leqq\left|\left\{x\,\Big|\sum_{k=K+1}^{\infty} 2^{krs_1}|(F^{-1}\varphi_k Ff)(x)|^r\geqq\frac{t^r}{2}\right\}\right|$$
$$\leqq|\{x\mid \sup_{k=0,1,2,\ldots}|(F^{-1}\varphi_k Ff)(x)|\geqq ct2^{-Ks_1}\}|\,.$$

We have $t2^{-Ks_1}\sim t^{1-s_1\frac{p_1}{n}}\sim t^{\frac{p_1}{p_0}}$. So it follows that

$$\int_{(2C)^{\frac{1}{r}}}^{\infty} t^{p_1-1}\left|\left\{x\,\Big|\sum_{k=0}^{\infty} 2^{krs_1}|(F^{-1}\varphi_k Ff)(x)|^r>t^r\right\}\right|\,dt$$
$$\leqq\int_{c(2C)^{\frac{1}{r}}}^{\infty} t^{p_1-1}|\{x\mid \sup_{k=0,1,2,\ldots}|(F^{-1}\varphi_k Ff)(x)|\geqq t^{\frac{p_1}{p_0}}\}|\,dt \tag{9}$$
$$\leqq c'\int_0^{\infty} t^{p_0-1}\,|\{x\mid \sup_{k=0,1,2,\ldots}|(F^{-1}\varphi_k Ff)(x)|\geqq t\}|\,dt$$
$$\leqq c''\|f\mid F^0_{p_0,\infty}(R_n)\|^{p_0}=c''\,.$$

Now (7), (8) and (9) prove that

$$\|f\mid F^{s_1}_{p_1,r}(R_n)\|^{p_1}\leqq c\|f\mid F^0_{p_0,\infty}(R_n)\|^{p_1} \tag{10}$$

$\|f\mid F^0_{p_0,\infty}(R_n)\|=1$. A homogeneity argument shows that (10) holds for all $\in F^0_{p_0,\infty}(R_n)$. This proves (2).

Remark 1. If $s>0$, $1\leqq p<\infty$ and $1\leqq q\leqq\infty$, then $\Lambda^s_{p,q}(R_n)=B^s_{p,q}(R_n)$ are the classical Besov spaces, cf. (2.5.7/5) or (2.5.12/21). In these cases, (1) is a well-known embedding theorem, cf. e.g. [I, Theorem 2.8.1]. The extension to the above values of the parameters is due to J. Peetre [4,8]. The proof is based on inequality (1.3.2/5) for entire analytic functions of exponential type. This is Nikol'skij's idea, cf. S. M. Nikol'skij [3] and the references given there. (2) and the above proof are due to B. Jawerth [1]. Some special cases of (2), if $1<p_0<p_1<\infty$ and $1<q<r<\infty$, have been proved in [I, 2.8.1]. However Jawerth's result is more general and his proof is simpler. Of peculiar interest is the fact, that the numbers q and r in (2) are independent of each other.

Remark 2. We recall that $\mathcal{C}^s(R_n)=B^s_{\infty,\infty}(R_n)$ with $s>0$ are the Zygmund spaces, cf. (2.5.7/6) or (2.5.12/22). Furthermore, if $m=0, 1, 2, \ldots$, it follows from (2.5.7/2) and (2.2.2/1) that

$$B^m_{\infty,1}(R_n)\subset C^m(R_n)\,. \tag{11}$$

Now it follows from (1) and (2.3.2/7) that

$$B_{p,q}^{s+\frac{n}{p}}(R_n)\subset\mathcal{C}^s(R_n) \quad \text{if} \quad s>0, \quad 0<p\leqq\infty \quad \text{and} \quad 0<q\leqq\infty \tag{12}$$

and

$$B_{p,q}^{m+\frac{n}{p}}(R_n)\subset C^m(R_n) \quad \text{if} \quad 0<p\leqq\infty, \quad 0<q\leqq 1 \quad \text{and} \quad m=0,1,2,\ldots \tag{13}$$

In other words, if the smoothness index s in $B^s_{p,q}(R_n)$ is large enough, then $f\in B^s_{p,q}(R_n)$ has some classical derivatives.

Remark 3. By (2) we have

$$F^{s_0}_{p_0,q}(R_n)\subset F^{s_1}_{p_1,p_1}(R_n)=B^{s_1}_{p_1,p_1}(R_n) \tag{14}$$

if $0<p_0<p_1<\infty$, $0<q\leqq\infty$ and $s_0-\frac{n}{p_0}=s_1-\frac{n}{p_1}$. B. Jawerth [1] proved that (14) can be strengthened by

$$F^{s_0}_{p_0,q}(R_n)\subset B^{s_1}_{p_1,p_0}(R_n) \quad \text{if} \quad 0<p_0<p_1<\infty, \tag{15}$$

$-\infty<s_1<s_0<\infty$, $0<q\leqq\infty$ and $s_0-\frac{n}{p_0}=s_1-\frac{n}{p_1}$. We recall that $F^s_{p,2}(R_n)=H^s_p(R_n)$, with $-\infty<s<\infty$ and $1<p<\infty$, are the Bessel-potential spaces (with the Sobolev spaces as specia cases), cf. Theorem 2.5.6. Hence it follows from (15) that

$$H^{s_0}_{p_0}(R_n)\subset B^{s_1}_{p_1,p_0}(R_n) \quad \text{if} \quad 1<p_0<p_1<\infty, \tag{16}$$

$-\infty<s_1<s_0<\infty$ and $s_0-\frac{n}{p_0}=s_1-\frac{n}{p_1}$. This is a well-known embedding theorem, cf. e.g. [I, 2.8.1/18)].

2.7.2. Traces

If $x=(x_1,\ldots,x_n)$ then we put $x'=(x_1,\ldots,x_{n-1})$. We ask for the trace of $f\in B^s_{p,q}(R_n)$ or $f\in F^s_{p,q}(R_n)$ on the hyperplane $\{x \mid x=(x',0)\}$. This hyperplane will be denoted by R_{n-1} in the sequel. It is clear that a clarification of this problem is of crucial interest for boundary value problems of elliptic differential operators. There are two types of embedding theorems.

(i) Direct embedding theorems. Let $A_j(R_{n-1})$ with $j=0,\ldots,r$ be $r+1$ quasi-normed spaces on R_{n-1} (with the above agreement that $R_{n-1}=\{x \mid x=(x',0)\}$ is a hyperplane in R_n). Let $A(R_n)$ be a quasi-normed space on R_n. If $\mathfrak{R}$ is given by

$$\mathfrak{R}f=\left\{f(x',0), \frac{\partial f}{\partial x_n}(x',0),\ldots,\frac{\partial^r f}{\partial x_n^r}(x',0)\right\}, \tag{1}$$

then we ask whether $\mathfrak{R}$ is a continuous linear mapping from $A(R_n)$ into $\prod_{j=0}^r A_j(R_{n-1})$. In other words, we ask whether there exists a positive constant c such that

$$\sum_{j=0}^r \left\|\frac{\partial^j f}{\partial x_n^j}(x',0) \mid A_j(R_{n-1})\right\|\leqq c\|f \mid A(R_n)\| \tag{2}$$

holds for all $f\in A(R_n)$. At first glance it seems reasonable to restrict (2) to functions $f\in S(R_n)$, where $\frac{\partial^j f}{\partial x_n^j}(x',0)$ has an obvious meaning. But this is not necessary in our case because the spaces $A_j(R_{n-1})$ and $A(R_n)$ are always of type

$B^s_{p,q}$ or $F^s_{p,q}$, and we shall be concerned with analytic functions $F^{-1}\varphi Ff$, where $\varphi \in S(R_n)$ has compact support. Of course, these functions have natural traces on R_{n-1}, and it makes sense to ask whether (2) is valid. Subsequently, one extends (2) to all $f \in A(R_n)$ via a limiting procedure. All formulations of type (2) should be understood in this sense.

(ii) Inverse embedding theorems. If $A_j(R_{n-1})$and $A(R_n)$ are the above spaces and if (2) is valid, then we ask whether there exists a continuous linear operator $\mathfrak{S}$ from $\prod_{j=0}^{r} A_j(R_{n-1})$ into $A(R_n)$ such that

$$\mathfrak{R}\mathfrak{S} = E \quad \left(\text{identity in } \prod_{j=0}^{r} A_j(R_{n-1})\right). \tag{3}$$

If $\mathfrak{S}$ exists, then $\mathfrak{R}$ maps $A(R_n)$ onto $\prod_{j=0}^{r} A_j(R_{n-1})$, in other words, $\prod_{j=0}^{r} A_j(R_{n-1})$ is the exact trace of $f \to \left\{f(x', 0), \ldots, \frac{\partial^r f}{\partial x_n^r}(x', 0)\right\}$. We introduce the following notations. If $\mathfrak{R}$ is a continuous linear operator from A(R_n) into $\prod_{j=0}^{r} A_j(R_{n-1})$ and if there exists a continuous linear operator $\mathfrak{S}$ from $\prod_{j=0}^{r} A_j(R_{n-1})$ into $A(R_n)$ satisfying (3), then $\mathfrak{R}$ is called a retraction and $\mathfrak{S}$ a (corresponding) coretraction. Hence, the statement that $\mathfrak{R}$ is a retraction includes both direct and inverse embedding theorems on the hyperplane R_{n-1}. The embedding theorems in [I, 2.9] are formulated in that language.

Theorem. Let $\mathfrak{R}$ be given by (1), where $r = 0, 1, 2, \ldots$

(i) If $0 < p \leq \infty$, $0 < q \leq \infty$ and $s > r + \frac{1}{p} + \max\left(0, (n-1)\left(\frac{1}{p} - 1\right)\right)$, then $\mathfrak{R}$ is a retraction from

$$B^s_{p,q}(R_n) \quad \text{onto} \quad \prod_{j=0}^{r} B_{p,q}^{s-\frac{1}{p}-j}(R_{n-1}). \tag{4}$$

(ii) If $0 < p < \infty$, $0 < q \leq \infty$ and $s > r + \frac{1}{p} + \max\left(0, (n-1)\left(\frac{1}{p} - 1\right)\right)$, then $\mathfrak{R}$ is a retraction from

$$F^s_{p,q}(R_n) \quad \text{onto} \quad \prod_{j=0}^{r} B_{p,p}^{s-\frac{1}{p}-j}(R_{n-1}). \tag{5}$$

Proof. Step 1. We prove a simple but important inequality. Let $f \in S'(R_n)$ and $\{\varphi_k(x)\}_{k=0}^{\infty} \in \Phi(R_n)$, cf. 2.3.1. Let $(\varphi_k^* f)(x)$ be the maximal function from Definition 2.3.6/2. If $x = (x', x_n)$ with $2^{-k-1} \leq x_n \leq 2^{-k}$ then

$$|(F^{-1}\varphi_k Ff)(x', 0)| \leq c(\varphi_k^* f)(x),$$

where c is independent of k. Let $0 < p < \infty$. Integration yields

$$\|(F^{-1}\varphi_k Ff)(x', 0) \mid L_p(R_{n-1})\|^p \leq c 2^k \int_{2^{-k-1}}^{2^{-k}} \|(\varphi_k^* f)(x) \mid L_p(R_{n-1})\|^p \, dx_n. \tag{6}$$

The counterpart of (6) for $p = \infty$ reads as follows:

$$\|(F^{-1}\varphi_k Ff)(x', 0) \mid L_\infty(R_{n-1})\| \leq c \|(\varphi_k^* f)(x) \mid L_\infty(R_n)\|. \tag{7}$$

Step 2. We prove a second inequality. Let $\{\varphi_k(x)\}_{k=0}^{\infty} \in \Phi(R_n)$ and $\{\varphi_k'(x')\}_{k=0}^{\infty} \in \Phi(R_{n-1})$. Temporarily we denote the Fourier transform on R_{n-1} and its inverse by F' and F'^{-1}, respectively, (F and F^{-1} are the Fourier transform and its inverse on R_n, respectively). We have

$$F'^{-1}\varphi'_k F' f = \sum_{l=k-1}^{\infty} F'^{-1}\varphi'_k F' F^{-1}\varphi_l F f \tag{8}$$

with $\varphi_{-1}=0$ (convergence in $S'(R_n)$). Of course, we extend F', F'^{-1} and φ'_k in (8) in an obvious way from R_{n-1} to R_n, e.g. $\varphi'_k(x)=\varphi'_k(x')$. In this sense we have $F^{-1}\varphi'_k F = F'^{-1}\varphi'_k F'$. We may assume without loss of generality, that $\varphi'_k(x')=\varphi'_1(2^{-k+1}x')$ if $k=2,3,\ldots$ Then

$$\|F'^{-1}\varphi'_k \mid L_1(R_{n-1})\| = 2^{(k-1)(n-1)}\|(F'^{-1}\varphi'_1)(2^{k-1}\cdot) \mid L_1(R_{n-1})\| = c < \infty \tag{9}$$

if $k=1,2,\ldots$ Let $1 \leqq p \leqq \infty$. Then (8) and (9) yield

$$\begin{aligned} &\|(F'^{-1}\varphi'_k F' f)(\cdot\,,0) \mid L_p(R_{n-1})\| \\ &\leqq \sum_{l=k-1}^{\infty} \|F'^{-1}\varphi'_k * (F^{-1}\varphi_l F f)(\cdot\,,0) \mid L_p(R_{n-1})\| \\ &\leqq \sum_{l=k-1}^{\infty} \|(F^{-1}\varphi_l F f)(\cdot,0) \mid L_p(R_{n-1})\|\,. \end{aligned} \tag{10}$$

If $0<p<1$, then (8) gives

$$\begin{aligned} &\|(F'^{-1}\varphi'_k F' f)(x',0) \mid L_p(R_{n-1})\|^p \\ &\leqq \sum_{l=k-1}^{\infty} \|F'^{-1}\varphi'_k * (F^{-1}\varphi_l F f)(\cdot,0) \mid L_p(R_{n-1})\|^p\,. \end{aligned} \tag{11}$$

It is easy to see that

$$\operatorname{supp} F'[(F^{-1}\varphi_l F f)(\cdot,0)] \subset \{y' \mid |y'| \leqq 2^{l+1}\} \tag{12}$$

(in R_{n-1}). We apply the $(n-1)$-dimensional version of (1.5.3/3) to the right-hand side of (11) to obtain

$$\begin{aligned} &\|(F'^{-1}\varphi'_k F' f)(x',0) \mid L_p(R_{n-1})\|^p \\ &\leqq c \sum_{l=k-1}^{\infty} 2^{l(n-1)\left(\frac{1}{p}-1\right)p} \|F'^{-1}\varphi_k \mid L_p(R_{n-1})\|^p \, \|(F^{-1}\varphi_l F f)(\cdot,0) \mid L_p(R_{n-1})\|^p\,. \end{aligned} \tag{13}$$

As in (9) we have

$$\begin{aligned} &\|F'^{-1}\varphi_k \mid L_p(R_{n-1})\|^p \\ &= 2^{(k-1)(n-1)p}\|(F'^{-1}\varphi'_1)(2^{k-1}\cdot) \mid L_p(R_{n-1})\|^p = c2^{(k-1)(n-1)(p-1)}\,. \end{aligned} \tag{14}$$

Substituting (14) into (13), we obtain

$$\begin{aligned} &\|(F'^{-1}\varphi'_k F' f)(x',0) \mid L_p(R_{n-1})\|^p \\ &\leqq c \sum_{l=k-1}^{\infty} 2^{(l-k)(n-1)(1-p)}\|(F^{-1}\varphi_l F f)(\cdot,0) \mid L_p(R_{n-1})\|^p\,. \end{aligned} \tag{15}$$

Step 3. Let $f \in B^s_{p,q}(R_n)$ with $1 \leqq p \leqq \infty$, $0<q \leqq \infty$ and $s>\frac{1}{p}$. Let ε be a positive number such that $s-\frac{1}{p}-\varepsilon>0$. Then (10) and (6) (resp. (7)) give

$$\begin{aligned} &2^{k\left(s-\frac{1}{p}\right)}\|(F'^{-1}\varphi'_k F' f)(x',0) \mid L_p(R_{n-1})\| \\ &\leqq c \sum_{l=k-1}^{\infty} 2^{-(l-k)\left(s-\frac{1}{p}-\varepsilon\right)} 2^{ls-(l-k)\varepsilon}\|\varphi_l{}^* f \mid L_p(R_n)\| \\ &\leqq c' \left(\sum_{l=k-1}^{\infty} 2^{qls-q\varepsilon(l-k)}\|\varphi_l{}^* f \mid L_p(R_n)\|^q\right)^{\frac{1}{q}} \end{aligned} \tag{16}$$

(modification if $q=\infty$). Now it follows that

$$\left(\sum_{k=0}^{\infty} 2^{k\left(s-\frac{1}{p}\right)q}\|(F'^{-1}\varphi'_k F'f)(x',0) \mid L_p(R_{n-1})\|^q\right)^{\frac{1}{q}} \leq c\left(\sum_{l=0}^{\infty} 2^{qls}\|\varphi_l^* f \mid L_p(R_n)\|^q\right)^{\frac{1}{q}} \tag{17}$$

(modification if $q=\infty$). By Theorem 2.3.6 we have

$$\|f(x',0) \mid B_{p,q}^{s-\frac{1}{p}}(R_{n-1})\| \leq c\|f \mid B_{p,q}^s(R_n)\| . \tag{18}$$

Let $f\in B_{p,q}^s(R_n)$ with $0<p<1$, $0<q\leq\infty$ and $s>\frac{1}{p}+(n-1)\left(\frac{1}{p}-1\right)$. Then we use (15) instead of (10). This has the consequence that we must replace $(l-k)\times\left(s-\frac{1}{p}-\varepsilon\right)$ in the first estimate in (16) by $(l-k)\left(s-\frac{1}{p}-\varepsilon-(n-1)\left(\frac{1}{p}-1\right)\right)$. The rest is the same as in the case $1\leq p\leq\infty$. Hence, (18) holds also for $0<p<1$ (under the above hypotheses for s). Now we assume that $f\in F_{p,q}^s(R_n)$ with $1\leq p<\infty$, $0<q\leq\infty$, and $s>\frac{1}{p}$. Let again $\varepsilon>0$ and $s-\frac{1}{p}-\varepsilon>0$. Then (10) and (6) yield

$$\begin{aligned}
&2^{k\left(s-\frac{1}{p}\right)}\|(F'^{-1}\varphi'_k F'f)(x',0) \mid L_p(R_{n-1})\| \\
&\leq c\sum_{l=k-1}^{\infty} 2^{-(l-k)\left(s-\frac{1}{p}-\varepsilon\right)}2^{-\varepsilon(l-k)}2^{ls}\left(\int_{2^{-l-1}}^{2^{-l}}\|(\varphi_l^* f)(x) \mid L_p(R_{n-1})\|^p\,\mathrm{d}x_n\right)^{\frac{1}{p}} \\
&\leq c'\left(\sum_{l=k-1}^{\infty} 2^{-\varepsilon p(l-k)}2^{lsp}\int_{2^{-l-1}}^{2^{-l}}\|(\varphi_l^* f)(x) \mid L_p(R_{n-1})\|^p\,\mathrm{d}x_n\right)^{\frac{1}{p}} .
\end{aligned} \tag{19}$$

By Theorem 2.3.6 we have

$$\begin{aligned}
&\sum_{k=0}^{\infty} 2^{k\left(s-\frac{1}{p}\right)p}\|(F'^{-1}\varphi'_k F'f)(x',0) \mid L_p(R_{n-1})\|^p \\
&\leq c\sum_{l=0}^{\infty} 2^{lsp}\int_{2^{-l-1}}^{2^{-l}}\|(\varphi_l^* f)(x) \mid L_p(R_{n-1})\|^p\,\mathrm{d}x_n \\
&\leq c\int_{R_n}\sup_{l=0,1,\ldots} 2^{lsp}|(\varphi_l^* f)(x)|^p\,\mathrm{d}x \\
&\leq c'\|f \mid F_{p,\infty}^s(R_n)\|^p \leq c''\|f \mid F_{p,q}^s(R_n)\|^p .
\end{aligned} \tag{20}$$

This proves that

$$\|f(x',0) \mid B_{p,p}^{s-\frac{1}{p}}(R_{n-1})\| \leq c\|f \mid F_{p,q}^s(R_n)\| . \tag{21}$$

Finally let $f\in F_{p,q}^s(R_n)$ with $0<p<1$, $0<q\leq\infty$ and $s>\frac{1}{p}+(n-1)\left(\frac{1}{p}-1\right)$. Then we use (15) instead of (10). This has the consequence that we must replace $(l-k)\times\left(s-\frac{1}{p}-\varepsilon\right)$ in the first estimate in (19) by $(l-k)\left(s-\frac{1}{p}-\varepsilon-(n-1)\left(\frac{1}{p}-1\right)\right)$. The rest is the same as in the case $1\leq p<\infty$. Hence, (21) holds also for $0<p<1$.

Step 4. If $f\in B^s_{p,q}(R_n)$ with $0<p\leqq\infty$, $0<q\leqq\infty$ and

$$s>j+\frac{1}{p}+\max\left(0,\,(n-1)\left(\frac{1}{p}-1\right)\right).$$

then we have $\frac{\partial^j f}{\partial x_n^j}\in B^{s-j}_{p,q}(R_n)$, cf. Theorem 2.3.8. Now (18) yields

$$\left\|\frac{\partial^j f}{\partial x_n^j}(x',0)\mid B^{s-j-\frac{1}{p}}_{p,q}(R_{n-1})\right\|\leqq c\left\|\frac{\partial^j f}{\partial x_n^j}\mid B^{s-j}_{p,q}(R_n)\right\|\leqq c'\|f\mid B^s_{p,q}(R_n)\|\,. \quad (22)$$

Similarly for the spaces $F^s_{p,q}(R_n)$. This proves that $\mathfrak{R}$ is a continuous linear mapping

$$\text{from}\quad B^s_{p,q}(R_n)\quad\text{into}\quad \prod_{j=0}^{r} B^{s-\frac{1}{p}-j}_{p,q}(R_{n-1})$$

and

$$\text{from}\quad F^s_{p,q}(R_n)\quad\text{into}\quad \prod_{j=0}^{r} B^{s-\frac{1}{p}-j}_{p,p}(R_{n-1})\,,$$

respectively (under the hypotheses of the theorem).

Step 5. Let $\{\varphi'_k(x')\}_{k=0}^\infty\in\Phi(R_{n-1})$. Furthermore, let $\psi_0(\lambda)\in S(R_1)$ and $\psi(\lambda)\in S(R_1)$ with

$$\operatorname{supp}\psi_0\subset(-1,1)\quad\text{and}\quad\operatorname{supp}\psi\subset(1,2)\,. \quad (23)$$

Let $\psi_l(\lambda)=\psi(2^{-l}\lambda)$ if $l=1,2,\ldots$ We denote the Fourier transform on R_1 and its inverse by F_1 and F_1^{-1}, respectively. We assume that

$$(F_1^{-1}\psi_0)(0)=(F_1^{-1}\psi)(0)=1\,. \quad (24)$$

Then

$$(F_1^{-1}\psi_j)(\lambda)\,|_{\lambda=0}=(F_1^{-1}\psi(2^{-j}\cdot))(\lambda)\,|_{\lambda=0}=2^j(F_1^{-1}\psi)(2^j\lambda)\,|_{\lambda=0}=2^j \quad (25)$$

if $j=1,2,3,\ldots$ Let $f\in B^s_{p,q}(R_{n-1})$ with $0<p\leqq\infty$, $0<q\leqq\infty$ and $-\infty<s<\infty$. Set

$$g(x',x_n)=\sum_{k=0}^{\infty}2^{-k}(F_1^{-1}\psi_k)(x_n)\,(F'^{-1}\varphi'_kF'f)(x')\,. \quad (26)$$

First we note that

$$g(x',0)=f(x'). \quad (27)$$

Hence, $f(x')$ is the trace of $g(x',x_n)$. We want to prove that $g\in B^{s+\frac{1}{p}}_{p,q}(R_n)$. Let $\{\varphi_k(x)\}_{k=0}^\infty\in\Phi(R_n)$. Then it follows that

$$\begin{aligned}\|g\mid B^{s+\frac{1}{p}}_{p,q}(R_n)\|&=\|2^{\left(s+\frac{1}{p}\right)k}F^{-1}\varphi_kFg\mid l_q(L_p(R_n))\|\\&\leqq c\|2^{\left(s+\frac{1}{p}-1\right)k}(F_1^{-1}\psi_k)(x_n)\,(F'^{-1}\varphi'_kF'f)(x')\mid l_q(L_p(R_n))\|\,.\end{aligned} \quad (28)$$

As in (25), we have

$$\|F_1^{-1}\psi_k\mid L_p(R_1)\|=2^{k\left(1-\frac{1}{p}\right)}\|F_1^{-1}\psi\mid L_p(R_1)\| \quad (29)$$

if $k=1,2,3,\ldots$ Now (28) and (29) yield

$$\begin{aligned}\|g\mid B^{s+\frac{1}{p}}_{p,q}(R_n)\|&\leqq c\|2^{sk}F'^{-1}\varphi'_kF'f\mid l_q(L_p(R_{n-1}))\|\\&=c\|f\mid B^s_{p,q}(R_{n-1})\|\,.\end{aligned} \quad (30)$$

Hence, $g=\mathfrak{S}f$ has the desired properties of a coretraction in the case of the $B^s_{p,q}$-spaces if $r=0$ in the sense of the theorem.

Step 6. Let $f\in B^s_{p,p}(R_{n-1})$ with $0<p<\infty$ and $-\infty<s<\infty$. We want to prove that g from (26) belongs to $F^{s+\frac{1}{p}}_{p,q}(R_n)$, where $0<q\leqq\infty$. Of course, we may assume that $0<q<p$. Let

$$a_k(x')=(F'^{-1}\varphi'_kF'f)(x') \quad\text{and}\quad b_k=\|a_k \mid L_p(R_{n-1})\| . \tag{31}$$

Let again $\{\varphi_k(x)\}_{k=0}^\infty\in\Phi(R_n)$. Then it follows from Theorem 1.6.3 that

$$\begin{aligned}
\|g \mid F^{s+\frac{1}{p}}_{p,q}(R_n)\|^p &= \|2^{k\left(s+\frac{1}{p}\right)}F^{-1}\varphi_kFg \mid L_p(R_n,l_q)\|^p\\
&\leqq c\|2^{\left(s+\frac{1}{p}-1\right)k}(F_1^{-1}\psi_k)(x_n)\,(F'^{-1}\varphi'_kF'f)(x') \mid L_p(R_n,l_q)\|^p\\
&= c\int_{R_n}\left(\sum_{k=0}^\infty |2^{-k}(F_1^{-1}\psi_k)(x_n)\,2^{k\left(s+\frac{1}{p}\right)}a_k(x')|^q\right)^{\frac{p}{q}}\mathrm{d}x\\
&= c\int_{R_1}\left\|\sum_{k=0}^\infty|\ldots|^q \mid L_{\frac{p}{q}}(R_{n-1})\right\|^{\frac{p}{q}}\mathrm{d}x_n\\
&\leqq c\int_{R_1}\left(\sum_{k=0}^\infty 2^{-kq}|(F_1^{-1}\psi_k)(x_n)|^q\,2^{k\left(s+\frac{1}{p}\right)q}b_k^q\right)^{\frac{p}{q}}\mathrm{d}x_n .
\end{aligned}\tag{32}$$

We split the last integral in (32) into

$$\int_{R_1}\ldots\mathrm{d}x_n=\int_I\ldots\mathrm{d}x_n+\sum_{l=0}^\infty\int_{I_l}\ldots\mathrm{d}x_n , \tag{33}$$

where $I=(-\infty,-1)\cup(1,\infty)$ and $I_l=(-2^{-l},-2^{-l-1})\cup(2^{-l-1},2^{-l})$. By (25) we have

$$2^{-k}|(F_1^{-1}\psi_k)(x_n)|\leqq c\,(1+|2^kx_n|)^{-\sigma} , \tag{34}$$

where σ is an arbitrary positive number (c in (34) depends on σ, but not on k). If σ is chosen sufficiently large, it follows that

$$\int_I\ldots\mathrm{d}x_n\leqq c\int_I\left(\sum_{k=0}^\infty\frac{2^{k\left(s+\frac{1}{p}\right)q}}{2^{k\sigma q}}b_k^q\right)^{\frac{p}{q}}\frac{\mathrm{d}x_n}{|x_n|^{\sigma p}}\leqq c'\sum_{k=0}^\infty 2^{ksp}b_k^p . \tag{35}$$

In order to estimate the corresponding integrals over I_l we again use (34) and obtain

$$\begin{aligned}
\sum_{l=0}^\infty\int_{I_l}\ldots\mathrm{d}x_n &\leqq c\sum_{l=0}^\infty 2^{-l}\left(\sum_{k=0}^l 2^{k\left(s+\frac{1}{p}\right)q}b_k^q\right)^{\frac{p}{q}}\\
&\quad+c\sum_{l=0}^\infty 2^{-l}\left(\sum_{k=l+1}^\infty 2^{k\left(s+\frac{1}{p}\right)q}2^{-(k-l)\sigma q}b_k^q\right)^{\frac{p}{q}}=A_1+A_2 .
\end{aligned}\tag{36}$$

If $0<\varepsilon<1$, we have

$$\begin{aligned} A_1 &\leq c \sum_{l=0}^{\infty} 2^{-l} \sum_{k=0}^{l} 2^{\varepsilon(l-k)} 2^{k\left(s+\frac{1}{p}\right)p} b_k^p \\ &= c \sum_{k=0}^{\infty} 2^{-\varepsilon k + k\left(s+\frac{1}{p}\right)p} b_k^p \sum_{l=k}^{\infty} 2^{-l+\varepsilon l} = c' \sum_{k=0}^{\infty} 2^{ksp} b_k^p . \end{aligned} \tag{37}$$

Furthermore,

$$\begin{aligned} A_2 &\leq c \sum_{l=0}^{\infty} 2^{-l} \sum_{k=l+1}^{\infty} 2^{k\left(s+\frac{1}{p}\right)p} 2^{-(k-l)(\sigma-1)p} b_k^p \\ &\leq c \sum_{k=0}^{\infty} 2^{ksp} b_k^p \sum_{l=0}^{k} 2^{-(k-l)(\sigma-1)p+k-l} \leq c' \sum_{k=0}^{\infty} 2^{ksp} b_k^p , \end{aligned} \tag{38}$$

if σ is sufficiently large. Now (32), (33), and (35)–(38) yield

$$\|g \mid F_{p,q}^{s+\frac{1}{p}}(R_n)\|^p \leq c \sum_{k=0}^{\infty} 2^{ksp} b_k^p = c \|f \mid B_{p,p}^s(R_{n-1})\|^p . \tag{39}$$

Hence, $g=\mathfrak{S}f$ has the desired properties of a coretraction in the case of the $F_{p,q}^s$-spaces if $r=0$ in the sense of the theorem.

Step 7. Let $r=0, 1, 2, \ldots$ be an arbitrary number. If $\mathfrak{R}$ has the meaning expressed in the theorem, we wish to construct a corresponding coretraction $\mathfrak{S}$. For that purpose we modify g from (26). Let $\varrho_j(\lambda)\in S(R_1)$ with $j=0, \ldots, r$ be non-negative

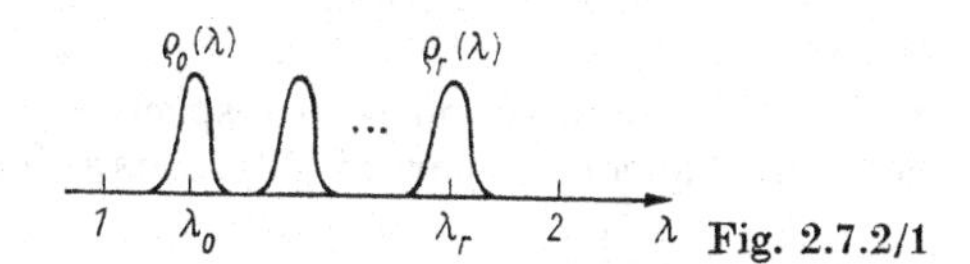

Fig. 2.7.2/1

non-trivial functions on the real line with supp $\varrho_j\subset(1, 2)$ if $j=0, \ldots, r$ and

$$\operatorname{supp} \varrho_j \cap \operatorname{supp} \varrho_k = \emptyset \quad \text{if} \quad j \neq k .$$

We assume that the supports of ϱ_j are concentrated near λ_j, cf. Fig. 2.7.2/1. Let d be a given number with $d=0, \ldots, r$. We determine real numbers $a_{l,d}$ such that

$$\psi^{(d)}(\lambda) = \sum_{l=0}^{r} a_{l,d} \varrho_l(\lambda)$$

and

$$\begin{aligned} & i^m (2\pi)^{-\frac{1}{2}} \int_{R_1} \lambda^m \psi^{(d)}(\lambda) \, d\lambda \\ &= i^m (2\pi)^{-\frac{1}{2}} \sum_{l=0}^{r} a_{l,d} \int_{R_1} \lambda^m \varrho_l(\lambda) \, d\lambda = \delta_{m,d} , \end{aligned} \tag{40}$$

where $m=0, \ldots, r$ (of course, $\delta_{m,d}$ is Kronecker's symbol). For fixed d, (40) has a unique solution because

$$\begin{aligned} \det \left(\int_{R_1} \lambda^m \varrho_l(\lambda) d\lambda \right) &\sim \det \left(\lambda_l^m \int_{R_1} \varrho_l(\lambda) \, d\lambda\right) \\ &\sim \prod_{l=0}^{r} \int_{R_1} \varrho_l(\lambda) \, d\lambda \cdot \det (\lambda_l^m) \neq 0 \end{aligned}$$

(Vandermonde's determinant). Now we replace the function ψ in the fifth step by $\psi^{(d)}$, in particular $\psi_j^{(d)}(\lambda)=\psi^{(d)}(2^{-j}\lambda)$ if $j=1, 2, \ldots$ Instead of (25) we have

$$\begin{aligned}2^{-j(1+m)}\frac{\mathrm{d}^m}{\mathrm{d}\mu^m}(F_1^{-1}\psi_j^{(d)})(\mu)\,|_{\mu=0}&=\left[\frac{\mathrm{d}^m}{\mathrm{d}\mu^m}F_1^{-1}\psi^{(d)}\right](0)\\ &=\mathrm{i}^m(2\pi)^{-\frac{1}{2}}\int_{R_1}\lambda^m\psi^{(d)}(\lambda)\,\mathrm{d}\lambda=\delta_{m,d}\quad\text{where}\quad m=0,\ldots,r\,.\end{aligned}\tag{41}$$

In a similar way we construct functions $\psi_0^{(d)}(\lambda)$. Let

$$\mathfrak{S}\{f_0,\ldots,f_r\}=\sum_{l=0}^{r}\sum_{k=0}^{\infty}2^{-k(1+l)}(F_1^{-1}\psi_k^{(l)})(x_n)\,(F'^{-1}\varphi_k'F'f_l)(x').\tag{42}$$

This is the counterpart of (26). Let

$$\{f_0,\ldots,f_r\}\in\prod_{l=0}^{r}B_{p,q}^{s-\frac{1}{p}-l}(R_{n-1})\,,$$

where $0<p\leqq\infty$, $0<q\leqq\infty$ and $-\infty<s<\infty$. Then the counterpart of (30) reads as follows:

$$\|\mathfrak{S}\{f_0,\ldots,f_r\}\mid B_{p,q}^s(R_n)\|\leqq c\sum_{l=0}^{r}\|f_l\mid B_{p,q}^{s-\frac{1}{p}-l}(R_{n-1})\|\,.\tag{43}$$

On the other hand, (41) yields

$$\frac{\partial^m}{\partial x_n^m}\mathfrak{S}\{f_0,\ldots,f_r\}\,(x',0)=\sum_{k=0}^{\infty}(F'^{-1}\varphi_k'F'f_m)(x')=f_m(x')\,,\tag{44}$$

where $m=0,\ldots,r$. This proves that $\mathfrak{S}$ is a coretraction with respect to $\mathfrak{R}$, if $\mathfrak{R}$ has the meaning of part (i) of the theorem. The counterpart of (39) reads as follows:

$$\|\mathfrak{S}\{f_0,\ldots,f_r\}\mid F_{p,q}^s(R_n)\|\leqq c\sum_{l=0}^{r}\|f_l\mid B_{p,p}^{s-\frac{1}{p}-l}(R_{n-1})\|\tag{45}$$

if $0<p<\infty$, $0<q\leqq\infty$ and $-\infty<s<\infty$. Now (44) and (45) prove that $\mathfrak{S}$ is a coretraction with respect to $\mathfrak{R}$, if $\mathfrak{R}$ has the meaning of part (ii) of the theorem. The proof is complete.

Remark 1. If $1\leqq p<\infty$, $1\leqq q\leqq\infty$ and $s>r+\frac{1}{p}$, then part (i) of the theorem is the well-known classical statement for the Besov spaces, cf. S. M. Nikol'skij [3] or [I, 2.9.3]. Furthermore, recalling that $H_p^s(R_n)=F_{p,2}^s(R_n)$ with $1<p<\infty$ are the Bessel-potential spaces, cf. Theorem 2.5.6, then (5) means that $\mathfrak{R}$ is a retraction from

$$H_p^s(R_n)\quad\text{onto}\quad\prod_{j=0}^{r}B_{p,p}^{s-\frac{1}{p}-j}(R_{n-1})\tag{46}$$

if $1<p<\infty$ and $s>r+\frac{1}{p}$. This statement is also well-known, cf. S. M. Nikol'skij [3] or [I, 2.9.3.].

Remark 2. Traces on hyperplanes of dimension $1, 2, \ldots, n-2$ can be obtained by iterative application of the above theorem. The method is clear, we shall not go into detail.

Remark 3. Some of the main ideas of the proof of the above theorem are due to B. Jawerth [1] (cf. also B. Jawerth [2]), who proved the above theorem for $r=0$. The most interesting assertion is the independence of the trace of $F_{p,q}^s(R_n)$ on R_{n-1} of q. This fact has also been observed (independently) by M. L. Gol'dman [4] and G. A. Kaljabin [3] (if $1<p<\infty$ and $1\leq q\leq\infty$). However, it is not a big surprise that the trace of $F^s_{p,q}(R_n)$ is independent of q: if $1<p<\infty$ and $\min(p,2)\leq q\leq\max(p,2)$, then

$$F^s_{p,\min(p,2)}(R_n)\subset F^s_{p,q}(R_n)\subset F^s_{p,\max(p,2)}(R_n)$$

and the two spaces on the left- and right-hand sides coincide with $B^s_{p,p}(R_n)$ and $H^s_p(R_n)$. In that case, the classical assertions (4), with $1<p<\infty$ and $p=q$, and (46) prove that the trace of $F^s_{p,q}(R_n)$ is independent of q.

Remark 4. The different formulations in the above theorem for the case $1\leq p\leq\infty$ on the one hand and the case $0<p<1$ on the other look a little curious at first glance. Nevertheless, these conditions are natural. Let us consider the spaces $B^s_{p,q}(R_n)$ and the trace $\frac{\partial^r f}{\partial x_n^r}(x',0)$ where $r=0,1,2,\ldots$ If $s>r+\frac{1}{p}+\max\left(0,(n-1)\left(\frac{1}{p}-1\right)\right)$, it follows from the above theorem that $\frac{\partial^r f}{\partial x_n^r}(x',0)$ makes sense and belongs to $B^{s-\frac{1}{p}-r}_{p,q}(R_{n-1})$, where this space is considered as a subspace of $S'(R_{n-1})$. It has been observed by J. Peetre [7] that the traces $\frac{\partial^r f}{\partial x_n^r}(x',0)$ of elements of $B^s_{p,q}(R_n)$ do not make sense in the framework of $S'(R_{n-1})$ if $s<r+\frac{1}{p}+$ $+\max\left(0,(n-1)\left(\frac{1}{p}-1\right)\right)$. On the other hand, B. Jawerth [2] proved that $\frac{\partial^r f}{\partial x_n^r}(x',0)$ exists if $f\in B^s_{p,q}(R_n)$ and $0<p<1$, $r+\frac{1}{p}<s<r+\frac{1}{p}+(n-1)\left(\frac{1}{p}-1\right)$, however not as an element of

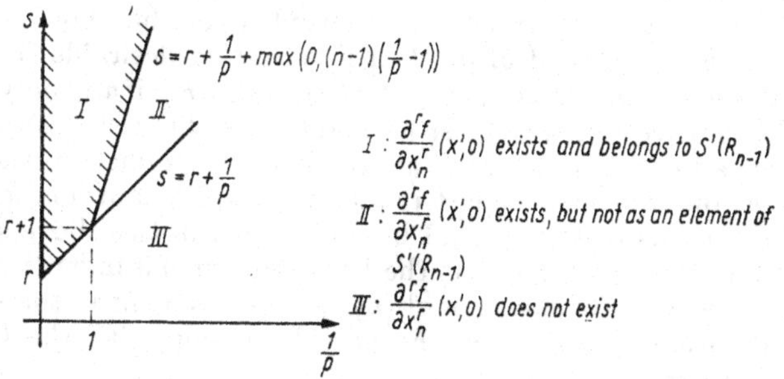

Fig. 2.7.2/2

$S'(R_{n-1})$: it belongs to appropriate subspaces of $L_p(R_{n-1})$. Finally, in [S, p. 115], we proved that $\frac{\partial^r f}{\partial x_n^r}(x',0)$ does not exist if $0<p\leq\infty$ and $s<r+\frac{1}{p}$.

Remark 5. The following limiting case is of some interest. If $1\leq p<\infty$ then the exact trace of $B^{\frac{1}{p}}_{p,1}(R_n)$ on R_{n-1} is $L_p(R_{n-1})$. This has been proved independently by J. Peetre [7], M. L. Gol'dman [3], and V. I. Burenkov, M. L. Gol'dman [1]. The latter papers contain also generalizations to traces on m-dimensional hyperplanes with $m<n$ and to the anisotropic case. However there does not exist a bounded linear extension operator from $L_p(R_{n-1})$ into $B^{\frac{1}{p}}_{p,1}(R_n)$. The corresponding problem for the trace of $W^1_1(R_n)$ on R_{n-1} (which is $L_1(R_{n-1})$) has been treated by J. Peetre [9].

2.8. Pointwise Multipliers

2.8.1. Definition and Preliminaries

Section 2.8. deals with multiplication properties of the spaces $B^s_{p,q}(R_n)$ and $F^s_{p,q}(R_n)$. If g is a given function on R_n, then we ask whether $f \to gf$ (pointwise multiplication) yields a bounded linear mapping from $B^s_{p,q}(R_n)$ into itself or from $F^s_{p,q}(R_n)$ into itself. Such a function g is called a pointwise multiplier for the corresponding spaces. For the sake of brevity, we shall use the notion "multiplier" instead of "pointwise multiplier" and in contrast to "Fourier multiplier". In other words: "multiplier" always means "pointwise multiplier" and never "Fourier multiplier". If $g \in S'(R_n)$ and $f \in S'(R_n)$, it is not clear in general what is meant by gf (pointwise multiplication). Our approach is the following. Let $\varphi(x) \in S(R_n)$ with compact support on R_n and with $\varphi(x)=1$ if $|x| \leqq 1$. If $g \in S'(R_n)$, then

$$g_j(x)=F^{-1}\varphi(2^{-j}\cdot)\, Fg \quad \text{with} \quad j=0, 1, 2, \ldots \tag{1}$$

is a sequence of entire analytic functions. $g_j(x)$ and all its derivatives have at most polynomial growth, cf. Theorem 1.2.1/2. Hence $g_j(x)\, f$ makes sense (pointwise multiplication of a tempered distribution). If $\{g_j f\}_{j=0}^{\infty}$ is a fundamental sequence in a space $B^{\sigma}_{u,v}(R_n)$ or $F^{\sigma}_{u,v}(R_n)$, we denote the limit element by gf (pointwise multiplication). If gf has an immediate meaning, e.g. if $g \in S(R_n)$ and $f \in S'(R_n)$, or if $g \in L_2(R_n)$ and $f \in L_2(R_n)$, then the above definition coincides with the usual one. Obviously, in order to make the above definition of gf reasonable, we must ensure that the limit element of $\{g_j f\}_{j=0}^{\infty}$ is independent of the choice of φ. Also, if f_j is given by (1) with f instead of g, then it would be desirable that $\{g_j f\}_{j=0}^{\infty}$ is a fundamental sequence in a given space $B^{\sigma}_{u,v}(R_n)$ or $F^{\sigma}_{u,v}(R_n)$ if and only if $\{g f_j\}_{j=0}^{\infty}$ be a fundamental sequence in the same space and the two limit elements coincide. As a matter of fact, all constructions in this section are symmetric with respect to f and g, and, as a consequence, the last requirement will be automatically satisfied. All products gf in this section should be understood in the above sense, and it will be clear from our calculations that the limit element gf is independent of the chosen function φ. However, we shall not always stress this point in the sequel. In particular, we shall usually work from the very beginning with the function g instead of the sequence $\{g_j\}_{j=0}^{\infty}$.

2.8.2. General Multipliers

We recall that $\mathcal{C}^{\varrho}(R_n)=B^{\varrho}_{\infty,\infty}(R_n)$ with $\varrho > 0$ are the Zygmund spaces, cf. (2.5.12/22) or (2.5.7/6).

Theorem. (i) Let $-\infty < s < \infty$, $0 < p \leqq \infty$, $0 < q \leqq \infty$ and

$$\varrho > \max\left(s, \frac{n}{p}-s\right). \tag{1}$$

Then $g \in \mathcal{C}^{\varrho}(R_n)$ is a multiplier for $B^s_{p,q}(R_n)$. In other words, $f \to gf$ yields a bounded linear mapping from $B^s_{p,q}(R_n)$ into itself and there exists a positive constant c such that

$$\|gf \mid B^s_{p,q}(R_n)\| \leqq c \|g \mid \mathcal{C}^{\varrho}(R_n)\|\, \|f \mid B^s_{p,q}(R_n)\| \tag{2}$$

holds for all $g \in \mathcal{C}^{\varrho}(R_n)$ and all $f \in B^s_{p,q}(R_n)$.

(ii) Let $-\infty<s<\infty$, $0<p<\infty$, $0<q\leqq\infty$ and

$$\varrho>\max\left(s,\ \frac{n}{\min(p,q)}-s\right). \tag{3}$$

Then $g\in\mathcal{C}^\varrho(R_n)$ is a multiplier for $F^s_{p,q}(R_n)$. In other words, $f\to gf$ yields a bounded linear mapping from $F^s_{p,q}(R_n)$ into itself and there exists a positive constant c such that

$$\|gf \mid F^s_{p,q}(R_n)\|\leqq c\|g \mid \mathcal{C}^\varrho(R_n)\|\ \|f \mid F^s_{p,q}(R_n)\| \tag{4}$$

holds for all $g\in\mathcal{C}^\varrho(R_n)$ and all $f\in F^s_{p,q}(R_n)$.

Proof. Step 1. Let $\{\varphi_k(x)\}_{k=0}^\infty\in\Phi(R_n)$, cf. Definition 2.3.1/1. We may assume that $\varphi_k(x)=\varphi_1(2^{-k+1}x)$ if $k=1, 2, 3, \ldots$ Let

$$b_k(x)=(F^{-1}\varphi_kFg)(x) \quad\text{and}\quad c_k(x)=(F^{-1}\varphi_kFf)(x)\,, \tag{5}$$

where $k=0, 1, 2, \ldots$ We assume temporarily that Fg has compact support (this is in accordance with our remarks in 2.8.1. about the definition of gf). In that case, the following formulas make sense. We have

$$g=\sum_{k=0}^\infty b_k \quad\text{and}\quad f=\sum_{k=0}^\infty c_k \quad (\text{in } S'(R_n))\,. \tag{6}$$

It follows that

$$\begin{aligned}[F^{-1}\varphi_kF(gf)](x) &= \int_{R_n}(F^{-1}\varphi_k)(y)\,(gf)(x-y)\,dy\\ &=2^{(k-1)n}\int_{R_n}(F^{-1}\varphi_1)(2^{k-1}y)(gf)(x-y)\,dy\\ &=\int_{R_n}(F^{-1}\varphi_1)(y)\sum_{l,j=0}^\infty b_j(x-2^{-k+1}y)\,c_l(x-2^{-k+1}y)\,dy\end{aligned} \tag{7}$$

if $k=1, 2, 3, \ldots$ and

$$\begin{aligned}[F^{-1}\varphi_kF(gf)](x) &= \int_{R_n}\varphi_k(y)\sum_{l,j=0}^\infty [F^{-1}(b_j\cdot c_l)](-y)\,e^{ixy}dy\\ &=\int_{R_n}\varphi_k(y)\,e^{ixy}\sum_{l,j=0}^\infty [F(b_jc_l)](y)\,dy\end{aligned} \tag{8}$$

f $k=0, 1, 2, \ldots$ The intersection of the supports of

$$\varphi_k(y) \quad\text{and}\quad [F(b_jc_l)](y)=\int_{R_n}(Fb_j)(z)\,(Fc_l)(y-z)\,dz \tag{9}$$

is empty if the non-negative integers j and l do not belong to one of the following three cases:

(i) $k-2\leqq l\leqq k+2$ and $j=0, \ldots, k$;
(ii) $k-2\leqq j\leqq k+2$ and $l=0, \ldots, k$;
(iii) $l\geqq k$, $j\geqq k$ and $|l-k|\leqq 3$.

Hence it follows from (8) that we may restrict ourselves in (7) to these three cases. In particular, the sums in (7) and (8) are finite and consequently, these formulas make sense. It will be sufficient to consider the following three model cases:

$$(\Sigma_k' f)(x)=\sum_{j=0}^k\int_{R_n}(F^{-1}\varphi_1)(y)\,b_j(x-2^{-k+1}y)\,c_k(x-2^{-k+1}y)\,dy\,, \tag{10}$$

$$(\Sigma_k''f)(x)=\sum_{l=0}^{k}\int_{R_n}(F^{-1}\varphi_1)(y)\,b_k(x-2^{-k+1}y)\,c_l(x-2^{-k+1}y)\,dy\,, \tag{11}$$

$$(\Sigma_k'''f)(x)=\sum_{l=k}^{\infty}\int_{R_n}(F^{-1}\varphi_1)(y)\,b_l(x-2^{-k+1}y)\,c_l(x-2^{-k+1}y)\,dy\,, \tag{12}$$

where $k=1, 2, \ldots$ All the other sums are treated in the same way, including the corresponding sums with $k=0$.

Step 2. We introduce the maximal functions

$$b_j^*(x)=\sup_{y\in R_n}\frac{|b_j(x-y)|}{(1+|2^jy|)^{a_1}}\,,\quad c_l^*(x)=\sup_{y\in R_n}\frac{|c_l(x-y)|}{(1+|2^ly|)^{a_2}} \tag{13}$$

from Definition 2.3.6/2. Let $a_1>0$ and

$$a_2>\frac{n}{p}\quad\text{if}\quad f\in B^s_{p,q}(R_n)\quad\text{and}\quad a_2>\frac{n}{\min(p,q)}\quad\text{if}\quad f\in F^s_{p,q}(R_n)\,. \tag{14}$$

Let

$$C=\int_{R_n}|(F^{-1}\varphi_1)(y)|\,(1+|2y|)^{a_1+a_2}dy\,.$$

Then (10)–(12) can be estimated by

$$|(\Sigma_k'f)(x)|\leqq Cc_k^*(x)\sum_{j=0}^{k}b_j^*(x)\,, \tag{15}$$

$$|(\Sigma_k''f)(x)|\leqq Cb_k^*(x)\sum_{l=0}^{k}c_l^*(x)\,, \tag{16}$$

$$|(\Sigma_k'''f)(x)|\leqq C\sum_{l=k}^{\infty}2^{(l-k)(a_1+a_2)}b_l^*(x)\,c_l^*(x)\,. \tag{17}$$

We prove (4). Let $f\in F^s_{p,q}(R_n)$ and let $g\in \mathcal{C}^\varrho(R_n)$, where ϱ satisfies (3). We use the abbreviations from 2.3.1. Let ε and ε' be positive numbers with $0<\varepsilon'<\varepsilon$. Then

$$2^{ks}|(\Sigma_k'f)(x)|\leqq C2^{ks}c_k^*(x)\|b_j^*\mid l_1(L_\infty(R_n))\|\,, \tag{18}$$

$$\begin{aligned}2^{ks}|(\Sigma_k''f)(x)|&\leqq C2^{k(|s|+\varepsilon)}b_k^*(x)\sum_{l=0}^{k}2^{l(s-\varepsilon)}c_l^*(x)\\&\leqq C'2^{k|s|+k\varepsilon}\|b_k^*\mid L_\infty(R_n)\|\left(\sum_{l=0}^{k}2^{lsq}c_l^{*q}(x)\right)^{\frac{1}{q}},\end{aligned} \tag{19}$$

$$\begin{aligned}2^{ks}|(\Sigma_k'''f)(x)|&\leqq C\sum_{l=k}^{\infty}2^{(l-k)(a_1+a_2+s-s)}b_l^*(x)\,2^{ls-\varepsilon(l-k)}c_l^*(x)\\&\leqq C'\|2^{j\max(0,a_1+a_2+\varepsilon-s)}b_j^*\mid l_\infty(L_\infty(R_n))\|\left(\sum_{l=k}^{\infty}2^{-\varepsilon'(l-k)q}2^{lsq}c_l^{*q}(x)\right)^{\frac{1}{q}},\end{aligned} \tag{20}$$

(modification if $q=\infty$). It follows that

$$\begin{aligned}&\|2^{ks}(\Sigma_k'f+\Sigma_k''f+\Sigma_k'''f)(\cdot)\mid L_p(R_n,l_q)\|\leqq c\|2^{ks}c_k^*\mid L_p(R_n,l_q)\|\\&\times[\|2^{j(|s|+\varepsilon)}b_j^*\mid l_\infty(L_\infty(R_n))\|+\|2^{j(a_1+a_2+\varepsilon-s)}b_j^*\mid l_\infty(L_\infty(R_n))\|]\,,\end{aligned} \tag{21}$$

where $\varepsilon>0$ and $c>0$. The above choice of a_1 and a_2 and Theorem 2.3.6 prove that the right-hand side of (21) can be estimated from above by

$$c\|f \mid F^s_{p,q}(R_n)\|[\|g \mid \mathcal{C}^{|s|+\varepsilon}(R_n)\|+\|g \mid \mathcal{C}^{a_1+a_2+\varepsilon-s}(R_n)\|]\,. \tag{22}$$

This proves (4). The proof of (2) follows the same lines (obvious modifications of the above calculations).

Step 3. We assumed temporarily that Fg had compact support. Now let $g\in\mathcal{C}^\varrho(R_n)$ and $f\in F^s_{p,q}(R_n)$ or $f\in B^s_{p,q}(R_n)$, respectively, be arbitrary functions, where the parameters involved satisfy the hypotheses of the theorem. Let again $\{\varphi_k(x)\}_{k=0}^\infty\in\Phi(R_n)$ and $g_j=\sum_{l=0}^{j}F^{-1}\varphi_l Fg$. Then it follows from (4) (and Step 2) that

$$\begin{aligned}\|fg_j-fg_k \mid F^s_{p,q}(R_n)\| &\leq c\|f \mid F^s_{p,q}(R_n)\|\left\|\sum_{l=j+1}^{k}F^{-1}\varphi_l Fg \mid \mathcal{C}^\varrho(R_n)\right\|\\ &\leq c\varepsilon\|f \mid F^s_{p,q}(R_n)\|\;\|g \mid \mathcal{C}^{\varrho+\delta}(R_n)\|\end{aligned} \tag{23}$$

if $k\geq j\geq j(\varepsilon)$ and $\delta>0$. This proves that $\{fg_j\}_{j=0}^\infty$ is a fundamental sequence in $F^s_{p,q}(R_n)$. Similarly for $B^s_{p,q}(R_n)$. The limit element gf satisfies (4), cf. the considerations in 2.8.1. The proof is complete.

Remark 1. The question arises whether the restrictions for ϱ in (1) and (3) are natural. If s is large, then we have $\varrho>s$, and this is natural. However, if s is near zero, then the restrictions for ϱ are too strong. This can be seen easily if one considers $F^0_{p,2}(R_n)=L_p(R_n)$ with $1\leq p<\infty$. We discuss this problem in the following corollary, where we formulate natural restrictions for ϱ. Of course, the restrictions for ϱ in (1) and (3) depend on the technique of maximal functions. A detailed examination of the above proof shows that we need this technique only in the case of the space $F^s_{p,q}(R_n)$. As for the spaces $B^s_{p,q}(R_n)$, this technique is not necessary. Let e.g. $s>0$, $1\leq p\leq\infty$ and $0<q\leq\infty$. Then we apply simply Hölder's inequality to (10), (11), and (12) and obtain that

$$\|gf \mid B^s_{p,q}(R_n)\|\leq c\|g \mid B^s_{\infty,q}(R_n)\|\;\|f \mid B^s_{p,q}(R_n)\|\,, \tag{24}$$

where c is independent of g and f. Of course, this is an essential improvement of part (i) of the theorem, provided that $s>0$, $1\leq p\leq\infty$ and $0<q\leq\infty$. (24) is due to J. Peetre [8, pp. 144/145] (his proof is similar to the one above). It seems to be possible to extend this direct method for the spaces $B^s_{p,q}(R_n)$ to $s<0$ and/or $p<1$. We shall not go into detail because we obtain a rather definitive result for the spaces $B^s_{p,q}(R_n)$ in the following corollary on the basis of other methods. Further assertions for the spaces $B^s_{p,q}(R_n)$ may be found in J. Peetre [8, Chapter 7].

Corollary. (i) Let $-\infty<s<\infty$, $0<p\leq\infty$, $0<q\leq\infty$ and

$$\varkappa^s_p=\max\left(s,\,n\left(\frac{1}{\min(p,1)}-1\right)-s\right). \tag{25}$$

If $\varrho>\varkappa^s_p$, then every $g\in\mathcal{C}^\varrho(R_n)$ is a multiplier for $B^s_{p,q}(R_n)$ and (2) holds for all $g\in\mathcal{C}^\varrho(R_n)$ and all $f\in B^s_{p,q}(R_n)$. If $\varrho<\varkappa^s_p$, then there exists a function $g\in\mathcal{C}^\varrho(R_n)$ which is not a multiplier for $B^s_{p,q}(R_n)$.

(ii) Let $-\infty<s<\infty$, $0<p<\infty$, $0<q<\infty$ and

$$\varkappa^s_{p,q}=\max\left(s,\,n\left(\frac{1}{\min(p,q,1)}-1\right)-s\right). \tag{26}$$

If $\varrho>\varkappa^s_{p,q}$, then every $g\in\mathcal{C}^\varrho(R_n)$ is a multiplier for $F^s_{p,q}(R_n)$ and (4) holds for all $g\in\mathcal{C}^\varrho(R_n)$ and all $f\in F^s_{p,q}(R_n)$. If $p\leq q$ and $\varrho<\varkappa^s_{p,q}=\varkappa^s_p$, then there exists a function $g\in\mathcal{C}^\varrho(R_n)$ which is not a multiplier for $F^s_{p,q}(R_n)$.

Proof. (Outline). *Step 1.* We use Corollary 2.5.11 and Theorem 2.5.12 (in particular (2.5.12/4)). Then it follows by elementary calculations that $g \in \mathscr{C}^\varrho(R_n)$ is a multiplier both for $F^s_{p,q}(R_n)$ and $B^s_{p,q}(R_n)$ if $\varrho > s > 0$, $1 < p \leqq \infty$ and $1 \leqq q \leqq \infty$ (with $p < \infty$ and $q < \infty$ in the case of $F^s_{p,q}(R_n)$). We note that $f \to gf$ is a self-adjoint operation. Hence it follows by duality, cf. Theorem 2.11.2, that g is also a multiplier for $B^s_{p,q}(R_n)$ with $-\varrho < s < 0$, $1 < p \leqq \infty$, $1 < q \leqq \infty$ and for $F^s_{p,q}(R_n)$ with $-\varrho < s < 0$, $1 < p < \infty$, $1 < q < \infty$. We use the real interpolation formula (2.4.2/1) and the interpolation property from Proposition 2.4.1 to obtain that g is a multiplier for $B^s_{p,q}(R_n)$ if $|s| < \varrho$, $1 < p \leqq \infty$ and $0 < q \leqq \infty$. Next we wish to apply the complex interpolation formula (2.4.7/3). We may assume that Fg has compact support (in the sense of 2.8.1.; arbitrary g's are considered later in the same way as in Step 3 of the proof of the above theorem). Then it follows easily from the considerations in Step 1 of the proof of the above theorem that $f(z) \to gf(z)$ maps an $S'(R_n)$-analytic function in A in the sense of 2.4.4. into an $S'(R_n)$-analytic function in A. In other words, the multiplication operator $f \to gf$ has the interpolation property for the complex method from 2.4.4. It follows from (2.4.7/3) that the above function g is a multiplier for $F^s_{p,q}(R_n)$ if $|s| < \varrho$, $1 < p < \infty$, $1 < q < \infty$.

Step 2. We prove the first statement of part (i). If $1 < p \leqq \infty$, it follows from Step 1 that $g \in \mathscr{C}^\varrho(R_n)$, with $\varrho > |s| = \varkappa^s_p$, is a multiplier for $B^s_{p,q}(R_n)$, with $0 < q \leqq \infty$. Let $p < 1$. Let $0 < p_1 < p < 1 < p_0$ and $\frac{1}{p} = \frac{1-\Theta}{p_0} + \frac{\Theta}{p_1}$. Let

$$g(z) = F^{-1}(1+|\xi|^2)^{-\frac{z}{2}\varkappa} Fg \quad \text{with} \quad 0 \leqq \operatorname{Re} z \leqq 1 \quad \text{and} \quad g \in \mathscr{C}^\varepsilon(R_n) \tag{27}$$

where $\varepsilon > 0$. Then $g(it) \in \mathscr{C}^\varepsilon(R_n)$ is a multiplier for $B^0_{p_0,q_0}(R_n)$, where $t \in R_1$. Furthermore, $g(it+1) \in \mathscr{C}^{\varkappa+\varepsilon}(R_n)$ is a multiplier for $B^{s_1}_{p_1,q_1}(R_n)$ if $\varkappa = \max\left(s_1, \frac{n}{p_1} - s_1\right)$, cf. the above theorem. We may again assume (without loss of generality) that Fg has compact support. Then $f(z) \to g(z) f(z)$ maps an $S'(R_n)$-analytic function in A into an $S'(R_n)$-analytic function in A (cf. Step 1). Hence it follows from (2.4.7/2) and the interpolation property that $g(\Theta) \in \mathscr{C}^{\varepsilon+\Theta\varkappa}(R_n)$ is a multiplier for $B^s_{p,q}(R_n)$ if

$$s = \Theta s_1, \quad \frac{1}{p} = \frac{1-\Theta}{p_0} + \frac{\Theta}{p_1} \quad \text{and} \quad 0 < q \leqq \infty$$

(the value of q is not interesting). However,

$$\Theta\varkappa = \max\left(s, \frac{n}{p} - n\frac{1-\Theta}{p_0} - s\right) \to \max\left(s, \frac{n}{p} - n - s\right)$$

f $p_0 \downarrow 1$, $p_1 \to 0$ and (consequently) $\Theta \to 0$. This proves the first assertion of (i).

Step 3. The first statement of part (ii) is proved similarly. If $1 < p < \infty$ and $1 < q < \infty$, the desired assertion follows from Step 1. If $q > 1$ and $p \leqq 1$, we use (2.4.7/3) with $q_0 = q_1 = q$, p_0 near 1, $p_1 \to 0$, $s_0 = 0$, and the above theorem. Similarly if $p > 1$ and $q \leqq 1$. If $p \leqq 1$ and $q \leqq 1$, we use (2.4.7/3) with $s_0 = 0$, p_0 and q_0 near 1. The rest is the same as in Step 2.

Step 4. We prove the second statement of part (i). Let $\varrho > 0$ be given.

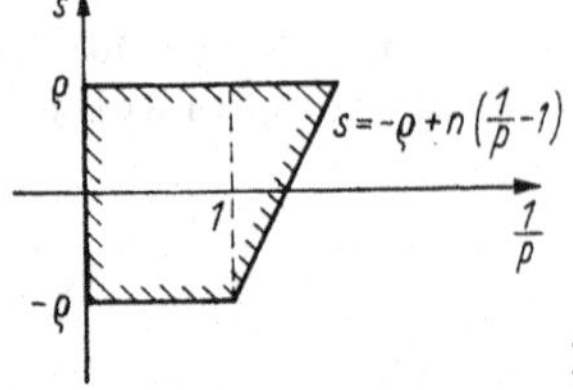

Fig. 2.8.2

Then it follows from the above considerations that $g \in \mathscr{C}^\varrho(R_n)$ is a multiplier for $B^s_{p,q}(R_n)$ if $0 < q \leqq \infty$ and if $\left(s, \frac{1}{p}\right)$ belongs to the interior of the shaded area (including the left) boundary which corresponds to $p=\infty$). Now we assume that a given function $g\in\mathcal{C}^\varrho(R_n)$ is a multiplier for a space $B^s_{p,q}(R_n)$ with $|s|>\varrho$. It follows by interpolation, cf. (2.4.7/2), and duality, cf. (2.11.2/1), that g is also a multiplier for certain spaces $B^\sigma_{u,u}(R_n)$ with $\sigma>\varrho$, $1<u<\infty$. By (2.5.12/4), it is not hard to see that g belongs at least locally to $B^\sigma_{u,u}(R_n)$, but neither is it complicated to construct a function $g\in\mathcal{C}^\varrho(R_n)$ which does not belong locally to $B^\sigma_{u,u}(R_n)$. This is not a multiplier for the above space $B^s_{p,q}(R_n)$. Let $\left(s,\frac{1}{p}\right)$ be on the right-hand side of the line $s=-\varrho+$ $+n\left(\frac{1}{p}-1\right)$. It follows from Corollary 2.8.5 (cf. also Fig. 2.8.5) that there exist functions $g\in\mathcal{C}^\varrho(R_n)$ (we put $\varrho=\alpha$ in comparison with Corollary 2.8.5) which are not multipliers for $B^s_{p,q}(R_n)$. This proves the second statement of part (i). Let $p\leqq q$ and let $g\in\mathcal{C}^\varrho(R_n)$ be a multiplier for $F^s_{p,q}(R_n)$, where $\left(s,\frac{1}{p}\right)$ is outside of the shaded area. Then it follows by complex interpolation, cf. (2.4.7/3), and afterwards by real interpolation, cf. (2.4.2/2), that g is a multiplier for certain spaces $B^\sigma_{u,u}(R_n)$ with $\left(\sigma,\frac{1}{u}\right)$ outside of the shaded area. Consequently, the second statement of part (ii) can be reduced to the corresponding assertion of part (i).

Remark 2. The corollary shows that $\varkappa^s_p$ and $\varkappa^s_{p,q}$ with $q\geqq p$ are natural constants in connection with the multiplier problem. Probably $\varkappa_{p,q}$ is also natural if $q<p$, but we do not have final results as in the other cases. Of particular interest are the local Hardy spaces $h_p(R_n)=F^0_{p,2}(R_n)$ if $0<p\leqq 1$, cf. 2.5.8. In that case we have

$$\mathcal{C}^\varrho(R_n)\cdot h_p(R_n)\subset h_p(R_n) \quad \text{if} \quad 0<p\leqq 1 \quad \text{and} \quad \varrho>n\left(\frac{1}{p}-1\right).$$

On the other hand, $\mathcal{C}^\varrho(R_n)$ is not a multiplier space (it is clear what this means) for $h_p(R_n)$ with $0<p\leqq 1$ if $\varrho<n\left(\frac{1}{p}-1\right)$.

Remark 3. Besides the almost classical paper by R. S. Strichartz [1] we refer to V. G. Maz'ja, T. O. Šapošnikova [1–8]. The latter authors obtained far-reaching multiplier theorems in function spaces, partly of necessary and sufficient character.

2.8.3. Multiplication Algebras

The pointwise product gf of two elements $f\in B^s_{p,q}(R_n)$ and $g\in B^s_{p,q}(R_n)$ has the same meaning as in 2.8.1. (if it exists). Similarly if $f\in F^s_{p,q}(R_n)$ and $g\in F^s_{p,q}(R_n)$. Our problem is to find conditions under which this product gf is an element of $B^s_{p,q}(R_n)$ or $F^s_{p,q}(R_n)$, respectively. Then $B^s_{p,q}(R_n)$ or $F^s_{p,q}(R_n)$, respectively, is called a multiplication algebra, in contrast to the convolution algebras from 2.6.5., where the pointwise multiplication was replaced by the convolution.

Definition. Let either $A(R_n)=B^s_{p,q}(R_n)$ with $-\infty<s<\infty$, $0<p\leqq\infty$, $0<q\leqq\infty$ or $A(R_n)=$ $=F^s_{p,q}(R_n)$ with $-\infty<s<\infty$, $0<p<\infty$, $0<q<\infty$. Then $A(R_n)$ is said to be a multiplication algebra if for all $g\in A(R_n)$ and all $f\in A(R_n)$ the product gf exists (in the sense of 2.8.1.) and belongs to $A(R_n)$, and if there exists a positive constant c such that

$$\|gf \mid A(R_n)\|\leqq c\|g \mid A(R_n)\|\cdot\|f \mid A(R_n)\| \tag{1}$$

holds for all $g\in A(R_n)$ and all $f\in A(R_n)$.

We recall that $C(R_n)$ is the space of all bounded and uniformly continuous functions on R_n, cf. 2.2.2. Furthermore, "$\subset$" always stands for linear and continuous embedding.

Theorem. Let $-\infty<s<\infty$, $0<p\leqq\infty$ and $0<q\leqq\infty$. The following three statements are pairwise equivalent:

(i) $B^s_{p,q}(R_n)$ is a multiplication algebra,

(ii) $B^s_{p,q}(R_n) \subset C(R_n)$,

(iii) There holds $\begin{cases} \text{either} \quad 0<p\leqq\infty, \quad 0<q\leqq\infty, \quad \text{and} \quad s>\dfrac{n}{p} \\ \text{or} \quad 0<p\leqq\infty, \quad 0<q\leqq 1, \quad \text{and} \quad s=\dfrac{n}{p}. \end{cases}$

Remark 1. A proof of this theorem can be found in [S, pp. 133–137]. It is a modification of the proof of Theorem 2.8.2 (in the sense of (2.8.2/24) without the explicit use of maximal functions). Of course, the equivalence of (ii) and (iii) is an embedding theorem. It can be proved in the same way as in Theorem 2.7.1 (i). As a special case of the above theorem we remark that the Zygmund spaces $\mathcal{C}^s(R_n)=B^s_{\infty,\infty}(R_n)$, with $-\infty<s<\infty$, are multiplication algebras if and only if $s>0$.

Corollary. $F^s_{p,q}(R_n)$ is a multiplication algebra if

$$\begin{cases} \text{either} \quad 0<p<q\leqq\infty \quad \text{and} \quad s>\dfrac{n}{p} \\ \text{or} \quad 0<q\leqq p<\infty \quad \text{and} \quad s>\dfrac{n}{2}\left(\dfrac{1}{p}+\dfrac{1}{q}\right). \end{cases} \tag{2}$$

Remark 2. A proof of this corollary has been given in [S, pp. 137–138]. It is a modification of the proof of Theorem 2.8.2, although the corollary does not have the same final character as the theorem above. A satisfactory answer for the cases $1<p<\infty$ and $1<q<\infty$ has been given recently by G. A. Kaljabin [4, 6]. He proved that $F^s_{p,q}(R_n)$ with $1<p<\infty$ and $1<q<\infty$ is a multiplication algebra if and only if

$$F^s_{p,q}(R_n) \subset C(R_n). \tag{3}$$

The "only if"-part has also been observed in [S, p. 139]. This fits quite well with the corresponding condition for the spaces $B^s_{p,q}(R_n)$. The problem is to extend Kaljabin's result to $p\leqq 1$ and/or $q\leqq 1$. A partial solution of this problem has been obtained recently by P. Nilsson [2], who proved that $F^s_{p,q}(R_n)$ with $0<p<\infty$, $0<q<\infty$ and $s>n\left(\dfrac{1}{\min(p,q,1)}-1\right)$ is a multiplication algebra if an only if (3) holds. It is easy to see that (3) holds if $0<p<\infty$, $0<q\leqq\infty$ and $s>\dfrac{n}{p}$. Then it follows from Kaljabin's result and the above corollary that $F^s_{p,q}(R_n)$ is a multiplication algebra if $1<p<\infty$, $1<q<\infty$ and $s>\dfrac{n}{p}$.

Remark 3. We give some references as far as the classical Bessel-potential spaces $H^s_p(R_n)= =F^s_{p,2}(R_n)$ with $1<p<\infty$ and the Besov spaces $B^s_{p,q}(R_n)$ with $1\leqq p\leqq\infty$, $1\leqq q\leqq\infty$, are concerned. R. S. Strichartz [1] proved that $H^s_p(R_n)$ with $1<p<\infty$ and $s>\dfrac{n}{p}$ is a multiplication algebra (if s is an integer, then this statement is a simple consequence of certain embedding theorems, cf. R. A. Adams [1, p. 115]). Some results in this direction for the spaces $B^s_{p,q}(R_n)$ with $1\leqq p\leqq\infty$ and $1\leqq q\leqq\infty$, respectively their homogeneous and periodic counterparts may be found in J. Peetre [8, p. 147], C. S. Herz [1], C. Bennett, J. E. Gilbert [1] and R. Johnson [4]. Cf. also J.-L. Zolésio [1] and M. Murata [1].

2.8.4. The Classes $P_{p,\alpha}(R_n)$

In Theorem 2.8.2 and Corollary 2.8.2 we formulated conditions under which $f\to gf$ with $g\in\mathcal{C}^\varrho(R_n)$ yields a bounded mapping from $B^s_{p,q}(R_n)$ or $F^s_{p,q}(R_n)$ into itself, respectively. The restrictions for ϱ in Corollary 2.8.2 are quite natural. But for our later purposes these results are not sufficient. Of particular interest is the clarification of the problem under which conditions the characteristic function of a half-space is a multiplier for $B^s_{p,q}(R_n)$ or for $F^s_{p,q}(R_n)$. Statements of this type will be needed in the sequel in a very decisive way. We discuss this problem from a slightly more general point of view in this subsection and introduce for that purpose the classes $P_{p,\alpha}(R_n)$. In particular, the characteristic function of a half-space belongs to $P_{p,0}(R_n)$. The results of this subsection are needed later on, but the proofs are not very typical, and they may be skipped at first glance. Let

$$\mathcal{C}^\infty(R_1) = \bigcap_{\sigma>0} \mathcal{C}^\sigma(R_1), \tag{1}$$

where $\mathcal{C}^\sigma(R_1) = B^\sigma_{\infty,\infty}(R_1)$ are the Zygmund spaces (one-dimensional case), cf. 2.5.7. or 2.5.12.

Definition. Let $0<p\leqq\infty$ and $\alpha\geqq 0$. Then $P_{p,\alpha}(R_n)$ is the set of all bounded complex-valued functions $g(x)$ on R_n which can be represented as

$$g(x)=G(x_n) \quad \text{and} \quad G(t)\in B^{\frac{1}{p}+\alpha}_{p,\infty}(R_1)+\mathcal{C}^\infty(R_1). \tag{2}$$

Remark 1. In particular, $g(x)\in P_{p,\alpha}(R_n)$ depends explicitly only on x_n. Furthermore, if $\alpha>0$, then it follows from (2.7.1/1) that

$$B^{\frac{1}{p}+\alpha}_{p,\infty}(R_1)\subset B^\alpha_{\infty,\infty}(R_1)=\mathcal{C}^\alpha(R_1)\subset C(R_1). \tag{3}$$

In particular, every $g(x)$ which satisfies (2) is bounded on R_n. In other words, the additional assumption above that $g(x)$ is bounded on R_n is not necessary if $\alpha>0$: in that case, (2) is sufficient. On the other hand, if $0<p\leqq\infty$, then there exist essentially unbounded functions $G(t)\in B^{\frac{1}{p}}_{p,\infty}(R_1)$. This follows from the equivalence of (ii) and (iii) in Theorem 2.8.3 (which also holds if one replaces $C(R_n)$ in (ii) by $L_\infty(R_n)$). Consequently, if $\alpha=0$, then (2) does not ensure automatically that $g(x)$ is bounded on R_n.

Lemma. The following inclusions hold:

$$P_{p,\alpha}(R_n)\subset P_{q,\alpha}(R_n) \quad \text{if} \quad 0<p\leqq q\leqq\infty \quad \text{and} \quad \alpha\geqq 0, \tag{4}$$

and

$$P_{p,\alpha}(R_n)\subset P_{p,\beta}(R_n) \quad \text{if} \quad 0\leqq\beta\leqq\alpha<\infty \quad \text{and} \quad 0<p\leqq\infty. \tag{5}$$

Proof. (5) is obvious and (4) follows from

$$B^{\frac{1}{p}+\alpha}_{p,\infty}(R_1)\subset B^{\frac{1}{q}+\alpha}_{q,\infty}(R_1),$$

cf. (2.7.1/1).

If α is a real number and if a and b are complex numbers, then we introduce

$$h^{a,b}_\alpha(t)=\begin{cases} at^\alpha & \text{if } t\geqq 0, \\ b|t|^\alpha & \text{if } t<0. \end{cases} \tag{6}$$

The Fourier transform on R_1 and its inverse will be denoted by F_1 and F_1^{-1}, respectively.

Proposition. (i) Let $0<p\leqq\infty$. Then the characteristic function of $R_n^+=\{x \mid x\in R_n, x_n>0\}$ belongs to $P_{p,0}(R_n)$.
(ii) Let $\sigma(t)$ be an infinitely differentiable function on R_1 with

$$\sigma(t)=1 \quad \text{if} \quad |t|\geqq c \quad \text{and} \quad \sigma(t)=0 \quad \text{if} \quad |t|\leqq\varepsilon, \tag{7}$$

where $c>\varepsilon>0$. Let $0<p\leqq\infty$, $\alpha\geqq 0$, a and b complex. Then $g(x)=G(x_n)$ with

$$G(t)=(F_1{}^{-1}\sigma F_1 h^{a,b}_\alpha)(t) \tag{8}$$

belongs to $P_{p,\alpha}(R_n)$.

(iii) Let $\varkappa(t)\in S(R_1)$, $\alpha\geqq 0$, a and b complex, $0<p\leqq\infty$. Then $g(x)=G(x_n)$ with

$$G(t)=\varkappa(t)\ h_\alpha^{a,b}(t) \tag{9}$$

belongs to $P_{p,\alpha}(R_n)$.

Proof. Step 1. Let $-1<\alpha<\infty$. We calculate the restriction of $F_1h_\alpha^{a,b}$ to $R_1-\{0\}$. In other words, we consider $F_1h_1^{a,b}$ temporarily as an element of $D'(R_1-\{0\})$. If $\alpha=0$, then

$$F_1h_0^{a,b}=\frac{t}{t}\ F_1h_0^{a,b}=\frac{c}{t}\ F_1(h_0^{a,b})'=\frac{a-b}{t}\ c=\frac{c'}{t}\in D'(R_1-\{0\}).$$

If $\alpha=m=1, 2, 3, \ldots$, then

$$F_1h_\alpha^{a,b}=F_1t^mh_0^{a',b'}=c(F_1h_0^{a',b'})^{(m)}=\frac{c'}{t^{1+m}}. \tag{10}$$

Finally, if $\alpha=m+\beta$ with $m=0, 1, 2, \ldots$ and $-1<\beta<0$, then we claim that

$$F_1h_\alpha^{a,b}=h_{-\alpha-1}^{a',b'}\quad(\text{on } R_1-\{0\}), \tag{11}$$

where a' and b' are appropriate numbers. In the same way as in (10) we may assume (without loss of generality) that $m=0$, i.e. $\alpha=\beta$. The function $h_\alpha(t)$ is homogeneous, i.e. we have $h_\alpha(\lambda t)=\lambda^\alpha h_\alpha(t)$ if $\lambda>0$. By standard properties of F_1 it follows that F_1h_α is also homogeneous (in the sense of distributions) and that

$$(F_1h_\alpha)(\lambda\,\cdot\,)=\lambda^{-1-\alpha}(F_1h_\alpha)(\,\cdot\,),\quad\lambda>0. \tag{12}$$

If $-1<\alpha<-\frac{1}{2}$, then $h_\alpha\in L_1(R_1)+L_2(R_1)$ and consequently, $F_1h_\alpha\in L_\infty(R_1)+L_2(R_1)$ is a regular distribution. Then (11) follows from (12), provided that $-1<\alpha<-\frac{1}{2}$ (it is not hard to see that $(F_1h_\alpha)(\lambda t)=\lambda^{-1-\alpha}(F_1h_\alpha)(t)$ if t is a Lebesgue point of the regular distribution F_1h_α). Of course, (11) is also valid if one replaces F_1 by F_1^{-1}. Then it follows that

$$F_1h_{-\alpha-1}^{a',b'}=h_\alpha^{a,b}\quad\text{if}\quad-1<\alpha<-\frac{1}{2}. \tag{13}$$

If a and b run independently through the complex plane, than a' and b' also run independently through the complex plane. Thus (13) proves (11) with $-\frac{1}{2}<\alpha<0$. Finally, a limiting argument proves (11) with $\alpha=-\frac{1}{2}$.

Step 2. We prove (ii). Let $\{\varphi_k(t)\}_{k=0}^\infty\in\Phi(R_1)$, cf. 2.3.1, where we may assume that $\varphi_k(t)=\varphi_1(2^{-k+1}t)$ if $k=2, 3, \ldots$ Since $\sigma(t)$ vanishes near the origin we may apply the considerations from Step 1 and obtain

$$F_1^{-1}\varphi_kF_1G=F_1^{-1}\varphi_k\sigma F_1h_\alpha^{a,b}=F_1^{-1}\varphi_k\sigma h_{-\alpha-1}^{a',b'}\in S(R_1). \tag{14}$$

On the other hand, if k is large, then

$$\begin{aligned}\|F_1^{-1}\varphi_kF_1G\mid L_p(R_1)\|&=\|F_1^{-1}(\varphi_kh_{-\alpha-1}^{a',b'})\mid L_p(R_1)\|\\&=2^{-(k-1)(\alpha+1)}\|F_1^{-1}\varphi_1(2^{-k+1}\,\cdot\,)\,h_{-\alpha-1}^{a',b'}(2^{-k+1}\,\cdot\,)\mid L_p(R_1)\|\\&=2^{-\alpha(k-1)}\|(F_1^{-1}\varphi_1h_{-\alpha-1}^{a',b'})\,(2^{k-1}\,\cdot\,)\mid L_p(R_1)\|\\&=2^{-\left(\alpha+\frac{1}{p}\right)(k-1)}\|F_1^{-1}\varphi_1h_{-\alpha-1}^{a',b'}\mid L_p(R_1)\|.\end{aligned} \tag{15}$$

Now (14) and (15) prove that

$$G\in B_{p,\infty}^{\alpha+\frac{1}{p}}(R_1). \tag{16}$$

If $\alpha=0$, then

$$G(t)=h_0^{a,b}+F_1^{-1}(\sigma-1)\ F_1h_0^{a,b}=h_0^{a,b}+c\ (F_1^{-1}\ (1-\sigma))*h_0^{a,b}$$

is bounded. Hence, $g(x)=G(x_n)\in P_{p,\alpha}(R_n)$ in any case.

Step 3. We prove (i). Let $\chi=\chi(t)$ be the characteristic function of $[0,\infty)$. If $\sigma(t)$ has the same meaning as in part (ii) of the proposition then we have

$$(F_1^{-1}(1-\sigma)F_1\chi)(t)=\int_{R_1}\chi(t-\tau)\,[F_1^{-1}(1-\sigma)](\tau)\,d\tau = \int_{-\infty}^{t}[F_1^{-1}(1-\sigma)](\tau)\,d\tau\in C^\infty(R_1)\,. \tag{17}$$

(17) and the second step yield

$$\chi=F^{-1}\sigma F\chi+F^{-1}(1-\sigma)\,F\chi\in B^{\frac{1}{p}}_{p,\infty}(R_1)+C^\infty(R_1)\,. \tag{18}$$

This proves (i).

Step 4. We prove (iii). If G is given by (8), then

$$\varkappa h_\alpha^{a,b}=\varkappa G+\varkappa F_1^{-1}(1-\sigma)\,F_1 h_\alpha^{a,b}\,. \tag{19}$$

The second term on the right-hand side of (19) belongs to $S(R_1)$. This follows essentially from Theorem 1.2.1/2. Furthermore, (16) and Theorem 2.8.2 show that $\varkappa G\in B^{\alpha+\frac{1}{p}}_{p,\infty}(R_1)$. Consequently, $\varkappa h_\alpha^{a,b}\in B^{\alpha+\frac{1}{p}}_{p,\infty}(R_1)$ and $\varkappa h_\alpha^{a,b}$ is bounded. This proves (iii).

Remark 2. We add a few further results, which we proved in [S, pp. 142–145]. Let $\chi_{[c,d]}(t)$ be the characteristic function of the interval $[c,d]$, where $-\infty<c<d<\infty$. If $0<p\leqq\infty$ and $0<q<\infty$, then

$$\chi_{[c,d]}\in B^{\frac{1}{p}}_{p,\infty}(R_1)\quad\text{and}\quad \chi_{[c,d]}\notin B^{\frac{1}{p}}_{p,q}(R_1)\,. \tag{20}$$

The first statement is an easy consequence of the above considerations, in particular of Step 4 of the above proof. Furthermore, we remark that part (iii) of the proposition is sharp in the following sense. Let $\alpha>0$, $b\neq 0$, $\varkappa(t)\in S(R_1)$ with $\varkappa(0)\neq 0$. If $0<p\leqq\infty$ and $0<q<\infty$, then

$$\varkappa h_\alpha^{0,b}\in B^{\frac{1}{p}+\alpha}_{p,\infty}(R_1)\quad\text{and}\quad \varkappa h_\alpha^{0,b}\notin B^{\frac{1}{p}+\alpha}_{p,q}(R_1)\,. \tag{21}$$

The first statement of (21) has been proved in Step 4 of the above proof.

2.8.5. Special Multipliers for $B^s_{p,q}(R_n)$

We study the problem of determining conditions under which the operator

$$\mathcal{G}f=gf\quad\text{with}\quad g\in P_{r,\alpha}(R_n) \tag{1}$$

(pointwise multiplication in the sense of 2.8.1.) yields a bounded linear mapping from $B^s_{p,q}(R_n)$ into itself. We are looking for necessary and sufficient conditions. The theorem below contains (the more interesting, from the standpoint of the later application) sufficient conditions. In a corollary we state that these conditions are almost necessary (i.e. with the exception of some limiting cases). It is convenient to introduce the number $\bar{p}$ by

$$\bar{p}=p\quad\text{if}\quad 0<p\leqq 2\quad\text{and}\quad \bar{p}=p'\quad\text{if}\quad 2<p\leqq\infty,\quad \frac{1}{p}+\frac{1}{p'}=1\,. \tag{2}$$

Furthermore, $\varkappa_p^s$ is given by (2.8.2/25).

Theorem. Let $0<p\leqq\infty$, $0<q\leqq\infty$ and $\alpha\geqq 0$. Let $g\in P_{\bar{p},\alpha}(R_n)$ with $g(x)=G(x_n)$ and

$$G(t)=G_0(t)+G_1(t),\quad G_0(t)\in B^{\frac{1}{\bar{p}}+\alpha}_{\bar{p},\infty}(R_1)\cap L_\infty(R_1),\quad G_1(t)\in C^\infty(R_1)\,. \tag{3}$$

Let

$$\begin{cases} \frac{1}{p}-1-\alpha<s<\frac{1}{p}+\alpha & \text{if} \quad 1\leqq p\leqq\infty \\ n\left(\frac{1}{p}-1\right)-\alpha<s<\frac{1}{p}+\alpha & \text{if} \quad 0<p<1\,, \end{cases} \tag{4}$$

(i.e. the shaded area in Fig. 2.8.5). Then $\mathcal{G}$ with $\mathcal{G}f=gf$ yields a bounded linear mapping from $B^s_{p,q}(R_n)$ into itself. Furthermore, if $\varrho>\varkappa^s_p$, then there exists a positive constant c such that

$$\begin{aligned} &\|gf \mid B^s_{p,q}(R_n)\| \\ &\leqq c[\|G_0 \mid B^{\frac{1}{p}+\alpha}_{p,\infty}(R_1)\|+\|G_0 \mid L_\infty(R_1)\|+\|G_1 \mid \mathcal{C}^\varrho(R_1)\|]\,\|f \mid B^s_{p,q}(R_n)\| \end{aligned} \tag{5}$$

for all $g(x)=G(x_n)$ with (3) and all $f\in B^s_{p,q}(R_n)$.

Proof. Step 1. Let $b_k(x)$ and $c_k(x)$ be the functions from (2.8.2/5) and (2.8.2/6). As in the first step of the proof of Theorem 2.8.2 it is sufficient to deal with $\Sigma'_k f$, $\Sigma''_k f$ and $\Sigma'''_k f$ from (2.8.2/10)–(2.8.2/12). In this step we estimate $\Sigma'_k f$ and

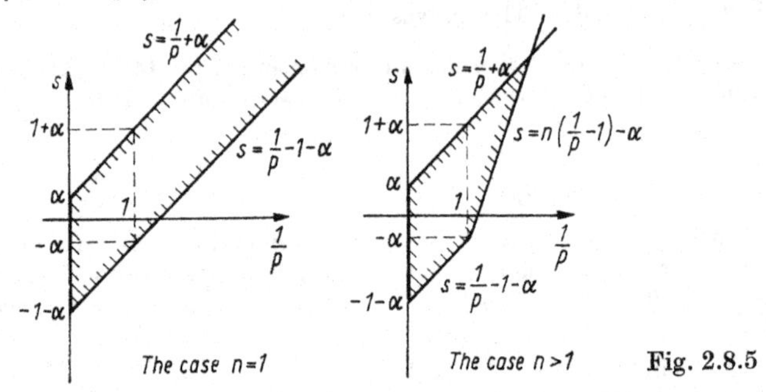

Fig. 2.8.5

$\Sigma'''_k f$. The estimate for $G_1(x_n)$ with $G_1(t)\in\mathcal{C}^\infty(R_1)$ follows immediately from Corollary 2.8.2. Consequently, we may assume in the sequel (without restriction of generality) that $G(t)=G_0(t)$ in (3). Let $\{\varphi_j(x)\}_{j=0}^\infty\in\Phi(R_n)$ be the system from the first step of the proof of Theorem 2.8.2. Then we have

$$\sum_{j=0}^{k}\varphi_j(x)=\varphi_0(2^{-k+1}x)+\varphi_1(2^{-k+1}x)=\varphi(2^{-k}x)$$

if $k=1, 2, \ldots$ Now (2.8.2/10) yields

$$(\Sigma'_k f)(x)=\int_{R_n}(F^{-1}\varphi_1)(y)\,c_k\,(x-2^{-k+1}y)\,(F^{-1}\varphi(2^{-k}\,\cdot)\,Fg)\,(x-2^{-k+1}y)\,dy\,. \tag{6}$$

The Fourier transform on R_1 and its inverse are again denoted by F_1 and F_1^{-1}, respectively. Hence it follows from (3) (with $G=G_0$) that

$$\begin{aligned} &\sup_{z\in R_n}|(F^{-1}\varphi(2^{-k}\,\cdot)\,Fg)(z)|=\sup_{t\in R_1}|(F_1^{-1}\varphi(0,\,2^{-k}\,\cdot)\,F_1G)(t)| \\ &=c\left|\int_{R_1}[F_1^{-1}\varphi(0,\,2^{-k}\,\cdot)](\tau)G(t-\tau)\,d\tau\right| \\ &\leqq c\|G \mid L_\infty(R_1)\|\int_{R_1}2^k|[F_1^{-1}\varphi(0,\,\cdot)]\,(2^k\tau)|\,d\tau=c'\|G \mid L_\infty(R_1)\|\,. \end{aligned} \tag{7}$$

(6) and (7) yield

$$|(\Sigma'_k f)(x)|\leqq c\|G \mid L_\infty(R_1)\|\int_{R_n}|(F^{-1}\varphi_1)(y)|\,|c_k\,(x-2^{-k+1}y)|\,dy\,. \tag{8}$$

We estimate $(\Sigma_k'''f)(x)$ from (2.8.2/12). Similarly, as above, we have

$$b_l(x)=(F^{-1}\varphi_l Fg)(x)=(F_1^{-1}\varphi_l(0,\cdot)\, F_1G)(x_n)=(\beta_l G)(x_n)\,. \tag{9}$$

However, $\{\varphi_l(0, x_n)\}_{l=0}^{\infty}\in\Phi(R_1)$. Because $G(t)\in B^{\alpha}_{\infty,\infty}(R_1)$, cf. (2.8.4/3), we have

$$|b_l(x)|=|(\beta_l G)(x_n)|\leqq 2^{-l\alpha}\,\|G \mid B^{\alpha}_{\infty,\infty}(R_1)\| \quad\text{if}\quad x\in R_n\,. \tag{10}$$

(10) and (2.8.2/12) yield

$$\begin{aligned}|(\Sigma_k'''f)(x)|&\leqq\|G \mid B^{\alpha}_{\infty,\infty}(R_1)\|\sum_{l=k}^{\infty}2^{-l\alpha}\int_{R_n}|(F^{-1}\varphi_1)(y)|\,|c_l\,(x-2^{-k+1}y)|\,dy\\ &\leqq\|G \mid B^{\alpha}_{\infty,\infty}(R_1)\|\sum_{l=k}^{\infty}2^{-(l-k)\alpha}\int_{R_n}|(F^{-1}\varphi_1)(y)|\,|c_l\,(x-2^{-k+1}y)\mid dy\,.\end{aligned} \tag{11}$$

The estimates in (8) and in (11) are of the same type. If $1\leqq p\leqq\infty$ and $0<q\leqq\infty$, it follows from (8) and (11) that

$$\begin{aligned}&2^{ks}\|\Sigma_k'f \mid L_p(R_n)\|+2^{ks}\|\Sigma_k'''f \mid L_p(R_n)\|\\ &\leqq c(\|G \mid L_\infty(R_1)\|+\|G \mid B^{\alpha}_{\infty,\infty}(R_1)\|)\,2^{ks}\sum_{l=k}^{\infty}2^{-(l-k)\alpha}\|c_k \mid L_p(R_n)\|\\ &\leqq c(\,\dots)\sum_{l=k}^{\infty}2^{-(l-k)(s+\alpha)}2^{ls}\|c_l \mid L_p(R_n)\|\\ &\leqq c'(\,\dots)\left(\sum_{l=k}^{\infty}2^{-(l-k)(s+\alpha-\varepsilon)q}2^{lsq}\|c_l \mid L_p(R_n)\|^q\right)^{\frac{1}{q}}\end{aligned}$$

(modification if $q=\infty$), where $\varepsilon>0$ is an arbitrary number. Now it follows that

$$\begin{aligned}&\|2^{ks}\Sigma_k'f \mid l_q(L_p(R_n))\|+\|2^{ks}\Sigma_k'''f \mid l_q(L_p(R_n))\|\\ &\leqq c(\|G \mid L_\infty(R_1)\|+\|G \mid B^{\alpha}_{\infty,\infty}(R_1)\|)\,\|f \mid B^{s}_{p,q}(R_n)\|\\ &\text{if } 1\leqq p\leqq\infty,\ 0<q\leqq\infty \text{ and } s>-\alpha.\end{aligned} \tag{12}$$

Let $0<p<1$ and $0<q\leqq\infty$. Let $h_l(y)=(F^{-1}\varphi_1)(y)\;c_l\,(x-2^{-k+1}y)$, where $x\in R_n$ is fixed. We have

$$(Fh_l)(x)=\int_{R_n}\varphi_1\,(z-\zeta)\,[F^{-1}c_l\,(x-2^{-k+1}\,\cdot)]\,(\zeta)\,d\zeta \tag{13}$$

and

$$\begin{aligned}&\operatorname{supp} F^{-1}c_l\,(x-2^{-k+1}\,\cdot)=\operatorname{supp} Fc_l\,(2^{-k+1}\,\cdot)\\ &=\operatorname{supp}\,(Fc_l)\,(2^{k-1}\,\cdot)=\operatorname{supp}\varphi_l(2^{k-1}\,\cdot)\,(Ff)\,(2^{k-1}\,\cdot)\subset\{z \mid |z|\leqq 2^{l-k+2}\}\,.\end{aligned}$$

Then it follows from (13) that

$$\operatorname{supp} Fh_l\subset\{z \mid |z|\leqq c2^{l-k}\}\,, \tag{14}$$

where c is an appropriate positive number. By (1.3.2/5) and Remark 1.4.1/4 we have

$$\begin{aligned}&\|(F^{-1}\varphi_1)\,(\cdot)\,c_l\,(x-2^{-k+1}\,\cdot) \mid L_1(R_n)\|\\ &\leqq c2^{(l-k)\left(\frac{1}{p}-1\right)n}\,\|(F^{-1}\varphi_1)\,(\cdot)\,c_l\,(x-2^{-k+1}\,\cdot) \mid L_p(R_n)\|\,,\end{aligned} \tag{15}$$

where c is independent of l and k. We put (15) in (8) and (11) to obtain

$$\begin{aligned}&|(\Sigma_k'f)(x)|^p+|(\Sigma_k'''f)(x)|^p\leqq c(\|G \mid L_\infty(R_1)\|+\|G \mid B^{\alpha}_{\infty,\infty}(R_1)\|)^p\\ &\times\sum_{l=k}^{\infty}2^{-(l-k)p\alpha+(l-k)np\left(\frac{1}{p}-1\right)}\,\|(F^{-1}\varphi_1)\,(\cdot)\,c_l\,(x-2^{-k+1}\,\cdot) \mid L_p(R_n)\|^p\,.\end{aligned} \tag{16}$$

Now the rest is the same as in the case $1 \leqq p \leqq \infty$, and we obtain

$$\|2^{ks}\Sigma_k' f \mid l_q(L_p(R_n))\| + \|2^{ks}\Sigma_k''' f \mid l_q((L_p(R_n))\| \leqq c(\|G \mid L_\infty(R_1)\| + \|G \mid B^\alpha_{\infty,\infty}(R_1)\|) \|f \mid B^s_{p,q}(R_n)\| \tag{17}$$

if $0<p<1$, $0<q\leqq\infty$ and $s>-\alpha+n\left(\frac{1}{p}-1\right)$.

Step 2. We estimate $\Sigma_k'' f$ from (2.8.2/11). We have

$$|(\Sigma_k'' f)(x)| \leqq \sum_{l=0}^{k} \int_{R_n} |(F^{-1}\varphi_1)(y)| \sup_{z\in R_1} |c_l(x'-2^{-k+1}y', z)| (\beta_k G)(x_n - 2^{-k+1}y_n)\, dy\,, \tag{18}$$

where $\beta_k G$ has the same meaning as in (9). If $1\leqq p\leqq\infty$, it follows that

$$\|\Sigma_k'' f \mid L_p(R_n)\| \leqq c \sum_{l=0}^{k} \|\sup_{z\in R_1} |c_l(\cdot, z)| \mid L_p(R_{n-1})\| \, \|\beta_k \mid L_p(R_1)\|\,. \tag{19}$$

Let $x'\in R_{n-1}$ be fixed. Then the support of

$$[F_1 c_l(x', \cdot)](z) = (F'^{-1}\varphi_l F f)(x', z)$$

is contained in the interval $[-c2^l, c2^l]$, where c is an appropriate positive number and F' is the $(n-1)$-dimensional Fourier transform. Then it follows from the one-dimensional case of (1.3.2/5) and Remark 1.4.1/4 that

$$\sup_{z\in R_1} |c_l(x', z)| \leqq c 2^{\frac{l}{p}} \|c_l(x', \cdot) \mid L_p(R_1)\|\,. \tag{20}$$

(20), (19) and (3) yield

$$2^{ks}\|\Sigma_k'' f \mid L_p(R_n)\| \leqq c 2^{-\alpha k} \|G \mid B_{p,\infty}^{\alpha+\frac{1}{p}}(R_1)\| \sum_{l=0}^{k} 2^{(k-l)\left(s-\frac{1}{p}\right)} 2^{ls}\|c_l \mid L_p(R_n)\| \leqq c\|G \mid B_{p,\infty}^{\alpha+\frac{1}{p}}(R_1)\| \left(\sum_{l=0}^{k} 2^{(k-l)\left(s-\frac{1}{p}-\alpha+\varepsilon\right)q} 2^{lsq}\|c_l \mid L_p(R_n)\|^q\right)^{\frac{1}{q}}, \tag{21}$$

where $\varepsilon>0$ is an arbitrary positive number (modification if $q=\infty$). Now it follows that

$$\|2^{sk}\Sigma_k'' f \mid l_q(L_p(R_n))\| \leqq c\|G \mid B_{p,\infty}^{\alpha+\frac{1}{p}}(R_1)\| \, \|f \mid B^s_{p,q}(R_n)\| \tag{22}$$

if $1\leqq p\leqq\infty$, $0<q\leqq\infty$ and $s<\alpha+\frac{1}{p}$.

Let $0<p<1$. We modify $h_l(y)$ from (13) and (14) by

$$h_l(y) = (F^{-1}\varphi_1)(y)\, b_k(x-2^{-k+1}y)\, c_l(x-2^{-k+1}y)$$

with $l=0, \ldots, k$. Then

$$\operatorname{supp} Fh_l \subset \{z \mid |z|\leqq c\}\,,$$

where c is independent of k and $l=0, 1, \ldots, k$. This shows that $\Sigma_k'' f$ from (2.8.2/11) can be estimated by

$$|(\Sigma_k'' f)(x)| \leqq c \sum_{l=0}^{k} \|(F^{-1}\varphi_1)(\cdot)\, b_k(x-2^{-k+1}\cdot)\, c_l(x-2^{-k+1}\cdot) \mid L_p(R_n)\|\,. \tag{23}$$

The rest is the same as above. In other words, (22) is also valid if

$$0<p<1, \quad 0<q\leqq\infty \quad \text{and} \quad s<\alpha+\frac{1}{p}. \tag{24}$$

Step 3. We prove the theorem. (12) and (22) show that

$$\begin{aligned} &\|\mathcal{G}f \mid B^s_{p,q}(R_n)\| \leqq c(\|G \mid L_\infty(R_1)\| + \|G \mid B^{\alpha+\frac{1}{p}}_{p,\infty}(R_1)\|)\, \|f \mid B^s_{p,q}(R_n)\| \\ &\text{if} \quad 1\leqq p\leqq\infty, \quad 0<q\leqq\infty \quad \text{and} \quad -\alpha<s<\alpha+\frac{1}{p}. \end{aligned} \tag{25}$$

We have used the fact that

$$B^{\frac{1}{p}+\alpha}_{p,\infty}(R_1)\subset B^{\frac{1}{r}+\alpha}_{r,\infty}(R_1) \quad \text{if} \quad 0<p\leqq r\leqq\infty, \tag{26}$$

cf. (2.7.1/1). On the other hand, (17) and (22) (with (24)) show that

$$\begin{aligned} &\|\mathcal{G}f \mid B^s_{p,q}(R_n)\| \leqq c(\|G \mid L_\infty(R_n)\| + \|G \mid B^{\alpha+\frac{1}{p}}_{p,\infty}(R_1)\|)\, \|f \mid B^s_{p,q}(R_n)\| \\ &\text{if} \quad 0<p<1, \quad 0<q\leqq\infty \quad \text{and} \quad -\alpha+n\left(\frac{1}{p}-1\right)<s<\alpha+\frac{1}{p}. \end{aligned} \tag{27}$$

We always have $\bar{p}\leqq p$. Thus (5) is a consequence of (25) and (27), respectively. (25) and (27) cover all cases of the theorem with the exception of

$$1<p\leqq\infty, \quad 0<q\leqq\infty, \quad \frac{1}{p}-1-\alpha<s\leqq-\alpha \tag{28}$$

(with $s<-\alpha$ if $\alpha=0$ and $p=\infty$). If $\frac{1}{p}+\frac{1}{p'}=1$, then $1\leqq p'<\infty$, where p is given by (28). We have $G\in B^{\alpha+\frac{1}{p'}}_{p',\infty}(R_1)\cap L_\infty(R_1)$. This follows from $\bar{p}=p'$ if $2\leqq p\leqq\infty$, and it is a consequence of $p\leqq p'$ and (26) if $1<p<2$. This shows that (25) holds if one replaces p by p' and if s is restricted by $-\alpha<s<\alpha+\frac{1}{p'}$. We use the duality formula

$$(B^s_{p',q'}(R_n))' = B^{-s}_{p,q}(R_n) \tag{29}$$

if $-\infty<s<\infty$, $1\leqq p'<\infty$, $1<q<\infty$ and $\frac{1}{q}+\frac{1}{q'}=1$, cf. 2.11.2. The operator $\mathcal{G}f=gf$ is formally self-adjoint, so it follows from the above modification of (25) that (5) holds if

$$1<p\leqq\infty, \quad 1<q<\infty, \quad -\alpha+\frac{1}{p}-1<s<\alpha. \tag{30}$$

It follows by real interpolation, cf. (2.4.2/1), that (5) holds if

$$1<p\leqq\infty, \quad 0<q\leqq\infty, \quad -\alpha+\frac{1}{p}-1<s<\alpha.$$

If $\alpha=0$, then a second real interpolation covers also the remaining case $s=0$ and $p<\infty$. The proof is complete.

Remark 1. The question arises whether one can replace $\bar{p}$ in (5) by p. If $0<p\leqq2$, then this is obvious by definition. If $2<p\leqq\infty$ and $-\alpha<s<\alpha+\frac{1}{p}$, then a corresponding statement follows from (25).

Remark 2. Let $\bar{p}$ be given by (2). By the above technique (but in a more direct and simpler way) one can prove the following result: if $0<p\leqq\infty$, $0<q\leqq\infty$ and $n\left(\frac{1}{p}-1\right)<s<\frac{n}{p}$, then there exists a positive constant c such that

$$\|gf \mid B^s_{p,q}(R_n)\| \leqq c[\|g \mid L_\infty(R_n)\| + \|g \mid B^{\frac{n}{\bar{p}}}_{\bar{p},\infty}(R_n)\|] \, \|f \mid B^s_{p,q}(R_n)\|$$

holds for all $f\in B^s_{p,q}(R_n)$ and all $g\in L_\infty(R_n)\cap B^{\frac{n}{\bar{p}}}_{\bar{p},\infty}(R_n)$.

Corollary. Let $g^\alpha(x)=G^\alpha(x)$ with

$G^0(t)=\chi_{[0,1]}(t)$ (characteristic function of the interval $[0, 1)$),

$G^\alpha(t)=\varkappa(t)\, h^{0,b}_\alpha(t)$ if $\alpha>0$,

where $\varkappa(t)$ and $h^{0,b}_\alpha(t)$ have the same meaning as in Proposition 2.8.4. Let $b\neq 0$ and $\varkappa(0)\neq 0$. Let $0<p\leqq\infty$ and $0<q\leqq\infty$. Then $g^\alpha(x)$ with $g^\alpha(x)=G^\alpha(x_n)$ is a multiplier for $B^s_{p,q}(R_n)$ if s is restricted by (4), and $g^\alpha(x)$ is not a multiplier for $B^s_{p,q}(R_n)$ if

$$\begin{cases} \text{either } \frac{1}{p}-1-\alpha>s \quad \text{or} \quad s>\frac{1}{p}+\alpha \quad \text{in the case} \quad 1\leqq p\leqq\infty \\ \text{either } n\left(\frac{1}{p}-1\right)-\alpha>s \quad \text{or} \quad s>\frac{1}{p}+\alpha \quad \text{in the case} \quad 0<p<1\,. \end{cases} \tag{31}$$

Remark 3. By Proposition 2.8.4 we have $G^\alpha(x)\in P_{\bar{p},\alpha}(R_n)$. This shows that the first statement of the corollary (the one which claims that g^α is a multiplier) is a special case of the above theorem. A proof of the rest of the corollary has been given in [S, pp. 150–152]. That proof uses essentially Remark 2.8.4/2. The corollary shows that the restrictions (4) for s are natural. What about the limiting cases, i.e. if s equals one of the limiting numbers in (4)? A few remarks in that direction may be found in [S, p. 153] and in H. Triebel [5].

Remark 4. By Proposition 2.8.4 and the above theorem, the characteristic function of a half-space is a multiplier for $B^s_{p,q}(R_n)$ if $0<q\leqq\infty$ and

$$\begin{cases} 1\leqq p\leqq\infty\,, \quad \frac{1}{p}-1<s<\frac{1}{p}\,, \\ 0<p<1\,, \quad n\left(\frac{1}{p}-1\right)<s<\frac{1}{p}\,. \end{cases} \tag{32}$$

For the classical Bessel-potential spaces $H^s_p(R_n)=F^s_{p,2}(R_n)$ with $1<p<\infty$ and the classical Besov spaces $\Lambda^s_{p,q}(R_n)=B^s_{p,q}(R_n)$ with $1\leqq p<\infty$ and $1\leqq q\leqq\infty$, the problem whether the characteristic function χ of a half-space is a multiplier has been studied in several papers. E. Shamir [1] and R. S. Strichartz [1] proved that χ is a multiplier for $H^s_p(R_n)$ if $1<p<\infty$ and $0\leqq s<\frac{1}{p}$. The corresponding problem for "parabolic" spaces $H^s_p(R_n)$ has been treated in R. J. Bagby [1]. The statement that χ is a multiplier for the special Besov spaces $B^s_{p,p}(R_n)$ with $1<p<\infty$ and $0\leqq s<\frac{1}{p}$ is due to J. L. Lions, E. Magenes [1] (the important case $p=2$ may also be found in J. L. Lions, E. Magenes [2, I; 1.11.3]); cf. also [I, 2.10.2]. A systematic treatment of the multiplier problem in $H^s_p(R_n)$ with $1<p<\infty$ and $s\geqq 0$ has been given in R. S. Strichartz [1]. Concerning multipliers for $B^s_{p,q}(R_n)$ we refer also to J. Peetre [8, Chapter 7].

2.8.6. Two Propositions

Our main aim is to extend some results from 2.8.5. for the spaces $B^s_{p,q}(R_n)$ to the spaces $F^s_{p,q}(R_n)$. Statements for the spaces $F^s_{p,q}(R_n)$ are always more difficult than corresponding ones for the spaces $B^s_{p,q}(R_n)$. For that reason we restrict ourselves to a special but very important case. We study the problem of determinging conditions under which the characteristic function of a half-space is a multiplier for $F^s_{p,q}(R_n)$. This subsection deals with two propositions of auxiliary character (the somewhat special proofs can be skipped at first glance; the proof of the second proposition uses a lot of special assertions from interpolation theory). The main results will be proved in the following subsection. Let again $x=(x', x_n)$.

Proposition 1. Let χ be the characteristic function of the half-space $\{x \mid x_n \geqq 0\}$. Let $1 \leqq p < \infty$, $1 \leqq q < \infty$ and $0 < s < 1$. Then $f \to \chi f$ yields a bounded linear mapping from $F^s_{p,q}(R_n)$ into itself if and only if there exists a positive constant c such that

$$\||x_n|^{-s} f \mid L_p(R_n)\| \leqq c\|f \mid F^s_{p,q}(R_n)\| \tag{1}$$

holds for all $f \in F^s_{p,q}(R_n)$.

Proof. Step 1. Let $f \to \chi f$ be a bounded operator from $F^s_{p,q}(R_n)$ into itself. Let $f \in F^s_{p,q}(R_n)$. We use (2.5.11/16) with $M=1$, replacing the unit ball B_n by the unit cube $Q_n = \{x \mid |x_l| \leqq 1$ if $l=1, \ldots, n\}$ (this change is completely immaterial). If $x_n > 0$ and $r > 0$, then

$$\int_{Q_n} |(\chi f)(x+rh)-(\chi f)(x)| \, dh \geqq \int_{Q_{n-1}} \left(\int_{-1}^{-\frac{x_n}{r}} |f(x) \mid dh_n \right)_+ dh' = |f(x)| \left(1-\frac{x_n}{r}\right)_+ \tag{2}$$

with $(a)_+ = a$ if $a > 0$ and $(a)_+ = 0$ if $a \leqq 0$. This estimate and (2.5.11/16) yield

$$\left[\int_{R_n} |f(x)|^p \left(\int_0^\infty r^{-sq} \left(1-\frac{x_n}{r}\right)_+^q \frac{dr}{r} \right)^{\frac{p}{q}} dx \right]^{\frac{1}{p}} \leqq c\|\chi f \mid F^s_{p,q}(R_n)\| \leqq c'\|f \mid F^s_{p,q}(R_n)\| . \tag{3}$$

(1) follows from (3) and $(\ldots) = c x_n^{-sq}$ with $c > 0$ (and a symmetry argument $x_n \to -x_n$).

Step 2. Let (1) be satisfied. We recall that $S(R_n)$ is dense in $F^s_{p,q}(R_n)$, cf. Theorem 2.3.3. Furthermore, we may assume that $sp < 1$, otherwise (1) does not make sense. Then it follows from (2.3.2/9) and Remark 2.8.5/4 that χf belongs to $F^s_{p,q}(R_n)$ if $f \in S(R_n)$. This shows that we may restrict ourselves to functions $f \in S(R_n)$ (the rest will be a matter of completion). Let $f \in S(R_n)$. If $x_n \geqq 0$, then we have as in the first step

$$\int_{Q_n} |(\chi f)(x+rh)-(\chi f)(x) \mid dh \leqq \int_{Q_n} |f(x+rh)-f(x) \mid dh + f|(x) \mid \left(1-\frac{x_n}{r}\right)_+ . \tag{4}$$

If $x_n < 0$, then

$$\int_{Q_n} |(\chi f)(x+rh)-(\chi f)(x) \mid dh = \int_{Q_{n-1}} \left(\int_{-\frac{x_n}{r}}^{1} |f(x+rh) \mid dh_n \right)_+ dh' \tag{5}$$

$$\leqq \int_{Q_n} |f(x+rh)-f(x) \mid dh + |f(x)| \left(1+\frac{x_n}{r}\right)_+ .$$

(4), (5) and (2.5.11/16) with $M=1$ (and Q_n instead of B_n) yield

$$\|\chi f \mid F^s_{p,q}(R_n)\| \leqq c\|f \mid F^s_{p,q}(R_n)\| + c\| |x_n|^{-s} f \mid L_p(R_n)\| .$$

By (1) this is the desired estimate.

Remark 1. We did not use the fact that $p \geqq 1$ and $q \geqq 1$. In other words, one can extend the above proposition to all values p, q and s for which (2.5.11/16) with $M=1$ is an equivalent quasinorm in $F^s_{p,q}(R_n)$, i.e.

$$0 < p < \infty, \quad 0 < q < \infty, \quad 1 > s > \tilde{\sigma}_{p,q} , \tag{6}$$

where $\tilde{\sigma}_{p,q}$ is given by (2.5.11/2).

Real interpolation of the $L_p(R_n)$-spaces yields the Lorentz spaces $L_{p,q}(R_n)$ with the Marcinkiewicz spaces $L_{p,\infty}(R_n)$ as special cases. More precisely, let $1<p_0<\infty$, $1<p_1<\infty$, with $p_0 \neq p_1$. Let $1 \leqq q_0 \leqq \infty$, $1 \leqq q_1 \leqq \infty$ and $1 \leqq q \leqq \infty$. If $0<$ $<\Theta<1$ and $\frac{1}{p}=\frac{1-\Theta}{p_0}+\frac{\Theta}{p_1}$, then

$$(L_{p_0,q_0}(R_n), L_{p_1,q_1}(R_n))_{\Theta,q}=L_{p,q}(R_n)\,, \tag{7}$$

cf. [I, p. 134], where one also finds the necessary definitions and explanations about the Lorentz spaces $L_{p,q}(R_n)$. In particular, we have

$$L_{p,p}(R_n)=L_p(R_n) \quad \text{if} \quad 1<p<\infty\,. \tag{8}$$

Proposition 2. Let $1<p<\infty$, $0<q<\infty$ and $0<s<\frac{1}{p}$. If $f(t) \in L_{\frac{1}{s},\infty}(R_1)$ and $g(x) \in$ $\in F^s_{p,q}(R_n)$, then $f(x_n)\, g(x) \in L_p(R_n)$. Furthermore, there exists a positive constant c such that

$$\|f(x_n)\, g(x) \mid L_p(R_n)\| \leqq c\|f \mid L_{\frac{1}{s},\infty}(R_1)\|\, \|g \mid F^s_{p,q}(R_n)\| \tag{9}$$

holds for all $f \in L_{\frac{1}{s},\infty}(R_1)$ and all $g \in F^s_{p,q}(R_n)$.

Proof. Step 1. Let $n=1$. In that case we prove the proposition for $0<q \leqq \infty$, including the limiting case $q=\infty$. By (2.3.2/6) it is clear that we may restrict ourselves to that limiting case $q=\infty$. Let $G_s(t)=(F^{-1}(1+|\tau|^2)^{-\frac{s}{2}})(t)$ be the well-known Bessel kernels, cf. S. M. Nikol'skij [3, 8.1]. We recall that under our hypotheses $G_s(t)$ can be estimated by

$$|G_s(t)| \leqq c|t|^{-(1-s)}, \tag{10}$$

cf. S. M. Nikol'skij [3, 8.1]. Then we have $G_s(t) \in L_{\frac{1}{1-s},\infty}(R_1)$, cf. e.g. [I. pp. 132/140]. On the other hand, for every $g \in F^s_{p,\infty}(R_1)$ we have

$$g=G_s * h, \quad h \in F^0_{p,\infty}(R_1)\,, \tag{11}$$

$$\|h \mid F^0_{p,\infty}(R_1)\| \sim \|g \mid F^s_{p,\infty}(R_1)\|\,, \tag{12}$$

where $*$ stands for convolution. This follows from $g=I_{-s}h$ and the lifting property from Theorem 2.3.8 (recall that $F(G_s * h)=c\, FG_s \cdot Fh$). Consequently, the one-dimensional case of (9) is equivalent to

$$\|f \cdot g \mid L_p(R_1)\| \leqq c\|f \mid L_{\frac{1}{s},\infty}(R_1)\|\, \|h \mid F^0_{p,\infty}(R_1)\|\,, \tag{13}$$

where h is given by (11).

Step 2. Let again $n=1$ and

$$\frac{1}{r}=\frac{1}{p}-s\,. \tag{14}$$

We claim that

$$L_{\frac{1}{1-s},\infty}(R_1) * F^0_{p,\infty}(R_1) \subset L_r(R_1)\,. \tag{15}$$

Let $G \in L_{\frac{1}{1-s},\infty}(R_1)$ (later on we identify G with G_s), $h \in F^0_{p,\infty}(R_1)$ and $\{\varphi_j\}_{j=0}^{\infty} \in \Phi(R_1)$, cf. 2.3.1. Let $\psi_j(t)=\sum_{l=-1}^{1} \varphi_{j+l}(t)$ with $t \in R_1$ and $\varphi_{-1} \equiv 0$. Then we have $\psi_j(t)=1$ if $t \in \operatorname{supp} \varphi_j$. Because $F(G * h)=c\, FG \cdot Fh$ it follows that

$$\begin{aligned} F^{-1}[\varphi_j F(G * h)](t) &= cF^{-1}[(\varphi_j FG)(\psi_j Fh)] \\ &= c' \int_{R_1} F^{-1}[\varphi_j FG](t-\tau)\, F^{-1}[\psi_j Fh](\tau)\, d\tau\,. \end{aligned}$$

Furthermore,

$$\left(\sum_{j=0}^{\infty} |F^{-1}[\varphi_j F(G * h)](t)|^2\right)^{\frac{1}{2}} \tag{16}$$

$$\leq \int_{R_1} \sup_k |F^{-1}[\psi_k Fh](\tau)| \left(\sum_{j=0}^{\infty} |F^{-1}[\varphi_j FG](t-\tau)|^2\right)^{\frac{1}{2}} d\tau .$$

We take the L_r-norm. Then

$$\|G * h \mid F^0_{r,2}(R_1)\| \leq c\|\sup_k \mid F^{-1}\psi_k Fh \|L_p(R_1)\| \tag{17}$$

$$\times \left\| \left(\sum_{j=0}^{\infty} |(F^{-1}\varphi_j FG)(\cdot)|^2\right)^{\frac{1}{2}} \middle| L_{\frac{1}{1-s},\infty}(R_1) \right\| .$$

This follows from (14) and well-known properties of convolution operators, cf. [I, p. 139, Theorem 2]. The left-hand side of (17) coincides with $\|G * h \mid L_r(R_1)\|$, cf. Theorem 2.5.6. The first term on the right-hand side can be estimated from above by $c\|h \mid F^0_{p,\infty}(R_1)\|$. We claim that the second term on the right-hand side is equivalent to $\|G \mid L_{\frac{1}{1-s},\infty}(R_1)\|$. This is a Littlewood-Paley theorem for the Marcinkiewicz spaces $L_{\frac{1}{1-s},\infty}(R_1)$. It can be proved by real interpolation, cf. (7) and (8), and the corresponding assertion for the spaces $L_p(R_1)=F^0_{p,2}(R_1)$ with $1<p<\infty$. Now (15) follows from (17).

Step 3. We prove (13). Hölder's inequality gives

$$L_{\frac{1}{s}}(R_1) \cdot L_r(R_1) \subset L_p(R_1) ,$$

where r is given by (14). By interpolation, cf. (7) and (8), it follows that

$$L_{\frac{1}{s},\infty}(R_1) \cdot L_r(R_1) \subset L_{p,\infty}(R_1) . \tag{18}$$

We apply this formula to $f \cdot g = f \cdot (G_s * h)$. Then (18) and (15) (cf. also (17)) yield

$$\|fg \mid L_{p,\infty}(R_1)\| \leq c\|f \mid L_{\frac{1}{s},\infty}(R_1)\| \, \|G_s * h \mid L_r(R_1)\| \tag{19}$$

$$\leq c'\|f \mid L_{\frac{1}{s},\infty}(R_1)\| \, \|h \mid F^0_{p,\infty}(R_1)\|$$

(where we used the fact that $G_s \in L_{\frac{1}{1-s},\infty}(R_1)$). Let p_0 and p_1 be two numbers, with $p_0<p<p_1$, which are near p. Let $\frac{1}{p}=\frac{1-\Theta}{p_0}+\frac{\Theta}{p_1}$. Then (19) holds with p_0 and p_1 instead of p, respectively. We use the interpolation formulas

$$(L_{p_0,\infty}(R_1), L_{p_1,\infty}(R_1))_{\Theta,p} = L_p(R_1) \tag{20}$$

and

$$(F^0_{p_0,\infty}(R_1), F^0_{p_1,\infty}(R_1))_{\Theta,p} = F^0_{p,\infty}(R_1) . \tag{21}$$

(20) follows from (7) and (8). A proof of (21) with $1<q<\infty$ instead of ∞ has been given in [I, p. 185]. This proof can also be used if $q=\infty$ and then we obtain (21). But the interpolation of (19), with the help of (20) and (21), yields (13), which, in turn, proves (9) with $n=1$.

Step 4. We prove (9) with $n>1$. Let $(\Delta^1_{h,n}g)(x)$ be the differences from (2.5.13/4). Let $g(x) \in F^s_{p,q}(R_n)$ with the above restrictions for p, q and, s (we may assume that $q>1$). Then it follows from the first step of the proof of Theorem 2.5.10 (the case $m<0$) and from the first step of the proof of Theorem 2.5.11 (the case $m \geq 0$ with $\lambda=1$ and with the one-dimensional version of the Hardy-Littlewood maximal function) and Corollary 2.5.11 that

$$\left\| \left[\int_0^\infty r^{-sq} \left(\int_{-1}^{1} |(\Delta_{rh}^1 g)(\cdot)|\, dh \right)^q \frac{dr}{r} \right]^{\frac{1}{q}} \,\Big|\, L_p(R_n) \right\| \leq c \|g \mid F_{p,q}^s(R_n)\| . \tag{22}$$

In particular, Corollary 2.5.11 gives

$$\| \|g(x', x_n) \mid F_{p,q}^s(R_1)\| \, L_p(R_{n-1})\| \leq c \|g \mid F_{p,q}^s(R_n)\| , \tag{23}$$

cf. (2.5.13/3) and the explanations in Remark 2.5.13/1. We apply the one-dimensional case of (9) to $g(x', x_n)$, where $x' \in R_{n-1}$ is fixed and x_n is variable. Afterwards we take the $L_p(R_{n-1})$-norm with respect to $x' \in R_{n-1}$ and then (9) follows immediately from (23).

Remark 2. A few modifications of the above proof yield the following result: let $1 < p < \infty$, $0 < q \leq \infty$ and $0 < s < \frac{n}{p}$. If $f \in L_{\frac{n}{s}, \infty}(R_n)$ and $g(x) \in F_{p,q}^s(R_n)$, then $fg \in L_p(R_n)$. Furthermore, there exists a positive constant c such that

$$\|fg \mid L_p(R_n)\| \leq c \|f \mid L_{\frac{n}{s}, \infty}(R_n)\| \, \|g \mid F_{p,q}^s(R_n)\| \tag{24}$$

holds for all $f \in L_{\frac{n}{s}, \infty}(R_n)$ and all $g \in F_{p,q}^s(R_n)$. A proof has been given in H. Triebel [14]. Steps 1–3 of the above proof coincide essentially with the one-dimensional case of that proof. If $q = 2$, then $F_{p,2}^s(R_n) = H_p^s(R_n)$ are the Bessel-potential spaces, cf. Theorem 2.5.6. In that case, (24) is due to R. S. Strichartz [1, p. 1049]. We carried over some ideas in the proof from R. S. Strichartz's paper.

2.8.7. Characteristic Functions as Multipliers

The problem of whether the characteristic function of a rectangle or of a polygon in R_n is a multiplier for $B_{p,q}^s(R_n)$ or for $F_{p,q}^s(R_n)$ can be reduced immediately to the question of whether the characteristic function χ of the half-space $\{x \mid x \in R_n, x_n \geq 0\}$ is a multiplier for the corresponding spaces. A clarification of this problem is not only of interest for its own sake but is crucial for some further considerations. Of course, the notion of a multiplier always must be understood in the sense of 2.8.1.

Theorem. (i) Let $0 < p \leq \infty$ and $0 < q \leq \infty$. If

$$\begin{cases} \text{either} & \frac{1}{p} - 1 < s < \frac{1}{p} & \text{in the case} \quad 1 \leq p \leq \infty \\ \text{or} & n\left(\frac{1}{p} - 1\right) < s < \frac{1}{p} & \text{in the case} \quad 0 < p < 1 \end{cases} \tag{1}$$

(cf. Fig. 2.8.7/1), then χ is a multiplier for $B_{p,q}^s(R_n)$.
(ii) Let $0 < p < \infty$ and $0 < q < \infty$. If

$$\begin{cases} \text{either} & \frac{1}{p} - 1 + n\left(\frac{1}{\min(q, 1)} - 1\right) < s < \frac{1}{p} & \text{in the case} \quad 1 \leq p < \infty \\ \text{or} & n\left(\frac{1}{\min(p, q)} - 1\right) < s < \frac{1}{p} & \text{in the case} \quad 0 < p < 1 \end{cases} \tag{2}$$

then χ is a multiplier for $F_{p,q}^s(R_n)$.

(iii) Let $0<p\leqq\infty$ (with $p<\infty$ in the case of the spaces $F^s_{p,q}(R_n)$) and $0<q\leqq\infty$. If χ is a multiplier for $B^s_{p,q}(R_n)$ or for $F^s_{p,q}(R_n)$, respectively, then

$$\begin{cases} \text{either} \quad \frac{1}{p}-1\leqq s\leqq\frac{1}{p} & \text{in the case} \quad p\geqq 1 \\ \text{or} \quad n\left(\frac{1}{p}-1\right)\leqq s\leqq\frac{1}{p} & \text{in the case} \quad 0<p<1 \end{cases} \tag{3}$$

(cf. Fig. 2.8.7/1).

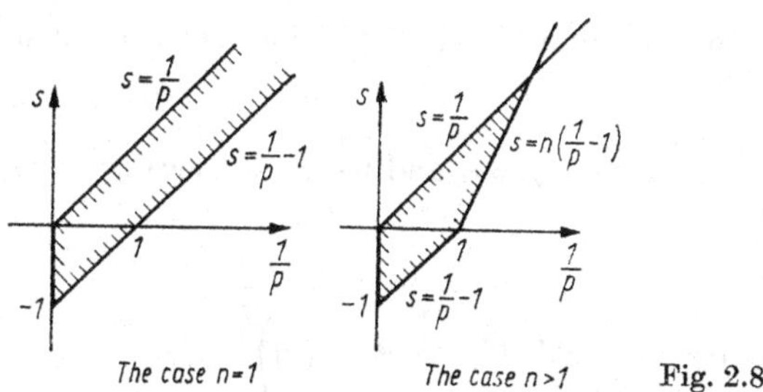

Fig. 2.8.7/1

Proof. Step 1. The statements for the spaces $B^s_{p,q}(R_n)$ in (i) and in (iii) follow immediately from Theorem 2.8.5 in cojunction with Proposition 2.8.4(i) (cf. also Remark 2.8.5/4) and from Corollary 2.8.5.

Step 2. Let $1<p<\infty$ and $1<q<\infty$. In order to prove that χ is a multiplier for $F^s_{p,q}(R_n)$ if $\frac{1}{p}-1<s<\frac{1}{p}$ we wish to apply Proposition 2.8.6/1 and Proposition 2.8.6/2. We recall of the well-known fact that $|t|^{-s}$ is an element of the Marcinkiewicz space $L_{\frac{1}{s},\infty}(R_1)$ if $0<s<1$, cf. e.g. [I, pp. 140 and 132]. Then Proposition 2.8.6/2 with $f(x_n)=|x_n|^{-s}$ and Proposition 2.8.6/1 prove that χ is a multiplier for $F^s_{p,q}(R_n)$ if $0<s<\frac{1}{p}$. Because $f\to\chi f$ is self-adjoint it follows by duality, cf. (2.11.2/2), that χ is also a multiplier if $\frac{1}{p}-1<s<0$. The corresponding statement for the remaining cases $F^0_{p,q}(R_n)$ follows by complex interpolation, cf. (2.4.7/3) (in the proof of Corollary 2.8.2 we discussed the problem that the multiplication operator $f\to gf$ has the interpolation property for our complex method; via a limiting argument this is also valid if $g=\chi$).

Step 3. In order to prove the remaining cases of (ii) we use the same method as in the proofs of the Theorems 2.8.2 and 2.8.5, respectively. In other words, we must estimate $\Sigma'_k f$, $\Sigma''_k f$ and $\Sigma'''_k f$ from (2.8.2/10)–(2.8.2/12). Again let

$$c_k(x)=(F^{-1}\varphi_k Ff)(x) \quad \text{with} \quad f\in F^s_{p,q}(R_n)$$

and

$$\beta_l(x_n)=(\beta_l G)(x_n)=F_1^{-1}\varphi_l(0,\cdot)\,F_1G \quad \text{with} \quad G=\chi_1\,,$$

cf. (2.8.2/5) and (2.8.5/9), where all symbols have the same meaning as above and χ_1 is the characteristic function of $[0,\infty)$. Furthermore, we recall that $c_k^*(x)$ and $\beta_l^*(x_n)$ are the corresponding maximal functions, cf. (2.8.2/13) or (2.3.6/2) (one-dimensional case as far as $\beta_l^*(x_n)$ is concerned). We replace b_k in (2.8.2/11) by β_k, cf. (2.8.5/9). Then it follows from (2.8.5/8), (2.8.2/11), and (2.8.2/12), respectively, that

$$|(\Sigma_k' f)(x)| \leq c c_k^*(x), \tag{4}$$

$$|(\Sigma_k'' f)(x)| \leq c \beta_k^*(x_n) \sum_{l=0}^{k} c_l^*(x), \tag{5}$$

$$|(\Sigma_k''' f)(x)| \leq c \sum_{l=k}^{\infty} 2^{a(l-k)} c_l^*(x), \tag{6}$$

with $a > \frac{n}{\min(p, q)}$. Of course, the estimate for $\Sigma_k' f$ is a special case of the one for $\Sigma_k''' f$. The crucial point of the following considerations is the estimate of $\Sigma_k'' f$ if $-\infty < s < \frac{1}{p}$.

Step 4. We estimate $\Sigma_k'' f$ if $s<0$, $0<p<\infty$ and $0<q\leq\infty$. Because $\beta_k^*(x_n)$ is uniformly bounded, (5) yields

$$2^{ks}|(\Sigma_k'' f)(x)| \leq c \sum_{l=0}^{k} 2^{(k-l)s} 2^{ls} c_l^*(x) \leq c' \left(\sum_{l=0}^{k} 2^{(k-l)(s+\varepsilon)q} 2^{lsq} c_l^{*q}(x) \right)^{\frac{1}{q}}$$

with $\varepsilon>0$, $s+\varepsilon<0$ and $0<q\leq\infty$ (modification if $q=\infty$). Hence

$$\sum_{k=1}^{\infty} 2^{ksq} |(\Sigma_k'' f)(x)|^q \leq c \sum_{l=0}^{\infty} 2^{lsq} c_l^{*q}(x)$$

and

$$\left\| \left(\sum_{k=1}^{\infty} 2^{ksq} |(\Sigma_k'' f)(x)|^q \right)^{\frac{1}{q}} \mid L_p(R_n) \right\| \leq c \| 2^{ls} c_l^* \mid L_p(R_n, l_q) \|,$$

so it follows from Theorem 2.3.6 that

$$\left\| \left(\sum_{k=1}^{\infty} 2^{ksq} |(\Sigma_k'' f)(x)|^q) \right)^{\frac{1}{q}} \mid L_p(R_n) \right\| \leq c \| f \mid F_{p,q}^s(R_n) \|. \tag{7}$$

Step 5. We estimate Σ_k'' if $0<s<\frac{1}{p}$, $p\leq 1$ and $\infty\geq q\geq p$. Since $k=1, 2, 3, \ldots$ and $\varphi_k(x)=\varphi_1(2^{-k+1}x)$ we have

$$\beta_k(x_n) = (F_1^{-1}\varphi_1(0, 2^{-k+1}\cdot)\, F_1\chi_1)(x_n) = \left(F_1^{-1}\varphi_1(0, 2^{-k+1}t)\, \frac{c}{t} \right)(x_n) \tag{8}$$

$$= c'(F_1^{-1}\varphi_1(0, t)\, t^{-1})\, (2^{k-1}x_n) = \psi(2^k x_n)$$

with $\psi \in S(R_1)$, cf. the first step of the proof of Proposition 2.8.4. Let

$$\psi^*(t) = \sup_{\tau \in R_1} \frac{|\psi(t-\tau)|}{1+|\tau|^d} \tag{9}$$

where $d>0$ is an arbitrary number. Then (8) yields $\beta_k^*(x_n) = \psi^*(2^k x_n)$ and we have

$$|\beta_k^*(x_n)| \leq \frac{c}{1+|2^k x_n|^d}, \tag{10}$$

where c is independent of k. If $|x_n|\geq 1$, then (5) and (10) with a big d yield

$$2^{ks}|(\Sigma_k'' f)(x)| \leq 2^{-k} \left(\sum_{l=0}^{k} 2^{lsq} c_l^{*q}(x) \right)^{\frac{1}{q}} \leq c 2^{-k} \left(\sum_{l=0}^{\infty} 2^{lsq} c_l^{*q}(x) \right)^{\frac{1}{q}}$$

and

$$\int_{\substack{x' \in R_{n-1} \\ |x_n| \geq 1}} \left(\sum_{k=1}^{\infty} 2^{ksq} |(\Sigma_k'' f)(\cdot)|^q \right)^{\frac{p}{q}} \mathrm{d}x' \mathrm{d}x_n \leq c \|2^{ls} c_l^* \mid L_p(R_n, l_q)\|^p . \tag{11}$$

Let $2^{-j-1} \leq |x_n| \leq 2^{-j}$ with $j=1, 2, \ldots$ We consider the following three cases of (5): (i) $k=1, \ldots, j$, (ii) $k>j$ and $l=0, \ldots, j$, (iii) $k>j$ and $l \geq j$. In the last two cases, we write $\Sigma_k'' f = \Sigma_{k,1}'' f + \Sigma_{k,2}'' f$, where $\Sigma_{k,1}'' f$ and $\Sigma_{k,2}'' f$ contain the summands of (5) with $k \geq l > j$ and $0 \leq l \leq j$, respectively. We start with the last case. Then (10) and $p \leq 1$ yield

$$\begin{aligned} 2^{ksp} |(\Sigma_{k,1}'' f)(x)|^p &\leq c 2^{-kdp} 2^{jdp} \sum_{l=j}^{k} c_l^{*p}(x)\, 2^{ksp} \\ &\leq c 2^{-kdp+ksp} 2^{jdp} \sum_{l=j}^{k} 2^{-lsp} \left(\sum_{r=0}^{\infty} 2^{rsq} c_r^{*q}(x) \right)^{\frac{p}{q}} \\ &\leq c' 2^{(k-j)(s-d)p} \left(\sum_{r=0}^{\infty} 2^{rsq} c_r^{*q}(x) \right)^{\frac{p}{q}} . \end{aligned} \tag{12}$$

Integration of (12) over the strip

$$S_j = \{x \mid x = (x', x_n),\ x' \in R_{n-1},\ 2^{-j-1} \leq |x_n| \leq 2^{-j}\}$$

and summation over $k>j$ and $j=1, 2, \ldots$ yield

$$\sum_{j=1}^{\infty} \sum_{k=j}^{\infty} \int_{S_j} 2^{ksp} |(\Sigma_{k,1}'' f)(x)|^p \mathrm{d}x \leq c \int_{R_{n-1} \times [-1,1]} \left(\sum_{r=0}^{\infty} 2^{rsq} c_r^{*q}(x) \right)^{\frac{p}{q}} \mathrm{d}x , \tag{13}$$

if one chooses $d>s$. Next we consider the case $k>j$ and $l=0, \ldots, j$. In particular we have $l \leq j$. Let $x=(x', x_n)$ (recall that $2^{-j-1} \leq |x_n| \leq 2^{-j}$). Then it follows from the definition of $c_l^*(x)$ that

$$c_l^*(x) \leq c c_l^*(x', 0) \leq c' 2^{\frac{l}{p}} \left(\int_{2^{-l-1}}^{2^{-l}} c_l^{*p}(x', t)\, \mathrm{d}t \right)^{\frac{1}{p}} . \tag{14}$$

Now (5), (10) and (14) yield

$$\begin{aligned} 2^{ksp} |(\Sigma_{k,2}'' f)(x)|^p &\leq c 2^{-kdp+jdp} \sum_{l=0}^{j} 2^l \int_{2^{-l-1}}^{2^{-l}} c_l^{*p}(x', t)\, \mathrm{d}t \cdot 2^{ksp} \\ &\leq c 2^{-kdp+jdp} 2^{ksp} \sum_{l=0}^{j} 2^{l-lsp} \int_{2^{-l-1}}^{2^{-l}} \left(\sum_{r=0}^{\infty} 2^{rsq} c_r^{*q}(x', t) \right)^{\frac{p}{q}} \mathrm{d}t . \end{aligned} \tag{15}$$

Integration of (15) over the above strip S_j (which gives the factor 2^{-j} on the right-hand side) and summation over $k>j$ and $j=1, 2, \ldots$ yield

$$\begin{aligned} &\sum_{j=1}^{\infty} \sum_{k=j}^{\infty} \int_{S_j} 2^{ksp} |(\Sigma_{k,2}'' f)(x)|^p \, \mathrm{d}x \\ &\leq c \sum_{j=1}^{\infty} \sum_{l=0}^{j} 2^{-j+jdp} 2^{l-lsp} \int_{R_{n-1}} \int_{2^{-l-1}}^{2^{-l}} (\ldots)^{\frac{p}{q}} \mathrm{d}x \sum_{k=j}^{\infty} 2^{-kp(d-s)} . \end{aligned} \tag{16}$$

Let $d>s$. Then the last sum equals $c2^{-jpd}2^{jps}$, and so the factor in front of the integral has the form $2^{(j-l)(sp-1)}$. Because $sp<1$ we have

$$\sum_{j=1}^{\infty}\sum_{k=j}^{\infty}\int_{S_j} 2^{ksp}|(\Sigma_{k,2}''f)(x)|^p dx \leqq c \int_{R_{n-1}\times[0,1]}\left(\sum_{r=0}^{\infty} 2^{rsq}c_r^{*q}(x)\right)^{\frac{p}{q}} dx\,. \tag{17}$$

It follows from (13) and (17) that

$$\sum_{j=1}^{\infty}\sum_{k=j}^{\infty}\int_{S_j} 2^{ksp}|(\Sigma_k''f)(x)|^p\, dx \leqq c\|2^{rs}c_r^* \mid L_p(R_n, l_q)\|^p\,. \tag{18}$$

Finally we consider the case $k\leqq j$. Then we have also $l\leqq j$ and (14) is applicable. It follows that

$$\begin{aligned} 2^{ksp}|(\Sigma_k''f)(x)|^p &\leqq c\sum_{l=0}^{k} 2^{lsp}c_l^{*p}(x)\, 2^{ksp-lsp} \\ &\leqq c'\sum_{l=0}^{k} 2^{ksp-lsp+l}\int_{2^{-l-1}}^{2^{-l}}\left(\sum_{r=0}^{\infty} 2^{rsq}c_r^{*q}(x',t)\right)^{\frac{p}{q}} dt\,. \end{aligned} \tag{19}$$

Integration of (19) over the strip S_j and summation over $k\leqq j$ and $j=1,2,\ldots$ yield

$$\begin{aligned} &\sum_{j=1}^{\infty}\sum_{k=1}^{j}\int_{S_j} 2^{ksp}|(\Sigma_k''f)(x)|^p\, dx \\ &\leqq c\sum_{j=1}^{\infty}\sum_{l=0}^{j} 2^{-j}2^{-lsp+l}\int_{R_{n-1}}\int_{2^{-l-1}}^{2^{-l}}(\ldots)^{\frac{p}{q}}\, dx\sum_{k=l}^{j} 2^{ksp}\,. \end{aligned}$$

The last sum can be estimated by $c2^{jsp}$. So the factor in front of the integral has the form $2^{(j-l)(sp-1)}$. This shows that

$$\sum_{j=1}^{\infty}\sum_{k=1}^{j}\int_{S_j} 2^{ksp}|(\Sigma_k''f)(x)|^p\, dx \leqq c \int_{R_{n-1}\times[0,1]}\left(\sum_{r=0}^{\infty} 2^{rsq}c_r^{*q}(x)\right)^{\frac{p}{q}} dx \tag{20}$$

(recall that $sp<1$). But (18) and (20) prove that

$$\int_{R_{n-1}\times[-1,1]}\sum_{k=1}^{\infty} 2^{ksp}|(\Sigma_k''f)(x)|^p\, dx \leqq c\|2^{rs}c_r^* \mid L_p(R_n, l_q)\|^p \tag{21}$$

and, because $q\geqq p$

$$\int_{R_{n-1}\times[-1,1]}\left(\sum_{k=1}^{\infty} 2^{ksq}|(\Sigma_k''f)(x)|^q\right)^{\frac{p}{q}} dx = c\|2^{sk}c_k^* \mid L_p(R_n, l_q)\|^p \tag{22}$$

(11), (22) and Theorem 2.3.6 yield

$$\begin{aligned} \left\|\left(\sum_{k=1}^{\infty} 2^{ksq}|(\Sigma_k''f)(x)|^q\right)^{\frac{1}{q}} \mid L_p(R_n)\right\| &\leqq c\|2^{sk}c_k^* \mid L_p(R_n, l_q)\| \\ &\leqq c'\|f \mid F_{p,q}^s(R_n)\|\,. \end{aligned} \tag{23}$$

Step 6. We introduce the operators

$$T_1f=\sum_{k=1}^{\infty}(\Sigma_k''f)(x),\quad T_2f=\sum_{k=1}^{\infty}[(\Sigma_k'f)(x)+(\Sigma_k'''f)(x)]\,. \tag{24}$$

Then it follows from the considerations in the first step of the proof of Theorem 2.8.2 that

$$\chi f = T_1 f + T_2 f + \dots, \tag{25}$$

where $+\dots$ indicates operators of the same type as T_1 and T_2 (and unimportant terms with $k=0$). Essentially we split the operator $Tf=\chi f$ into T_1 and T_2 in the sense of the model cases from Step 1 of the proof of Theorem 2.8.2. Step 4 and Step 5 prove that T_1 yields a bounded linear mapping from $F^s_{p,q}(R_n)$ into $F^s_{p,q}(R_n)$ if

$$\begin{cases} \text{either} & 0<p<\infty, \quad 0<q\leqq\infty, \quad s<0 \\ \text{or} & p\leqq 1, \quad p\leqq q\leqq\infty, \quad 0<s<\frac{1}{p}. \end{cases} \tag{26}$$

We wish to apply the complex interpolation formula (2.4.7/3). It is not hard to see that both T_1 and T_2 have the interpolation property, cf. Step 1 of the proof

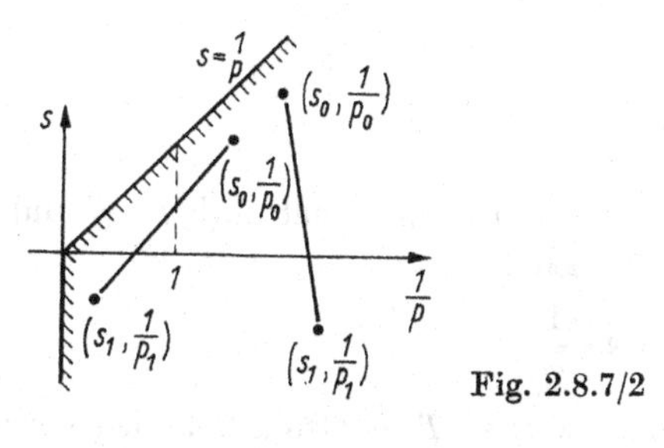

Fig. 2.8.7/2

of Corollary 2.8.2. We apply (2.4.7/3), where $\left(s_0, \frac{1}{p_0}\right)$ is given by the second case in (26), and $\left(s_1, \frac{1}{p_1}\right)$ by the first one. The interpolation is characterized by the lines between the two couples, and all couples $\left(s, \frac{1}{p}\right)$ in the shaded area in Fig. 2.8.7/2 are covered. If we choose $q_0=\infty$ then $\frac{1}{q}=\frac{\Theta}{q_1}$, cf. (2.4.7/1). If Θ with $0<\Theta<1$ and q with $0<q\leqq\infty$ are given, then we choose $q_1=\Theta q$ and obtain $F^s_{p,q}(R_n)$ as an interpolation space in the sense of (2.4.7/3). By the interpolation property it follows that T_1 yields a bounded linear mapping from $F^s_{p,q}(R_n)$ into itself if

$$0<p<\infty, \quad 0<q\leqq\infty, \quad -\infty<s<\frac{1}{p}. \tag{27}$$

Step 7. We estimate T_2. It follows from (6) that

$$\begin{aligned} 2^{ks}|(\Sigma_k''' f)(x)| &\leqq c \sum_{l=k}^{\infty} 2^{(a-s)(l-k)} 2^{ls} c_l^*(x) \\ &\leqq c' \left(\sum_{l=k}^{\infty} 2^{(a+\varepsilon-s)(l-k)q} 2^{lsq} c_l^{*q}(x)\right)^{\frac{1}{q}} \end{aligned} \tag{28}$$

with $a>\frac{n}{\min(p,q)}$ and $\varepsilon>0$. If $s>\frac{n}{\min(p,q)}$, we determine a and ε such that $s>a+\varepsilon$. Then it follows from (28) that

$$\left(\sum_{k=1}^{\infty} 2^{ksq}|(\Sigma_k''' f)(x)|^q\right)^{\frac{1}{q}} \leqq c \left(\sum_{l=1}^{\infty} 2^{lsq} c_l^{*q}(x)\right)^{\frac{1}{q}}.$$

We take the L_p-quasi-norm. Then it follows from (4), (5), (24) and Theorem 2.3.6 that

$$\|T_2 f \mid F^s_{p,q}(R_n)\| \leqq c\|f \mid F^{s'}_{p,q}(R_n)\| \tag{29}$$

if

$$0<p<\infty, \quad 0<q\leqq\infty \quad \text{and} \quad s>\frac{n}{\min(p,q)}. \tag{30}$$

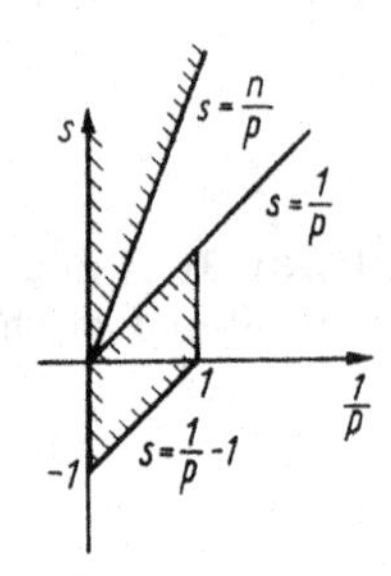

Fig. 2.8.7/3

On the other hand it follows from Step 2 and Step 6 that both $f\to\chi f$ and T_1 are bounded operators from $F^s_{p,q}(R_n)$ into $F^s_{p,q}(R_n)$ if

$$1<p<\infty, \quad 1<q<\infty, \quad \frac{1}{p}-1<s<\frac{1}{p}. \tag{31}$$

Then (25) and the following remarks show that T_2 is also a bounded operator in these spaces. Hence, T_2 is bounded in the two shaded areas in Fig. 2.8.7/3, which are characterized by (30) and (31). We again apply the complex interpolation formula (2.4.7/3) in the same way as in Step 6. As a result we obtain that T_2 is a bounded operator from $F^s_{p,q}(R_n)$ into $F^s_{p,q}(R_n)$ if

$$\begin{cases} \text{either} & \dfrac{1}{p}-1+n\left(\dfrac{1}{\min(q,1)}-1\right)<s, \quad 1\leqq p<\infty, \quad 0<q<\infty \\ \text{or} & n\left(\dfrac{1}{\min(p,q)}-1\right)<s, \quad 0<p<1, \quad 0<q<\infty. \end{cases} \tag{32}$$

Now, part (ii) of the theorem is a consequence of (25), the restriction (27) for T_1, and the restriction (32) for T_2.

Step 8. We prove (iii) for the spaces $F^s_{p,q}(R_n)$. Let χ be a multiplier for $F^s_{p,q}(R_n)$, where $\left(s, \frac{1}{p}\right)$ is outside of the shaded area in Fig. 2.8.7/1 given by (3). By complex interpolation via (2.4.7/3) between $F^s_{p,q}(R_n)$ and spaces $F^{s_0}_{p_0,q_0}(R_n)$ with $\left(s_0, \frac{1}{p_0}\right)$ inside of the shaded area in Fig. 2.8.7/1 we obtain new spaces for which χ is a multiplier. In particular, we may assume without restriction of generality that χ is a multiplier for all spaces $F^\sigma_{p,q}(R_n)$ with $s-\varepsilon<\sigma<s+\varepsilon$ where $\varepsilon>0$ is an appropriate number (s and p have the above meaning). Hence it follows by real interpolation, cf. (2.4.2/2), that χ is also a multiplier for $B^s_{p,q}(R_n)$. But this is a contradiction to Step 1. The proof is complete.

Remark 1. In Remark 2.8.5/4 we gave some references about the problem of whether χ is a multiplier in certain function spaces. The one-dimensional case of the above theorem holds a key position in later considerations, in particular if $p\leqq 1$. In that case we have a rather final answer (besides some limiting cases) if $q \geqq p$. The problem is whether the conditions (2) (in the n-dimensional case) are also natural if $q<1$ or $q<p$ in the cases $p \geqq 1$ and $p<1$, respectively.

Remark 2. What about the limiting cases of (1) and (2), i.e. if $s=\frac{1}{p}$ or $s=\frac{1}{p}-1$ or $s=n\left(\frac{1}{p}-1\right)$, respectively? One can give a few partial answers. In [S, p. 153] we remarked that χ is not a multiplier for $B^{\frac{1}{p}}_{p,q}(R_n)$ if $0<p\leqq\infty$ and $0<q<\infty$. Furthermore, it follows immediately from Proposition 2.8.6/1 that χ is not a multiplier for $F^{\frac{1}{p}}_{p,q}(R_n)$ if $1\leqq p<\infty$ and $1\leqq q<\infty$. By duality one can obtain further assertions, at least if $p\geqq 1$ and $q \geqq 1$.

2.8.8. Further Multipliers

We describe a method which yields further multipliers for $B^s_{p,q}(R_n)$ and $F^s_{p,q}(R_n)$, respectively, if one combines Theorem 2.8.7 and Theorem 2.8.2. Since the results for the spaces $B^s_{p,q}(R_n)$ are covered by Theorem 2.8.5, we deal with the spaces $F^s_{p,q}(R_n)$ only. Furthermore, for the sake of simplicity we consider the one-dimensional case, i.e. $n=1$, and assume that either $\min(p,q)\geqq 1$ or $\infty\geqq q\geqq p>0$ with $p<1$. Then (2.8.7/2) coincides with (2.8.7/1). However, there is no doubt that our method can be applied to the remaining couples (p, q) in the sense of Theorem 2.8.7, too. Let $C^{\infty,-1}(R_1)$ be the set of all complex-valued functions $g(x)$ on R_1 for which there exists a finite number of points $-\infty<a_1<\ldots<a_N$ such that $g(x)$ is infinitely differentiable in $(-\infty, a_1]$, $[a_j, a_{j+1}]$ with $j=1,\ldots,N-1$, and in $[a_N,\infty)$ (extension of $g(x)$ and its derivatives from the interior of these intervals to its closures) and such that $g(x)$ and all its derivatives are bounded and uniformly continuous in $(-\infty, a_1]$ and $[a_N, \infty)$. In other words, $C^{\infty,-1}(R_1)$ contains piecewise infinitely differentiable functions. Cf. with $\mathfrak{C}^\infty(R_1)$ from (2.8.4/1). Furthermore, $C^k(R_1)$ has the meaning of (2.2.2/1). Then we define

$$C^{\infty,k}(R_1)=C^{\infty,-1}(R_1)\cap C^k(R_1) \quad \text{if} \quad k=0,1,2,\ldots \tag{1}$$

Constructions of this type are typical in the theory of spline-functions.

Proposition. Let $0<p<\infty$, $0<q\leqq\infty$ and

$$\begin{cases} \text{either} & \min(p,q)\geqq 1 \\ \text{or} & p<1 \quad \text{and} \quad q\geqq p\,. \end{cases} \tag{2}$$

Let $k=-1,0,1,2,\ldots$ Then $g(x)\in C^{\infty,k}(R_1)$ is a multiplier for $F^s_{p,q}(R_1)$ if

$$\frac{1}{p}-1-(k+1)<s<\frac{1}{p}+k+1\,. \tag{3}$$

Proof. Step 1. We claim that $g\in C^{\infty,-1}(R_1)$ is a multiplier for $F^s_{p,q}(R_n)$ if and only if the characteristic function χ of the interval $[0,\infty)$ is a multiplier for $F^s_{p,q}(R_1)$. One direction is trivial because $\chi\in C^{\infty,-1}(R_1)$. On the other hand, if χ is a multiplier, then it follows from Theorem 2.8.2 that $h(x)\,\chi_{[a,b]}$ is also a multiplier, where $\chi_{[a,b]}$ is the characteristic function of the interval $[a,b]$ and $h\in\mathfrak{C}^\infty(R_1)$. Summation shows that any $g\in C^{\infty,-1}(R_1)$ is a multiplier, too. Hence, Theorem 2.8.7 proves the proposition if $k=-1$.

Step 2. (Outline). If $k=0,1,2,\ldots$ and $g\in C^{\infty,k}(R_1)$, then $g^{(j)}(x)\in C^{\infty,-1}(R_1)$ with $j=0,\ldots,k+1$. Thus it follows from Step 1 and Theorem 2.3.8 that g is a multiplier for $F^s_{p,q}(R_1)$ if

$$\frac{1}{p}-1<s<\frac{1}{p}+k+1 \quad \text{with (2)}\,. \tag{4}$$

Now it follows by duality that g is also a multiplier for $F^s_{p,q}(R_1)$ if $1<p<\infty$, $1<q<\infty$ and $\frac{1}{p}-1-(k+1)<s<\frac{1}{p}$. The rest is again a matter of complex interpolation in the same way as in the proof of Theorem 2.8.7.

Remark. If (2) is not satisfied, then one has to use (2.8.7/2) and one obtains similar assertions. This method can be extended to $n>1$. However, there is a difficulty. If $n>1$ and $n\left(\frac{1}{p}-1\right)>\frac{1}{p}$, then there does not exist a space $F^s_{p,q}(R_n)$ for which the characteristic function of a half-space is a multiplier (and this was our starting point). Probably one can overcome this difficulty if one uses the one-dimensional case, Fubini type theorems in the sense of 2.5.13. (provided that k in $g\in C^{\infty,k}(R_1)$ is sufficiently large) and complex interpolation (which covers all numbers k).

2.9. Extensions

2.9.1. The Spaces $B^s_{p,q}(R_n^+)$ and $F^s_{p,q}(R_n^+)$

Recall that R_n always denotes the real n-dimensional Euclidean space. If $x\in R_n$, we put $x=(x', x_n)$ with $x'\in R_{n-1}$. Let

$$R_n^+=\{x \mid x=(x', x_n)\in R_n \quad \text{and} \quad x_n>0\}\,. \tag{1}$$

If $g\in B^s_{p,q}(R_n)$ or $g\in F^s_{p,q}(R_n)$, respectively, then the restriction of g on R_n^+ should be understood in the sense of distributions on R_n^+, i.e. as an element of $D'(R_n^+)$.

Definition. Let $-\infty<s<\infty$ and $0<q\leqq\infty$.
(i) Let $0<p\leqq\infty$. Then $B^s_{p,q}(R_n^+)$ is the collection of all restrictions of elements of $B^s_{p,q}(R_n)$ on R_n^+. If $f\in B^s_{p,q}(R_n^+)$, then

$$\|f \mid B^s_{p,q}(R_n^+)\|=\inf \|g \mid B^s_{p,q}(R_n)\|\ , \tag{2}$$

where the infimum is taken over all $g\in B^s_{p,q}(R_n)$ whose restriction on R_n^+ coincides with f.
(ii) Let $0<p<\infty$. Then $F^s_{p,q}(R_n^+)$ is the collection of all restrictions of elements of $F^s_{p,q}(R_n)$ on R_n^+. If $f\in F^s_{p,q}(R_n^+)$, then

$$\|f \mid F^s_{p,q}(R_n^+)\|=\inf \|g \mid F^s_{p,q}(R_n)\|\ , \tag{3}$$

where the infimum is taken over all $g\in F^s_{p,q}(R_n)$ whose restriction on R_n^+ coincides with f.

Remark 1. Of course, one can extend part (ii) of the definition to $p=\infty$, where $F^s_{\infty,q}(R_n$ has the meaning of 2.3.4.

Proposition 1. (i) $B^s_{p,q}(R_n^+)$ with $-\infty<s<\infty$, $0<p\leqq\infty$ and $0<q\leqq\infty$ is a quasi-Banach space (Banach space if $p\geqq 1$ and $q\geqq 1$).
(ii) $F^s_{p,q}(R_n^+)$ with $-\infty<s<\infty$, $0<p<\infty$ and $0<q\leqq\infty$ is a quasi-Banach space (Banach space if $p\geqq 1$ and $q\geqq 1$).

Proof. It is easy to see that both $B^s_{p,q}(R_n^+)$ and $F^s_{p,q}(R_n^+)$ are quasi-normed (resp. normed) linear spaces. The completeness follows by standard arguments.

The Extension Problem. Let R, with $Rf=f/R_n^+$ if $f\in S'(R_n)$, be the restriction operator (from R_n onto R_n^+). Obviously, R is a bounded linear operator from $B^s_{p,q}(R_n)$ onto $B^s_{p,q}(R_n^+)$ and from $F^s_{p,q}(R_n)$ onto $F^s_{p,q}(R_n^+)$, respectively (we use the symbol R also for restrictions of R to linear subspaces of $S'(R_n)$). The problem is to find a bounded linear operator S from $B^s_{p,q}(R_n^+)$ or $F^s_{p,q}(R_n^+)$ into $B^s_{p,q}(R_n)$ or $F^s_{p,q}(R_n)$, respectively, such that

$$RS=E \quad \text{(identity in } B^s_{p,q}(R_n^+) \text{ or } F^s_{p,q}(R_n^+)\text{, resp.)} \tag{4}$$

In that case, S is called a bounded linear extension operator. If (4) holds, we shall say that the restriction operator R is a retraction, and the extension operator S is a corresponding coretraction. Cf. (2.7.2/3), where we studied a similar problem.

Remark 2. We are looking for common coretractions S in the sense of (4) for the restriction operator R in a given family of spaces. Let e.g. $\{B^s_{p,q}(R_n^+)$ with $s_0<s<s_1, p_0<p<q_1, q_0<q<q_1\}$ be such a family, where s_0, s_1, p_0, p_1, q_0, and q_1 are given. Then we ask whether there exists a linear operator S, which is defined on the union of all spaces of the family, such that its restrictions on $B^s_{p,q}(R_n^+)$ has the desired property. In such a case we shall speak of a common coretraction S.

Proposition 2. Let $0<p<\infty, 0<q_0\leqq\infty, 0<q_1\leqq\infty$ and $-\infty<s_0<s_1<\infty$. Let the restriction operator R be a retraction for $F^{s_0}_{p,q_0}(R_n)$ and for $F^{s_1}_{p,q_1}(R_n)$ with a common coretraction S in the sense of (4). Then R is a retraction for all spaces $B^s_{p,q}(R_n)$ with $s_0<s<s_1$ and $0<q\leqq\infty$ with the common coretraction S.

Proof. The property (4), including the boundedness of the operators R and S, is preserved under real interpolation. Hence the proposition follows immediately from (2.4.2/2).

Remark 3. By the last proposition we may restrict ourselves to the spaces $F^s_{p,q}(R_n)$ in the sequel.

2.9.2. The Case $\min(p, q) > 1$

As was mentioned at the end of the last subsection, it is sufficient to deal with the spaces $F^s_{p,q}(R_n)$.

Theorem. Let $r=0, 1, 2, \ldots$ Then the restriction operator R from 2.9.1. is a retraction for all spaces $F^s_{p,q}(R_n)$ with

$$1<p<\infty,\quad 1<q<\infty,\quad \text{and}\quad -r+\frac{1}{p}-1<s<r+\frac{1}{p}. \tag{1}$$

and there exists a common coretraction S in the sense of (2.9.1/4) (which depends on r).

Proof. Step 1. Let $1<p<\infty, 1<q<\infty$ and $\frac{1}{p}-1<s<\frac{1}{p}$ (i.e.. $r=0$ in (1)). Then the characteristic function of R_n^+ is a multiplier for $F^s_{p,q}(R_n)$, cf. Theorem 2.8.7 (ii). Thus it follows that $S=S_0$, with

$$(S_0f)(x)=\begin{cases} f(x) & \text{if}\quad x\in R_n^+\,, \\ 0 & \text{if}\quad x\in R_n-R_n^+, \end{cases} \tag{2}$$

has the desired properties (if f is smooth, then (2) makes sense immediately, the rest is a matter of completion).

Step 2. In order to deal with the cases $r=1, 2, \ldots$ we need some preparations. If $r=1, 2, \ldots$ and $\frac{1}{p}-1+r<s<\frac{1}{p}+r$ (we always assume that $1<p<\infty$ and $1<q<\infty$), then

$$\mathring{F}^s_{p,q}(R_n^+)=\left\{f \mid f\in F^s_{p,q}(R_n^+),\, f(x', 0)=\ldots=\frac{\partial^{r-1}f}{\partial x_n^{r-1}}(x', 0)=0\right\} \tag{3}$$

makes sense, cf. Theorem 2.7.2 (ii). We claim that S_0 from (2) yields a bounded linear extension operator from $\mathring{F}^s_{p,q}(R_n^+)$ into $F^s_{p,q}(R_n)$. For that purpose we note that

$$\frac{\partial^j S_0 f}{\partial x_n^j}(x)=\begin{cases}\dfrac{\partial^j f}{\partial x_n^j} & \text{if} \quad x\in R_n^+\\ 0 & \text{if} \quad x\in R_n-R_n^+\end{cases}=S_0\,\frac{\partial^j f}{\partial x_n^j}(x) \tag{4}$$

if $j=0,\ldots,r$ and $f\in\mathring{F}^s_{p,q}(R_n^+)$ (we may again assume that f is smooth, so (4) follows by standard arguments). Furthermore, by Theorem 2.3.8 we have

$$\|S_0 f \mid F^s_{p,q}(R_n)\|\leq c\sum_{|\alpha'|+j\leq r}\left\|\frac{\partial^j}{\partial x_n^j}D^{\alpha'}S_0 f \mid F^{s-r}_{p,q}(R_n)\right\|, \tag{5}$$

with $\alpha'=(\alpha_1,\ldots,\alpha_{n-1},0)$. But (4) yields

$$\frac{\partial^j}{\partial x_n^j}D^{\alpha'}S_0 f=S_0\left(\frac{\partial^j}{\partial x_n^j}D^{\alpha'}f\right) \quad \text{if} \quad |\alpha'|+j\leq r\,. \tag{6}$$

We put (6) in (5). Then it follows from Step 1 that S_0 is an extension operator from $\mathring{F}^s_{p,q}(R_n^+)$ into $F^s_{p,q}(R_n)$.

Step 3. Again let $r=1, 2, \ldots$ (and $1<p<\infty$, $1<q<\infty$). and let

$$0<\lambda_1<\ldots<\lambda_{2r}<\infty\,.$$

If $\varphi\in S(R_n)$, put

$$(S_r^1\varphi)(x)=\sum_{j=1}^{2r}b_j\varphi(x',-\lambda_j x_n)\,,\quad (S_r^2\varphi)(x)=-\sum_{j=1}^{2r}\frac{b_j}{\lambda_j}\varphi\left(x',-\frac{x_n}{\lambda_j}\right). \tag{7}$$

We determine the numbers b_j such that

$$\frac{\partial^j S_r^1\varphi(x',0)}{\partial x_n^j}=\frac{\partial^j S_r^2\varphi(x',0)}{\partial x_n^j}=\frac{\partial^j\varphi(x',0)}{\partial x_n^j} \quad \text{if} \quad j=0,\ldots,r-1 \tag{8}$$

(this is possible in a unique way). The function $\varphi-S_r^2\varphi$ and its derivatives up to order $r-1$ vanish on the hyperplane $x_n=0$. Now, if s is given by (1), we define S_r via

$$(S_r f)(\varphi)=f(S_0(\varphi-S_r^2\varphi))\,,\quad \varphi\in S(R_n)\,. \tag{9}$$

If f is smooth (and again we may assume that this is valid), then we have

$$(S_r f)(x)=\begin{cases}f(x) & \text{if} \quad x\in R_n^+\,,\\ (S_r^1 f)(x) & \text{if} \quad x\in R_n-R_n^+\,.\end{cases} \tag{10}$$

(8) proves that we have the following counterparts of (4) and (6).

$$D^\alpha(S_r f)(x)=\begin{cases}(D^\alpha f)(x) & \text{if} \quad x\in R_n^+\\ (D^\alpha S_r^1 f)(x) & \text{if} \quad x\in R_n-R_n^+\end{cases} \quad \text{if} \quad |\alpha|\leq r\,. \tag{11}$$

Let $\frac{1}{p}-1+r<s<r+\frac{1}{p}$. Then it follows from Theorem 2.3.8 and elementary calculations that

$$\begin{aligned}\|S_r f \mid F^s_{p,q}(R_n)\| &\leq c\sum_{|\alpha|\leq r}\|D^\alpha S_r f \mid F^{s-r}_{p,q}(R_n)\|\\ &\leq c'\sum_{|\alpha|\leq r}\|S_0 D^\alpha f \mid F^{s-r}_{p,q}(R_n)\|\,.\end{aligned} \tag{12}$$

Step 1 and (12) prove that S_r has the desired properties if $\frac{1}{p}-1+r<s<r+\frac{1}{p}$. Next we consider the case $\frac{1}{p}-1-r<s<-r+\frac{1}{p}$ (and again $1<p<\infty$, $1<q<\infty$). We use the duality formula

$$(F^s_{p,q}(R_n))' = F^{-s}_{p',q'}(R_n), \quad \frac{1}{p}+\frac{1}{p'}=\frac{1}{q}+\frac{1}{q'}=1\,, \tag{13}$$

cf. (2.11.2/2). Then we can apply the argument of Step 2 to the function $S_0(\varphi - S^2_r\varphi)$, considered as an element of $F^{-s}_{p',q'}(R_n)$. Now (9) yields

$$\begin{aligned}\|S_r f \mid F^s_{p,q}(R_n)\| &= \sup |(S_r f)(\varphi)| = \sup |f\,(S_0(\varphi - S^2_r\varphi))| \\ &\leq \|f \mid F^s_{p,q}(R_n)\| \sup \|S_0\,(\varphi - S^2_r\varphi) \mid F^{-s}_{p',q'}(R_n)\|\,,\end{aligned} \tag{14}$$

where the supremum is taken over all $\varphi \in S(R_n)$ with $\|\varphi \mid F^{-s}_{p',q'}(R_n)\| \leq 1$. It follows from Step 2 and the explicit form of S^2_r that the last factor on the right-hand side of (14) is bounded. Thus (10) and (14) prove that S_r has the desired properties.

Step 4. Again let $r=1, 2, \ldots$ (and $1<p<\infty$, $1<q<\infty$). We know that S_r is a coretraction for the restriction operator R in $F^s_{p,q}(R_n)$ if

$$\begin{cases} \text{either} & \frac{1}{p}-1+r<s<\frac{1}{p}+r \\ \text{or} & \frac{1}{p}-1-r<s<\frac{1}{p}-r\,. \end{cases} \tag{15}$$

Thus it follows from the complex interpolation formula (2.4.7/3) that S_r has the desired properties for the full range of the parameters given by (1) if S_r has the interpolation property. Besides simple affine transformations of R_n of the type $x \to (x', \lambda x_n)$, the operator S_r from (10) can essentially be represented as a finite sum of multiplication operators with characteristic functions of half-spaces. But such multiplication operators have the interpolation property, cf. the proof of Corollary 2.8.2 and the end of Step 2 of the proof of Theorem 2.8.7. Hence, S_r has the interpolation property. The proof is complete.

Remark 1. The above proof is based on the fact that the characteristic function of a half-space is a pointwise multiplier for $F^s_{p,q}(R_n)$ if $1<p<\infty$, $1<q<\infty$ and $\frac{1}{p}-1<s<\frac{1}{p}$. By Theorem 2.8.7 (ii) this fact holds also for other spaces $F^s_{p,q}(R_n)$ and hence, one can generalize the above considerations. However, if $n>1$, then it is not possible to prove the counterpart of the above theorem for all values of p (and this is one of the main goals of our theory). So we consider the case $n=1$ in the following subsection and reduce later on the n-dimensional case to the one-dimensional one.

Remark 2. At the end of the above proof one can also use the classical complex interpolation method $[\cdot,\cdot]_\Theta$ in the sense of Remark 2.4.7/2 instead of $(\cdot,\cdot)_\Theta$. In that case the interpolation property is satisfied automatically. But the disadvantage is that the classical interpolation method cannot be extended to the spaces $F^s_{p,q}(R_n)$ with $p<1$ or $q<1$ (and we shall need this extension in the sequel).

Remark 3. Cf. [I, pp. 218/219] as far as Step 2 and Step 3 of the above proof are concerned.

2.9.3. The Case $0<p\leqq q<\infty$ and $n=1$

It is our aim to extend Theorem 2.9.2 to certain spaces $F^s_{p,q}(R_n)$ with $\min(p,q)\leqq 1$. In this subsection we consider the one-dimensional case, i.e. $n=1$. Furthermore, we assume that

$$0<p<\infty, \quad 0<q<\infty \quad \text{and} \quad \min(p,q)\geqq \min(p,1). \tag{1}$$

This means $q\geqq 1$ if $p\geqq 1$ and $q\geqq p$ if $p<1$. The first case, i.e. $p\geqq 1$, is essentially covered by Theorem 2.9.2 (besides some limiting cases). Furthermore, we assume that S_r has the same meaning as in (2.9.2/9) and (2.9.2/10) with $n=1$.

Theorem. Let p and q be given by (1). Then the restriction operator R from 2.9.1. with $n=1$ is a retraction for all spaces $F^s_{p,q}(R_1)$ with $-\infty<s<\infty$. Let $r=0, 1, 2, \ldots$ Then S_r is a common coretraction in the sense of (2.9.1/4) for all spaces $F^s_{p,q}(R_1)$ with

$$-r+\frac{1}{p}-1<s<r+\frac{1}{p}. \tag{2}$$

Proof. Step 1. If (1) is satisfied, then Theorem 2.8.7(ii) with $n=1$ shows that the characteristic function of the interval $(0,\infty)$ is a multiplier for $F^s_{p,q}(R_1)$ if $\frac{1}{p}-1<s<\frac{1}{p}$. Hence it follows in the same way as in Step 1 and as in the first part of Step 3 of the proof of Theorem 2.9.2 that S_r has the desired properties if

$$r+\frac{1}{p}-1<s<r+\frac{1}{p}. \tag{3}$$

Step 2. If (1) is satisfied, it follows from Step 1 and from Theorem 2.9.2 with $S=S_r$ that S_r is an extension operator (coretraction for the restriction operator R) if $\left(s, \frac{1}{p}\right)$ belongs to the shaded area in Fig. 2.9.3. We apply complex interpolation in the same way as in Step 4 of the proof of Theorem 2.9.2. Then (2.4.7/3) covers all cases of the theorem (inclusively $q=1$ if $p>1$). The proof is complete.

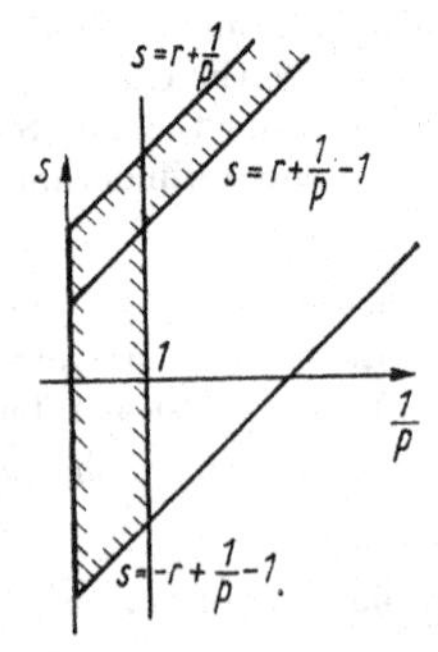

Fig. 2.9.3

Remark 1. The restriction (1) is convenient but not necessary for our method. On the one hand one can start with the more general cases of Theorem 2.8.7(ii). On the other, one obtains by complex interpolation (if one starts with the restrictions (1)) some couples (p,q) which are not covered by (1). We return to this problem in 2.9.5.

Corollary. Let $0<p\leqq\infty$ and $0<q\leqq\infty$. Then the restriction operator R from 2.9.1. with $n=1$ is a retraction for all spaces $B^s_{p,q}(R_1)$ with $-\infty<s<\infty$. Let $r=0, 1, 2, \ldots$ Then S_r is a common coretraction in the sense of (2.9.1/4) for all spaces $B^s_{p,q}(R_1)$ with (2).

Proof. If $p<\infty$, the corollary follows immediately from the above theorem and Proposition 2.9.1/2. If $p=\infty$, we use the duality formula $B^s_{\infty,\infty}(R_1)=(B^{-s}_{1,1}(R_1))'$, cf. (2.11.2/1). We obtain in a similar way as in (2.9.2/14) that S_r has the desired properties if $-r-1<s<-r$. If $r-1<s<r$, then $B^s_{\infty,\infty}(R_n)=\mathcal{C}^s(R_n)$ are the Zygmund spaces, cf. (2.5.12/22). If one uses the explicit norm (2.5.12/20) with $p=q=\infty$, $m=r-1$ and $M=1$, then it follows by simple direct calculations that S_r is an extension operator in the above sense. Finally, the cases $B^s_{\infty,q}(R_1)$ with $-r-1<s<r$ and $0<q\leqq\infty$ can be obtained by real interpolation, cf. (2.4.2/1).

Remark 2. One can give a direct proof of this corollary by beginning with Theorem 2.8.5 and Proposition 2.8.4(i) (one-dimensional case) and proceeding as in the proof of Theorem 2.9.2.

2.9.4. The Extension Theorem

In this subsection we prove the main assertion of Section 2.9.

Theorem. (i) If $0<p\leqq\infty$, $0<q\leqq\infty$ and $-\infty<s<\infty$, then the restriction operator R from 2.9.1. is a retraction for the spaces $B^s_{p,q}(R_n)$. If N is a natural number, then there exists a common coretraction $S=S^N$ in the sense of (2.9.1/4) for all spaces $B^s_{p,q}(R_n)$ with

$$|s|<N, \quad \frac{1}{N}<p\leqq\infty \quad \text{and} \quad 0<q\leqq\infty. \tag{1}$$

(ii) If $0<p<\infty$, $0<q<\infty$,

$$\min(p, q)\geqq\min(p, 1) \tag{2}$$

and $-\infty<s<\infty$, then the restriction operator R from 2.9.1. is a retraction for the space $F^s_{p,q}(R_n)$. If N is a natural number, then the operator $S=S^N$ from part (i) is also a common coretraction in the sense of (2.9.1/4) for all spaces $F^s_{p,q}(R_n)$ with

$$|s|<N, \quad \frac{1}{N}<p<\infty, \quad 0<q<\infty \quad \text{and} \quad (2). \tag{3}$$

Proof. Step 1. Let us assume that part (ii) is proven with $S^N=S_r$, where $r=r(N)$ is sufficiently large and S_r has the same meaning as in the proof of Theorem 2.9.2, cf. (2.9.2/9) and (2.9.2/10). If $p<\infty$, then part (i) follows immediately from Proposition 2.9.1/2. If $p=\infty$, we argue in the same way as in the proof of Corollary 2.9.3 or as in Remark 2.9.3/2.

Step 2. We prove (ii). Let S_r be the same operator as above. Let $0<p<\infty$ and

$$r+\frac{1}{p}>s>\max\left(\frac{n}{\min(p, q)}, 4+\frac{3}{\min(p, q)}\right)=s_0(p, q) \tag{4}$$

Then it follows from Theorem 2.5.13 that $F^s_{p,q}(R_n)$ has the Fubini property, cf. Definition 2.5.13. The explicit form of S_r (in particular the fact that S_r is essentially a one-dimensional affair) shows that the extension problem for $F^s_{p,q}(R_n)$ reduces to the corresponding problem for $F^s_{p,q}(R_1)$. Hence, Theorem 2.9.3 proves that S_r has the desired properties. Consequently, by Theorem 2.9.2 the operator

S_r is an extension operator if either (2.9.2/1) holds or if p, q, s satisfy $0<p<\infty$, (2), (4), cf. the shaded area in Fig. 2.9.4. We again apply the complex interpolation formula (2.4.7/3), cf. Step 4 of the proof of Theorem 2.9.2. As a result we obtain that S_r is an extension operator in the convex hull of the shaded area in Fig. 2.9.4, where q satisfies (2). But if $r\to\infty$, then this procedure covers all couples $\left(s, \frac{1}{p}\right)$. This proves the theorem.

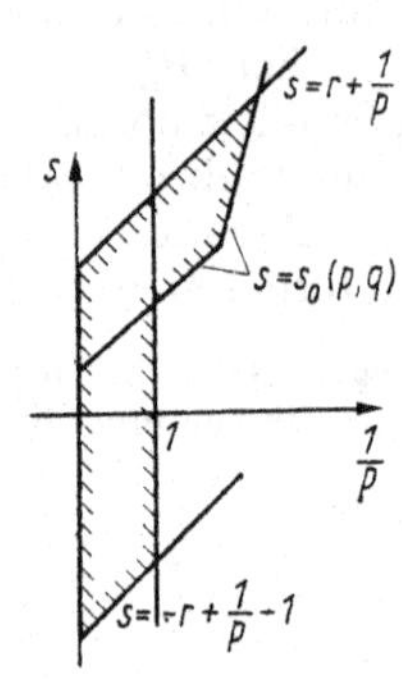

Fig. 2.9.4

2.9.5. The Case $q<p$

In the preceding subsections we started with Theorem 2.8.7(ii) under the additional hypothesis (2.9.3/1) (one-dimensional case). Then we obtained a quite satisfactory solution of the extension problem, both in the one-dimensional and in the general n-dimensional case. Of course, (2.8.7/2) contains also some cases other than these in (2.9.3/1) and one can hope to find affirmative solutions of the extension problem for more general spaces $F^s_{p,q}(R_n)$. We deal with the one-dimensional case. We can translate these results afterwards by the method of 2.9.4. via the Fubini property of certain spaces $F^s_{p,q}(R_n)$ to the n-dimensional case. The method is clear, and we shall not give detailed formulations. But even in the one-dimensional case we cannot remove by our method the disturbing assumption $\min(p, q) \geqq \min(p, 1)$ in 2(.9.3/1) completely. However, it is possible to weaken this restriction by

$$\begin{cases} \text{either } 1\leqq p<\infty & \text{and } \frac{1}{2}<q<\infty \\ \text{or } \quad 0<p<1 & \text{and } 0<\frac{1}{q}<\frac{1}{p}+1\,. \end{cases} \tag{1}$$

Theorem. Let p and q be given by (1). Then the restriction operator R from 2.9.1. with $n=1$ is a retraction for all spaces $F^s_{p,q}(R_1)$ with $-\infty<s<\infty$. If N is a natural number, then there exists a common corretraction S^N in the sense of (2.9.1/4) for all spaces $F^s_{p,q}(R_1)$ with $|s|<N$ and

$$\begin{cases} \text{either } 1\leqq p<\infty & \text{and } \frac{1}{2}+\frac{1}{N}<q<\infty \\ \text{or } \quad \frac{1}{N}<p<1 & \text{and } 0<\frac{1}{q}<\frac{1}{p}+\frac{1}{N}+1\,. \end{cases} \tag{2}$$

Proof. If $0<p<\infty$ is fixed, the characteristic function of the interval $(0, \infty)$ is a multiplier for $F^s_{p,q}(R_1)$ if $\left(s, \frac{1}{p}\right)$ belongs to the shaded areas in Fig. 2.9.5. This follows immediately from Theorem 2.8.7(ii). The rest is the same as in the proofs of Theorem 2.9.3 and Theorem 2.9.2. In particular we use the complex interpolation formula (2.4.7/3) with fixed $p_0=p_1=p$.

Problem. Do all spaces $F^s_{p,q}(R_n)$ with $0<p<\infty$, $0<q\leqq\infty$ and $-\infty<s<\infty$ have the extension property, i.e. is R from 2.9.1. a retraction? Of course, the conjecture is yes. One must find a proof which is not based on Theorem 2.8.7.

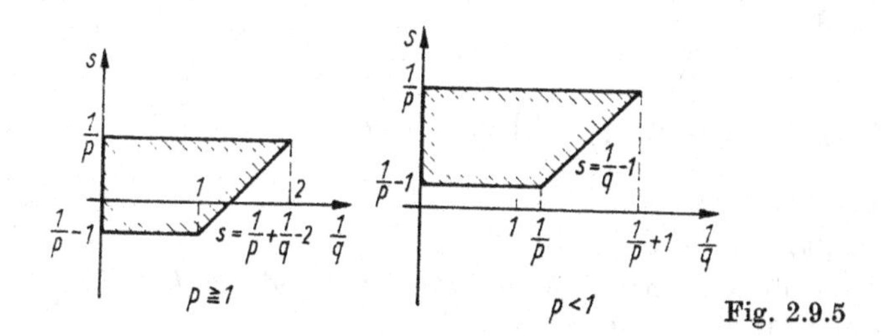

Fig. 2.9.5

2.10. Diffeomorphic Maps

2.10.1. Preliminaries

The problem is the following. Let

$$y=\psi(x)\,,\quad \text{i.e.}\quad y_j=\psi_j(x)\quad \text{if}\quad j=1,\ldots,n\,, \tag{1}$$

be an infinitely differentiable one-to-one mapping from R_n onto R_n, i.e. a diffeomorphism. We assume in the sequel that $\psi(x)=x$ if $|x|$ is large. This is not necessary for our considerations, but convenient. The problem is to prove that the linear operator

$$f(x)\to f(\psi(x)) \tag{2}$$

maps $B^s_{p,q}(R_n)$ onto itself, and $F^s_{p,q}(R_n)$ onto itself. Of course, one must interpret (2) with $f\in S'(R_n)$ in the sense of distributions. For some spaces it is easy to see that this problem has an affirmative solution. Let $W^m_p(R_n)=F^m_{p,2}(R_n)$ with $1<p<\infty$ and $m=0, 1, 2, \ldots$ be the usual Sobolev spaces, cf. (2.2.2/7). Then it follows by straightforward calculations that (2) yields a one-to-one mapping from $W^m_p(R_n)$ onto itself. By duality (recall that $F^{-m}_{p',2}(R_n)=(F^m_{p,2}(R_n))'$ with $\frac{1}{p}+\frac{1}{p'}=1$, cf. (2.11.2/2)) and certain multiplication properties in the sense of Theorem 2.8.2 it follows that $F^{-m}_{p,2}(R_n)$ also has the diffeomorphism property. By real and complex interpolation, cf. (2.4.2/2) and Remark 2.4.7/2, one can extend this assertion to all spaces $B^s_{p,q}(R_n)$ and $H^s_p(R_n)=F^s_{p,2}(R_n)$ with $1<p<\infty$, $0<q\leqq\infty$ and $-\infty<s<\infty$. In the same way one proves a corresponding statement also for $B^s_{\infty,q}(R_n)$ with $0<q\leqq\infty$ and $-\infty<s<\infty$. Unfortunately, however, it is not possible to cover all spaces $B^s_{p,q}(R_n)$ or $F^s_{p,q}(R_n)$ in this simple way, and the problem for these general spaces is much harder. We start with a simple proposition which will be useful in the sequel.

Proposition. Let $x\in R_1$, $h\in R_1$, $(\Delta^1_h f)(x)=f(x+h)-f(x)$ and $\Delta^l_h f=\Delta^{l-1}_h\Delta^1_h f$. If M and N are natural numbers with $1\leqq N<M$, then

$$\begin{aligned}&M\,(M-1)\ldots(M-N+1)\,(\Delta^{M-N}_h f)(x)\\&=\sum_{l=N}^{M} l\,(l-1)\ldots(l-N+1)\binom{M}{l}(-1)^{M-l}f(x+(l-N)\,h)\,.\end{aligned} \tag{3}$$

Proof. We have

$$(t-1)^M=\sum_{l=0}^{M}\binom{M}{l}(-1)^{M-l}t^l, \quad t\in R_1 .$$

Differentiation yields

$$M(M-1)\dots(M-N+1)(t-1)^{M-N} = \sum_{l=N}^{M} l(l-1)\dots(l-N+1)\binom{M}{l}(-1)^{M-l}t^{l-N} . \tag{4}$$

Moreooer, we remark that

$$[F(\Delta_h^{M-N}f)](y)=(e^{iyh}-1)^{M-N}Ff , \tag{5}$$

where F is the one-dimensional Fourier transform and e.g. $f\in S(R_1)$. Now (3) follows from (4) and (5) with $t=e^{iyh}$.

Remark 1. The method shows that every identity for polynomials has a corresponding counterpart for differences. One can extend this procedure to the n-dimensional case.

Remark 2. The following consequence of (3) is of interest later on,

$$\sum_{l=0}^{M} l^N\binom{M}{l}(-1)^{M-l}f(x+lh)=\sum_{r=0}^{N} c_r(\Delta_h^{M-N+r}f)(x+(N-r)h) , \tag{6}$$

where c_r are appropriate constants.

2.10.2. The Main Theorem

The diffeomorphic map $y=\psi(x)$ from R_n onto R_n has the same meaning as in the preceding subsection.

Theorem. (i) Let $0<p\leqq\infty$, $0<q\leqq\infty$ and $-\infty<s<\infty$. Then (2.10.1/2) yields a one-to-one mapping from $B^s_{p,q}(R_n)$ onto itself.
(ii) Let $0<p<\infty$, $0<q\leqq\infty$ and $-\infty<s<\infty$. Then (2.10.1/2) yields a one-to-one mapping from $F^s_{p,q}(R_n)$ onto itself.

Proof. Step 1. We consider the spaces $F^s_{p,q}(R_n)$. The proof for the spaces $B^s_{p,q}(R_n)$ is the same. In this step we prove (ii) for $0<p<\infty$, $0<q\leqq\infty$ and s sufficiently large, where we assume for sake of simplicity that $n=1$. However, the proof works also for $n>1$ (obvious technical modifications). Furthermore, we may assume that $f\in F^s_{p,q}(R_n)$ is a real function. If s is sufficiently large, we can use the equivalent quasi-norms for $F^s_{p,q}(R_1)$ from (2.5.9/17) and (2.5.9/49). If $|h|<2$ and $x\in R_1$, then

$$[\Delta_h^M f(\psi(\cdot))](x)=\sum_{l=0}^{M}(-1)^{M-l}\binom{M}{l}f(\psi(x+lh)) . \tag{1}$$

Moreover, if L is a natural number, then Taylor's formula yields

$$\psi(x+lh)=\sum_{l=0}^{L}\frac{1}{j!}\psi^{(j)}(x)(lh)^j+\frac{1}{L!}\int_0^{lh}(lh-t)^L\psi^{(L+1)}(x+t)\,dt .$$

Consequently,

$$f(\psi(x+lh))=f\left(\sum_{j=0}^{L}\frac{1}{j!}\psi^{(j)}(x)(lh)^j\right) + f'\left(\sum_{j=0}^{L}\dots+\vartheta\frac{1}{L!}\int_0^{lh}\dots\right)\frac{1}{L!}\int_0^{lh}(lh-t)^L\psi^{(L+1)}(x+t)\,dt , \tag{2}$$

where $0 \leqq \vartheta \leqq 1$. The last integral vanishes if $|x|$ is large (recall that $\psi(x)=x$ for large values of $|x|$ and $L \geqq 1$). We put (2) in (1) and the result in (2.5.9/17) with $f(\psi(x))$ instead of $f(x)$. Then the terms with $f'(\ldots)$ can be estimated from above by

$$c\|f' \mid L_\infty(R_1)\| \left(\sum_{k=0}^{\infty} 2^{-k(L+1)q} 2^{skq}\right)^{\frac{1}{q}} \leqq c'\|f \mid F^s_{p,q}(R_1)\|, \tag{3}$$

where we have used $L+1>s$ and $F^s_{p,q}(R_1) \subset C^1(R_1)$ if $s-\frac{1}{p}>1$. The last inclusion from Theorem 2.7.1, cf. (2.7.1/11) and (2.7.1/14). The estimate $L+1>s$ means that we determine L step by step (sufficiently large). In other words, we can restrict our attention to the first term in (2). If s is large, then we find an integer K such that

$$K+\frac{1}{p}+1<K+G_{p,q}<s<2K, \tag{4}$$

where $G_{p,q}$ has been defined in (2.5.9/7). We have

$$\begin{aligned} f\left(\sum_{j=0}^{L}{}' \frac{1}{j!}\psi^{(j)}(x)\,(lh)^j\right) &= f(\psi(x)+lh\psi'(x)) \\ &+\sum_{r=1}^{K}\frac{1}{r!} f^{(r)}(\psi(x)+lh\psi'(x))\left(\sum_{j=2}^{L}\frac{1}{j!}\psi^{(j)}(x)\,l^jh^j\right)^r \\ &+\frac{1}{K!}\int_0^{\sum_{j=2}^{L}\cdots}\left(\sum_{j=2}^{L}\cdots-t\right)^K f^{(K+1)}(\psi(x)+\psi'(x)\,lh+t)\,\mathrm{d}t. \end{aligned} \tag{5}$$

We put (5) in (1) and the result in (2.5.9/17). Of interest is the influence of the last term in (5), the one with the integral. The corresponding terms in (2.5.9/17) can be estimated from above by

$$c\|f^{(K+1)} \mid L_\infty(R_1)\| \left(\sum_{k=0}^{\infty} 2^{ksq} 2^{-2Kkq}\right)^{\frac{1}{q}} \leqq c'\|f \mid F^s_{p,q}(R_1)\|. \tag{6}$$

We recall that the last term in (5) (and also the middle term) vanishes if $|x|$ is large. The last estimate in (6) follows from (4) and $F^s_{p,q}(R_1) \subset C^{K+1}(R_1)$ ⇆ $s>K+1+\frac{1}{p}$, cf. again Theorem 2.7.1, (2.7.1/11) and (2.7.1/14). Consequently, only the first and the second term in (5) are of interest in the sequel. If we put the first term of (5) in (1), then we have $(\Delta^M_{h\psi'(x)}f)(\psi(x))$. Now replace h by $2^{-k}h$. Then the corresponding maximal function, in the sense of (2.5.9/6) and (2.5.9/17), can be estimated from above by

$$\sup_{\substack{1 \leqq |h| \leqq 2 \\ y \in R_n}} \frac{|(\Delta^M_{2^{-k}h\psi'(y)}f)(\psi(y))|}{1+|x-y|^a\, 2^{ka}} \leqq c \sup_{\substack{|h| \leqq 2 \\ y \in R_n}} \frac{|(\Delta^M_{2^{-k}h}f)(\psi(y))|}{1+|\psi(x)-\psi(y)|^a\, 2^{ka}}. \tag{7}$$

In the last estimate we assumed without restriction of generality that $|\psi'(x)| \leqq 1$ (otherwise we must replace $|h| \leqq 2$ by $|h| \leqq c$, and such a modification is completely immaterial). The right-hand side of (7) can be estimated by

$$c \sup_{|h| \leqq 2} (D^M_{2^{-k}h}f)(\psi(x)). \tag{8}$$

If s is sufficiently large (and we assume that it is), we can apply (2.5.9/49). This shows that the corresponding part of (2.5.9/17) (with $f(\psi(x))$ instead of $f(x)$) which systems from the first term of (5) can be estimated from above by $c\|f\,|\,F^s_{p,q}(R_1)\|$. Finally, we must estimate the term in (2.5.9/17) (with $f(\psi(x))$ instead of $f(x)$) which systems from the second term of (5). We put the second term of (5) in (1) and apply Proposition 2.10.1 and (2.10.1/6). Then we obtain terms of the type

$$\begin{aligned} &h^N \sum_{l=0}^{M} (-1)^{M-l} \binom{M}{l} l^N f^{(r)}(\psi(x)+lh\psi'(x)) \\ &= h^N \sum_{m=0}^{N} c_r(\Delta^{M-N+m}_{h\psi'(x)} f^{(r)})(\psi(x)+(N-m)\,h\psi'(x)) \end{aligned} \tag{9}$$

with $N=2r, \ldots, Lr$ and $r=1, \ldots, K$, where L and K have the above meaning. Recall that the number M is at our disposal (we must choose $M>s$ in any case). If L and K are fixed and if M is sufficiently large, we estimate the maximal functions of the corresponding terms in (2.5.9/17) in the same way as in (7) and (8). As a result we see that the part in (2.5.9/17) which stems from the second term of (5) can be estimated from above by

$$c\|f^{(r)}\,|\,F^{s-r}_{p,q}(R_1)\|\,, \quad \text{where} \quad r=1, \ldots, K \tag{10}$$

(of course, we assumed again that $s-K$ is sufficiently large and applied again (2.5.9/49)). By Theorem 2.3.8 we estimate (10) from above by $c\|f\,|\,F^s_{p,q}(R_1)\|$. If we put together all terms, then it follows that the quasi-norm in (2.5.9/17), with $f(\psi(x))$ instead of $f(x)$, can be estimated from above by $c\|f\,|\,F^s_{p,q}(R_1)\|$. The proof is complete (provided that s is a large number).

Step 2. We must remove the additional assumption of Step 1 that for given p and q the number s must be large. We shall do this in the n-dimensional case. Let $0<p<\infty$, $0<q\leqq\infty$ and $-\infty<s<\infty$. Let $A=(1-\Delta)^m$, where m is a natural number such that Step 1 is applicable to $F^{s+2m}_{p,q}(R_n)$. Let $f\in F^{s+2m}_{p,q}(R_n)$. Then it follows from the multiplication properties of the spaces $F^s_{p,q}(R_n)$ from Theorem 2.8.2 that

$$\begin{aligned} \|(Af)(\psi(\cdot))\,|\,F^s_{p,q}(R_n)\| &\leqq c \sum_{|\alpha|\leqq 2m} \|D^\alpha[f(\psi(\cdot))]\,|\,F^s_{p,q}(R_n)\| \\ &\leqq c'\|f(\psi(\cdot))\,|\,F^{s+2m}_{p,q}(R_n)\|\,, \end{aligned}$$

cf. also Theorem 2.3.8. By Step 1, the right-hand side can be estimated from above by $c\|f\,|\,F^{s+2m}_{p,q}(R_n)\|$. We have $A=I_{2m}$, where I_{2m} has the meaning of (2.3.8/1). Hence, A yields a one-to-one mapping from $F^{s+2m}_{p,q}(R_n)$ onto $F^s_{p,q}(R_n)$. This proves

$$\|(Af)(\psi(\cdot))\,|\,F^s_{p,q}(R_n)\| \leqq c\|Af\,|\,F^s_{p,q}(R_n)\|\,.$$

But every $g\in F^s_{p,q}(R_n)$ can be represented by $g=Af$ with $f\in F^{s+2m}_{p,q}(R_n)$, so this proves the theorem.

2.11. Dual Spaces

2.11.1. Preliminaries

The aim of this section is the study of the dual spaces of $B^s_{p,q}(R_n)$ and $F^s_{p,q}(R_n)$. If $-\infty<s<\infty$, $0<p<\infty$, and $0<q<\infty$, then $S(R_n)$ is a dense subset of $B^s_{p,q}(R_n)$ and of $F^s_{p,q}(R_n)$, cf. Theorem 2.3.3. Consequently, in that case a continuous linear functional on $B^s_{p,q}(R_n)$ or $F^s_{p,q}(R_n)$, respectively, can be interpreted in the usual way as an element of $S'(R_n)$, the dual of $S(R_n)$. More precisely, $g \in S'(R_n)$ belongs to the dual space $(B^s_{p,q}(R_n))'$ of the space $B^s_{p,q}(R_n)$ with $-\infty<s<\infty$, $0<p<\infty$, $0<q<\infty$, if and only if there exists a positive number c such that

$$|g(\varphi)| \leqq c\|\varphi \mid B^s_{p,q}(R_n)\| \quad \text{for all} \quad \varphi \in S(R_n) . \tag{1}$$

Similarly for $F^s_{p,q}(R_n)$ with $-\infty<s<\infty$, $0<p<\infty$ and $0<q<\infty$. All statements in the subsections 2.11.2. and 2.11.3. must be understood in this sense.

In this subsection we determine the dual spaces of $L_p(R_n, l_q)$ and $l_q(L_p(R_n))$, which are denoted by $(L_p(R_n, l_q))'$ and $(l_q(L_p(R_n)))'$, respectively. We recall that $L_p(R_n, l_q)$, or $l_q(L_p(R_n))$, is the set of all sequences $\{f_k(x)\}_{k=0}^\infty$ of Lebesgue-measurable functions on R_n for which the quasi-norm (1.2.2/5) or (1.2.2/6), respectively, is finite. We indicate "R_n" in these quasi-norms now. Dual means always the topological dual and the quasi-norm $\|g\|$ of a continuous linear functional g is calculated in the usual way. We shall need the determination of these duals later on. However, the proof is not typical and a reader who is ready to accept the result may skip its proof at first glance. If $1 \leqq q<\infty$, then q' is determined in the usual way by $\frac{1}{q}+\frac{1}{q'}=1$. If $0<q<1$, we put $q'=\infty$. Similarly for p if $1 \leqq p<\infty$.

Proposition. Let $1 \leqq p<\infty$ and $0<q<\infty$.

(i) Then $g \in (L_p(R_n, l_q))'$ if and only if it can be represented uniquely as

$$g(f)=\sum_{j=0}^{\infty} \int_{R_n} g_j(x)\, f_j(x)\, dx \tag{2}$$

for every $f=\{f_j(x)\}_{j=0}^\infty \in L_p(R_n, l_q)$, where

$$\{g_j(x)\}_{j=0}^\infty \in L_{p'}(R_n, l_{q'}) \quad \text{and} \quad \|g\|=\|g_j \mid L_{p'}(R_n, l_{q'})\| . \tag{3}$$

(ii) Then $g \in (l_q(L_p(R_n)))'$ if and only if it can be represented uniquely as

$$g(f)=\sum_{j=0}^{\infty} \int_{R_n} g_j(x)\, f_j(x)\, dx \tag{4}$$

for every $f=\{f_j(x)\}_{j=0}^\infty \in l_q(L_p(R_n))$, where

$$\{g_j(x)\}_{j=0}^\infty \in l_{q'}(L_{p'}(R_n)) \quad \text{and} \quad \|g\|=\|g_j \mid l_{q'}(L_{p'}(R_n))\| . \tag{5}$$

Proof. Step 1. By Hölder's inequality and the monotonicity properties of the l_q-spaces we have in any case that

$$L_{p'}(R_n, l_{q'}) \subset (L_p(R_n, l_q))', \quad l_{q'}(L_{p'}(R_n)) \subset (l_q(L_p(R_n)))' \tag{6}$$

and corresponding inequalities for the quasi-norms, respectively. We must prove the reverse assertion. If $1<p<\infty$ and $1<q<\infty$ then we have the well-known classical case, cf. e.g. R. E. Edwards [1, 8.20.5]. Hence we may restrict our attention to the remaining cases.

Step 2. Let $1<p<\infty$ and $0<q \leqq 1$. Let $g \in (L_p(R_n, l_q))'$ and

$$g_j(f)=g(\{0, \ldots, 0, f_j, 0, \ldots\}), \quad f=\{f_j\} \in L_p(R_n, l_q) . \tag{7}$$

By the usual representation of elements of $(L_p(R_n))'$ we have

$$g_j(f)=\int_{R_n} g_j(x)\, f_j(x)\, dx, \quad \|g_j\|=\|g_j(\cdot) \mid L_{p'}(R_n)\| , \tag{8}$$

where g_j in the last formula is interpreted as an element of $(L_p(R_n))'$. It follows that

$$\int_{R_n} \sum_{j=0}^{N} g_j(x)\, f_j(x)\, dx \leqq \|g\|\, \|\{f_j\}_{j=0}^N \mid L_p(R_n, l_q)\| . \tag{9}$$

Let

$$f_j(x)=\sup_{l=0,\dots,N} |g_l(x)|^{p'-1}\,\delta_{j,k}\,\mathrm{sgn}\,\overline{g_k(x)}\,, \tag{10}$$

where sgn $\overline{g_k(x)}$ is the argument of the complex number $\overline{g_k(x)}$ and $k=k(x)$ is an integer between 0 and N with $|g_k(x)|=\max_{l=0,\dots,N}|g_l(x)|$. We put (10) in (9) and obtain

$$\int_{R_n}\sup_{l=0,\dots,N}|g_l(x)|^{p'}\,dx\le\|g\|\left(\int_{R_n}\sup_{l=0,\dots,N}|g_l(x)|^{(p'-1)p}\,dx\right)^{\frac{1}{p}}. \tag{11}$$

Recall that $(p'-1)\,p=p'$. Then letting $N\to\infty$ in (11) yields the desired result. The case $1<p<\infty$, $0<q\le 1$ and $g\in(l_q(L_p(R_n)))'$ can be treated in the same way.

Step 3. Let $p=1$ and $1<q<\infty$. Let $g\in(L_1(R_n,l_q))'$. We use (7)–(9) and replace (10) by

$$f_j(x)=\chi(x)\,|g_j(x)|^{q'-1}\,\mathrm{sgn}\,\overline{g_j(x)}\,, \tag{12}$$

where $\chi(x)$ is the characteristic function of a set M in R_n with finite, positive Lebesgue measure such that

$$\left(\sum_{j=0}^{N}|g_j(x)|^{q'}\right)^{\frac{1}{q'}}\ge(1-\varepsilon)\operatorname*{ess\,supp}_{y\in R_n}\left(\sum_{j=0}^{N}|g_j(y)|^{q'}\right)^{\frac{1}{q'}}\quad\text{if}\quad x\in M$$

with $\varepsilon>0$. This makes sense because $g_j(y)\in L_\infty(R_n)$. We put (12) in (9) to obtain

$$\begin{aligned}(1-\varepsilon)^{q'}\,|M|\,\left\|\sum_{j=0}^{N}|g_j(\cdot)|^{q'}\mid L_\infty(R_n)\right\|&\le\int_{R_n}\chi(x)\sum_{j=0}^{N}|g_j(x)|^{q'}\,dx\\ &\le\|g\|\,\|\{f_j\}_{j=0}^{N}\mid L_1(R_n,l_q)\|\le\|g\|\,|M|\left\|\sum_{j=0}^{N}|g_j(\cdot)|^{q'}\mid L_\infty(R_n)\right\|^{\frac{1}{q}}.\end{aligned} \tag{13}$$

Letting $N\to\infty$ and $\varepsilon\downarrow 0$ yields the desired result. The case $p=1$, $1<q<\infty$ and $g\in(l_q(L_1(R_n)))'$ can be treated in the same way.

Step 4. Let $p=1$ and $0<q<1$. Let $g\in(L_1(R_n,l_q))'$ or $g\in(l_q(L_1(R_n)))'$. We again use (7)–(9) and replace (10) by

$$f_j(x)=\chi(x)\,\delta_{j,k}\,\mathrm{sgn}\,\overline{g_k(x)}\,, \tag{14}$$

where k and $\chi(x)$ have the same meaning as in (10) and (12), respectively. We put (14) in (9) and obtain that

$$(1-\varepsilon)\,|M|\,\|\sup_{j=0,\dots,N}|g_j(x)|\mid L_\infty(R_n)\|\le\|g\|\,|M|\,.$$

Again letting $N\to\infty$ and $\varepsilon\downarrow 0$ yields the desired result. The proof is complete.

2.11.2. The Case $1\le p<\infty$

Recall that duality formulas must always be interpreted in the sense of 2.11.1., and that $\frac{1}{q}+\frac{1}{q'}=1$ if $1<q<\infty$ and $q'=\infty$ if $0<q\le 1$. Similarly for p'. The space $F^s_{\infty,q'}(R_n)$ has the meaning of 2.3.4.

Theorem. (i) Let $-\infty<s<\infty$, $1\le p<\infty$ and $0<q<\infty$. Then

$$(B^s_{p,q}(R_n))'=B^{-s}_{p',q'}(R_n)\,. \tag{1}$$

(ii) Let $-\infty<s<\infty$, $1\le p<\infty$ and $1<q<\infty$. Then

$$(F^s_{p,q}(R_n))'=F^{-s}_{p',q'}(R_n)\,. \tag{2}$$

Proof. Step 1. Let $-\infty<s<\infty$, $1\leqq p<\infty$ and $0<q<\infty$. We prove in this step that

$$F^{-s}_{p',q'}(R_n)\subset(F^s_{p,q}(R_n))' \quad\text{and}\quad B^{-s}_{p',q'}(R_n)\subset(B^s_{p,q}(R_n))'\,. \tag{3}$$

Note that $1<p'\leqq\infty$ and $1<q'\leqq\infty$. Let $f\in F^{-s}_{p',q'}(R_n)$. We apply Definition 2.3.4. In any case we find systems $\varphi=\{\varphi_k(x)\}_{k=0}^{\infty}\in\Phi(R_n)$ and $\{f_k(x)\}_{k=0}^{\infty}\subset L_{p'}(R_n)$ such that

$$f=\sum_{k=0}^{\infty}F^{-1}\varphi_kFf_k \quad\text{in}\quad S'(R_n) \tag{4}$$

and

$$\|2^{-sk}f_k\mid L_{p'}(R_n,l_{q'})\|\leqq 2\|f\mid F^{-s}_{p',q'}(R_n)\| \tag{5}$$

(if $1<p'<\infty$, we may choose $f_k=\sum_{r=-1}^{1}F^{-1}\varphi_{k+r}Ff$, cf. also Proposition 2.3.4/1). If $\varphi\in S(R_n)$, then

$$\begin{aligned}|f(\varphi)|&=\left|\sum_{k=0}^{\infty}f(F\varphi_kF^{-1}\varphi)\right|=\left|\sum_{k=0}^{\infty}\sum_{r=-1}^{1}(F^{-1}\varphi_{k+r}Ff_{k+r})\,(F\varphi_kF^{-1}\varphi)\right|\\&=\left|\sum_{l=0}^{\infty}\sum_{r=-1}^{1}f_l(F\varphi_l\varphi_{l+r}F^{-1}\varphi)\right|\\&\leqq c\sum_{r=-1}^{1}\|2^{-sk}f_k\mid L_{p'}(R_n,l_{q'})\|\,\|2^{sk}F\varphi_k\varphi_{k+r}F^{-1}\varphi\mid L_p(R_n,l_q)\|\\&\leqq c'\|f\mid F^{-s}_{p',q'}(R_n)\|\,\|\varphi\mid F^s_{p,q}(R_n)\|\,.\end{aligned} \tag{6}$$

This proves the first part of (3). The second part can be proved in the same way.

Step 2. We prove (ii). Let $-\infty<s<\infty$, $1\leqq p<\infty$ and $1<q<\infty$. Let again $\{\varphi_k(x)\}_{k=0}^{\infty}\in\Phi(R_n)$. Because

$$F^s_{p,q}(R_n)\ni f\to\{2^{sk}F^{-1}\varphi_kFf\}_{k=0}^{\infty}$$

is a one-to-one mapping from $F^s_{p,q}(R_n)$ onto a subspace of $L_p(R_n,l_q)$, every functional $g\in(F^s_{p,q}(R_n))'$ can be interpreted as a functional on that subspace. By the Hahn-Banach theorem, g can be extended to a continuous linear functional on $L_p(R_n,l_q)$, where the norm of g is preserved. If $\varphi\in S(R_n)$, then Proposition 2.11.1 (i) yields

$$g(\varphi)=\int_{R_n}\sum_{k=0}^{\infty}g_k(x)\,(F^{-1}\varphi_kF\varphi)(x)\,dx\,, \tag{7}$$

$$\|2^{-sk}g_k\mid L_{p'}(R_n,l_{q'})\|\sim\|g\|\,. \tag{8}$$

(7) can be written as

$$g(\varphi)=\sum_{k=0}^{\infty}(F\varphi_kF^{-1}g_k)(\varphi)\,. \tag{9}$$

But (9) and (8) is a representation of type (2.3.4/1) and (2.3.4/4), respectively. This proves $g\in F^{-s}_{p',q'}(R_n)$ and a corresponding inequality for the norms. The proof of (ii) is complete.

Step 3. We prove (i). Let $-\infty<s<\infty$, $1\leqq p<\infty$, $1\leqq q<\infty$, and $g\in(B^s_{p,q}(R_n))'$. Then we can argue in the same way as in Step 2. The result is (8), (9) with $l_{q'}(L_{p'}(R_n))$ instead of $L_{p'}(R_n,l_{q'})$. But it is not hard to see that there exists a constant c such that

$$\|F^{-1}\varphi_kFg\mid L_{p'}(R_n)\|\leqq c\sum_{r=-1}^{1}\|g_{k+r}\mid L_{p'}(R_n)\|\,, \tag{10}$$

where c is independent of k. Thus it follows from (8) and (10) that g belongs to $B^{-s}_{p',q'}(R_n)$. This proves (1) provided $1 \leqq q < \infty$. Now let $0 < q < 1$ (and $-\infty < s < \infty$, $1 \leqq p < \infty$). Then $B^s_{p,q}(R_n) \subset B^s_{p,1}(R_n)$ gives

$$B^{-s}_{p',\infty}(R_n) = (B^s_{p,1}(R_n))' \subset (B^s_{p,q}(R_n))' . \tag{11}$$

On the other hand, if $g \in (B^s_{p,q}(R_n))'$ with $-\infty < s < \infty$, $1 \leqq p < \infty$ and $0 < q < 1$ and if $\{\varphi_k(x)\}_{k=0}^\infty \in \Phi(R_n)$, then we have

$$|(F^{-1}\varphi_k Fg)(\varphi)| = |g(F\varphi_k F^{-1}\varphi)|$$
$$\leqq c\|g\| 2^{sk} \|F\varphi_k F^{-1}\varphi \mid L_p(R_n)\| \leqq c'\|g\|\, 2^{sk} \|\varphi \mid L_p(R_n)\| , \quad \varphi \in S(R_n) .$$

Of course, c and c' are independent of k. Consequently,

$$\sup_k 2^{-sk} \|F^{-1}\varphi_k Fg \mid L_{p'}(R_n)\| \leqq c'\|g\| .$$

It follows that $g \in B^{-s}_{p',\infty}(R_n)$. This proves (i) if $0 < q < 1$.

Remark 1. One can extend part (ii) of the theorem and its proof to $q = 1$ (if $-\infty < s < \infty$ and $1 < p < \infty$). Then $(F^s_{p,1}(R_n))'$ equals the space on the right-hand side of (2.3.4/4) if one replaces there s, p, q by $-s, p', \infty$, respectively. But this space (probably) does not coincide with $F^{-s}_{p',\infty}(R_n)$. Furthermore, the determination of $(F^s_{p,q}(R_n))'$ with $-\infty < s < \infty$, $1 < p < \infty$, $0 < q < 1$ is an open problem. One can prove that

$$(F^s_{1,q}(R_n))' = B^{-s}_{\infty,\infty}(R_n) \quad \text{if} \quad -\infty < s < \infty,\ 0 < q \leqq 1 ,$$

cf. [S, p. 121].

Remark 2. The restrictions $p < \infty$ and $q < \infty$ in the theorem are natural, because, if either $p = \infty$ or $q = \infty$, then $S(R_n)$ is not dense in $B^s_{p,q}(R_n)$ and in $F^s_{p,q}(R_n)$, respectively, and the density of $S(R_n)$ in the considered spaces was the basis of our interpretation of the dual spaces in 2.11.1. Let us define $\mathring{B}^s_{p,q}(R_n)$ with $-\infty < s < \infty$, $0 < p \leqq \infty$, $0 < q \leqq \infty$, as the completion of $S(R_n)$ in $B^s_{p,q}(R_n)$ (of course, only the limiting cases $\max(p, q) = \infty$ are of interest). Then we can ask for $(\mathring{B}^s_{p,q}(R_n))'$ in the sense of the interpretation in 2.11.1. In [S, pp. 121/122] we proved that

$$(\mathring{B}^s_{p,q}(R_n))' = B^{-s}_{p',q'}(R_n) \quad \text{if} \quad -\infty < s < \infty, \quad 1 \leqq p \leqq \infty, \quad \text{and} \quad 0 < q \leqq \infty , \tag{12}$$

where again p' and q' have the above meaning (of course, only the cases with $\max(p, q) = \infty$ are new in comparison with (1)). Recall that $\mathcal{C}^s(R_n) = B^s_{\infty,\infty}(R_n)$ with $s > 0$ are the Zygmund spaces. Then we have

$$(\mathring{\mathcal{C}}^s(R_n))' = B^{-s}_{1,1}(R_n) \quad \text{with} \quad s > 0 .$$

Remark 3. If $1 < p < \infty$ and $1 \leqq q < \infty$, then (1) is well-known, cf. [I, 2.6.1.] and the references given there. The extension to $0 < q < 1$ is also known, cf. T. M. Flett [1] and J. Peetre [11]. (2) with $1 < p < \infty$ and $1 < q < \infty$ is due to H. Triebel [2]. Some special cases of (12) may be found in [I, 2.6.1.].

2.11.3. The Case $0 < p < 1$

We recall that $\frac{1}{q} + \frac{1}{q'} = 1$ if $1 < q < \infty$ and $q' = \infty$ if $0 < q \leqq 1$.

Theorem. (i) Let $-\infty < s < \infty$, $0 < p < 1$ and $0 < q < \infty$. Then

$$(B^s_{p,q}(R_n))' = B^{-s+n\left(\frac{1}{p}-1\right)}_{\infty,q'}(R_n) . \tag{1}$$

(ii) Let $-\infty < s < \infty$, $0 < p < 1$ and $0 < q < \infty$. Then

$$(F^s_{p,q}(R_n))' = B^{-s+n\left(\frac{1}{p}-1\right)}_{\infty,\infty}(R_n) . \tag{2}$$

Proof. Step 1. We have

$$B^s_{p,q}(R_n) \subset B^{s-n\left(\frac{1}{p}-1\right)}_{1,q}(R_n) \quad \text{if} \quad 0<p<1, \quad 0<q<\infty,$$

cf. (2.7.1/1). Then it follows from (2.11.2/1) that

$$B^{-s+n\left(\frac{1}{p}-1\right)}_{\infty,q'}(R_n) \subset (B^s_{p,q}(R_n))' \text{ with } -\infty<s<\infty,\ 0<p<1,\ 0<q<\infty. \quad (3)$$

Step 2. Let $-\infty<s<\infty$, $0<p<1$ and $0<q<\infty$. Let $g\in(B^s_{p,q}(R_n))'$, and $\{\varphi_k(x)\}_{k=0}^\infty\in\Phi(R_n)$. Then

$$|(F^{-1}\varphi_k Fg)(x)| = |g\left((F^{-1}\varphi_k)(x-\cdot)\right)| \leq \|g\|\ \|(F^{-1}\varphi_k)(x-\cdot) \mid B^s_{p,q}(R_n)\|. \quad (4)$$

We may assume that $\varphi_k(x)=\varphi_1(2^{-k+1}x)$ if $k=1,2,\ldots$ Then

$$\|(F^{-1}\varphi_k)(x-\cdot) \mid B^s_{p,q}(R_n)\| \leq c2^{sk+kn\left(1-\frac{1}{p}\right)}. \quad (5)$$

We put (5) in (4) and have

$$2^{-sk+kn\left(\frac{1}{p}-1\right)}\|F^{-1}\varphi_k Fg \mid L_\infty(R_n)\| \leq c\|g\|. \quad (6)$$

If $0<q\leq 1$, then (3) and (6) prove (1).

Step 3. Let $-\infty<s<\infty$, $0<p<1$ and $1<q<\infty$. Let $g\in(B^s_{p,q}(R_n))'$. We choose $x_k\in R_n$ such that

$$|(F^{-1}\varphi_k Fg)(x_k)| \sim \|F^{-1}\varphi_k Fg \mid L_\infty(R_n)\|, \quad (7)$$

cf. (6). Let a_k be arbitrary complex numbers with $k=0,\ldots,N$ and let

$$\varphi(x) = \sum_{k=0}^N a_k (F^{-1}\varphi_k)(x_k-x)\, 2^{-sk+nk\left(\frac{1}{p}-1\right)}.$$

Of course, $\varphi\in S(R_n)$. We use (4) and (5) to obtain

$$\begin{aligned} &\left|\sum_{k=0}^N a_k 2^{-sk+nk\left(\frac{1}{p}-1\right)}(F^{-1}\varphi_k Fg)(x_k)\right| = |g(\varphi)| \\ &\leq \|g\|\ \|\varphi \mid B^s_{p,q}(R_n)\| \leq c\|g\| \left(\sum_{k=0}^N |a_k|^q\right)^{\frac{1}{q}}, \end{aligned} \quad (8)$$

where c is independent of g, N and a_k. But (7) and (8) prove that

$$\left[\sum_{k=0}^N \left(2^{-sk+nk\left(\frac{1}{p}-1\right)}\|F^{-1}\varphi_k Fg \mid L_\infty(R_n)\|\right)^{q'}\right]^{\frac{1}{q'}} \leq c\|g\|, \quad (9)$$

where c is independent of g and N. If N tends to infinity, then (9) and (3) prove (1).

Step 4. We prove (ii). Because $0<p<1$ it follows from (2.3.2/9) and (2.7.1/2) that

$$B^s_{p,\min(p,q)}(R_n) \subset F^s_{p,q}(R_n) \subset F^{s-n\left(\frac{1}{p}-1\right)}_{1,1}(R_n) = B^{s-n\left(\frac{1}{p}-1\right)}_{1,1}(R_n).$$

Then (2.11.2/1) and (1) show that

$$B^{-s+n\left(\frac{1}{p}-1\right)}_{\infty,\infty}(R_n) \subset (F^s_{p,q}(R_n))' \subset B^{-s+n\left(\frac{1}{p}-1\right)}_{\infty,\infty}(R_n).$$

This is the desired assertion.

Remark 1. We recall that $F^0_{p,2}(R_n)=h_p(R_n)$ with $0<p<1$ are the local Hardy spaces, cf. (2.2.2/12) and Theorem 2.5.8. On the other hand, $\mathcal{C}^s(R_n)=B^s_{\infty,\infty}(R_n)$ with $s>0$ are the Zygmund spaces. Hence we have as a special case of (2) that

$$(h_p(R_n))'=\mathcal{C}^{n\left(\frac{1}{p}-1\right)}(R_n)\,,\quad 0<p<1\,. \tag{10}$$

Remark 2. Formula (1) for the 1-torus is due to T. M. Flett [1], at least if $0<q\leq 1$. J. Peetre [4 and 8, p. 249] extended this result to R_n. The case $1<q<\infty$ may be found in B. Jawerth [1]. Formula (10) for the 1-torus (classical Hardy spaces) is due to P. L. Duren, B. W. Romberg, A. L. Shields [1]. The extension to R_n for the Hardy spaces $H_p(R_n)$ from Remark 2.2.2/2 has been given by T. Walsh [1]. Concerning (10) we refer to D. Goldberg [1]. (2) is due to B. Jawerth [1].

Remark 3. Let $\mathring{B}^s_{p,q}(R_n)$ be the spaces from Remark 2.11.2/2. Then

$$(\mathring{B}^s_{p,\infty}(R_n))'=B_{\infty,1}^{-s+n\left(\frac{1}{p}-1\right)}(R_n)\quad\text{if}\quad -\infty<s<\infty\quad\text{and}\quad 0<p<1\,,$$

cf. [S, p. 124].

2.12. Further Properties

2.12.1. Characterizations of $F^s_{p,q}(R_n)$ via Lusin and Littlewood-Paley Functions

Lusin and Littlewood-Paley functions play an important role in the theory of the spaces $L_p(R_n)$ with $1<p<\infty$, cf. E. M. Stein [1, Chapter IV]. They are of particular interest in connection with Fourier multiplier theorems. Furthermore, they are also one of the main tools for the study of Hardy spaces, cf. in particular A. P. Calderón, A. Torchinsky [1]. On the basis of that paper, L. Päivärinta [1–3] extended this approach to the spaces $F^s_{p,q}(R_n)$. In this subsection we give a brief description of his results. Let $B(r,x)=\{y\mid y\in R_n,\ |x-y|\leq r\}$, where $r>0$ and $x\in R_n$.

Definition. Let $0<q<\infty$, $-\infty<s<\infty$, $\{\varphi_k(x)\}_{k=0}^\infty\in\Phi(R_n)$, $f\in S'(R_n)$, and $f_k(x)=2^{sk}F^{-1}\varphi_kFf$ if $k=0,1,2,\ldots$

(i) (*Lusin function*). Let $-\infty<a<\infty$ and $x\in R_n$. Then

$$S^s_a(f,q,x)=\left(\sum_{k=0}^\infty |B(2^{a-k},x|^{-1}\int_{B(2^{a-k},x)}|f_k(y)|^q dy\right)^{\frac{1}{q}}. \tag{1}$$

(ii) (*Littlewood-Paley function*). Let $0<\lambda<\infty$ and $x\in R_n$. Then

$$G^s_\lambda(f,q,x)=\left(\sum_{k=0}^\infty 2^{kn}\int_{R_n}\frac{|f_k(y)|^q}{(1+2^k\,|x-y|)^{q\lambda}}\,\mathrm{d}y\right)^{\frac{1}{q}}. \tag{2}$$

Remark 1. These are essentially L. Päivärinta's [1, p.123] definitions. If $q=2$, they coincide essentially with corresponding functions in A. P. Calderón, A. Torchinsky [1, I, p. 17]. These are new types of maximal functions reminiscent of the classical Hardy-Littlewood maximal function from (1.2.3/1) and the maximal function from Definition 2.3.6/2, which were of great service to us.

Theorem. Let $-\infty<s<\infty$, $0<p<\infty$ and $0<q<\infty$.

(i) If a is small, then

$$F^s_{p,q}(R_n)=\{f\mid f\in S'(R_n),\ \|S^s_a(f,q,\cdot)\mid L_p(R_n)\|<\infty\}\,, \tag{3}$$

where $\|S^s_a(f,q,\cdot)\mid L_p(R_n)\|$ is an equivalent quasi-norm in $F^s_{p,q}(R_n)$.

(ii) If $\lambda>0$ is sufficiently large, then

$$F^s_{p,q}(R_n)=\{f \mid f\in S'(R_n),\ \|G^s_\lambda(f,q,\cdot) \mid L_p(R_n)\|<\infty\}, \tag{4}$$

where $\|G^s_\lambda(f,q,\cdot) \mid L_p(R_n)\|$ is an equivalent quasi-norm in $F^s_{p,q}(R_n)$.

Remark 2. A proof has been given by L. Päivärinta [1–3]. One can use this theorem in order to prove new Fourier multiplier theorems. Corresponding ideas for the spaces $L_p(R_n)$ with $1<p<\infty$ and the spaces $H_p(R_n)$ with $0<p\leq 1$ may be found in E. M. Stein [1, Chapter IV] and in A. P. Calderón, A. Torchinsky [1].

2.12.2. Characterizations via Gauss-Weierstrass and Cauchy-Poisson Semi-Groups

It is well-known that the classical Besov spaces $B^s_{p,q}(R_n)$ with $1<p<\infty$ and $1\leq q\leq\infty$ can be characterized as traces of harmonic functions in the half-space $R^+_{n+1}=\{(x,t) \mid x\in R_n, t\geq 0\}$ on the hyperplane $\{t=0\}$ (which is identified with R_n). Furthermore, we recall that the Hardy spaces $H_p(R_n)$ were considered originally in the same way as quasi-normed linear spaces of harmonic functions in R^+_{n+1}. As far as the classical Besov spaces are concerned we refer e.g. to [I, pp. 192–196]. For the Hardy spaces cf. E. M. Stein, G. Weiss [1, 2], E. M. Stein [1], and in particular C. Fefferman, E. M. Stein [2]. One of the main aims of the latter paper is to show that there is a wide variety of other possible characterizations of $H_p(R_n)$, cf. Remark 2.2.2/2. Let $\{F^{-1}\varphi_t Ff\}_{t>0}$ be the family of functions from Remark 2.2.2/2. Of course,

$$(F^{-1}\varphi_t Ff)(x)=c\int_{R_n}(F^{-1}\varphi_t)(x-y)\,f(y)\,dy, \tag{1}$$

if the integral makes sense. Recall that $\varphi_t(x)=\varphi(tx)$. In other words, we have characterizations via

$$(G_t f)(x)=(F^{-1}\varphi_t Ff)(x), \tag{2}$$

which can be represented in the form of the convolution integral (1) with the kernel

$$(F^{-1}\varphi_t)(x)=[F^{-1}\varphi(t\,\cdot)](x)=t^{-n}(F^{-1}\varphi)\left(\frac{x}{t}\right). \tag{3}$$

If we choose $\varphi(x)=e^{-|x|}$, then G_t from (2) is the well-known Cauchy-Poisson semi-group and $(F^{-1}\varphi_t Ff)(x)$ is a harmonic function in R^+_{n+1} with the Poisson kernel

$$(F^{-1}e^{-|\xi|})(x)=\frac{c}{(1+|x|^2)^{\frac{n+1}{2}}}.$$

In that case, (1) reads as follows,

$$[P(t)\,f](x)=c\int_{R_n}\frac{t}{(|x-y|^2+t^2)^{\frac{n+1}{2}}}\,f(y)\,dy, \tag{4}$$

where $x\in R_n$ and $t>0$, i.e. $(x,t)\in R^+_{n+1}$. Cf. [I, pp. 192–196]. The above remarks about harmonic functions are based on these formulas. Another distinguished choice is $\varphi(x)=e^{-|x|^2}$. In that case, G_t from (2) is the well-known Gauss-Weierstrass semi-group and $(F^{-1}\varphi_t Ff)(x)$ is the (uniquely determined) solution of the Cauchy problem of the heat equation,

$$\frac{\partial u(x,t)}{\partial t}=\Delta u(x,t)\quad\text{in}\quad (x,t)\in R^+_{n+1},\quad u(x,0)=f(x)\quad\text{in } R_n. \tag{5}$$

Recall that $F(e^{-\frac{|\xi|^2}{2}})(x)=e^{-\frac{|x|^2}{2}}$. In that case, (1) reads as follows,

$$[W(t)\,f](x)=u(x,t)=(4\pi t)^{-\frac{n}{2}}\int_{R_n}e^{-\frac{|x-y|^2}{4t}}\,f(y)\,dy. \tag{6}$$

where $x \in R_n$ and $t>0$, i.e. $(x, t) \in R^+_{n+1}$. The formulas (4) and (6) are the basis for further considerations. Since $e^{-|x|^2} \in S(R_n)$, formula (6) makes sense for every $f \in S'(R_n)$ (of course, in that general case, (6) must be interpreted as $f(\psi)$ with the test function $\psi(y)=(4\pi t)^{-\frac{n}{2}} e^{-\frac{|x-y|^2}{4t}}$, $x \in R_n$). Similarly for (4) with a few modifications. A characterization of $B^s_{p,q}(R_n)$, $1<p<\infty$, with the help of (4) and (6) has been given in [I, pp. 190–196]. This approach is based on the theory of semi-groups. If $p<1$, this method breaks down, and the question arises whether the spaces $B^s_{p,q}(R_n)$ with $p<1$ can also be characterized by means of (4) and (6). A first result for the homogeneous counterpart of $B^s_{p,q}(R_n)$ and with some restrictions on the parameters involved is due to J. Peetre [8, p. 256]. Recently, a satisfactory answer for the above spaces $B^s_{p,q}(R_n)$ has been given by Bui Huy Qui [2] (for the Gauss-Weierstrass semi-group). We formulate here a result which has been obtained recently by H. Triebel [24] and which includes also the spaces $F^s_{p,q}(R_n)$.

Theorem. (i) Let $0<p\leqq\infty$, $0<q\leqq\infty$ and $s>\max\left(0, n\left(\frac{1}{p}-1\right)\right)$. If k and m are natural numbers with $2m>s$ and

$$k > \begin{cases} s & \text{if } 1\leqq p\leqq\infty\,, \\ s+n\left(\frac{1}{p}-1\right) & \text{if } 0<p<1\,, \end{cases} \tag{7}$$

then

$$\|f \mid L_p(R_n)\| + \left(\int_0^\infty t^{\left(m-\frac{s}{2}\right)q} \left\| \frac{\partial^m W(t) f}{\partial t^m}(\cdot) \mid L_p(R_n)\right\|^q \frac{dt}{t}\right)^{\frac{1}{q}} \tag{8}$$

and

$$\|f \mid L_p(R_n)\| + \left(\int_0^\infty t^{(k-s)q} \left\| \frac{\partial^k P(t) f}{\partial t^k}(\cdot) \mid L_p(R_n)\right\|^q \frac{dt}{t}\right)^{\frac{1}{q}} \tag{9}$$

are equivalent quasi-norms in $B^s_{p,q}(R_n)$ (modification if $q=\infty$).

(ii) Let $0<p<\infty$, $0<q\leqq\infty$ an $s>\frac{n}{\min(p, q)}$. If m is a sufficiently large natural number, then

$$\|f \mid L_p(R_n)\| + \left\| \left(\int_0^\infty t^{\left(m-\frac{s}{2}\right)q} \left| \frac{\partial^m W(t) f}{\partial t^m}(\cdot)\right|^q \frac{dt}{t}\right)^{\frac{1}{q}} \middle| L_p(R_n)\right\| \tag{10}$$

and

$$\|f \mid L_p(R_n)\| + \left\| \left(\int_0^\infty t^{(m-s)q} \left| \frac{\partial^m P(t) f}{\partial t^m}(\cdot)\right|^q \frac{dt}{t}\right)^{\frac{1}{q}} \middle| L_p(R_n)\right\| \tag{11}$$

are equivalent quasi-norms in $F^s_{p,q}(R_n)$ (modifications if $q=\infty$).

Remark. In H. Triebel [24] we proved a corresponding theorem for more general functions $\varphi(x)$ in the sense of (1)–(3). As we mentioned, some results of this type for the spaces $B^s_{p,q}(R_n)$ are known, cf. [I, 2.5.2] and the references given there, and Bui Huy Qui [2]. As far as the spaces $F^s_{p,q}(R_n)$ with $1<p<\infty$ and $1<q<\infty$ are concerned cf. also G.A. Kaljabin [2,5].

2.12.3. Characterizations via Spline Functions

In Remark 2.5.5/2 we defined (unconditional) Schauder bases in quasi-Banach spaces. We constructed in 2.5.5. Schauder bases of analytic functions for $B^s_{p,q}(R_n)$ if $1<p<\infty$ and $0<q<\infty$. What about other distinguished systems of functions as Schauder bases in $B^s_{p,q}(R_n)$ and

$F^s_{p,q}(R_n)$? In this connection, much attention has been paid to spline functions. There are some results for the one-dimensional case, i.e. $n=1$, and in particular for the spaces $B^s_{p,q}(I)$ and $F^s_{p,q}(I)$ on the interval $I=(0,1)$. We refer to Z. Ciesielski [1–3], P. Oswald [2, 3, 5, 6], R. Ropela [1, 2], P. Sjölin, J.- O. Strömberg [1], and H. Triebel [18]. As for the general n-dimensional case, there are many open problems. The most remarkable progress is probably the announcement by Z. Ciesielski, T. Figiel [1], (cf. also Z. Ciesielski, J. Domsta [1] as far as the Sobolev spaces are concerned). In this subsection we give a brief description of some statements which have been obtained in H. Triebel [18] via real and complex interpolation of the results of Z. Ciesielski, P. Oswald, and R. Ropela.

Spaces on I. Let $I=(0,1)$ be the unit interval. Let $D'(I)$ be the collection of all distributions on I. Then

$$B^s_{p,q}(I)=\{f \mid f\in D'(I),\quad \exists g\in B^s_{p,q}(R_1)\quad \text{with}\quad g/I=f\}\,, \tag{1}$$

where g/I means the restriction of g to I. Let

$$\|f \mid B^s_{p,q}(I)\|=\inf \|g \mid B^s_{p,q}(R_1)\|\,, \tag{2}$$

where the infimum is taken over all $g\in B^s_{p,q}(R_1)$ with $g/I=f$. Similarly for $F^s_{p,q}(I)$. We assume that $-\infty<s<\infty$, $0<p<\infty$ and $0<q<\infty$.

Splines. We follow Z. Ciesielski [2]. The system of Haar functions $\{\chi_j(x)\}_{j=1}^{\infty}$ on I is defined as follows. $\chi_1(x)=1$ in I,

$$\chi_{2^\mu+\nu}(x)=\begin{cases} 2^{\frac{\mu}{2}} & \text{if}\quad x\in\left[\dfrac{\nu-1}{2^\mu},\dfrac{2\nu-1}{2^{\mu+1}}\right) \\ -2^{\frac{\mu}{2}} & \text{if}\quad x\in\left[\dfrac{2\nu-1}{2^{\mu+1}},\dfrac{\nu}{2^\mu}\right) \\ 0 & \text{elsewhere in I,} \end{cases} \tag{3}$$

where $\mu\geqq 0$ and $1\leqq\nu\leqq 2^\mu$. Let

$$(Gf)(x)=\int_0^x f(y)\,\mathrm{d}y\,,\quad (Hf)(x)=\int_x^1 f(y)\,\mathrm{d}y\,,\qquad x\in I\,.$$

Let $m\geqq -1$. We apply Schmidt's orthonormalization method (with respect to $L_2(I)$) to the system $\{1, x, \ldots, x^{m+1}, G^{m+1}\chi_j\}_{j=2}^{\infty}$ to obtain the orthonormal system $\{f_j^{(m)}\}_{j=-m}^{\infty}$ of splines of order m. For $0\leqq k\leqq m+1$ and $j\geqq k-m$ we put

$$f_j^{(m,k)}=\frac{\mathrm{d}^k}{\mathrm{d}x^k}f_j^{(m)}\,,\quad g_j^{(m,k)}=H^k f_j^{(m)}\,.$$

Furthermore, let

$$h_j^{(m,k)}(x)=f_j^{(m,k)}(x)\,\|f_j^{(m,k)} \mid L_2(I)\|^{-1}\quad\text{if}\quad 0\leqq k\leqq m+1\quad\text{and}\quad j\geqq k-m\,,$$

$$h_j^{(m,k)}(x)=g_j^{(m,-k)}(x)\,\|f_j^{(m,-k)} \mid L_2(I)\|\quad\text{if}\quad 0\leqq -k\leqq m+1\quad\text{and}\quad j\geqq |k|-m.$$

The two systems $\{h_j^{(m,k)}(x)\}_{j=|k|-m}^{\infty}$ and $\{h_j^{(m,-k)}(x)\}_{j=|k|-m}^{\infty}$ are biorthogonal, i.e.

$$\int_0^1 h_j^{(m,k)}(x)\,h_i^{(m,-k)}(x)\,\mathrm{d}x=\delta_{j,i}\,,\quad j,i\geqq |k|-m$$

and $|k|\leqq m+1$.

The problem is whether a distribution $f\in D'(I)$ which belongs to $B^s_{p,q}(I)$ or $F^s_{p,q}(I)$ can be represented as

$$f=\sum_{j=|k|-m}^{\infty} a_j h_j^{(m,k)}\quad\text{with}\quad a_j=(f,h_j^{(m,-k)})\,, \tag{4}$$

where $(\cdot\,,\cdot)$ is the scalar product in $L_2(I)$. We formulate two results which are not in a final form, but which shed some light upon the question raised.

Theorem 1. Let m and k be two integers with $m \geqq -1$ and $|k| \leqq m+1$. Let $l=m-k+1$.
(i) Let $1<p<\infty$, $1 \leqq q<\infty$ and $0<s<2l \min\left(\frac{1}{p}, \frac{1}{p'}\right)$. Then $\{h_j^{(m,k)}\}_{j=|k|-m}^{\infty}$ is an unconditional Schauder basis in $B_{p,q}^s(I)$. If $f \in B_{p,q}^s(I)$ is given by (4), then

$$\sum_{j=|k|-m}^{1} |a_j| + \left[\sum_{r=0}^{\infty} 2^{r\left(s+\frac{1}{2}-\frac{1}{p}\right)q}\left(\sum_{j=2^r+1}^{2^{r+1}} |a_j|^p\right)^{\frac{q}{p}}\right]^{\frac{1}{q}} \tag{5}$$

and

$$\sum_{j=|k|-m}^{1} |a_j| + \left[\sum_{r=0}^{\infty} 2^{rsq}\left\|\sum_{j=2^r+1}^{2^{r+1}} a_j h_j^{(m,k)} \mid L_p(I)\right\|^q\right]^{\frac{1}{q}} \tag{6}$$

are equivalent norms in $B_{p,q}^s(I)$.
(ii) Let $1<p<\infty$ and $1<q<\infty$. Let $\left(\frac{1}{p}, \frac{1}{q}\right)$ be a point of the shaded area in Fig. 2.12.3. If l is sufficiently large, then $\{h_j^{(m,k)}\}_{j=|k|-m}^{\infty}$ is an unconditional Schauder basis in $F_{p,q}^s(I)$. If $f \in F_{p,q}^s(I)$ is given by (4), then

$$\left\|\left(\sum_{j=|k|-m}^{\infty} (1+|j|^s)^q \, |a_j h_j^{(m,k)}|^q\right)^{\frac{1}{q}} \mid L_p(I)\right\| \tag{7}$$

is an equivalent norm in $F_{p,q}^s(I)$.

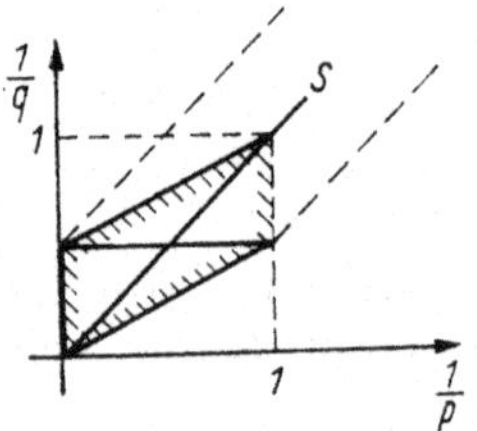

Fig. 2.12.3

Remark 1. A proof has been given in H. Triebel [18]. As far as the Sobolev spaces $W_p^N(I) = F_{p,2}^N(I)$ with $N=0, 1, 2, \ldots$ and the Besov spaces $B_{p,q}^s(I)$ are concerned we refer again to the cited papers by Z. Ciesielski and S. Ropela. Probably the above theorem can be strengthened. We conjecture that $\{h_j^{(m,k)}\}_{j=|k|-m}^{\infty}$ is an unconditional Schauder basis at least in $B_{p,q}^s(I)$ if $1<p<\infty$, $1 \leqq q<\infty$, and $\frac{1}{p}-1<s<l+\frac{1}{p}$ and in $F_{p,q}^s(I)$ if $1<p<\infty$, $1<q<\infty$ and $\frac{1}{p}-1<s<l+\frac{1}{p}$. The cases with $\frac{1}{p}-1<s \leqq 0$ can be obtained by duality and interpolation, cf. [I, p. 332].

Theorem 2. Let m be an integer with $m \geqq -1$.
(i) Let $0<p<\infty$, $0<q<\infty$ and $\frac{1}{p}-1<s<m+1+\frac{1}{p}$. Then $\{h_j^{(m,0)}\}_{j=-m}^{\infty}$ is a Schauder basis in $B_{p,q}^s(I)$.
(ii) Let $0<p<\infty$ and $0<q<\infty$. Let $\left(\frac{1}{p}, \frac{1}{q}\right)$ be a point of the strip S in Fig. 2.12.3 (including the shaded area). If m is sufficiently large, then $\{h_j^{(m,0)}\}_{j=-m}^{\infty}$ is a Schauder basis in $F_{p,q}^s(I)$.

Remark 2. Part (i) of the theorem with $1 \leqq p<\infty$ and $1 \leqq q<\infty$, including (5), has been proved by S. Ropela [1]. The extension to $p<1$ and $q<1$ is due to P. Oswald [2, 3,7]. Part (ii) has been proved in H. Triebel [18] by complex interpolation. Probably this theorem can also be strengthened. We conjecture that $\{h_j^{(m,0)}\}_{j=-m}^{\infty}$ is a Schauder basis in $B^s_{p,q}(I)$ and in $F^s_{p,q}(I)$ if $0<p<<\infty$, $0<q<\infty$ and $\frac{1}{p}-1<s<m+1+\frac{1}{p}$.

Remark 3. In the case of the Haar system $\{h_j^{(-1,0)}\}_{j=1}^{\infty}$ one can extend the results for $B^s_{p,q}(I)$ to $B^s_{p,q}(R_1)$. There are also some results for n-dimensional Haar systems as Schauder bases in $B^s_{p,q}(R_n)$ and in $B^s_{p,q}(Q)$, where Q is a cube in R_n. We do not go into detail and refer to [I, *pp.* 338–342], H. Triebel [8], and Tran Duc Long, H. Triebel [1]. Concerning higher spline functions as Schauder bases in $B^s_{p,q}(R_n)$ (and corresponding spaces on manifolds) we refer to Z. Ciesielski, T. Figiel [1, 2], cf. also I. P. Irodova [1]. As far as unconditional bases in the classical Hardy spaces H_1 (boundary values of analytic functions in the unit disc in the complex plane) are concerned, we refer also to B. Maurey [1], L. Carleson [1] and P. Wojtaszczyk [1]. Unconditional spline bases in H_p with $p<1$ have been constructed by P. Wojtaszczyk [2] and J.-O. Strömberg [1], cf. also P. Sjölin, J.-O. Strömberg [2].

3. Function Spaces on Domains

3.1. Preliminaries, Motivations and Methods

3.1.1. Motivations

In Subsection 2.2.1. we spoke about the history of function spaces from our point of view, cf. in particular Fig. 2.2.1. The theory of function spaces was closely connected from the very beginning with applications, in particular to partial differential equations, cf. S. L. Sobolev [4]. In this context, spaces on bounded domains or on the half-space R_n^+ are more interesting than corresponding spaces on the Euclidean n-space R_n. We have in mind boundary value problems for elliptic differential equations and initial value problems for parabolic and hyperbolic differential equations. For some basic spaces, such as spaces of differentiable functions, Hölder spaces, and Sobolev spaces no difficulties arise in defining these spaces on domains directly. For more complicated spaces, e.g. Bessel-potential spaces or local Hardy spaces, the situation is quite different. At least for the latter spaces it is not only convenient, but it seems to be necessary, to begin with corresponding spaces on R_n and to define spaces on domains and on half-spaces by restrictions. Even a detailed study of some aspects of Sobolev spaces on domains, e.g. interpolation properties, leads almost automatically to corresponding spaces on R_n. Consequently, there are two possibilities in dealing with function spaces on domains Ω in R_n:
(i) The direct way, where function spaces on Ω are defined as subspaces of $D'(\Omega)$ (the collection of all complex-valued distributions on Ω) via explicit norms or quasi-norms.
(ii) Definition of spaces on Ω as restrictions of corresponding spaces on R_n.
Our way is the second one. It has many advantages, but also few disadvantages. The advantages will be clear if one looks at Chapter 2. Our approach to spaces on R_n was based on Fourier transforms, maximal inequalities etc. in R_n, and there is no counterpart of that method for arbitrary domains in R_n. The definition of some of the spaces we listed in 2.2.2. depends heavily on such an approach, cf. (2.2.2/11) and (2.2.2/12). The treatment of the two scales $B^s_{p,q}(R_n)$ and $F^s_{p,q}(R_n)$ from 2.3.1. is completely based on Fourier analysis. We shall define the spaces $B^s_{p,q}(\Omega)$ and $F^s_{p,q}(\Omega)$ on the domain Ω in R_n as restrictions of $B^s_{p,q}(R_n)$ and $F^s_{p,q}(R_n)$ to Ω, respectively. It is our aim to carry over many of the properties from the spaces on R_n to the corresponding spaces on Ω. In order to do so one is faced with a few key problems, which will be described in more detail in 3.1.3. In any case this approach is successful if Ω is a bounded smooth domain. Two questions arise:
(i) What about domains with non-smooth boundaries?
(ii) What about inner descriptions of the spaces on domains which have nothing to do with the restriction procedure?
As far as the first question is concerned we recall that even for the Sobolev spaces $W^m_p(\Omega)$ there is an essential difference according to whether the underlying domain Ω has a smooth boundary or not. For non-smooth domains the familiar embedding theorems for the Sobolev spaces look qui tedifferent. Much attention has been given to the study of Sobolev spaces in non-smooth domains and their non-standard behaviour, cf. R. A. Adams [1], W. Mazja [1], and D. E. Edmunds

H. Triebel, *Theory of Function Spaces*, Modern Birkhäuser Classics,
DOI 10.1007/978-3-0346-0416-1_3, © Birkhäuser Verlag 1983

[1] (cf. also the remarks in [I, 4.6.2]). So it is quite natural to apply the restriction procedure sketched above only to smooth domains. In the next subsection we make a few comments about the second question.

3.1.2. The Problem of Inner Descriptions

Let Ω be a domain in the Euclidean n-space R_n and let $D'(\Omega)$ be the collection of all complex-valued distributions on Ω. We recall that "domain" always stands for "open set". Let $1<p<\infty$ and $m=0, 1, 2, \ldots$ Then the Sobolev space $W_p^m(\Omega)$ is defined as follows:

$$W_p^m(\Omega)=\{f \mid f\in L_p(\Omega), \|f \mid W_p^m(\Omega)\| = \sum_{|\alpha|\leq m} \|D^\alpha f \mid L_p(\Omega)\|<\infty\}, \tag{1}$$

cf. (2.2.2/7). Of course, $L_p(\Omega)$ is the collection of all complex-valued Lebesgue-measurable functions $g(x)$ on Ω such that

$$\|g(x) \mid L_p(\Omega)\| = \left(\int_\Omega |g(x)|^p \, dx\right)^{\frac{1}{p}} < \infty . \tag{2}$$

We consider $W_p^m(\Omega)$ as a subspace of $D'(\Omega)$, in particular the derivatives $D^\alpha f$ must be understood in the sense of distributions. We call (1) an inner description because only points $x\in\Omega$ are involved in the definition. If Ω is a bounded domain in R_n with smooth boundary, then $W_p^m(\Omega)$ coincides with the restriction of $W_p^m(R_n)=$ $=F_{p,2}^m(R_n)$ to Ω, i.e.

$$W_p^m(\Omega)=\{f \mid f\in D'(\Omega), \exists g\in W_p^m(R_n) \quad \text{with} \quad g \mid \Omega = f\}, \tag{3}$$

where $g \mid \Omega$ means the restriction of g to Ω in the sense of $D'(\Omega)$. Furthermore,

$$\|f \mid W_p^m(\Omega)\|^* = \inf \|g \mid W_p^m(R_n)\|, \tag{4}$$

where the infimum is taken over all g in the sense of (3), is an equivalent norm in $W_p^m(\Omega)$. Later on we begin with definitions of type (3), (4) and then there arises the problem of finding an inner description of type (1). For the Sobolev spaces the above situation is quite satisfactory if Ω is a bounded domain with smooth boundary (it is well-known that rather weak assumptions on the boundary of the bounded domain Ω are sufficient in order to prove that (1) on the one hand, and (3)–(4) on the other are equivalent to each other, cf. [I, 4.2.3]). What about other spaces, if one replaces W_p^m in (3) and (4) by $B_{p,q}^s$ or $F_{p,q}^s$ with the full range of admissible parameters? In the general case one cannot expect to find a simple inner description. On the other hand, for the classical Besov spaces $\Lambda_{p,q}^s(\Omega)=B_{p,q}^s(\Omega)$ with $s>0$, $1\leq p<\infty$, $1\leq q\leq\infty$ and the (Hölder)-Zygmund spaces $\mathcal{C}^s(\Omega)=B_{\infty,\infty}^s(\Omega)$ with $s>0$ there are satisfactory inner descriptions (we always assume that Ω is a bounded domain with smooth boundary), cf. [I, 4.4.2, 4.5.2]. We give an example. Let $s>0$ with $s=[s]+\{s\}$, where $[s]$ is an integer and $0<\{s\}<1$. Let $1\leq p=q\leq\infty$. Then

$$\|f \mid B_{p,p}^s(\Omega)\|^* = \|f \mid L_p(\Omega)\| + \sum_{|\alpha|=[s]} \left(\int_{\Omega\times\Omega} \frac{|D^\alpha f(x)-D^\alpha f(y)|^p}{|x-y|^{n+\{s\}p}} \, dx dy \right)^{\frac{1}{p}} \tag{5}$$

is an equivalent norm in $B_{p,p}^s(\Omega)$ (with the usual modification if $p=q=\infty$). Of course, this is a nice inner description. Can it be extended to values $p<1$? In a few cases we shall describe affirmative answers later on, cf. 3.4.2., but in general there are no satisfactory inner descriptions of the spaces $B_{p,q}^s(\Omega)$ with $p<1$ at the moment. Similarly, there are no nice inner descriptions for the general spaces $F^s_{p,q}(\Omega)$ with $0<p<\infty$ and $0<q\leqq\infty$. Perhabs some progress can be made in the future on the basis of the subsections 2.5.9.–2.5.12., cf. in particular Remark 2.5.9/2. In any case, s in $B^s_{p,q}(\Omega)$ and in $F^s_{p,q}(\Omega)$ must be large (at least positive) if one wishes to find inner descriptions in that way. For $s<0$, there are no inner descriptions, even in the classical cases. This situation justifies our approach: we define the spaces $B^s_{p,q}(\Omega)$ and $F^s_{p,q}(\Omega)$ as in (3)–(4), and derive all properties from that definition.

3.1.3. The Localization Method

As we remarked at the end of the last subsection, we define the spaces $B^s_{p,q}(\Omega)$ and $F^s_{p,q}(\Omega)$ on bounded smooth domains Ω as in (3.1.2/3) and (3.1.2/4), and it is our aim to carry over properties of the spaces $B^s_{p,q}(R_n)$ and $F^s_{p,q}(R_n)$ to the corresponding spaces on Ω. In order to establish such a programme we use the method of local coordinates. In this subsection we give a rough description and list some key properties for the spaces $B^s_{p,q}(R_n)$ and $F^s_{p,q}(R_n)$ which we shall need for a successful approach. Precise formulations will be given later on. Let Ω be a bounded domain in R_n with smooth boundary $\partial\Omega$. Let $A(R_n)$ be either $B^s_{p,q}(R_n)$ or $F^s_{p,q}(R_n)$ and let $A(\Omega)$ be the restriction of $A(R_n)$ to Ω, i.e. $A(\Omega)$ is defined via (3.1.2/3) – (3.1.2/4), with A instead of W^m_p.
(i) *Localization.* Let $K_1, \ldots, K_N$ be a finite number of open balls in R_n which cover $\bar{\Omega}$. Let $\{\psi_j\}_{j=1}^N$ be a corresponding resolution of unity. The first goal is to decompose $f\in A(\Omega)$ in $f=\sum_{j=1}^N f\psi_j$ with $\operatorname{supp} f\psi_j\subset\bar{\Omega}\cap K_j$ and $f\psi_j\in A(\Omega)$. For that purpose we must be sure that $g\to\psi g$, with $g\in A(R_n)$ and $\psi\in\mathcal{C}^s(R_n)$, is a bounded linear operator in $A(R_n)$. But this is the multiplication property from 2.8.2.
(ii) *Rectification.* By (i) we may assume without restriction of generality that the support of $f\in A(\Omega)$ is a subset of $\bar{\Omega}\cap K_1$. If $\bar{K}_1\subset\Omega$, then we may consider $f\in A(\Omega)$ as an element of $A(R_n)$. If $\partial\Omega\cap K_1\neq\emptyset$, then we use a rectification procedure. We are looking for an infinitely differentiable diffeomorphic map $y=\psi(x)$ of R_n onto itself such that the image of $\partial\Omega\cap K_1$ becomes a part of the hyperplane $\{y \mid y_n=0\}$. The question is whether $f(x)\to f(\psi(x))$ is a one-to-one map from $A(R_n)$ onto $A(R_n)$. This is the diffeomorphism property from 2.10.2. The affirmative answer from Theorem 2.10.2 reduces the spaces $A(\Omega)$ to corresponding spaces $A(R_n^+)$, which are defined on $R_n^+=\{x \mid x=(x', x_n),\, x_n>0\}$, cf. also 2.9.1.
(iii) *Extension.* Finally, we wish to reduce $A(R_n^+)$ to $A(R_n)$. But this is the extension problem from 2.9.1.

This is a very rough description of our method: localization, rectification and extension reduces the spaces A(Ω) to the corresponding spaces $A(R_n)$.

Remark. Besides spaces on bounded smooth domains Ω we are also interested in spaces on $\partial\Omega$, which is a compact C^∞-manifold. It is almost clear that the above programme (i) and (ii) can also be applied to the spaces $A(\partial\Omega)$.

3.2. Definitions and Basic Properties

3.2.1. C^∞-Domains

We consider three domains in the sequel: the Euclidean n-space R_n, the half-space

$$R_n^+ = \{x \mid x=(x', x_n) \in R_n,\ x' \in R_{n-1},\ x_n > 0\} \tag{1}$$

and bounded C^∞-domains Ω in R_n. We recall that "domain" always stands for "open set". The boundary of Ω is denoted by $\partial\Omega$.

Definition. Let Ω be a bounded domain in R_n. Then Ω is said to be a C^∞-domain if there exist N open balls $K_1, \ldots, K_N$ such that

$$\bigcup_{j=1}^{N} K_j \supset \partial\Omega \quad \text{and} \quad K_j \cap \partial\Omega \neq \emptyset \quad \text{if} \quad j=1, \ldots, N \tag{2}$$

with the following property. For every ball K_j there exists an infinitely differentiable real vector-function

$$f^{(j)}(x) = (f_1^{(j)}(x), \ldots, f_n^{(j)}(x)),$$

which is defined on $\overline{K_j}$, such that $y = f^{(j)}(x)$ yields a one-to-one mapping from K_j onto a bounded domain in R_n, where $\partial\Omega \cap K_j$ is mapped onto an $(n-1)$-dimensional domain in the hyper-plane $\{y \mid y \in R_n,\ y_n = 0\}$ and $\Omega \cap K_j$ is mapped onto a simply connected domain in R_n^+. Furthermore,

$$\frac{\partial(f_1^{(j)}, \ldots, f_n^{(j)})}{\partial(x_1, \ldots, x_n)} \neq 0 \quad \text{if} \quad x \in \overline{K_j}. \tag{3}$$

Remark 1. Of course, (3) is the Jacobian. This is the somewhat complicated definition of a C^∞-domain. But the basic idea is clear: local rectification of the boundary via C^∞-vector-functions. If we replace C^∞ by C^k with $k=1, 2, 3, \ldots$, we have the definition of a bounded C^k-domain.

Diffeomorphic Maps. The following observation will be of some interest later on. First we describe how to find balls K_j and convenient mappings $y = f^{(j)}(x)$ with the desired properties. We may assume without restriction of generality, that $0 \in \partial\Omega$ and that the boundary $\partial\Omega$ in a neighbourhood of 0 is given by

$$x_n = g(x'), \quad g(0) = 0, \quad \frac{\partial g}{\partial x_k}(0) = 0 \quad \text{if} \quad k = 1, \ldots, n-1, \tag{4}$$

where x' is an element of an $(n-1)$-dimensional neighbourhood ω of 0 in the hyper-plane $\{y \mid y_n = 0\}$. Furthermore, $g(x')$ is a C^∞-function in ω. Let $y = h(x) = (h_1(x), \ldots, h_n(x))$ with

$$h_k(x) = x_k \quad \text{if} \quad k = 1, \ldots, n-1 \quad \text{and} \quad h_n(x) = x_n - g(x'). \tag{5}$$

If $K(=K_j)$ is a sufficiently small ball centered at the origin, then $f^{(j)}(x) = h(x)$ has the desired properties. The explicit construction above has the following advantage. It is not hard to see that $h(x)$ can be extended outside of K in such a way that the extended vector-function, which is also denoted by $h(x)$, yields a diffeomorphic mapping from R_n onto itself with $h(x) = x$ for large values of $|x|$. We need this property in order to apply Theorem 2.10.2. In the general case one must combine a mapping $h(x)$ of the above type with an affine mapping of R_n onto itself. However, it is easy to see that the spaces $B_{p,q}^s(R_n)$ and $F_{p,q}^s(R_n)$ are preserved under affine mappings of R_n onto itself. In this sense we may assume that the vector-functions $f^{(1)}, \ldots, f^{(N)}$ from the above definition can be extended to R_n, in such a way that the extended vector-functions yield diffeomorphic mappings from R_n onto itself for which Theorem 2.10.2 is applicable.

Resolution of Unity. The balls K_j are as in the above definition. We find a C^∞-domain K_0 such that

$$\overline{K_0} \subset \Omega \quad \text{and} \quad \overline{\Omega} \subset \bigcup_{j=0}^{N} K_j \,. \tag{6}$$

Then there exists a resolution of unity $\{\psi_j(x)\}_{j=0}^N$ with the following properties: $\psi_j(x)$ is an infinitely differentiable function in R_n with

$$\operatorname{supp} \psi_j \subset K_j \quad \text{if} \quad j=0, \ldots, N \,. \tag{7}$$

Furthermore,

$$\sum_{j=0}^{N} \psi_j(x) = 1 \tag{8}$$

in a neighbourhood of $\overline{\Omega}$. This is a well-known basic fact from the theory of distributions, cf. e.g. H. Triebel [1, p. 47]. Obviously, the restriction of $\psi_1(x), \ldots, \psi_N(x)$ to $\partial\Omega$ is a resolution of unity with respect to $\partial\Omega$.

Remark 2. $\partial\Omega$ is a compact $(n-1)$-dimensional C^∞-manifold. $(f^{(j)}, \partial\Omega \cap k_j)$ with $j=1, \ldots, N$ are local charts. For the definition of $B^s_{p,q}(\partial\Omega)$ and $F^s_{p,q}(\partial\Omega)$ we shall need only this fact (an atlas with local charts). In other words, the definition of spaces on $\partial\Omega$ can be extended immediately to arbitrary compact and para-compact C^∞-manifolds.

3.2.2. Definitions

We use the notations of the preceding subsection. In particular, Ω stands for a bounded C^∞-domain with the boundary $\partial\Omega$. Furthermore, the domain K_0, the balls $K_1, \ldots, K_N$ and the corresponding functions $f^{(j)}$ and ψ_j are as described above. If $y=f^{(j)}(x)$, then $x={f^{(j)}}^{-1}(y)$ is the inverse mapping. We recall that $D'(\Omega)$ is the collection of all complex-valued distributions on Ω. Similarly $D'(\partial\Omega)$ stands for the distributions on the compact C^∞-manifold $\partial\Omega$. If $g \in S'(R_n)$ then there striction of g to Ω is an element of $D'(\Omega)$ which will be denoted by $g \mid \Omega$.

Definition 1. (i) Let $0<p\leqq\infty$, $0<q\leqq\infty$ and $-\infty<s<\infty$. Then

$$B^s_{p,q}(\Omega) = \{f \mid f \in D'(\Omega),\ \exists g \in B^s_{p,q}(R_n) \quad \text{with} \quad g \mid \Omega = f\}\,, \tag{1}$$

$$\|f \mid B^s_{p,q}(\Omega)\| = \inf \|g \mid B^s_{p,q}(R_n)\|\,, \tag{2}$$

where the infimum is taken over all $g \in B^s_{p,q}(R_n)$ in the sense of (1).

(ii) Let $0<p<\infty$, $0<q\leqq\infty$ and $-\infty<s<\infty$. Then

$$F^s_{p,q}(\Omega) = \{f \mid f \in D'(\Omega),\ \exists g \in F^s_{p,q}(R_n) \quad \text{with} \quad g \mid \Omega = f\}\,, \tag{3}$$

$$\|f \mid F^s_{p,q}(\Omega)\| = \inf \|g \mid F^s_{p,q}(R_n)\|\,, \tag{4}$$

where the infimum is taken over all $g \in F^s_{p,q}(R_n)$ in the sense of (3).

Definition 2. (i) Let $0<p\leqq\infty$, $0<q\leqq\infty$ and $-\infty<s<\infty$. Then

$$B^s_{p,q}(\partial\Omega) = \{f \mid f \in D'(\partial\Omega),\ (\psi_j f)({f^{(j)}}^{-1}(y', 0)) \in B^s_{p,q}(R_{n-1}) \text{ if } j=1, \ldots, N\}, \tag{5}$$

$$\|f \mid B^s_{p,q}(\partial\Omega)\| = \sum_{j=1}^{N} \|(\psi_j f)({f^{(j)}}^{-1}(\cdot, 0)) \mid B^s_{p,q}(R_{n-1})\|\,. \tag{6}$$

(ii) Let $0<p<\infty$, $0<q\leqq\infty$ and $-\infty<s<\infty$. Then

$$F^s_{p,q}(\partial\Omega)=\{f\mid f\in D'(\partial\Omega),\,(\psi_j f)(f^{(j)-1}(y',0))\in F^s_{p,q}(R_{n-1})\text{ if } j=1,\ldots,N\}\,, \tag{7}$$

$$\|f\mid F^s_{p,q}(\partial\Omega)\|=\sum_{j=1}^{N}\|(\psi_j f)(f^{(j)-1}(\cdot,0))\mid F^s_{p,q}(R_{n-1})\|\,. \tag{8}$$

Remark 1. In (5)–(8) the distribution $(\psi_j f)(f^{(j)-1}(y',0))$ is extended by zero outside of $f^{(j)}(\partial\Omega\cap K_j)$. This must be understood in the sense of distributions on R_{n-1}. Of course, the quasi-norms in (6) and (8) depend on the choice of $\{\psi_j\}$ and $\{f^{(j)}\}$ but we do not indicate this dependence because all admissible quasi-norms (6) are equivalent to each other, cf. the following subsection. Similarly for (8). Cf. also [I, 3.6.1.].

Remark 2. There is no difficulty in extending the definitions of $F^s_{p,q}(\Omega)$ and $F^s_{p,q}(\partial\Omega)$ to $F^s_{\infty,q}(\Omega)$ and $F^s_{\infty,q}(\partial\Omega)$ with $0<q<\infty$, respectively. One must start with the spaces $F^s_{\infty,q}(R_n)$, resp. $F^s_{\infty,q}(R_{n-1})$, from 2.3.4.

3.2.3. Quasi-Banach Spaces

The notations have the same meaning as in the two preceding subsections. Furthermore, the spaces $B^s_{p,q}(R^+_n)$ and $F^s_{p,q}(R^+_n)$ have been defined in 2.9.1. If $f\in B^s_{p,q}(\Omega)$, then $\psi_0 f$ is extended outside of K_0 by zero, and $(\psi_j f)(f^{(j)-1}(\cdot))$ is extended from $f^{(j)}(K_j\cap\Omega)$ to $R^+_n-f^{(j)}(K_j\cap\Omega)$ by zero if $j=1,\ldots,N$. In this sense we consider $\psi_0 f$ as an element of $B^s_{p,q}(R_n)$, and $\psi_j f$ with $j=1,\ldots,N$ as an element of $B^s_{p,q}(R^+_n)$. Similarly if $f\in F^s_{p,q}(\Omega)$.

Proposition. (i) Let $0<p\leqq\infty$, $0<q\leqq\infty$ and $-\infty<s<\infty$. Then $B^s_{p,q}(\Omega)$ is a quasi-Banach space (Banach space if $1\leqq p\leqq\infty$ and $1\leqq q\leqq\infty$). Furthermore,

$$\|\psi_0 f\mid B^s_{p,q}(R_n)\|+\sum_{j=1}^{N}\|(\psi_j f)(f^{(j)-1}(\cdot))\mid B^s_{p,q}(R^+_n)\| \tag{1}$$

is an equivalent quasi-norm in $B^s_{p,q}(\Omega)$.
(ii) Let $0<p\leqq\infty$, $0<q\leqq\infty$ and $-\infty<s<\infty$. Then $B^s_{p,q}(\partial\Omega)$ is a quasi-Banach space (Banach space if $1\leqq p\leqq\infty$ and $1\leqq q\leqq\infty$). It is independent of the choice of the domain K_0, the balls $K_1,\ldots,K_N$, the vector-functions $f^{(1)},\ldots,f^{(N)}$ and the resolution of unity from (3.2.1/7) and (3.2.1/8).
(iii) Let $0<p<\infty$, $0<q\leqq\infty$ and $-\infty<s<\infty$. The statements of part (i) and part (ii) remain valid if one replaces $B^s_{p,q}$ by $F^s_{p,q}$.

Proof. Step 1. We prove (i). Definition 3.2.2/1(i) is the well-known construction of a factor space. It is clear that $B^s_{p,q}(\Omega)$ is a linear space, and it is easy to see that $\|f\mid B^s_{p,q}(\Omega)\|$ is a quasi-norm. Let

$$\|g_1+g_2\mid B^s_{p,q}(R_n)\|\leqq c\,(\|g_1\mid B^s_{p,q}(R_n)\|+\|g_2\mid B^s_{p,q}(R_n)\|)\,, \tag{2}$$

where $c\geqq 1$ is independent of g_1 and g_2. Let $\{f_k\}_{k=1}^{\infty}$ be a fundamental sequence in $B^s_{p,q}(\Omega)$. We may assume, without restriction of generality, that

$$\|f_{k+1}-f_k\mid B^s_{p,q}(\Omega)\|\leqq(3c)^{-k}\,,\quad k=1,2,\ldots, \tag{3}$$

where c has the same meaning as in (2) (otherwise we choose a subsequence). Let $g_k\in B^s_{p,q}(R_n)$ be corresponding elements in the sense of (3.2.2/1) and (3.2.2/2) with

$$\|g_{k+1}-g_k\mid B^s_{p,q}(R_n)\|\leqq(2c)^{-k}\,.$$

Then we have

$$g=\lim_{k\to\infty} g_k=g_1+\sum_{k=1}^{\infty}(g_{k+1}-g_k) \tag{4}$$

with

$$\left\| \sum_{k=1}^{\infty}(g_{k+1}-g_k) \mid B^s_{p,q}(R_n)\right\| \leqq \sum_{k=1}^{\infty} c^k \, \|g_{k+1}-g_k \mid B^s_{p,q}(R_n)\| \leqq \sum_{k=1}^{\infty} 2^{-k}<\infty. \tag{5}$$

Let $f=g \mid \Omega$. Then it follows easily from (4) and (5) that f is the limit element of f_k. Hence $B^s_{p,q}(\Omega)$ is a quasi-Banach space. We prove that (1) is an equivalent quasi-norm in $B^s_{p,q}(\Omega)$. Let Ω_1 be a bounded domain with

$$\bar{\Omega}_1 \subset \left\{ x \mid x\in R_n,\ \sum_{j=0}^{N} \psi_j(x)=1\right\}$$

and $\bar{\Omega}\subset \Omega_1$. Let $f\in B^s_{p,q}(\Omega)$. If we restrict the infimum in (3.2.2/2) to $g\in B^s_{p,q}(R_n)$ with

$$g \mid \Omega=f \quad \text{and} \quad \operatorname{supp} g\subset \Omega_1\,, \tag{6}$$

then we obtain a new equivalent quasi-norm in $B^s_{p,q}(\Omega)$. This follows from Theorem 2.8.2 if one multiplies an arbitrary element $g\in B^s_{p,q}(R_n)$ with a fixed infinitely differentiable function $\varphi(x)$ with

$$\varphi(x)=1 \quad \text{if} \quad x\in\Omega \quad \text{and} \quad \operatorname{supp}\varphi\subset\Omega_1\,. \tag{7}$$

For elements $g\in B^s_{p,q}(R_n)$ with (6),

$$\sum_{k=0}^{N} \|\psi_k g \mid B^s_{p,q}(R_n)\| \tag{8}$$

is an equivalent quasi-norm. This is also a consequence of Theorem 2.8.2. Next we apply Theorem 2.10.2 to $g(x)\to g(f^{(j)}(x))$, cf. 3.2.1. Then

$$\|\psi_0 g \mid B^s_{p,q}(R_n)\| + \sum_{k=1}^{N} \|(\psi_k g)(f^{(k)^{-1}}(\cdot)) \mid B^s_{p,q}(R_n)\| \tag{9}$$

is an equivalent quasi-norm for all $g\in B^s_{p,q}(R_n)$ with (6). But the infimum over all admissible g with (6) yields (1). The proof of (i) is complete.

Step 2. We prove (ii). Let $K_1,\ldots,K_N$ and $K'_1,\ldots,K'_{N'}$ be two coverings of $\partial\Omega$ in the sense of 3.2.1. and 3.2.2. Let $f^{(1)},\ldots,f^{(N)}$ and $f'^{(1)},\ldots,f'^{(N')}$ be the corresponding vector-functions and $\{\psi_j\}_{j=1}^{N}$ and $\{\psi'_j\}_{j=1}^{N'}$ the corresponding resolutions of unity. We consider the finer covering $K_j\cap K'_k$ where $j=1,\ldots,N$ and $k=1,\ldots,N'$ with the resolution of unity $\{\psi_j\psi'_k\}_{j,k}$. We can choose either $f^{(1)},\ldots,f^{(N)}$ or $f'^{(1)},\ldots,f'^{(N')}$ as vector-functions. Now it is not complicated to compare these two possibilities on the basis of Theorem 2.10.2 (with $B^s_{p,q}(R_{n-1})$ instead of $B^s_{p,q}(R_n)$) in the same way as in the first step. Afterwards one reduces the original coverings to their refinements on the basis of Theorem 2.8.2.

Step 3. The proof of (iii) is the same.

Remark. In 3.1.2. we described the problem of an inner description of the corresponding spaces. The above proposition shows that this is essentially a local problem.

3.2.4. Elementary Embeddings

Let again Ω be a bounded C^∞-domain with boundary $\partial\Omega$. We formulate some elementary assertions for the spaces $B^s_{p,q}(\Omega)$ and $F^s_{p,q}(\Omega)$. Obviously, corresponding statements hold for the spaces $B^s_{p,q}(\partial\Omega)$ and $F^s_{p,q}(\partial\Omega)$. Let $C^\infty(\bar{\Omega})$ be the collection of all infinitely differentiable complex-valued functions $f(x)$ in Ω, such that every derivative $D^\alpha f(x)$ can be extended continuously to $\bar{\Omega}$ (the spaces $C^\infty(\bar{\Omega})$ can also be defined as the restriction of $S(R_n)$ to Ω). Recall that "$\subset$" always stands for continuous embedding.

Proposition. (i) Let $0<q_0\leqq q_1\leqq\infty$ and $-\infty<s<\infty$. Then

$$C^\infty(\bar{\Omega})\subset B^s_{p,q_0}(\Omega)\subset B^s_{p,q_1}(\Omega)\subset D'(\Omega) \quad \text{if} \quad 0<p\leqq\infty \tag{1}$$

and

$$C^\infty(\bar{\Omega})\subset F^s_{p,q_0}(\Omega)\subset F^s_{p,q_1}(\Omega)\subset D'(\Omega) \quad \text{if} \quad 0<p<\infty\,. \tag{2}$$

If $0<p<\infty$ and $0<q<\infty$, then $C^\infty(\bar{\Omega})$ is dense in $B^s_{p,q}(\Omega)$ and in $F^s_{p,q}(\Omega)$.
(ii) Let $0<q\leqq\infty$, $0<p<\infty$ and $-\infty<s<\infty$. Then

$$B^s_{p,\min(p,q)}(\Omega)\subset F^s_{p,q}(\Omega)\subset B^s_{p,\max(p,q)}(\Omega)\,. \tag{3}$$

Proof. The proof follows immediately from the definition of $B^s_{p,q}(\Omega)$ and $F^s_{p,q}(\Omega)$ on the one hand and from Proposition 2.3.2/2 and Theorem 2.3.3 on the other.

Remark. In the same way one can carry over (2.3.2/7) and (2.3.2/8) from R_n to Ω, cf. also 3.3.1.

3.3. Main Properties

3.3.1. Embeddings

We again recall that "$\subset$" always stands for continuous embedding. Furthermore, Ω always denotes a bounded C^∞-domain in R_n. We begin with a preliminary.

Lemma. Let $\varphi(x)$ be an infinitely differentiable function on R_n with compact support.
(i) Let $0<p_0\leqq p_1\leqq\infty$, $0<q\leqq\infty$ and $-\infty<s<\infty$. Then $f\to\varphi(x)f$ yields a continuous linear mapping from $B^s_{p_1,q}(R_n)$ into $B^s_{p_0,q}(R_n)$.
(ii) Let $0<p_0\leqq p_1<\infty$, $0<q<\infty$ and $-\infty<s<\infty$. Then $f\to\varphi(x)f$ yields a continuous linear mapping from $F^s_{p_1,q}(R_n)$ into $F^s_{p_0,q}(R_n)$.

Proof. Step 1. In the first three steps we prove (i). We apply Theorem 2.5.12. It is easy to see that the integration over R_n in the second term on the right-hand side of (2.5.12/4) can be replaced by an integration over the unit ball $\{h \mid |h|\leqq 1\}$. We apply this modified equivalent quasi-norm in $B^s_{p_0,q}(R_n)$ to $\varphi(x)\,f(x)$ instead of $f(x)$ (we assume that $s>\tilde{\sigma}_{p_0}$). If $|h|\leqq 1$, then $\Delta^M_h\varphi f$ has compact support and we have

$$\|\Delta^M_h\varphi f \mid L_{p_0}(R_n)\|\leqq c\|\Delta^M_h\varphi f \mid L_{p_1}(R_n)\|\,, \tag{1}$$

where c is independent of h with $|h|\leqq 1$ and $f\in B^s_{p_1,q}(R_n)$. We put (1) in the modified formula (2.5.12/4). Furthermore, $\varphi(x)$ is a pointwise multiplier in $B^s_{p_1,q}(R_n)$, cf. Theorem 2.8.2. Thus we have the statement of part (i) of the lemma provided that

$$0<p_0\leqq p_1\leqq\infty,\quad 0<q\leqq\infty \quad \text{and} \quad s>\tilde{\sigma}_{p_0}=n\left(\frac{1}{\min(p_0,1)}-1\right). \tag{2}$$

Step 2. If $1 \leqq p_0 \leqq p_1 \leqq \infty$, $1 \leqq q \leqq \infty$ and $s > 0$, it follows from step 1 that $f \to \varphi f$ has the desired property. But $f \to \varphi f$ is a self-adjoint operation. Consequently, $f \to \varphi f$ is also a continuous linear mapping from $(B^s_{p_0,q}(R_n))'$ into $B^s_{p_1,q}(R_n))'$. We use Theorem 2.11.2 to obtain that $f \to \varphi f$ is a continuous linear mapping from $B^s_{p_1,q}(R_n)$ into $B^s_{p_0,q}(R_n)$ if

$$1 < p_0 \leqq p_1 \leqq \infty, \quad 1 < q \leqq \infty, \quad s < 0 . \tag{3}$$

Step 3. We apply the complex interpolation formula (2.4.7/2) in a similar way as e.g. in the proof of Corollary 2.8.2. The "corner" spaces in (2.4.7/2) are characterized by (2) and (3), respectively. Every space $B^s_{p_0,q}(R_n)$ and $B^s_{p_1,q}(R_n)$ can be obtained in this way. This proves part (i) of the lemma.

Step 4. In order to prove (ii) we begin with (2.5.10/2). We replace f by φf. For large values of H we have

$$\begin{aligned}\left(\int_{|h| \geqq H} |h|^{-sq} \, |(\Delta_h^M \varphi f)(x)|^q \frac{dh}{|h|^n}\right)^{\frac{1}{q}} &\leqq \delta \sup_{|h| \geqq H} |h|^{-s+\varepsilon} |(\Delta_h^M \varphi f)(x)| \\ &\leqq c\delta \, (1+|x|)^{-s+\varepsilon} \sup_{y \in R_n} |(\varphi f)(y)| ,\end{aligned} \tag{4}$$

where $\delta > 0$ and $\varepsilon > 0$ are arbitrary numbers, and $H = H(\delta, \varepsilon)$. Let $(s-\varepsilon)p > n$ and put (4) in (2.5.10/2). We have $F^s_{p,q}(R_n) \subset C(R_n)$ under the hypotheses of Theorem 2.5.10, cf. Theorem 2.7.1 and (2.7.1/13). Hence

$$\begin{aligned}&\|\varphi f \mid F^s_{p,q}(R_n)\|^{(2)}_M \leqq c \|\varphi f \mid L_p(R_n)\| \\ &+ \left\| \left(\int_{|h| \leqq H} |h|^{-sq} \, |(\Delta_h^M \varphi f)(\cdot)|^q \frac{dh}{|h|^n}\right)^{\frac{1}{q}} \mid L_p(R_n) \right\| + \frac{1}{2} \|\varphi f \mid F^s_{p,q}(R_n)\|^{(2)}_M\end{aligned} \tag{5}$$

if δ in (4) is chosen sufficiently small. In other words, if φ is fixed, then

$$\|\varphi f \mid L_p(R_n)\| + \left\| \left(\int_{|h| \leqq H} |h|^{-sq} \, |(\Delta_h^M \varphi f)(\cdot)|^q \frac{dh}{|h|^n}\right)^{\frac{1}{q}} \mid L_p(R_n) \right\|$$

is equivalent to $\|\varphi f \mid F^s_{p,q}(R_n)\|$. Now we can argue in the same way as in step 1–step 3: duality and complex interpolation, cf. Theorem 2.11.2 and (2.4.7/3). It is not hard to see that all spaces $F^s_{p_0,q}(R_n)$ and $F^s_{p_1,q}(R_n)$ can be obtained in this way. This proves part (ii) of the lemma.

Remark 1. This lemma is an extension of a corresponding assertion in [I, 4.6.2.].

Theorem. (i) Let $0 < p_0 \leqq \infty$, $0 < p_1 \leqq \infty$, $0 < q_0 \leqq \infty$, $0 < q_1 \leqq \infty$ and $-\infty < s_1 < s_0 < \infty$. Then

$$B^{s_0}_{p_0,q_0}(\Omega) \subset B^{s_1}_{p_1,q_0}(\Omega) \quad \text{if} \quad s_0 - \frac{n}{p_0} = s_1 - \frac{n}{p_1} \tag{6}$$

and

$$B^{s_0}_{p_0,q_0}(\Omega) \subset B^{s_1}_{p_1,q_1}(\Omega) \quad \text{if} \quad s_0 - \frac{n}{p_0} > s_1 - \frac{n}{p_1} . \tag{7}$$

(ii) Let $0 < p_0 < \infty$, $0 < p_1 < \infty$, $0 < q_0 \leqq \infty$, $0 < q_1 \leqq \infty$ and $-\infty < s_1 < s_0 < \infty$. Then

$$F^{s_0}_{p_0,q_0}(\Omega) \subset F^{s_1}_{p_1,q_1}(\Omega) \quad \text{if} \quad s_0 - \frac{n}{p_0} \geqq s_1 - \frac{n}{p_1} . \tag{8}$$

(iii) Let $-\infty<s<\infty$. Then

$$B^s_{p_0,q}(\Omega)\subset B^s_{p_1,q}(\Omega) \quad\text{if}\quad 0<p_1\leqq p_0\leqq\infty \quad\text{and}\quad 0<q\leqq\infty \tag{9}$$

and

$$F^s_{p_0,q}(\Omega)\subset F^s_{p_1,q}(\Omega) \quad\text{if}\quad 0<p_1\leqq p_0<\infty \quad\text{and}\quad 0<q<\infty\,. \tag{10}$$

Proof. Step 1. We prove (iii). As we remarked in the proof of Proposition 3.2.3, it is sufficient to assume that g in (3.2.2/2) and (3.2.2/4) has the property (3.2.3/6). Then (9) and (10) follow from the above lemma.

Step 2. The definition of the spaces $B^s_{p,q}(\Omega)$ and $F^s_{p,q}(\Omega)$ on the one hand and Theorem 2.7.1 on the other hand prove (6) and (8) with $s_0-\frac{n}{p_0}=s_1-\frac{n}{p_1}$. Let $s_0-\frac{n}{p_0}>s_1-\frac{n}{p_1}$ and $p_0<p_1$. Let $s_0-\frac{n}{p_0}=s_1+\varepsilon-\frac{n}{p_1}$ with $\varepsilon>0$. Then we have (6) and (8) with $s_1+\varepsilon$ instead of s_1. The definition of the spaces $B^s_{p,q}(\Omega)$ and $F^s_{p,q}(\Omega)$ shows that (2.3.2/7) and (2.3.2/8) remain valid if we replace R_n by Ω. These modified formulas on the one hand and (6) and (8), with $s_1+\varepsilon$ instead of s_1, on the other prove (7) and (8) provided that $p_0<p_1$. Finally, let $p_1\leqq p_0$ and let $s_0-s_1=2\varepsilon$. Then (8) and (10) yield

$$F^{s_0}_{p_0,q_0}(\Omega)\subset F^{s_0-\varepsilon}_{p_0,p_0}(\Omega) \quad\text{and}\quad F^{s_0-\varepsilon}_{p_1,p_0}(\Omega)\subset F^{s_1}_{p_1,q_1}(\Omega) \tag{11}$$

and

$$F^{s_0-\varepsilon}_{p_0,p_0}(\Omega)\subset F^{s_0-\varepsilon}_{p_1,p_0}(\Omega)\,, \tag{12}$$

respectively. This proves (8) if $p_1\leqq p_0$. The proof of (7) with $p_1\leqq p_0$ is the same.

Remark 2. Of course, one can combine (6) and (7) with (8) or with (2.3.2/7) and (2.3.2/8), where R_n is replaced by Ω. The above theorem contains a lot of classical embedding theorems for different metrics, cf. [I, 4.6.1. and 4.6.2.]. Furthermore, the formulas (2.7.1/11)–(2.7.1/16) remain valid if one replaces R_n by Ω.

Remark 3. The theorem remains valid if n is replaced by $n-1$ and Ω by $\partial\Omega$.

3.3.2. Pointwise Multipliers

Let again Ω be a bounded C^∞-domain. Pointwise multipliers for spaces on R_n have been defined in 2.8.1. By restriction, every multiplier for a space on R_n is also a multiplier for a corresponding space on Ω. But more interesting than this almost obvious observation is the following fact.

Proposition. (i) The characteristic function of Ω is a multiplier for $B^s_{p,q}(R_n)$ if

$$\begin{cases}\text{either} & 1\leqq p\leqq\infty, \quad 0<q\leqq\infty, \quad \frac{1}{p}-1<s<\frac{1}{p}\\ \text{or} & 0<p<1, \quad 0<q\leqq\infty, \quad n\left(\frac{1}{p}-1\right)<s<\frac{1}{p}\,.\end{cases} \tag{1}$$

(ii) The characteristic function of Ω is a multiplier for $F^s_{p,q}(R_n)$ if

$$\begin{cases}\text{either} & 1\leqq p<\infty, \quad 0<q<\infty, \quad \frac{1}{p}-1+n\left(\frac{1}{\min(q,1)}-1\right)<s<\frac{1}{p}\\ \text{or} & 0<p<1\,, \quad 0<q<\infty, \quad n\left(\frac{1}{\min(p,q)}-1\right)<s<\frac{1}{p}\,.\end{cases} \tag{2}$$

Proof. By Theorem 2.8.7 the characteristic function of R^+_n is a multiplier for $B^s_{p,q}(R_n)$ and $F^s_{p,q}(R_n)$ if (1) and (2) are satisfied, respectively. We must prove that this assertion can be extended to characteristic functions χ_Ω of bounded C^∞-domains Ω. Let $f^{(1)}, \ldots, f^{(N)}$ be the vector-functions from 3.2.1. and $\psi_0, \psi_1, \ldots, \psi_N$ be a corresponding resolution of unity in the sense of 3.2.1. We recall that $f^{(1)}, \ldots, f^{(N)}$ are diffeomorphic maps from R_n onto R_n for which Theorem 2.10.2 is applicable. Furthermore, $\psi_0, \psi_1, \ldots, \psi_N$ are multipliers in $B^s_{p,q}(R_n)$ and $F^s_{p,q}(R_n)$, cf. Theorem 2.8.2. Let (1) be satisfied. Then

$$\begin{aligned}\|\chi_\Omega f \mid B^s_{p,q}(R_n)\| &\leqq c\|\psi_0 f \mid B^s_{p,q}(R_n)\| + c\sum_{j=1}^N \|\psi_j\chi_\Omega f \mid B^s_{p,q}(R_n)\| \\ &\leqq c\|\psi_0 f \mid B^s_{p,q}(R_n)\| + c'\sum_{j=1}^N \|(\psi_j\chi_\Omega f)(f^{(j)^{-1}}(\cdot)) \mid B^s_{p,q}(R_n)\| .\end{aligned} \tag{3}$$

However

$$(\psi_j\chi_\Omega)(f^{(j)^{-1}}(x)) = \psi_j(f^{(j)^{-1}}(x))\chi_+(x) ,$$

where χ_+ is the characteristic function of R_n^+. Hence, (3) yields

$$\begin{aligned}\|\chi_\Omega f \mid B^s_{p,q}(R_n)\| &\leqq c\|\psi_0 f \mid B^s_{p,q}(R_n)\| + c\sum_{j=1}^N \|(\psi_j f)(f^{(j)^{-1}}(\cdot)) \mid B^s_{p,q}(R_n)\| \\ &\leqq c\|\psi_0 f \mid B^s_{p,q}(R_n)\| + c'\sum_{j=1}^N \|\psi_j f \mid B^s_{p,q}(R_n)\| \leqq c''\|f \mid B^s_{p,q}(R_n)\| .\end{aligned}$$

This proves part (i) of the proposition. The proof of (ii) is the same.

Remark 1. If $\min(p, q) \geqq \min(p, 1)$, then (1) and (2) coincide (apart from the limiting cases $p=\infty$ and/or $q=\infty$). In Chapter 2 we used this restriction several times. Moreover, the third part of Theorem 2.8.7 can also be transformed from R_n to Ω. This shows that (1) is natural.

Our next aim is to extend Theorem 2.8.2 and Corollary 2.8.2. Let $\mathcal{C}^\varrho(\Omega) = B^\varrho_{\infty,\infty}(\Omega)$ if $\varrho > 0$. Furthermore, let $\varkappa^s_p$ and $\varkappa^s_{p,q}$ be given by (2.8.2/25) and (2.8.2/26), i.e.

$$\varkappa^s_p = \max\left(s,\, n\left(\frac{1}{\min(p, 1)} - 1\right) - s\right) \tag{4}$$

and

$$\varkappa^s_{p,q} = \max\left(s,\, n\left(\frac{1}{\min(p, q, 1)} - 1\right) - s\right) . \tag{5}$$

Theorem. (i) Let $-\infty < s < \infty$, $0 < p \leqq \infty$, $0 < q \leqq \infty$ and $\varrho > \varkappa^s_p$. Then every $g \in \mathcal{C}^\varrho(\Omega)$ is a multiplier for $B^s_{p,q}(\Omega)$, and there exists a positive constant c such that

$$\|gf \mid B^s_{p,q}(\Omega)\| \leqq c\|g \mid \mathcal{C}^\varrho(\Omega)\|\, \|f \mid B^s_{p,q}(\Omega)\| \tag{6}$$

holds for all $g \in \mathcal{C}^\varrho(\Omega)$ and all $f \in B^s_{p,q}(\Omega)$.
(ii) Let $-\infty < s < \infty$, $0 < p < \infty$, $0 < q < \infty$ and $\varrho > \varkappa^s_{p,q}$. Then every $g \in \mathcal{C}^\varrho(\Omega)$ is a multiplier for $F^s_{p,q}(\Omega)$, and there exists a positive constant c such that

$$\|gf \mid F^s_{p,q}(\Omega)\| \leqq c\|g \mid \mathcal{C}^\varrho(\Omega)\|\, \|f \mid F^s_{p,q}(\Omega)\| \tag{7}$$

holds for all $g \in \mathcal{C}^\varrho(\Omega)$ and all $f \in F^s_{p,q}(\Omega)$.

Proof. The proof follows immediately from the definitions of the spaces involved, $\mathcal{C}^\varrho(\Omega)$, $B^s_{p,q}(\Omega)$, $F^s_{p,q}(\Omega)$, and from Theorem 2.8.2 and Corollary 2.8.2.

Remark 2. The other assertions of Corollary 2.8.2 remain valid also if one replaces R_n by Ω. This shows at least that $\varkappa^s_p$ is a natural constant.

3.3.3. Traces

Again let Ω be a bounded C^∞-domain in R_n. We extend Theorem 2.7.2 from R_n to Ω. Let ν be the outward normal with respect to the boundary $\partial\Omega$ of Ω. As in (2.7.2/1) we define the trace operator $\mathfrak{R}$ by

$$\mathfrak{R}f = \left\{ f \mid \partial\Omega, \frac{\partial f}{\partial \nu}\bigg|\partial\Omega, \ldots, \frac{\partial^r f}{\partial \nu^r}\bigg|\partial\Omega \right\}, \tag{1}$$

where $\frac{\partial^j f}{\partial \nu^j}\Big|\partial\Omega$ denotes the restriction of $\frac{\partial^j f}{\partial \nu^j}$ to the boundary $\partial\Omega$. For any distribution $f \in D'(\Omega)$, the derivative $\frac{\partial^j f}{\partial \nu^j}$ makes sense near the boundary $\partial\Omega$ (and this is sufficient). As far as the meaning of $\frac{\partial^j f}{\partial \nu^j}\Big|\partial\Omega$ is concerned we have the following situation. We always assume that f belongs to $B^s_{p,q}(\Omega)$ or $F^s_{p,q}(\Omega)$, respectively. By the above localization method we reduce f to $(\varphi_j f)(f^{(j)^{-1}}(\cdot))$ in the sense of (3.2.3/1). Then we can ask for the trace of $(\varphi_j f)(f^{(j)^{-1}}(\cdot))$ on $R_{n-1} = \{y \mid y_n = 0\}$ in the sense of 2.7.2., cf. (2.7.2/1). In other words, our statements about traces should always be understood in the sense of this reduction to the traces in 2.7.2. The balls $K_1, \ldots, K_N$, the diffeomorphic maps $f^{(1)}, \ldots, f^{(N)}$ and the resolution of unity $\{\varphi_k\}_{k=0}^N$ have the above meaning, cf. 3.2.1. We may assume, without loss of generality, that the diffeomorphism $y = f^{(i)}(x)$ maps the ν-direction in $\partial\Omega \cap K_j$ onto the y_n-direction and that $\frac{\partial}{\partial \nu}$ is transformed into $\frac{\partial}{\partial y_n}$. This is convenient but immaterial. As for the general case, we add the following useful remark. Let e.g. $f \in B^s_{p,q}(R_n)$ with $s > r + \frac{1}{p} + \max\left(0, (n-1)\left(\frac{1}{p} - 1\right)\right)$, where $r = 0, 1, 2, \ldots$ Then $\frac{\partial^j f}{\partial x_n^j}(x', 0) \in B_{p,q}^{s-j-\frac{1}{p}}(R_{n-1})$ if $j = 0, \ldots, r$, cf. Theorem 2.7.2. It follows that

$$(D^\alpha f)(x', 0) \in B_{p,q}^{s-|\alpha|-\frac{1}{p}}(R_{n-1}) \quad \text{if} \quad 0 \leq |\alpha| \leq r.$$

Let $\mu(x') = (\mu_1(x'), \ldots, \mu_n(x'))$ be an arbitrary non-tangential C^∞-vector field on $R_{n-1} = \{x \mid x_n = 0\}$, i. e. $\mu_k(x')$ are C^∞-functions and $\mu_n(x') \neq 0$. Then we have

$$\frac{\partial^j f}{\partial \mu^j(x')}(x', 0) \in B_{p,q}^{s-j-\frac{1}{p}}(R_{n-1}) \quad \text{if} \quad j = 0, \ldots, r.$$

Vice versa, if $\frac{\partial^j f}{\partial \mu^j(x')}(x', 0)$ are known, $j = 0, \ldots, r$, then we can calculate $\frac{\partial^j f}{\partial x_n^j}(x', 0)$ with $j = 0, \ldots, r$, uniquely. In other words, we can replace $\frac{\partial^j f}{\partial x_n^j}(x', 0)$ in (2.7.2/1) and in Theorem 2.7.2 by $\frac{\partial^j f}{\partial \mu(x')^j}(x', 0)$. The above diffeomorphism $y = f^{(j)}(x)$ maps locally (i.e. in $\partial\Omega \cap K_j$) an arbitrary non-tangential C^∞-vector field on $\partial\Omega$ onto a corresponding vector field on $R_{n-1} = \{y \mid y_n = 0\}$. In this sense we can replace ν in (1) and in the following theorem by an arbitrary non-tangential C^∞-vector-field on $\partial\Omega$. In particular, no difficulties arise if $y = f^{(j)}(x)$ does not map the ν-direction onto the y_n-direction. Finally, we recall of the important notion of a retraction. Let $f \in B^s_{p,q}(\Omega)$ with $s > r + \frac{1}{p} + \max\left(0, (n-1)\left(\frac{1}{p} - 1\right)\right)$. Then $\mathfrak{R}$ is a retraction

$$\text{from} \quad B^s_{p,q}(\Omega) \quad \text{onto} \quad \prod_{j=0}^{r} B_{p,q}^{s-\frac{1}{p}-j}(\partial\Omega) \tag{2}$$

means the following: $\mathfrak{R}$ is a continuous linear map from $B^s_{p,q}(\Omega)$ onto $\prod_{j=0}^{r} B_{p,q}^{s-\frac{1}{p}-j}$ $(\partial\Omega)$ and there exists a continuous linear map $\mathfrak{S}$ from $\prod_{j=0}^{r} B_{p,q}^{s-\frac{1}{p}-j}(\partial\Omega)$ into $B^s_{p,q}(\Omega)$ such that $\mathfrak{R}\mathfrak{S}=E$ (identity in $\prod_{j=0}^{r} B_{p,q}^{s-\frac{1}{p}-j}(\partial\Omega)$). $\mathfrak{S}$ is called a corresponding coretraction. Similarly for $F^s_{p,q}(\Omega)$. Cf. 2.7.2., where we discussed this problem in greater detail. In particular, $\mathfrak{R}$ stands for a direct embedding theorem and $\mathfrak{S}$ for an inverse embedding theorem.

Theorem. Let $\mathfrak{R}$ be given by (1), where $r=0, 1, 2, \ldots$

(i) If $0<p\leqq\infty$, $0<q\leqq\infty$ and $s>r+\frac{1}{p}+\max\left(0, (n-1)\left(\frac{1}{p}-1\right)\right)$, then $\mathfrak{R}$ is a retraction

$$\text{from} \quad B^s_{p,q}(\Omega) \quad \text{onto} \quad \prod_{j=0}^{r} B_{p,q}^{s-\frac{1}{p}-j}(\partial\Omega)\,. \tag{3}$$

(ii) If $0<p<\infty$, $0<q\leqq\infty$ and $s>r+\frac{1}{p}+\max\left(0, (n-1)\left(\frac{1}{p}-1\right)\right)$, then $\mathfrak{R}$ is a retraction

$$\text{from} \quad F^s_{p,q}(\Omega) \quad \text{onto} \quad \prod_{j=0}^{r} B_{p,p}^{s-\frac{1}{p}-j}(\partial\Omega)\,. \tag{4}$$

Proof. We prove (i). The proof of (ii) is the same. We use (3.2.3/1), where the term with ψ_0 is unimportant because only the values of f near the boundary $\partial\Omega$ are of interest. We assume that $\psi_0 f\equiv 0$ and we extend $\frac{\partial^k f}{\partial \nu^k}$ from $\partial\Omega$ to a one-side-neighbourhood of $\partial\Omega$ in Ω in an appropriate way (by the above assumptions on $f^{(j)}$ we can take the curves $f^{(j)^{-1}}\{x_n=\text{const}\}$ locally as ν-directions). Now we replace f and s in (3.2.3/1) by $\frac{\partial^k f}{\partial \nu^k}$ and $s-k$, respectively and we apply Theorem 2.7.2 (with $r=0$). Then Definition 3.2.2/2 proves that $\mathfrak{R}$ maps $B^s_{p,q}(\Omega)$ into the space on the right-hand side of (3). Conversely let

$$\{g_0, \ldots, g_r\}\in \prod_{j=0}^{r} B_{p,q}^{s-\frac{1}{p}-j}(\partial\Omega)$$

be given. Let φ_j with $j=1, \ldots, N$ be infinitely differentiable functions on R_n with $\operatorname{supp}\varphi_j\subset K_j$ and $\varphi_j(x)=1$ if $x\in\operatorname{supp}\psi_j$. Let $\mathfrak{S}$ be the extension operator from (2.7.2/42) with the property (2.7.2/43). We put

$$\tilde{\mathfrak{S}}\{g_0, \ldots, g_r\}(x)=\sum_{j=1}^{N}\varphi_j(x)\,\mathfrak{S}\{\ldots, (\psi_j g_k)(f^{(j)^{-1}}(\cdot, 0), \ldots\}(f^{(j)}x) \tag{5}$$

if $x\in\Omega$ (with $\varphi_j(x)\,\mathfrak{S}\{\ldots\}(f^{(j)}x)=0$ outside of K_j). Recall that by our above assumption $y=f^{(j)}(x)$ transforms $\frac{\partial}{\partial\nu}$ into $\frac{\partial}{\partial y_n}$. Then it follows from Step 7 of the proof of Theorem 2.7.2 that $\tilde{\mathfrak{S}}$ from (5) is a coretraction with respect to $\mathfrak{R}$ from (1) with the desired properties. The proof is complete.

3.3.4. Extensions

Our next goal is the generalization of Theorem 2.9.4. An affirmative solution of the extension problem for the spaces $B^s_{p,q}(\Omega)$ and $F^s_{p,q}(\Omega)$ has far-reaching consequences. Some important properties of the $B^s_{p,q}$-$F^s_{p,q}$-spaces can be carried over from R_n to Ω if the existence of an extension operator is guaranteed. Again let Ω be a bounded C^∞-domain in R_n. Let R, with

$$Rf = f \mid \Omega\,, \tag{1}$$

be the restriction operator from $S'(R_n)$ onto $D'(\Omega)$. In other words, the restriction of every $f \in S'(R_n)$ to Ω will be interpreted as a distribution on Ω. Of course, R is a linear operator. The corresponding operator with R_n^+ instead of Ω has been defined in 2.9.1. We shall denote the restriction of R to linear subspaces of $S'(R_n)$, e.g. $B^s_{p,q}(R_n)$ or $F^s_{p,q}(R_n)$, by the same symbol. Of course, by Definition 3.2.2/1, R is a bounded linear operator from $B^s_{p,q}(R_n)$ and $F^s_{p,q}(R_n)$ onto $B^s_{p,q}(\Omega)$ and $F^s_{p,q}(\Omega)$, respectively. As in 2.9.1. we are looking for a bounded linear operator S from $B^s_{p,q}(\Omega)$ or $F^s_{p,q}(\Omega)$ into $B^s_{p,q}(R_n)$ or $F^s_{p,q}(R_n)$, respectively, such that

$$RS = E \quad (\text{identity in } B^s_{p,q}(\Omega) \text{ or } F^s_{p,q}(\Omega), \text{ resp.})\,. \tag{2}$$

In that case, S is a bounded linear extension operator. If (2) holds, then we shall say that the restriction operator R is a retraction, and the extension operator S is a corresponding coretraction. We are interested in common coretractions S in the sense of (2) for the restriction operator R in a given family of spaces. Let e.g.

$$\{B^s_{p,q}(\Omega) \text{ with } s_0 < s < s_1,\, p_0 < p < p_1,\, q_0 < q < q_1\}$$

be such a family, where s_0, s_1, p_0, p_1, q_0 and q_1 are given. Then we ask whether there exists a linear operator S, which is defined on the union of all spaces of the family, such that its restriction on $B^s_{p,q}(\Omega)$ has the desired property. In such a case we shall speak of a common coretraction S.

Theorem. (i) If $0 < p \leqq \infty$, $0 < q \leqq \infty$ and $-\infty < s < \infty$, then the restriction operator R from (1) is a retraction for the space $B^s_{p,q}(R_n)$. If N is a natural number then there exists a common coretraction $S = S^N$ in the sense of (2) for all spaces $B^s_{p,q}(R_n)$ with

$$|s| < N, \quad \frac{1}{N} < p \leqq \infty \quad \text{and} \quad 0 < q \leqq \infty\,. \tag{3}$$

(ii) If $0 < p < \infty$, $0 < q < \infty$,

$$\min(p, q) \geqq \min(p, 1)\,, \tag{4}$$

and $-\infty < s < \infty$, then the restriction operator R from (1) is a retraction for the space $F^s_{p,q}(R_n)$. If N is a natural number then the operator $S = S^N$ from part (i) is also a common coretraction in the sense of (2) for all spaces $F^s_{p,q}(R_n)$ with

$$|s| < N, \quad \frac{1}{N} < p < \infty, \quad 0 < q < \infty \quad \text{and} \quad (4)\,. \tag{5}$$

Proof. We use Proposition 3.2.3. Let φ_j with $j = 1, \ldots, N$ be the same functions as in the proof of Theorem 3.3.3. Temporarily, we denote the extension operator from Theorem 2.9.4 by S^N_+. Then

$$(S^N f)(x) = (\psi_0 f)(x) + \sum_{j=1}^{N} \{\varphi_j(f^{(j)^{-1}}(\cdot))\, S^N_+[(\psi_j f)(f^{(j)^{-1}}(\cdot))]\}\, (f^{(j)}(x)) \tag{6}$$

with $x \in R_n$ has the desired properties, cf. Theorem 2.9.4. Of course, (6) must be interpreted in the sense of distributions, and the distribution in the brackets $\{\ \}$ is extended by zero outside of $f^{(j)}(K_j)$. The rest follows from Proposition 3.2.3 and Theorem 2.9.4.

Remark. The above theorem is a copy of Theorem 2.9.4. However, in order to prove that S^N from (6) is an extension operator with the desired properties, one needs Theorem 2.8.2 (pointwise multiplication) and Theorem 2.10.2 (diffeomorphic maps). Both theorems are implicitely contained in Proposition 3.2.3. Of course, the remarks in 2.9.5. also have their counterparts when R_n^+ is replaced by Ω. In particular, Theorem 2.9.5 holds for arbitrary intervals on the real line. Furthermore, if Problem 2.9.5 has an affirmative solution, then all spaces $F^s_{p,q}(\Omega)$ with $0<p<\infty$, $0<q\leqq\infty$ and $-\infty<s<\infty$ have the extension property (i.e. there exists an extension operator).

3.3.5. Equivalent Quasi-Norms

The equivalent quasi-norms (2.3.8/2) and (2.3.8/3) for the spaces $B^s_{p,q}(R_n)$ and $F^s_{p,q}(R_n)$, respectively, are of great use to us. It is natural to ask whether there exists a counterpart for the spaces $B^s_{p,q}(\Omega)$ and $F^s_{p,q}(\Omega)$, where we always assume that Ω is a bounded C^∞-domain in R_n.

Theorem. (i) Let $0<p\leqq\infty$, $0<q\leqq\infty$, $-\infty<s<\infty$ and $m=1, 2, 3, \ldots$ Then

$$\sum_{|\alpha|\leqq m} \|D^\alpha f \mid B^{s-m}_{p,q}(\Omega)\| \tag{1}$$

is an equivalent quasi-norm in $B^s_{p,q}(\Omega)$.
(ii) Let $0<p<\infty$, $0<q<\infty$,

$$\min(p, q) \geqq \min(p, 1), \tag{2}$$

$-\infty<s<\infty$ and $m=1, 2, 3, \ldots$ Then

$$\sum_{|\alpha|\leqq m} \|D^\alpha f \mid F^s_{p,q}(\Omega)\| \tag{3}$$

is an equivalent quasi-norm in $F^s_{p,q}(\Omega)$.

Proof. We prove (ii), the proof of (i) being the same. If $f\in F^s_{p,q}(\Omega)$, then it follows easily from (2.3.8/3) and Definition 3.2.2/1 that

$$\sum_{|\alpha|\leqq m} \|D^\alpha f \mid F^{s-m}_{p,q}(\Omega)\| \leqq c\|f \mid F^s_{p,q}(\Omega)\|, \tag{4}$$

where c is independent of f (we need no additional assumptions for p, q and Ω). The problem is to prove the converse inequality. For that purpose we use the considerations in 2.9.2.–2.9.4. Again let $S_+^N=S_r$ be the extension operator from 2.9.4. which is given explicitly by (2.9.2/9) and (2.9.2/10) and let $|\alpha|\leqq m$. We replace b_j in (2.9.2/7) by $(-\lambda_j)^{|\alpha|}b_j$, where the b_j's are determined by (2.9.2/8). The corresponding operators $S^1_{r,|\alpha|}$, $S^2_{r,|\alpha|}$ and $S_{r,|\alpha|}$ from (2.9.2/7)–(2.9.2/10) have properties similar to S^1_r, S^2_r and S_r, respectively. In particular, $S_{r,|\alpha|}$ is an extension operator from $F^\sigma_{p,q}(R_n^+)$ into $F^\sigma_{p,q}(R_n)$ if σ is given and if r is sufficiently large, cf. Theorem 2.9.4 (and the preparations in 2.9.2. and 2.9.3.). By (2.9.2/10) we have

$$(D^\alpha S_r f)(x) = (S_{r,|\alpha|} D^\alpha f)(x)$$

(we always assume that r is large). Let $f\in F^s_{p,q}(R_n^+)$. Then

$$\|f \mid F^s_{p,q}(R_n^+)\| \leqq \|S_r f \mid F^s_{p,q}(R_n)\| \sim \sum_{|\alpha|\leqq m} \|D^\alpha S_r f \mid F^{s-m}_{p,q}(R_n)\| \tag{5}$$

$$= \sum_{|\alpha|\leqq m} \|S_{r,|\alpha|} D^\alpha f \mid F^{s-m}_{p,q}(R_n)\| \leqq c \sum_{|\alpha|\leqq m} \|D^\alpha f \mid F^{s-m}_{p,q}(R_n^+)\|.$$

This is the desired inequality with R_n^+ instead of Ω. The transformation from R_n^+ to Ω follows from Proposition 3.2.3 by standard arguments (pointwise multiplication property and diffeomorphism property of the spaces in question). The proof is complete.

Remark 1. The somewhat disturbing assumption (2) comes from Theorem 3.3.4 (or Theorem 2.9.4). The reason is that we used the existence of a (common) extension operator. Of course, one would conjecture that the above theorem is valid for all spaces $F^s_{p,q}(\Omega)$ with $0<p<\infty$, $0<q\leqq\infty$ and $-\infty<s<\infty$. Furthermore, it would be of interest to give explicit formulas for some special cases. We return to this problem in 3.4.2.

Remark 2. The above theorem remains valid if one replaces Ω by R_n^+. This follows from (5).

3.3.6. Interpolation

The real interpolation method $(\cdot,\cdot)_{\Theta,q}$ from 2.4.1.–2.4.3. for spaces on R_n has an immediate counterpart for spaces on R_n^+ and on Ω, where again Ω is a bounded C^∞-domain in R_n: Definition 2.4.1 works for arbitrary quasi-Banach spaces, in particular for $B^s_{p,q}(\Omega)$ and $F^s_{p,q}(\Omega)$. As far as the complex interpolation method $(\cdot,\cdot)_\Theta$ from 2.4.4. is concerned, the situation is different. Its definition depends on the Fourier transform on R_n, and there is no immediate counterpart on Ω. However, on the basis of Theorem 3.3.4, one can reduce the problem of complex interpolation for the spaces $B^s_{p,q}(\Omega)$ and $F^s_{p,q}(\Omega)$ to the complex interpolation of corresponding spaces on R_n. The idea is the following. Let R and $S=S^N$ be the operators from Theorem 3.3.4, where later on we assume that N is sufficiently large, such that S (more precisely: the restriction of S to the corresponding spaces) is the same for the spaces considered. Then (3.3.4/2) holds, i.e. $RS=E$. Furthermore, $P=SR$ is a projection, i.e. $P^2=P$ in the spaces considered. In other words, if $0<p\leqq\infty$, $0<q\leqq\infty$ and $-\infty<s<\infty$, then P maps $B^s_{p,q}(R_n)$ onto a complemented subspace of $B^s_{p,q}(R_n)$, which will be denoted by $PB^s_{p,q}(R_n)$. Similarly $PF^s_{p,q}(R_n)$, if the hypotheses of Theorem 3.3.4(ii) are satisfied. It is not hard to see that S yields an isomorphic mapping

$$\text{from } B^s_{p,q}(\Omega) \text{ onto } PB^s_{p,q}(R_n),\quad -\infty<s<\infty,\quad 0<p\leqq\infty,\quad 0<q\leqq\infty \tag{1}$$

and

$$\text{from } F^s_{p,q}(\Omega) \text{ onto } PF^s_{p,q}(R_n),\quad -\infty<s<\infty,\quad 0<p<\infty,\quad 0<q<\infty, \text{(3.3.4/4)}. \tag{2}$$

The inverse mapping from $PB^s_{p,q}(R_n)$ and $PF^s_{p,q}(R_n)$ onto $B^s_{p,q}(\Omega)$ and $F^s_{p,q}(\Omega)$, respectively, is the restriction operator R (restricted to the corresponding subspaces). On the other hand, $PB^s_{p,q}(R_n)$ and $PF^s_{p,q}(R_n)$ are subspaces of $S'(R_n)$, and the two definitions in 2.4.4. make sense if one replaces $B^s_{p,q}(R_n)$ by $PB^s_{p,q}(R_n)$ and $F^s_{p,q}(R_n)$ by $PF^s_{p,q}(R_n)$, provided that (2) is satisfied. R always has the meaning of 3.3.4. If $B(R_n)$ is a quasi-normed linear subspace of $S'(R_n)$, then $B(\Omega)=RB(R_n)$ is the restriction of $B(R_n)$ to Ω and

$$\|f \mid B(\Omega)\| = \inf \|g \mid B(R_n)\|, \tag{3}$$

where the infimum is taken over all $g\in B(R_n)$ with $g \mid \Omega=f$.

Definition. Let $-\infty<s_0<\infty$, $-\infty<s_1<\infty$ and $0<\Theta<1$.

(i) Let $0<p_0\leqq\infty$, $0<p_1\leqq\infty$, $0<q_0\leqq\infty$ and $0<q_1\leqq\infty$. Then

$$(B^{s_0}_{p_0,q_0}(\Omega), B^{s_1}_{p_1,q_1}(\Omega))_\Theta = R[(PB^{s_0}_{p_0,q_0}(R_n), PB^{s_1}_{p_1,q_1}(R_n))_\Theta]. \tag{4}$$

(ii) Let $0<p_0<\infty$, $0<p_1<\infty$, $0<q_0<\infty$, $0<q_1<\infty$,

$$\min(p_0, q_0) \geqq \min(p_0, 1) \quad\text{and}\quad \min(p_1, q_1) \geqq \min(p_1, 1). \tag{5}$$

Then

$$(F^{s_0}_{p_0,q_0}(\Omega), F^{s_1}_{p_1,q_1}(\Omega))_\Theta = R[(PF^{s_0}_{p_0,q_0}(R_n), PF^{s_1}_{p_1,q_1}(R_n))_\Theta]. \tag{6}$$

Remark 1. We recall that N, in $S=S^N$ is sufficiently large, such that (3.3.4/3) with s_0 and s_1 instead of s, respectively, holds. In this sense, the projection P is the same for all spaces in question. S has the interpolation property with respect to the complex interpolation method, cf. step 4 of the proof of Theorem 2.9.2 and step 2 of the proof of Theorem 2.9.4 as far as the extension from R_n^+ to R_n is concerned. This property is preserved if one replaces R_n^+ by Ω. Then P also has the interpolation property.

Theorem. (i) Let $-\infty<s_0<\infty$, $-\infty<s_1<\infty$, $0<p_0\leqq\infty$, $0<p_1\leqq\infty$, $0<q_0\leqq\infty$, $0<q_1\leqq\infty$, and $0<\Theta<1$. Let

$$s=(1-\Theta)s_0+\Theta s_1\,,\quad \frac{1}{p}=\frac{1-\Theta}{p_0}+\frac{\Theta}{p_1}\,,\quad \frac{1}{q}=\frac{1-\Theta}{q_0}+\frac{\Theta}{q_1}\,. \tag{7}$$

Then

$$(B^{s_0}_{p_0,q_0}(\Omega),\ B^{s_1}_{p_1,q_1}(\Omega))_\Theta=B^s_{p,q}(\Omega)\,.$$

(ii) Let $-\infty<s_0<\infty$, $-\infty<s_1<\infty$, $0<p_0<\infty$, $0<p_1<\infty$, $0<q_0<\infty$, $0<q_1<\infty$ and $0<\Theta<1$. Let (5) and (7) be satisfied. Then

$$(F^{s_0}_{p_0,q_0}(\Omega),\ F^{s_1}_{p_1,q_1}(\Omega))_\Theta=F^s_{p,q}(\Omega)\,. \tag{8}$$

(iii) Let $-\infty<s_0<\infty$ and $-\infty<s_1<\infty$ with $s_0\neq s_1$. Let $0<q_0\leqq\infty$, $0<q_1\leqq\infty$, $0<q\leqq\infty$. Let $0<\Theta<1$ and $s=(1-\Theta)s_0+\Theta s_1$. If $0<p\leqq\infty$, then

$$(B^{s_0}_{p,q_0}(\Omega),\ B^{s_1}_{p,q_1}(\Omega))_{\Theta,q}=B^s_{p,q}(\Omega) \tag{9}$$

and if $0<p<\infty$ then

$$(F^{s_0}_{p,q_0}(\Omega),\ F^{s_1}_{p,q_1}(\Omega))_{\Theta,q}=B^s_{p,q}(\Omega)\,. \tag{10}$$

Proof. Step 1. We prove (ii), the proof of (i) being the same. We begin with a remark. If

$$f(z)\in F(F^{s_0}_{p_0,q_0}(R_n),\ F^{s_1}_{p_1,q_1}(R_n))$$

in the sense of Definition 2.4.4/1, then it follows from Theorem 2.4.7 that $f(z)\in$ $\in F^\sigma_{\alpha,\beta}(R_n)$, where $0<\operatorname{Re} z<1$, σ is a number between s_0 and s_1, α is a number between p_0 and p_1, β is a number between q_0 and q_1. In particular $Pf(z)$ makes sense, where P is interpreted as in the above definition. Because P has the interpolation property, we have

$$Pf(z)\in F(PF^{s_0}_{p_0,q_0}(R_n),\ PF^{s_1}_{p_1,q_1}(R_n))\,. \tag{11}$$

Let $f\in(PF^{s_0}_{p_0,q_0}(R_n),\ PF^{s_1}_{p_1,q_1}(R_n))_\Theta$ and

$$f(z)\in F(PF^{s_0}_{p_0,q_0}(R_n),\ PF^{s_1}_{p_1,q_1}(R_n))\quad\text{with}\quad f(\Theta)=f\,. \tag{12}$$

Since $f(z)\in S'(R_n)$, with $0<\operatorname{Re} z<1$, is determined by the values of $f(it)$ and $f(1+it)$ with $t\in R_1$, it follows from (11) and (12) that $Pf(z)=f(z)$, and in particular $f=Pf$. By (2.4.7/3) we have

$$\|f\mid F^s_{p,q}(R_n)\|\leqq c\|f\mid(PF^{s_0}_{p_0,q_0}(R_n),\ PF^{s_1}_{p_1,q_1}(R_n))_\Theta\|\,. \tag{13}$$

On the other hand, if $Pf=f$ and

$$f(z)\in F(F^{s_0}_{p_0,q_0}(R_n),\ F^{s_1}_{p_1,q_1}(R_n))\quad\text{with}\quad f(\Theta)=f\,, \tag{14}$$

then it follows from $Pf=f$ and (11) that

$$\|f\mid(PF^{s_0}_{p_0,q_0}(R_n),\ PF^{s_1}_{p_1,q_1}(R_n))_\Theta\|\leqq\inf\|Pf(z)\mid F(F^{s_0}_{p_0,q_0}(R_n),\ F^{s_1}_{p_1,q_1}(R_n))\|\,,$$

where the infimum is taken over all $f(z)$ with (14). Since P has the interpolation property we obtain

$$\|f\mid(PF^{s_0}_{p_0,q_0}(R_n),\ PF^{s_1}_{p_1,q_1}(R_n))_\Theta\|\leqq c\|f\mid F^s_{p,q}(R_n)\|\,. \tag{15}$$

(13), (15), (6) and the definition of $F^s_{p,q}(\Omega)$ prove (8). As we remarked above, the proof of (i) is the same.

Step 2. We prove (9). By Theorem 3.3.4 (and the above remarks) $B^s_{p,q}(\Omega)$ is isomorphic to $PB^s_{p,q}(R_n)$ for all values of s, p, and q. Hence it follows from the interpolation property of the real interpolation method that

$$(B^{s_0}_{p_0,q_0}(\Omega), B^{s_1}_{p_1,q_1}(\Omega))_{\Theta,q} = R(PB^{s_0}_{p_0,q_0}(R_n), PB^{s_1}_{p_1,q_1}(R_n))_{\Theta,q}\,.$$

Now we have the same situation as in (4), so (9) is a consequence of (2.4.7/2). Finally, (10) follows from (9) and (3.2.4/3).

Remark 2. At first glance, part (ii) looks somewhat curious, because the restrictions (5) for the spaces $F^{s_0}_{p_0,q_0}(\Omega)$ and $F^{s_1}_{p_1,q_1}(\Omega)$ are not necessarily satisfied for $F^s_{p,q}(\Omega)$, cf. Fig. 3.3.6, where

$$\frac{1}{N} < p < \infty, \quad \min(p, q) \geqq \min(p, 1) \tag{16}$$

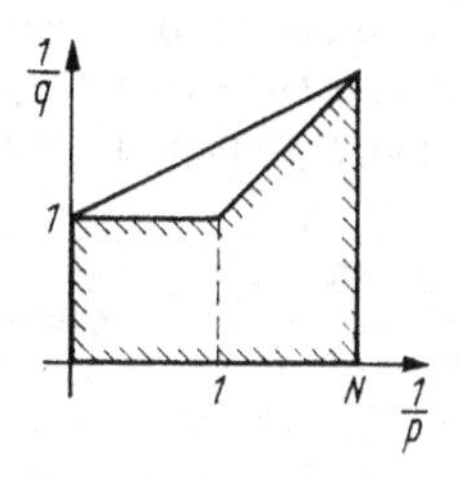

Fig. 3.3.6

is characterized by the shaded area. By interpolation, one obtains the convex hull of that area, which includes couples (p, q) for which (16) is not satisfied. This shows once more that the restriction (3.3.4/4) is not natural. One can weaken this restriction somewhat, cf. 2.9.5. In particular it follows that all spaces $F^s_{p,q}(\Omega)$ which can be obtained via (8) are isomorphic to $PF^s_{p,q}(R_n)$.

Again let Ω be a bounded C^∞-domain in R_n with boundary $\partial\Omega$. Our next goal is the interpolation of spaces on $\partial\Omega$. Again, the real interpolation $(\cdot,\cdot)_{\Theta,q}$ of spaces of type $B^s_{p,q}(\partial\Omega)$ or $F^s_{p,q}(\partial\Omega)$ is clear via Definition 2.4.1, but this is not so for the complex method $(\cdot,\cdot)_\Theta$. However, the situation is not as complicated as for spaces on the domain Ω. We use Definition 3.2.2/2. Let $-\infty < s_0 < \infty$, $-\infty < s_1 < \infty$, $0 < p_0 \leqq \infty$, $0 < p_1 \leqq \infty$, $0 < q_0 \leqq \infty$ and $0 < q_1 \leqq \infty$. If $0 < \Theta < 1$, then

$$\begin{aligned}&(B^{s_0}_{p_0,q_0}(\partial\Omega), B^{s_1}_{p_1,q_1}(\partial\Omega))_\Theta \\ &= \{f \mid f \in D'(\Omega),\ (\psi_j f)(f^{(j)-1}(y', 0)) \in (B^{s_0}_{p_0,q_0}(R_{n-1}), B^{s_1}_{p_1,q_1}(R_{n-1}))_\Theta\}\end{aligned} \tag{17}$$

and the quasi-norm in this interpolation space is given by the obvious counterpart of (3.2.2/6). Similarly one defines $(F^{s_0}_{p_0,q_0}(\partial\Omega), F^{s_1}_{p_1,q_1}(\partial\Omega))_\Theta$, where the parameters have the above meaning with $p_0 < \infty$ and $p_1 < \infty$ (we do not need (5)).

Proposition. (i). Part (i) and part (iii) of the above theorem remain valid if Ω is replaced by $\partial\Omega$.
(ii) Let $-\infty < s_0 < \infty$, $-\infty < s_1 < \infty$, $0 < q_0 \leqq \infty$, $0 < q_1 \leqq \infty$, $0 < p_0 < \infty$, and $0 < p_1 < \infty$. Let $0 < \Theta < 1$ and let s, p, and q be given by (7). Then

$$(F^{s_0}_{p_0,q_0}(\partial\Omega), F^{s_1}_{p_1,q_1}(\partial\Omega))_\Theta = F^s_{p,q}(\partial\Omega)\,. \tag{18}$$

Proof. The proposition is an easy consequence of the above definitions and of Theorem 2.4.7 and Theorem 2.4.2.

Remark 3. There is not difficulty in extending Theorem 2.4.3 to R_n^+, Ω and $\partial\Omega$.

Remark 4. The real interpolation method always has the interpolation property, cf. Proposition 2.4.1. As far as the above complex method is concerned, the situation is more complicated and one must check from time to time whether the operator considered has the interpolation property.

3.4. Further Properties

3.4.1. A Special Multiplication Property

One of the main aims of this book is the application of the theory of function spaces of the above type to boundary value problems for regular elliptic differential operators in bounded C^∞-domains Ω in R_n. In the preceding sections we proved a lot of fundamental properties of the spaces $B^s_{p,q}(\Omega)$ and $F^s_{p,q}(\Omega)$ which are necessary for that purpose. This section deals with a few more special properties. On the one hand they are helpful for our further studies, while on the other they are of self-contained interest. We start with a dilatation property and a special multiplication property for the spaces $B^s_{p,q}(R_n)$ and $F^s_{p,q}(R_n)$.

Proposition 1. (i) Let $0<p\leqq\infty$, $\infty>s>\max\left(0, n\left(\frac{1}{p}-1\right)\right)$ and $0<q\leqq\infty$. Then there exists a positive number c such that

$$\|f(\lambda\,\cdot)\mid B^s_{p,q}(R_n)\|\leqq c\lambda^{s-\frac{n}{p}}\|f\mid B^s_{p,q}(R_n)\| \tag{1}$$

holds for all λ with $1\leqq\lambda<\infty$ and all $f\in B^s_{p,q}(R_n)$.

(ii) Let $0<p<\infty$, $\infty>s>\max\left(0, n\left(\frac{1}{p}-1\right)\right)$ and $0<q\leqq\infty$. Then there exists a positive number c such that

$$\|f(\lambda\,\cdot)\mid F^s_{p,q}(R_n)\|\leqq c\lambda^{s-\frac{n}{p}}\|f\mid F^s_{p,q}(R_n)\| \tag{2}$$

holds for all λ with $1\leqq\lambda<\infty$ and all $f\in F^s_{p,q}(R_n)$.

Proof. Of course, $f(\lambda\,\cdot)$ must be interpreted in the sense of distributions. On the other hand, by Theorem 2.7.1, it follows that $f(x)$ is a regular distribution, and $f(\lambda x)$ makes also sense as a locally integrable function. We prove (2). The proof of (1) is the same. Let $\lambda=2^l$ with $l=1, 2, 3, \ldots$ and $\{\varphi_k(x)\}_{k=0}^\infty\in\Phi(R_n)$, cf. Definition 2.3.1. We may assume that $\varphi_k(x)=\varphi_1(2^{-k+1}x)$ if $k=1, 2, 3, \ldots$ Then it follows that

$$\begin{aligned}[F^{-1}\varphi_k Ff(2^l\,\cdot\,)](x)&=[F^{-1}\varphi_k(2^{-l}2^l\,\cdot\,)\,2^{-ln}(Ff)(2^{-l}\,\cdot\,)](x)\\&=[F^{-1}\varphi_1(2^{-k+1+l}\,\cdot\,)Ff](2^lx)\end{aligned} \tag{3}$$

if $k=1, 2, \ldots$ Similarly if $k=0$. Consequently,

$$\begin{aligned}&\left\|\left(\sum_{k=l+1}^\infty|[2^{sk}F^{-1}\varphi_k Ff(\lambda\,\cdot)]\,(\cdot)|^q\right)^{\frac{1}{q}}\mid L_p(R_n)\right\|\\&\leqq c\lambda^s\left\|\left(\sum_{j=1}^\infty|(2^{js}F^{-1}\varphi_j Ff)(2^l\cdot)|^q\right)^{\frac{1}{q}}\mid L_p(R_n)\right\|\leqq c'\lambda^{s-\frac{n}{p}}\|f\mid F^s_{p,q}(R_n)\|,\end{aligned} \tag{4}$$

cf. (2.3.1/6). For the remaining terms with $k=0, 1, \ldots, l$ we use Theorem 1.5.2 and obtain essentially (modification if $k=0$ or $k=l$) that

$$\begin{aligned}&\|F^{-1}\varphi_k Ff(2^l\,\cdot)\mid L_p(R_n)\|\\&\leqq c2^{-l\frac{n}{p}}\|\varphi_1(2^{-k+1+l}\,\cdot)\mid H^\sigma_2(R_n)\|\,\|F^{-1}\varphi_0 Ff\mid L_p(R_n)\|,\end{aligned} \tag{5}$$

where σ is an arbitrary number with $\sigma > n\left(\frac{1}{\min(p,1)} - \frac{1}{2}\right)$. It is easy to see that

$$\|\varphi_1(2^{-k+1+l}\cdot) \mid H_2^\sigma(R_n)\| \leqq c 2^{(-k+l)\sigma} 2^{(k-l)\frac{n}{2}} = c 2^{(l-k)\left(\sigma - \frac{n}{2}\right)}. \tag{6}$$

We may assume that $s > \sigma - \frac{n}{2}$. Then (5) and (6) yield

$$\left\| \left(\sum_{k=0}^{l} |[2^{ks} F^{-1} \varphi_k F f(\lambda\cdot)](\cdot)|^q \right)^{\frac{1}{q}} \mid L_p(R_n) \right\| \tag{7}$$

$$\leqq c \lambda^{s-\frac{n}{p}} \|F^{-1}\varphi_0 F f \mid L_p(R_n)\| + c\lambda^{s-\frac{n}{p}} \|F^{-1}\varphi_1 F f \mid L_p(R_n)\|.$$

Now, (4) and (7) prove (2), provided $\lambda = 2^l$. After minimal technical changes, the proof can be extended to arbitrary numbers λ with $\lambda \geqq 1$.

Remark 1. (3) and (4) show that $s - \frac{n}{p}$ in (1) and (2) is the best possible number. $s - \frac{n}{p}$ is denoted as the "differential dimension"; it plays an important role in embedding theorems, cf. e.g. Theorem 2.7.1.

Proposition 2. Let $a(x)$ be an infinitely differentiable function in R_n such that all derivatives of $a(x)$ are bounded. Let K_τ be a ball in R_n centered at $y \in R_n$ and of radius τ, where $0 < \tau \leqq 1$. Let $\varphi(x) \in S(R_n)$ with $\operatorname{supp} \varphi \subset K_\tau$ and

$$|D^\alpha \varphi(x)| \leqq \frac{c_\alpha}{\tau} \quad \text{for all multi-indices } \alpha. \tag{8}$$

(i) Let $0 < p \leqq \infty$, $0 < q \leqq \infty$ and $n\left(\frac{1}{p} - 1\right) \leqq s \leqq \frac{n}{p}$. If $\varepsilon > 0$, then there exists a constant c such that

$$\|(a(\cdot) - a(y))\varphi f \mid B_{p,q}^s(R_n)\| \leqq c \tau^{1-\varepsilon} \|f \mid B_{p,q}^s(R_n)\| \tag{9}$$

holds for all $f \in B_{p,q}^s(R_n)$. The constant c is independent of K_τ (in particular of y and τ).

(ii) Let $0 < p < \infty$, $0 < q < \infty$, $\min(p, q) \geqq \min(p, 1)$ and $n\left(\frac{1}{p} - 1\right) \leqq s \leqq \frac{n}{p}$. If $\varepsilon > 0$ then there exists a constant c such that

$$\|(a(\cdot) - a(y))\varphi f \mid F_{p,q}^s(R_n)\| \leqq c \tau^{1-\varepsilon} \|f \mid F_{p,q}^s(R_n)\| \tag{10}$$

holds for all $f \in F_{p,q}^s(R_n)$. The constant c is independent of K_τ (in particular of y and τ).

Proof. Step 1. We prove part (ii). Let $0 < p < \infty$, $0 < q < \infty$, $\min(p, q) > \min(p, 1)$ and $\sigma > \frac{n}{p}$. Then it follows from Corollary 2.8.3 and Remark 2.8.3/2 that

$$\|(a(\cdot) - a(y))\varphi f \mid F_{p,q}^\sigma(R_n)\| \leqq c \|(a(\cdot) - a(y))\varphi \mid F_{p,q}^\sigma(R_n)\| \, \|f \mid F_{p,q}^\sigma(R_n)\|. \tag{11}$$

By Proposition 1, we have

$$\|(a(\cdot) - a(y))\varphi \mid F_{p,q}^\sigma(R_n)\| \leqq c \tau^{-\left(\sigma - \frac{n}{p}\right)} \|(a(\tau\cdot) - a(y))\,\varphi(\tau\cdot) \mid F_{p,q}^s(R_n)\|. \tag{12}$$

Obviously, $\operatorname{supp} \varphi(\tau\cdot) \subset K_1$, where K_1 is an appropriate ball of radius 1 $\left(\text{centered at } \frac{y}{\tau}\right)$. By Lemma 3.3.1,

$$\|(a(\tau\cdot) - a(y))\,\varphi(\tau\cdot) \mid F_{p,q}^s(R_n)\| \tag{13}$$

$$\leqq c \sum_{|\alpha| \leqq m} \sup_{x \in R_n} |D^\alpha [(a(\tau x) - a(y))\,\varphi(\tau x)]| \leqq c'\tau,$$

where c and c' are independent of K_τ. We put (13) in (12) and the result in (11). Then it follows that

$$\|(a(\cdot) - a(y))\,\varphi f \mid F_{p,q}^\sigma(R_n)\| \leqq c \tau^{1-\varepsilon} \|f \mid F_{p,q}^\sigma(R_n)\| \tag{14}$$

with $\varepsilon=\sigma-\frac{n}{p}$. Let $\min(p,q)>1$. We remark that $f\to(a(x)-a(y))\varphi(x)f$ is a self-adjoint operator and use (2.11.2/2). It follows that (14) holds if

$$1<p<\infty,\quad 1<q<\infty \quad\text{and}\quad \sigma=n\left(\frac{1}{p}-1\right)-\varepsilon. \tag{15}$$

In order to use the complex interpolation formula (2.4.7/3) we remark that the operator $f\to$ $\to(a(x)-a(y))\,\varphi(x)f$ has the interpolation property, cf. the proof of Corollary 2.8.2. Interpolation of (14) with (15) and $1<p<\infty$, $1<q<\infty$, $\sigma=\varepsilon+\frac{n}{p}$ yields the desired assertion, provided $1<p<\infty$, $1<q<\infty$. A second application of (2.4.7/3) proves (ii).

Step 2. The proof of (i) is essentially the same. We use Theorem 2.8.3, Theorem 2.11.2 (i), and Theorem 2.4.7 (i). In order to cover all values of q one applies finally the real interpolation formula (2.4.2/1) and the proof is complete.

Remark 2. The proof shows that the restriction $n\left(\frac{1}{p}-1\right)\leqq s\leqq\frac{n}{p}$ is natural, otherwise (9) and (10) do not hold for arbitrary $\varepsilon>0$. The problem arises whether $\varepsilon=0$ is admissible in (9) and (10) (under the above restrictions for s). An affirmative answer for the spaces $B^s_{p,q}(R_n)$ with $n\left(\frac{1}{p}-1\right)<s<\frac{n}{p}$ has been given in H. Triebel [7, formula (52)].

3.4.2. Inner Descriptions, Equivalent Quasi-Norms

Again let Ω be a bounded C^∞-domain in R_n. In 3.1.2., we discussed the problem of the inner description of the spaces $B^s_{p,q}(\Omega)$ and $F^s_{p,q}(\Omega)$, i.e. it would be desirable to find explicit equivalent quasi-norms in these spaces, where only the values of $f(x)$ with $x\in\Omega$ are involved. It is well-known that there are nice inner descriptions for the Sobolev spaces

$$W^m_p(\Omega)=F^m_{p,2}(\Omega) \quad\text{with}\quad 1<p<\infty \quad\text{and}\quad m=0,1,2,\ldots, \tag{1}$$

the Zygmund spaces

$$\mathcal{C}^s(\Omega)=B^s_{\infty,\infty}(\Omega) \quad\text{with}\quad s>0, \tag{2}$$

and the classical Besov spaces

$$\Lambda^s_{p,q}(\Omega)=B^s_{p,q}(\Omega) \quad\text{with}\quad s>0,\quad 1\leqq p<\infty,\quad\text{and}\quad 1\leqq q\leqq\infty. \tag{3}$$

Concerning the corresponding spaces with R_n instead of Ω we refer to Theorem 2.5.6 and (2.2.2/7), (2.5.12/22) and (2.2.2/6), (2.5.12/21) and (2.2.2/9), (2.2.2/10), respectively. These equivalent norms in spaces on R_n have their counterparts for the corresponding spaces on Ω:

(i) If $1<p<\infty$ and $m=0,1,2,\ldots$ then

$$\|f\mid W^m_p(\Omega)\|=\sum_{|\alpha|\leqq m}\left(\int_\Omega |(D^\alpha f)(x)|^p\,dx\right)^{\frac{1}{p}} \tag{4}$$

is an equivalent norm in $W^m_p(\Omega)$, cf. [I, 4.4.2. and 4.2.4.].

(ii) If $h\in R_n$ and $l=1,2,3,\ldots$ then we put

$$\Omega_{h,l}=\bigcap_{j=0}^{l}\{x\mid x+jh\in\Omega\}. \tag{5}$$

Let $1\leqq p\leqq\infty$, $1\leqq q\leqq\infty$ and $s>0$. Let k and l be integers such that $0\leqq k<s$ and $l>s-k$. Then

$$\|f\mid \Lambda^s_{p,q}(\Omega)\|=\left(\int_\Omega |f(x)|^p\,dx\right)^{\frac{1}{p}}$$

$$+\sum_{|\alpha|\leqq k}\left(\int_{R_n}|h|^{-(s-k)q}\left(\int_{\Omega_{h,l}}|(\Delta^l_h D^\alpha f)(x)|^p\,dx\right)^{\frac{q}{p}}\frac{dh}{|h|^n}\right)^{\frac{1}{q}} \tag{6}$$

is an equivalent norm in $\Lambda^s_{p,q}(\Omega)$ (obvious modifications if either $p=\infty$ and/or $q=\infty$). Furthermore, we extended definition (3) to $p=\infty$. In particular $\Lambda^s_{\infty,\infty}(\Omega)=\mathcal{C}^s(\Omega)$. A proof of this statement may be found in [I, 4.4.2 and 4.4.1], where the limiting cases $p=1$ and $p=\infty$ can be included. Δ^l_h has the meaning of (2.5.9/1).

Obviously, (5) and (6) are satisfactory inner descriptions of the spaces $W^m_p(\Omega)$, $\Lambda^s_{p,q}(\Omega)$ and $\mathcal{C}^s(\Omega)$. Under the above restrictions for s, p and q, we have given in [I, 4.4.2] many other equivalent norms in $W^m_p(\Omega)$ and $\Lambda^s_{p,q}(\Omega)$. For the classical Bessel-potential spaces

$$H^s_p(\Omega)=F^s_{p,2}(\Omega), \quad -\infty<s<\infty, \quad 1<p<\infty,$$

the situation is more complicated and there are no satisfactory inner descriptions (with the exception of $s=m=0, 1, 2, \ldots$, where $H^s_p(\Omega)=W^m_p(\Omega)$). The main problem is to find nice inner descriptions of the spaces $B^s_{p,q}(\Omega)$ and $F^s_{p,q}(\Omega)$. The question is whether Theorem 2.5.10 (cf. also (2.5.10/21)) or Theorem 2.5.11 (cf. also Corollary 2.5.11) or Theorem 2.5.12 (cf. also (2.5.12/20)) have counterparts with Ω instead of R_n and with modifications in the sense of (6). There are no results at this moment, with the exception of the classical assertions we described above and the partial statement we shall formulate below.

Proposition. Let $0<p<\infty$ and

$$\max\left(0, n\left(\frac{1}{p}-1\right)\right)<s<1 \tag{7}$$

(the shaded area in Fig. 3.4.2). Then

$$\left(\int_\Omega |f(x)|^p\,dx\right)^{\frac{1}{p}}+\left(\int_{\Omega\times\Omega}\frac{|f(x)-f(y)|^p}{|x-y|^{n+sp}}\,dxdy\right)^{\frac{1}{p}} \tag{8}$$

is an equivalent quasi-norm in $B^s_{p,p}(\Omega)$.

Remark 1. A proof may be found in H. Triebel [11, Proposition 4; 7, Theorem 2]. Obviously, if $1\leq p<\infty$, then (8) is a special case of (6). Of interest is the extension of a special case of (6) to some values $p<1$. A combination of the above proposition with Theorem 3.3.5 yields further equivalent quasi-norms.

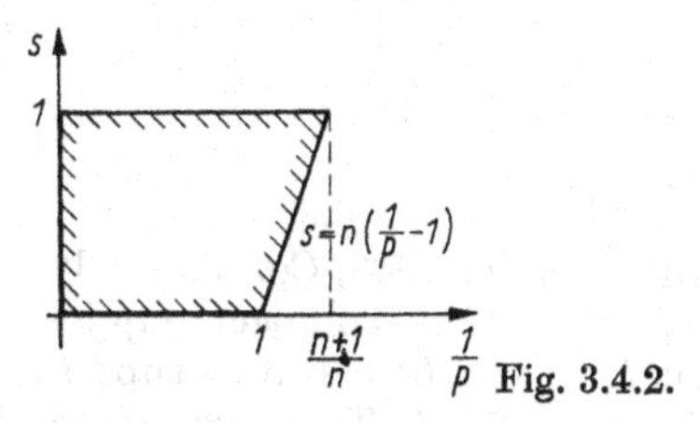

Fig. 3.4.2.

Remark 2. The above considerations show that the classical Sobolev spaces and Besov spaces on bounded smooth domains (and also on C^∞-manifolds) are special cases of our approach. If the underlying domain is not smooth, then the situation is different, cf. [I, 4.2.3] for further references and also D. E. Edmunds [1], and W. Mazja [1]. As far as Hardy spaces on (smooth) manifolds are concerned we refer also to R. S. Strichartz [2], and J. Peetre [10].

3.4.3. The Spaces $\mathring{B}^s_{p,q}(\Omega)$ and $\mathring{F}^s_{p,q}(\Omega)$

Let Ω be a bounded C^∞-domain. As usual, $D(\Omega)$ is the collection of all complex-valued infinitely differentiable functions $f(x)$ in R_n with $\operatorname{supp} f\subset\Omega$.

Definition. (i) Let $0<p\leq\infty$, $0<q\leq\infty$ and $-\infty<s<\infty$. Then $\mathring{B}^s_{p,q}(\Omega)$ is the completion of $D(\Omega)$ in $B^s_{p,q}(\Omega)$.

(ii) Let $0<p<\infty$, $0<q\leqq\infty$ and $-\infty<s<\infty$. Then $\mathring{F}^s_{p,q}(\Omega)$ is the completion of $D(\Omega)$ in $F^s_{p,q}(\Omega)$.

Remark 1. One can interpret $f(x)\in D(\Omega)$ either as a function on Ω or as a function on R_n (in the above way). In any case the above definition makes sense, cf. Definition 3.2.2/1 and Proposition 3.2.4.

Theorem. (i) If

$$\begin{cases} \text{either} & 1\leqq p<\infty, \quad 0<q<\infty, \quad \frac{1}{p}-1<s<\frac{1}{p} \\ \text{or} & 0<p<1, \quad 0<q<\infty, \quad n\left(\frac{1}{p}-1\right)<s<\frac{1}{p}, \end{cases} \tag{1}$$

then $B^s_{p,q}(\Omega)=\mathring{B}^s_{p,q}(\Omega)$.

(ii) If

$$\begin{cases} \text{either} & 1\leqq p<\infty, \quad 0<q<\infty, \quad \frac{1}{p}-1+n\left(\frac{1}{\min(q,1)}-1\right)<s<\frac{1}{p} \\ \text{or} & 0<p<1, \quad 0<q<\infty, \quad n\left(\frac{1}{\min(p,q)}-1\right)<s<\frac{1}{p}, \end{cases} \tag{2}$$

then $F^s_{p,q}(\Omega)=\mathring{F}^s_{p,q}(\Omega)$.

Proof. We prove (ii) (the proof of (i) being the same). By Proposition 3.2.3 it is sufficient to prove the following assertion: if $f\in F^s_{p,q}(R^+_n)$ with $\operatorname{supp} f\subset\{y \mid y\in R_n, |y|\leqq 1, y_n\geqq 0\}$, then f can be approximated in $F^s_{p,q}(R^+_n)$ by infinitely differentiable functions in R_n with compact support in R^+_n. First we extend f by zero to R_n and denote the extended function also by f. The characteristic function of R^+_n is a multiplier for $F^s_{p,q}(R_n)$, cf. Theorem 2.8.7. Hence it follows that

$$\|f \mid F^s_{p,q}(R^+_n)\|\leqq\|f \mid F^s_{p,q}(R_n)\| . \tag{3}$$

Next we remark that

$$f(x)\to f(x', x_n-h) \quad \text{with} \quad h>0$$

is an isomorphic operator in $F^s_{p,q}(R_n)$, and we have

$$\|f(\cdot)-f(\cdot, \cdot-h) \mid F^s_{p,q}(R_n)\|\to 0 \quad \text{if} \quad h\downarrow 0 . \tag{4}$$

This follows from the density assertion of Theorem 2.3.3. Consequently, we may assume, without restriction of generality, that f has compact support in R^+_n. Let $\varphi(x)\in S(R_n)$ with compact support in R^+_n and $\varphi(x)=1$ if $x\in\operatorname{supp} f$. Let $g_j\to f$ in $F^s_{p,q}(R_n)$ with $g_j\in S(R_n)$, cf. again Theorem 2.3.3. Then $\varphi g_j\to f$ in $F^s_{p,q}(R_n)$ is the desired approximation. The proof is complete.

Corollary. Let $m=1, 2, 3, \ldots$ and let ν be the outward normal of $\partial\Omega$.

(i) Under the hypotheses for s, p, and q of part (i) of the above theorem we have

$$\mathring{B}^{s+m}_{p,q}(\Omega)=\left\{f \,\middle|\, f\in B^{s+m}_{p,q}(\Omega), \left.\frac{\partial^j f}{\partial\nu^j}\right|\partial\Omega=0 \quad \text{if} \quad j=0,\ldots,m-1\right\}. \tag{5}$$

(ii) Under the hypotheses for s, p, and q of part (ii) of the above theorem we have

$$\mathring{F}^{s+m}_{p,q}(\Omega)=\left\{f \,\middle|\, f\in F^{s+m}_{p,q}(\Omega), \left.\frac{\partial^j f}{\partial\nu^j}\right|\partial\Omega=0 \quad \text{if} \quad j=0,\ldots,m-1\right\}. \tag{6}$$

Proof. In any case, the right-hand sides of (5) and (6) make sense, cf. Theorem 3.3.3. We may again assume that $f\in F^{s+m}_{p,q}(R^+_n)$ with $\operatorname{supp} f\subset\{y \mid y\in R_n, |y|\leqq 1,$

$y_n \geqq 0\}$ and $\frac{\partial^j f}{\partial x_n^j}(x', 0)=0$ if $j=0, \ldots, m-1$. We extend f by zero to R_n and denote the extended function also by f. Then $D^\alpha f$ in R_n is the extension by zero of $D^\alpha f$ in R_n^+, provided that $|\alpha| \leqq m$ (here we use that $\frac{\partial^j f}{\partial x_n^j}(x', 0)=0$ if $j=0, \ldots, m-1$). In particular, $D^\alpha f \in F^s_{p,q}(R_n)$ if $|\alpha| \leqq m$. Hence we have $f \in F^{s+m}_{p,q}(R_n)$ and

$$\|f \mid F^{s+m}_{p,q}(R_n^+)\| \leqq c' \|f \mid F^{s+m}_{p,q}(R_n)\| . \tag{7}$$

The rest is the same as in the proof of the theorem.

Remark 2. Essentially we proved that under the hypotheses of the corollary, the extension by zero from Ω to R_n, is a bounded linear operator

$$\text{from } \mathring{B}^{s+m}_{p,q}(\Omega) \text{ into } B^{s+m}_{p,q}(R_n)$$

and

$$\text{from } \mathring{F}^{s+m}_{p,q}(\Omega) \text{ into } F^{s+m}_{p,q}(R_n),$$

respectively. We have the same assertion with R_n^+ instead of Ω, including (5) with R_n^+ instead of Ω. This follows from the above proof.

Remark 3. The proof was based on Theorem 2.8.7 which for the spaces $B^s_{p,q}(R_n)$ covers also the limiting cases where $p=\infty$ and/or $q=\infty$. For density reasons, part (i) of the theorem and (5) cannot be valid for these limiting cases. Let $b^s_{p,q}(R_n)$ be the completion of $S(R_n)$ in $B^s_{p,q}(R_n)$ (we assume that $\max(p, q)=\infty$) and let $b^s_{p,q}(R_n^+)$ and $b^s_{p,q}(\Omega)$ be the corresponding restrictions to R_n^+ and Ω, respectively. Then it follows from the above considerations that parts (i) of the theorem and of the corollary remain valid if one replaces $B^s_{p,q}(\Omega)$ by $b^s_{p,q}(\Omega)$ and $B^{s+m}_{p,q}(\Omega)$ by $b^{s+m}_{p,q}(\Omega)$ (under the restrictions (1), where now $p=\infty$ and/or $q=\infty$ are included). Similarly, Remark 2 can be extended to these new spaces, including the possibility of replacing Ω by R_n^+.

4. Regular Elliptic Differential Equations

4.1. Definitions and Preliminaries

4.1.1. Introduction

Let Ω be a bounded C^∞-domain in R_n with boundary $\partial\Omega$. Let A,

$$(Au)(x)=\sum_{|\alpha|\leqq 2m} a_\alpha(x)\, D^\alpha u(x), \quad x\in\bar{\Omega}, \tag{1}$$

be a properly elliptic differential operator in $\bar{\Omega}$, and let $B_1, \ldots, B_m$,

$$(B_j u)\,(y)=\sum_{|\alpha|\leqq m_j} b_{j,\alpha}(y)\, D^\alpha u(y), \quad y\in\partial\Omega, \tag{2}$$

be m boundary operators such that $\{A;\, B_1, \ldots, B_m\}$ is regular elliptic (cf. the definition in 4.1.2.). The corresponding boundary value problem reads as follows,

$$(Au)(x)=f(x) \quad \text{if} \quad x\in\bar{\Omega} \quad \text{and} \quad (B_j u)(y)=g_j(y) \quad \text{if} \quad y\in\partial\Omega \tag{3}$$

and $j=1, \ldots, m$. This is the well-known boundary value problem for regular elliptic differential operators. Problems of this type have been studied for a long time, in particular in (Hölder)-Zygmund spaces $\mathcal{C}^s(\Omega)$ with $s>0$ and in $L_p(\Omega)$ with $1<p<\infty$. The L_p-theory includes also a corresponding theory in the Sobolev spaces $W_p^m(\Omega)$ with $1<p<\infty$ and $m=1, 2, \ldots$ Via the real and the classical complex interpolation method one can extend this theory to the Besov spaces $\Lambda^s_{p,q}(\Omega)$ with $s>0$, $1<p<\infty$, $1\leqq q\leqq\infty$ and to the Bessel-potential spaces $H^s_p(\Omega)$ with $s>0$ and $1<p<\infty$ (also to a few spaces of this type with $s\leqq 0$). A thorough description of this part of the theory has been given in [I, Chapter 5]. In [I] we have also given many references for the basic definitions and the main results. Furthermore, we added a few historical comments, cf. [I, Remark 5.2.1/5, Remark 5.3.4, 5.4.6, Remark 5.5.2/1]. Further extensive references about the classical theory may be found in J.-L. Lions, E. Magenes [2, I], Ju. M. Berezanskij [1], and C. Miranda [1]. A reader who is interested in historical comments and in additional remarks, which in particular shed more light on the basic definitions in the following subsection, is asked to consult the cited books. We only wish the mention the milestone paper by S. Agmon, A. Douglis, L. Nirenberg [1] (which somewhat marks the beginning of the up-to-date part of this theory) and the paper by L. Arkeryd [1], which was the basis of the considerations in [I, Chapter 5]. The main aim of Chapter 4 of this book is the extension of this theory to the spaces $B^s_{p,q}(\Omega)$ and $F^s_{p,q}(\Omega)$. Of particular interest are spaces with $p\leqq 1$. It was the main goal of the previous chapters to develop the theory of the spaces $B^s_{p,q}$ and $F^s_{p,q}$ in R_n and in Ω to such an extent that we have a sound basis for carrying over the methods from [I, Chapter 5] to the spaces $B^s_{p,q}(\Omega)$ and $F^s_{p,q}(\Omega)$. In this sense, Chapter 4 of this book may be considered as the continuation of [I, Chapter 5]. Although we follow the same lines as in [I] we shall try to make this chapter self-contained: we shall give detailed proofs of all statements pertaining to the spaces $B^s_{p,q}$ and $F^s_{p,q}$, in particular of the a priori estimates in Section 4.2. (although the arguments are sometimes simply a copy of corresponding considerations in [I] in the framework of an L_p-theory). In this sense, our treatment is an extension of the studies in H. Triebel [10, 14]. On the other hand, we shell need a few general

H. Triebel, *Theory of Function Spaces*, Modern Birkhäuser Classics,
DOI 10.1007/978-3-0346-0416-1_4, © Birkhäuser Verlag 1983

assertions about tempered distributions, analytic functions, algebraic polynomials etc., which have nothing to do with the underlying spaces. We formulate statements of such a type (cf. 4.1.3. and 4.1.4.) and refer for the proofs to [I].

4.1.2. Definitions

In this chapter, Ω always denotes a bounded C^∞-domain, cf. Definition 3.2.1. Furthermore, we recall that

$$R_n^+ = \{x \mid x = (x_1, \dots, x_n) \in R_n, x_n > 0\}.$$

As usual, $\bar{\Omega}$ and $\overline{R_n^+}$ is the closure of Ω and R_n^+, respectively. Finally, $C^\infty(\bar{\Omega})$ is the collection of all complex-valued infinitely differentiable functions $g(x)$ in Ω such that every derivative $D^\alpha g(x)$ can be extended continuously to $\bar{\Omega}$. Similarly, $C^\infty(\partial\Omega)$.

Definition 1. (i) The differential operator A,

$$Au = \sum_{|\alpha| \leqq 2m} a_\alpha(x) D^\alpha u, \quad a_\alpha(x) \in C^\infty(\bar{\Omega}) \quad \text{if} \quad |\alpha| \leqq 2m \tag{1}$$

is said to be properly elliptic in Ω if

$$a(x, \xi) = \sum_{|\alpha| \leqq 2m} a_\alpha(x)\, \xi^\alpha \neq 0 \quad \text{for every } 0 \neq \xi \in R_n \quad \text{and every} \quad x \in \bar{\Omega} \tag{2}$$

and if for every couple $\xi \in R_n$ and $\eta \in R_n$ of linearly independent vectors and for every $x \in \bar{\Omega}$ the polynomial $a(x, \xi + \tau\eta)$ in the complex variable τ has exactly m roots $\tau_k^+ = \tau_k^+(x; \xi, \eta)$ with positive imaginary part, $k = 1, \dots, m$.
(ii) Let Ω in (i) be replaced by R_n^+ and let $a_\alpha(x) = a_\alpha$ be complex constants. Then (i) defines a properly elliptic differential operator A in R_n^+ (with constant coefficients).

Remark 1. We recall that $\xi^\alpha = \prod_{j=1}^n \xi_j^{\alpha_j}$ if $\xi = (\xi_1, \dots, \xi_n) \in R_n$ and $\alpha = (\alpha_1, \dots, \alpha_n)$. Furthermore, the roots $\tau_1^+, \dots, \tau_m^+$ are enumerated with respect to their multiplicities. In particular, $a(x, \xi + \tau\eta)$ has also exactly m roots $\tau_k^- = \tau_k^-(x; \xi, \eta)$ with negative imaginary part, $k = 1, \dots, m$ (by (2) there do not exist real roots of $a(x, \xi + \tau\eta)$). Let

$$a^+(x; \xi, \eta, \tau) = \prod_{k=1}^m (\tau - \tau_k^+), \quad a^-(x; \xi, \eta, \tau) = \prod_{k=1}^m (\tau - \tau_k^-). \tag{3}$$

Remark 2. A discussion of this definition, in particular of the root condition, has been given in [I, 5.2.1.]. If $n \geqq 3$, the root condition is a consequence of (2).

Definition 2. (i) The system $\{B_j\}_{j=1}^k$ of the k differential operators

$$B_j u = \sum_{|\alpha| \leqq m_j} b_{j,\alpha}(x)\, D^\alpha u, \quad b_{j,\alpha}(x) \in C^\infty(\partial\Omega), \tag{4}$$

is said to be normal on $\partial\Omega$ if

$$0 \leqq m_1 < m_2 < \dots < m_k \tag{5}$$

and if

$$\sum_{|\alpha| = m_j} b_{j,\alpha}(x) \nu_x^\alpha \neq 0 \quad \text{for} \quad j = 1, \dots, k \tag{6}$$

and for every normal vector $\nu_x \neq 0$ on $\partial\Omega$, where $x \in \partial\Omega$.
(ii) Let Ω in (i) be replaced by R_n^+ (and $\partial\Omega$ by R_{n-1}) and let $b_{j,\alpha}(x) = b_{j,\alpha}$ be complex constants. Then (i) defines a normal system $\{B_j\}_{j=1}^k$ on R_{n-1}.

Remark 3. In part (ii) of the last definition we take $\nu_x=(0,\ldots,0,-1)$ (outward normal). The differential operators in (4) make sense if $u(x)$ is defined in $\overline{\Omega}$ or in $\overline{R_n^+}$, respectively (and if $u(x)$ has the necessary differentiability properties).

Definition 3. (i) Let A be the properly elliptic operator in Ω from Definition 1(i) and let $\{B_j\}_{j=1}^m$ be the normal system on $\partial\Omega$ from Definition 2(i) (with $k=m$, where m has the same meaning as in (1)). Then $\{B_j\}_{j=1}^m$ is said to be a complemented system on $\partial\Omega$ with respect to A if for every $x\in\partial\Omega$, every normal vector $\nu_x\neq 0$ and every tangential vector $\xi_x\neq 0$ in the point x with respect to $\partial\Omega$, the polynomials

$$b_j(x;\xi_x+\tau\nu_x)=\sum_{|\alpha|=m_j} b_{j,\alpha}(x)\,(\xi_x+\tau\nu_x)^\alpha \tag{7}$$

in the complex variable τ, are linearly independent modulo $a^+(x;\xi_x,\nu_x,\tau)$.
(ii) Let Ω in (i) be replaced by R_n^+ (with $\partial\Omega=R_{n-1}$) and let the above references to the parts (i) of the Definitions 1 and 2 be replaced by references to the corresponding parts (ii). Then (i) defines a complemented system $\{B_j\}_{j=1}^m$ on R_{n-1} with respect to A.

Remark 4. We recall that $a^+(x;\xi_x,\nu_x,\tau)$ has the meaning of (3). Furthermore, the polynomials in (7) are linearly independent modulo $a^+(x;\xi_x,\nu_x,\tau)$ (where $x\in\partial\Omega$, ξ_x and ν_x are fixed) if the identity

$$\sum_{j=1}^m c_j b_j(x;\xi_x+\tau\nu_x)=P(\tau)\,a^+(x;\xi_x,\nu_x,\tau) \tag{8}$$

in τ with the complex numbers $c_1,\ldots,c_m$ and the polynomial $P(\tau)$ has only the trivial solution $c_1=\ldots=c_m=0$ and $P(\tau)\equiv 0$. Finally we remark that one can replace $a^+(x;\xi_x,\nu_x,\tau)$ in the last definition by $a^-(x;\xi_x,\nu_x,\tau)$, cf. [I, Remark 5.2.1/3].

Definition 4. (i) $\{A;B_1,\ldots,B_m\}$ is said to be a regular elliptic system in Ω if
(a) A is properly elliptic in Ω in the sense of Definition 1(i),
(b) $\{B_j\}_{j=1}^m$ is a normal system in $\partial\Omega$ in the sense of Definition 2(i) with $k=m$ and $m_m\leq 2m-1$.
(c) $\{B_j\}_{j=1}^m$ is a complemented system on $\partial\Omega$ with respect to A in the sense of Definition 3(i).
(ii) Let Ω in (i) be replaced by R_n^+ (with $\partial\Omega=R_{n-1}$) and let the above references to the parts (i) of the Definitions 1–3 be replaced by references to the corresponding parts (ii). Then (i) defines a regular elliptic system $\{A;B_1,\ldots,B_m\}$ in R_n^+.

Example. We describe an interesting and important example. Let $\{\mu_x\}_{x\in\partial\Omega}$ be a non-tangential vector field on $\partial\Omega$ such that the components of the vectors μ_x belong to $C^\infty(\partial\Omega)$. Let k be a fixed integer with $k=0,\ldots,m$. If A is properly elliptic in Ω in the sense of Definition 1(i), then $\{A;B_1,\ldots,B_m\}$ with

$$B_j u=\frac{\partial^{k+j-1}u}{\partial\mu_x^{k+j-1}}+\sum_{|\alpha|<k+j-1} b_{j,\alpha}(x)\,D^\alpha u\,,\quad b_{j,\alpha}(x)\in C^\infty(\partial\Omega)\,, \tag{9}$$

is a regular elliptic system in Ω. Similarly if Ω is replaced by R_n^+, cf. [I, Remark 5.2.2/4]. Special cases of (9) yield Dirichlet's boundary value problem and Neumann's boundary value problem, respectively.

4.1.3. Basic Properties of Elliptic Operators

In Section 4.2. we reduce arbitrary regular elliptic systems in Ω to corresponding regular elliptic systems $\{A; B_1, \ldots, B_m\}$ in R_n^+ (which have, by definition, constant coefficients). Obviously, of particular interest are the top-order differential expressions of A and B_j, respectively. In this subsection we formulate some assertions for such an operator A, and in the following subsection for $\{A; B_1, \ldots, B_m\}$. Throughout this subsection, A,

$$Au = \sum_{|\alpha|=2m} a_\alpha D^\alpha u, \tag{1}$$

denotes a properly elliptic operator in R_n^+ in the sense of Definition 4.1.2/1(ii) having only top-order coefficients. The polynomials $a(\xi)=a(x,\xi)$, $a^+(\xi,\eta,\tau)=a^+(x;\xi,\eta,\tau)$ and $a^-(\xi,\eta,\tau)=a^-(x;\xi,\eta,\tau)$, and the roots τ_k^+ and τ_k^- have the same meaning as in (4.1.2/2) and (4.1.2/3), respectively. If $\xi=(\xi_1,\ldots,\xi_{n-1},0)=(\xi',0)$ with $\xi'\in R_{n-1}$ and if $\eta=(0,\ldots,0,1)$, then the corresponding roots of the polynomial $a(\xi+\tau\eta)$ in τ are denoted by $\tau_k^+(\xi')$ and $\tau_k^-(\xi')$, respectively.

Lemma 1. (i) There exist two positive numbers c_1 and c_2 such that

$$\begin{aligned} &|\tau_k^+(\xi')| \leqq c_1|\xi'|, \qquad |\tau_k^-(\xi')| \leqq c_1|\xi'|, \\ &|\operatorname{Im} \tau_k^+(\xi')| \geqq c_2|\xi'|, \quad |\operatorname{Im} \tau_k^-(\xi')| \geqq c_2|\xi'|. \end{aligned} \tag{2}$$

(ii) The coefficients $a_k^+(\xi')$ and $a_k^-(\xi')$ in

$$a^+(\xi',\tau) = \prod_{k=1}^m (\tau-\tau_k^+(\xi')) = \sum_{k=0}^m a_k^+(\xi')\,\tau^{m-k}, \tag{3}$$

$$a^-(\xi',\tau) = \prod_{k=1}^m (\tau-\tau_k^-(\xi')) = \sum_{k=0}^m a_k^-(\xi')\,\tau^{m-k} \tag{4}$$

are analytic functions of ξ' in $R_{n-1}-\{0\}$. Furthermore,

$$a_k^+(\lambda\xi') = \lambda^k a_k^+(\xi'),\ a_k^-(\lambda\xi') = \lambda^k a_k^-(\xi') \tag{5}$$

if $k=0,\ldots,m$ and $\lambda>0$. If $\xi'\neq 0$ and τ complex, then

$$a^-(-\xi',-\tau) = (-1)^m a^+(\xi',\tau). \tag{6}$$

Remark 1. A proof may be found in [I, 5.2.2].

Lemma 2. Let $\varphi(t)$ be an infinitely differentiable function on $[0,\infty)$ with

$$\varphi(t)=0 \quad \text{if} \quad 0\leqq t\leqq \frac{1}{2} \quad \text{and} \quad \varphi(t)=1 \quad \text{if} \quad 1\leqq t<\infty. \tag{7}$$

Let

$$g^+(\xi) = (1+|\xi|^2)^{-\frac{m}{2}} a^+(\xi',\xi_n), \quad g^-(\xi) = (1+|\xi|^2)^{-\frac{n}{2}} a^-(\xi',\xi_n)$$

with $\xi=(\xi',\xi_n)$. Then

$$\sup_{\xi\in R_n} (1+|\xi|^2)^{\frac{|\alpha|}{2}} \left(|D^\alpha[g^+(\xi)\,\varphi(\xi')]| + |D^\alpha[(g^+(\xi))^{-1}\varphi(\xi')]|\right) < \infty, \tag{8}$$

$$\sup_{\xi\in R_n} (1+|\xi|^2)^{\frac{|\alpha|}{2}} \left(|D^\alpha[g^-(\xi)\,\varphi(\xi')]| + |D^\alpha[(g^-(\xi))^{-1}\varphi(\xi')]|\right) < \infty \tag{9}$$

for every multi-index α.

Proof. (2) proves that

$$|a^+(\xi', \xi_n)| \geqq c\,(|\xi_n| + |\xi'|)^m \geqq c'\,(1+|\xi|^2)^{\frac{m}{2}}, \quad c'>0$$

if $|\xi'| \geqq \frac{1}{2}$. By (5) we have $a_k^+(\xi') = |\xi'|^k a_k^+\left(\frac{\xi'}{|\xi'|}\right)$, where a_k^+ has the needed differentiability properties. (8) is a consequence of these two formulas. Similarly one proves (9).

Remark 2. This is the heart of Lemma 2 in [I, 5.2.2.]. We prefered the above (apparently more complicated) formulation, because it is free of the underlying spaces (such as $L_p(R_n)$ etc.) The lemma shows that $g^+(\xi)\,\varphi(\xi')$ etc. are Fourier multipliers for $B^s_{p,q}(R_n)$ and $F^s_{p,q}(R_n)$, cf. Theorem 2.3.7.

Lemma 3. (i) Let $f \in S'(R_n)$, $\operatorname{supp} f \subset \overline{R_n^+}$ and $(Ff)(\xi', \xi_n) = 0$ of $|\xi'| \leqq 1$. Then

$$\operatorname{supp} F^{-1} a^+(\xi', \xi_n)\, Ff \subset \overline{R_n^+}\,.$$

(ii) Let $f \in S'(R_n)$, $\operatorname{supp} f \subset \overline{R_n^-}$ and $(Ff)(\xi', \xi_n) = 0$ if $|\xi'| \leqq 1$. Then

$$\operatorname{supp} F^{-1}[a^-(\xi', \xi_n)]^{-1}\, Ff \subset \overline{R_n^-}\,.$$

Remark 3. A proof may be found in [I, 5.2.2.]. Obviously, $\overline{R_n^-}$ is the closure of $R_n^- = \{x \mid x = (x_1, \ldots, x_n) \in R_n, x_n < 0\}$.

4.1.4. Basic Properties of Regular Elliptic Systems

We follow [I, 5.2.3.]. Again let A,

$$Au = \sum_{|\alpha|=2m} a_\alpha D^\alpha u \tag{1}$$

be a properly elliptic operator in R_n^+ in the sense of Definition 4.1.2/1(ii) having only top-order coefficients. Furthermore, let $\{B_j\}_{j=1}^m$ with

$$B_j u = \sum_{|\alpha|=m_j} b_{j,\alpha} D^\alpha u \tag{2}$$

be differential operators having only constant top-order coefficients. Let $\{A; B_1, \ldots, B_m\}$ be a regular elliptic system in R_n^+ in the sense of Definition 4.1.2/4(ii). We use the same notations as in the previous subsection. In particular, $\xi = (\xi_1, \ldots, \xi_{n-1}, 0) = (\xi', 0)$ with $\xi' \in R_{n-1}$ and $\eta = (0, \ldots.\, 0, 1)$. Furthermore, $a^+(\xi', \tau)$ has the meaning of (4.1.3/3) and $b_j(x; \xi + \tau\eta) = b_j(\xi + \tau\eta)$ is given by (4.1.2/7), where $b_{j,\alpha}(x) = b_{j,\alpha}$ are independent of $x \in R_{n-1}$. The complementing condition from Definition 4.1.2/3(ii) reads as follows. The polynomials in τ,

$$b_j(\xi + \tau\eta) = \sum_{|\alpha|=m_j} b_{j,\alpha}(\xi + \tau\eta)^\alpha = \sum_{l=0}^{m_j} b_{j,l}(\xi')\,\tau^l \tag{3}$$

are linearly independent modulo $a^+(\xi', \tau)$. It is easy to see that $b_{j,l}(\xi')$ is a homogeneous polynomial in $\xi' \in R_{n-1}$ of degree $m_j - l$. By Lemma 4.1.3/1 it follows that

$$\sum_{l=0}^{m_j} b_{j,l}(\xi')\,\tau^l = \sum_{l=0}^{m-1} b'_{j,l}(\xi')\,\tau^l + q_j(\xi', \tau)\, a^+(\xi', \tau)\,, \tag{4}$$

where $q_j(\xi', \tau)$ is a polynomial in τ (for every fixed ξ') and $b'_{j,l}(\xi')$ are analytic functions with respect to $\xi' \in R_{n-1} - \{0\}$ such that

$$b'_{j,l}(\lambda\xi') = \lambda^{m_j - l}\, b'_{j,l}(\xi') \quad \text{with} \quad \lambda > 0\,.$$

We need some properties of these functions $b'_{j,l}(\xi')$.

Lemma 1. Let $(b'_{j,l}(\xi'))_{\substack{j=1,\dots,m\\ l=0,\dots,m-1}}$ be the matrix of the functions $b'_{j,l}(\xi')$. Then

$$\det (b'_{j,l}(\xi'))_{\substack{j=1,\dots,m\\ l=0,\dots,m-1}} \neq 0 \quad \text{if} \quad \xi' \in R_{n-1} - \{0\} . \tag{5}$$

Let $(c_{j,l}(\xi'))_{\substack{j=1,\dots,m\\ l=0,\dots,m-1}}$ be the corresponding inverse matrix. Let $\varphi(t)$ be an infinitely differentiable function $[0, \infty)$ with $\varphi(t)=0$ if $0 \leqq t \leqq \frac{1}{2}$ and $\varphi(t)=1$ if $1 \leqq t < \infty$. Then

$$\sup_{\xi' \in R_{n-1}} (1+|\xi'|^2)^{\frac{|\alpha|}{2}} \, |D^\alpha[c_{j,l}(\xi')\,(1+|\xi'|^2)^{\frac{m_j-l}{2}} \varphi(|\xi'|)]| < \infty \tag{6}$$

for every multi-index α.

Remark 1. This lemma coincides essentially with [I, Lemma 5.2.3/1], where (6) follows from the proof of that lemma. The above lemma is the counterpart of Lemma 4.1.3/2. In particular, it follows from Theorem 2.3.7, that the functions $c_{j,l}(\xi')\,(1+|\xi'|^2)^{\frac{m_j-l}{2}} \varphi(|\xi'|)$ are Fourier multipliers for $B^s_{p,q}(R_{n-1})$ and $F^s_{p,q}(R_{n-1})$.

Temporarily we denote the $(n-1)$-dimensional Fourier transform with respect to the $n-1$ coordinates $x_1, \dots, x_{n-1}$ of $u(x_1, \dots, x_{n-1}, x_n)$ by F', i.e. $F'(u(\cdot, x_n)) = (F'u)\,(\xi', x_n)$ with $\xi' \in R_{n-1}$.

Lemma 2. Let $u(x)$ be a complex-valued function on R_n having classical derivatives up to order M such that

$$\sup_{x \in R_n} (1+|x|^2)^M \sum_{|\alpha| \leqq M} |D^\alpha u(x)| < \infty .$$

If M is sufficiently large and if

$$(Au)(x) = \sum_{|\alpha|=2m} a_\alpha D^\alpha u(x) = 0 \quad \text{for} \quad x_n \geqq 0 ,$$

then

$$(F'B_j u)(\xi', 0) = \mathrm{i}^{m_j} \sum_{l=0}^{m-1} b'_{j,l}(\xi') \frac{\partial^l}{\mathrm{i}^l \partial x_n^l} (F'u)\,(\xi', x_n)|_{x_n=0}$$

where $j=1, \dots, m$.

Remark 2. A proof of this lemma may be found in [I, 5.2.3.]. The functions $b'_{j,l}(\xi')$ have the meaning of (4). This formula is also the basis of the proof of the last lemma.

4.2. A Priori Estimates

4.2.1. Introduction and the Spaces $F^{s,r}_{p,q}(R_n^+)$

Our main goal in this chapter is the study of regular elliptic boundary value problems in bounded C^∞-domains Ω in R_n in the framework of the spaces $B^s_{p,q}(\Omega)$ and $F^s_{p,q}(\Omega)$. This problem can be reduced to a priori estimates for regular elliptic systems $\{A; B_1, \dots, B_m\}$ in R_n^+ in the sense of Definition 4.1.2/4 in the corresponding spaces on R_n^+. Spaces of type $B^s_{p,q}$(on R_n, R_n^+ or Ω) can be obtained by real interpolation of spaces of type $F^s_{p,q}$ and one can reduce the above problems for $B^s_{p,q}$ to corresponding problems for $F^s_{p,q}$ (we shall describe this method in detail later on). In this sense we restrict our attention in Section 4.2. to a priori estimates for regular elliptic systems $\{A; B_1, \ldots, B_m\}$ in R_n^+ and in Ω in the framework of the space $F^s_{p,q}(R_n^+)$ and $F^s_{p,q}(\Omega)$, respectively. As in [I, 5.3], our treatment in this section is based on the paper by L. Arkeryd [1]. This section may also be considered as an extended version of the brief announcements of the main results of this chapter in H. Triebel [10, 14]. In [I, 5.3] we proved a priori estimates for regular elliptic systems $\{A; B_1, \ldots, B_m\}$ in R_n^+ and in Ω in the framework of the spaces $L_p(R_n^+)$ and $L_p(\Omega)$, respectively, with $1<p<\infty$. The problem is to find a substitute for the "basic" spaces L_p if we deal with spaces of type $F^s_{p,q}$. It turns out that the spaces $F^s_{p,q}(\Omega)$ from Theorem 3.4.3 (and their counterparts on R_n and R_n^+) are the new "basic" spaces we are looking for. However, in order to simplify our considerations and to make them more transparent we assume in the later parts of Section 4.2. that

$$\min(p, q) \geqq \min(p, 1)\,. \tag{1}$$

In that case, (3.4.3/2) coincides with (3.4.3/1). However, we wish to emphasize that the theory which we develop in this chapter can be extended to values of p and q which are restricted by (3.4.3/2) (as far as the spaces $F^s_{p,q}$ are concerned; there will be no restrictions for q of that type for the spaces $B^s_{p,q}$). We remark that we used restrictions of type (1) many times in chapters 2 and 3.

For our purpose it will be convenient to generalize the spaces $F^s_{p,q}(R_n)$ from Definition 2.3.1(ii) and their restrictions to R_n^+ from Definition 2.9.1(ii) somewhat. Let $\xi=(\xi_1, \ldots, \xi_{n-1}, \xi_n)=(\xi', \xi_n)\in R_n$.

Definition. Let $0<p<\infty$, $0<q\leqq\infty$, $-\infty<s<\infty$ and $-\infty<r<\infty$. Then

$$F^{s,r}_{p,q}(R_n)=\{f \mid f\in S'(R_n),\ F^{-1}(1+|\xi'|^2)^{\frac{r}{2}} Ff\in F^s_{p,q}(R_n)\} \tag{2}$$

and $F^{s,r}_{p,q}(R_n^+)$ is the restriction of $F^{s,r}_{p,q}(R_n)$ to R_n^+.

Remark 1. Let F' and F'^{-1} be the $(n-1)$-dimensional Fourier transform and its inverse, respectively. Then the n-dimensional Fourier transform F and its inverse F^{-1} in (2) can be replaced by F' and F'^{-1}, respectively. If $r\neq 0$, then (2) defines an anisotropic space, where the differentiability properties in the directions $x_1, \ldots x_{n-1}$ are different from those in the direction x_n. One can study anisotropic spaces for their own sake, but this is not our aim. We are interested in the above spaces simply for technical reasons. Obviously, $F^{s,r}_{p,q}(R_n)$ is a quasi-Banach space, quasi-normed by

$$\|F^{-1}(1+|\xi'|^2)^{\frac{r}{2}} Ff \mid F^s_{p,q}(R_n)\|\,.$$

Furthermore, $F^{s,r}_{p,q}(R_n^+)$ is also a quasi-Banach space, endowed in the usual way with the quasi-norm

$$\|f \mid F^{s,r}_{p,q}(R_n^+)\| = \inf \|g \mid F^{s,r}_{p,q}(R_n)\|\,,$$

where the infimum is taken over all $g\in F^{s,r}_{p,q}(R_n)$ whose restrictions to R_n^+ coincide with f.

Lemma. Let A,

$$Au=\sum_{|\alpha|=2m} a_\alpha D^\alpha u\,,$$

be a properly elliptic differential operator in R_n^+ having only constant top-order coefficients. Let $0<p<\infty$, $0<q\leqq\infty$, $-\infty<s<\infty$ and $-\infty<r<\infty$. Then there exists a positive number c such that

$$\|u \mid F^{s,r}_{p,q}(R_n^+)\| \leqq c\,(\|Au \mid F^{s-2m,r}_{p,q}(R_n^+)\| + \|u \mid F^{s-1,r+1}_{p,q}(R_n^+)\|) \tag{3}$$

holds for every $u\in F^{s,r}_{p,q}(R_n^+)$.

Proof. Step 1. First we prove a few useful inequalities. Let k and j be two non-negative integers such that $k \geqq j$ and let σ and r be real numbers. We wish to prove that there exists a positive number c such that

$$\left\| \frac{\partial^j}{\partial x_n^j} F^{-1} (1+|\xi'|^2)^{\frac{k-j}{2}} Fu \mid F_{p,q}^{\sigma-k,r}(R_n^+) \right\| \leqq c \|u \mid F_{p,q}^{\sigma,r}(R_n^+)\| \tag{4}$$

holds for every $u \in F_{p,q}^{\sigma,r}(R_n^+)$. If $v \in S'(R_n)$ with $v \mid R_n^+ = u$, then

$$\frac{\partial^j}{\partial x_n^j} F^{-1} (1+|\xi'|^2)^{\frac{k-j}{2}} Fv \mid R_n^+ = \frac{\partial^j}{\partial x_n^j} F^{-1} (1+|\xi'|^2)^{\frac{k-j}{2}} Fu\,.$$

Hence, it follows easily that the left-hand side of (4) can be estimated from above by

$$\inf \left\| \frac{\partial^j}{\partial x_n^j} F^{-1} (1+|\xi'|^2)^{\frac{k-j}{2}} Fv \mid F_{p,q}^{\sigma-k,r}(R_n) \right\|, \tag{5}$$

where the infimum is taken over all $v \in F_{p,q}^{\sigma,r}(R_n)$ with $v \mid R_n^+ = u$. We have $\frac{\partial^j}{\partial x_n^j} F^{-1} = \mathrm{i}^j F^{-1} \xi_n^j$. Furthermore, the Fourier multipliers for the spaces $F_{p,q}^{\sigma-k,r}(R_n)$ are the same as for the space $F_{p,q}^{\sigma-k}(R_n)$. This follows from (2). In particular, $\xi_n^j (1+|\xi'|^2)^{\frac{k-j}{2}} \times (1+|\xi|^2)^{-\frac{k}{2}}$ is a Fourier multiplier for $F_{p,q}^{\sigma-k}(R_n)$, cf. Theorem 2.3.7. Hence it follows that (5) can be estimated from above by

$$c \inf \|F^{-1} (1+|\xi|^2)^{\frac{k}{2}} Fv \mid F_{p,q}^{\sigma-k,r}(R_n)\|\,.$$

Now (4) is a consequence of the lifting property from Theorem 2.3.8. In the same way it follows from Theorem 2.3.8 that

$$\sum_{\substack{|\alpha|=k \\ \alpha \neq (0,\dots,0,k)}} \|D^\alpha u \mid F_{p,q}^{\sigma,r}(R_n^+)\| \leqq c \|u \mid F_{p,q}^{\sigma+k-1,r+1}(R_n^+)\| \tag{6}$$

and

$$\|F^{-1} (1+|\xi'|^2)^{\frac{\varkappa}{2}} Fu \mid F_{p,q}^{\sigma,r}(R_n^+)\| = \|u \mid F_{p,q}^{\sigma,r+\varkappa}(R_n^+)\|\,, \tag{7}$$

where $\varkappa$ is an arbitrary real number.

Step 2. After these preliminaries we prove (3). Let J_σ,

$$J_\sigma f = F^{-1} (\mathrm{i}\xi_n - (1+|\xi'|^2)^{\frac{1}{2}})^\sigma Ff, \quad f \in S'(R_n)\,, \tag{8}$$

where σ is an arbitrary real number. It is easy to see that the lifting operator I_σ in Theorem 2.3.8 can be replaced by J_σ. In other words, J_σ is also a lifting operator, and it follows from the modified Theorem 2.3.8 and the above definition that J_σ maps $F_{p,q}^{s,r}(R_n)$ onto $F_{p,q}^{s-\sigma,r}(R_n)$. Obviously, J_σ maps also $S'(R_n)$ onto itself. Furthermore, in [I, 2.10.3.] (first step of the proof of Theorem 2.10.3 and Remark 2.10.3/3) we proved that

$$\operatorname{supp} J_\sigma f \subset \overline{R_n^-} \quad \text{if} \quad \operatorname{supp} f \subset \overline{R_n^-}\,, \tag{9}$$

where $R_n^- = \{x \mid x \in R_n,\ x_n < 0\}$. The above lifting property of J_σ and (9) yield that J_σ maps also $F_{p,q}^{s,r}(R_n^+)$ onto $F_{p,q}^{s-\sigma,r}(R_n^+)$. Let $\sigma = 2m$. Then the explicit from of J_{2m} shows that

$$\begin{aligned} \|u \mid F_{p,q}^{s,r}(R_n^+)\| \sim \|J_{2m} u \mid F_{p,q}^{s-2m,r}(R_n^+)\| \leqq c \Bigg(\left\| \frac{\partial^{2m} u}{\partial x_n^{2m}} \right| F_{p,q}^{s-2m,r}(R_n^+) \Bigg\| \\ + \sum_{j=0}^{2m-1} \left\| \frac{\partial^j}{\partial x_n^j} F^{-1} (1+|\xi'|^2)^{\frac{2m-1-j}{2}} FF^{-1} (1+|\xi'|^2)^{\frac{1}{2}} Fu \mid F_{p,q}^{s-2m,r}(R_n^+) \right\| \Bigg)\,. \end{aligned} \tag{10}$$

We apply (4) with $k=2m-1$ and then (7) with $\varkappa=1$. We have

$$\|u \mid F^{s,r}_{p,q}(R^+_n)\| \leqq c\left(\left\|\frac{\partial^{2m}u}{\partial x_n^{2m}} \,\middle|\, F^{s-2m,r}_{p,q}(R^+_n)\right\| + \|u \mid F^{s-1,r+1}_{p,q}(R^+_n)\|\right). \tag{11}$$

Since A is properly elliptic, the coefficient a_β with $\beta=(0, \ldots, 0, 2m)$ cannot be zero. Consequently,

$$\frac{\partial^{2m}u}{\partial x_n^{2m}} = cAu + \sum_{\substack{|\alpha|=2m \\ \alpha \neq \beta}} c_\alpha D^\alpha u. \tag{12}$$

We put (12) in (11) and apply (6). This proves (3).

Remark 2. The above proof is essentially a copy of a corresponding proof in [I, 5.3.1.].

4.2.2. A Priori Estimates. Part I: R_n^+, constant coefficients, Dirichlet problem

As has been mentioned at the beginning of 4.2.1., our first problem is to find a good substitute for the basic spaces L_p with $1<p<\infty$. We mentioned that the spaces $F^s_{p,q}$ from Theorem 3.4.3 (and their counterparts on R_n and R^+_n) have the desired properties, where we assume additionally that $\min(p, q) \geqq \min(p, 1)$ (mostly in order to avoid awkward formulations). In the first place we shall be interested in a priori estimates for the spaces $F^s_{p,q}$. However, these statements have immediate counterparts for the simpler spaces $B^s_{p,q}$, which we shall formulate as corollaries. In the sense of Theorem 3.4.3 we assume that

$$\begin{cases} \text{either} \quad 1 \leqq p < \infty, \quad 0<q<\infty, \quad \dfrac{1}{p}-1<\sigma<\dfrac{1}{p} \\ \text{or} \quad 0<p<1, \quad 0<q<\infty, \quad n\left(\dfrac{1}{p}-1\right)<\sigma<\dfrac{1}{p} \end{cases} \tag{1}$$

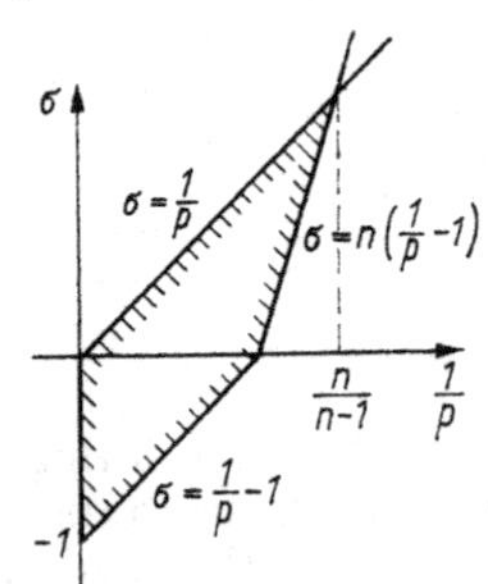

Fig. 4.2.2

(cf. Fig. 4.2.2) at least as far as the spaces $F^s_{p,q}$ are concerned. As for the spaces $B^s_{p,q}$, the case $p=\infty$ is also of some interest. We recall that $\xi=(\xi_1, \ldots, \xi_{n-1}, \xi_n)=(\xi', \xi_n)\in R_n$.

Lemma. Let A,

$$Au = \sum_{|\alpha|=2m} a_\alpha D^\alpha u, \tag{2}$$

be a properly elliptic differential operator on R^+_n having only constant top-order coefficients. Let p, q and σ be real numbers with (1) and $\min(p, q) \geqq \min(p, 1)$. Then there exists a positive number c such that

$$\|u \mid F^{\sigma+m}_{p,q}(R^+_n)\| \leqq c\|Au \mid F^{\sigma-m}_{p,q}(R^+_n)\| \tag{3}$$

for every $u \in F^{\sigma+m}_{p,q}(R_n)$ with

$$\operatorname{supp} u \subset \overline{R^+_n} \quad \text{and} \quad (Fu)(\xi', \xi_n)=0 \quad \text{if} \quad |\xi'| \leqq 1 . \tag{4}$$

Proof. Obviously, $f=Au \in F^{\sigma-m}_{p,q}(R_n)$. Let $S=S^N$ be the extension operator from Theorem 2.9.4(ii), where N is sufficiently large. S is defined via (2.9.2/9) and (2.9.2/10). We may assume that u is smooth. Then f is smooth and $\operatorname{supp} f \subset \overline{R^+_n}$. It follows that

$$(Sf)(x)=f(x)+\sum_{j=1}^{2r} b_j f(x', -\lambda_j x_n) , \tag{5}$$

where r is a sufficiently large natural number and $\lambda_j>0$. By Theorem 2.9.4 (ii) we have

$$\|Sf \mid F^{\sigma-m}_{p,q}(R_n)\| \leqq c\|f \mid F^{\sigma-m}_{p,q}(R^+_n)\| , \quad Sf \mid R^+_n = f . \tag{6}$$

Since

$$(Ff)(\xi', \xi_n)=(FAu)(\xi', \xi_n)=0 \quad \text{if} \quad |\xi'| \leqq 1 , \tag{7}$$

we have

$$(FSf)\,(\xi', \xi_n)=0 \quad \text{if} \quad |\xi'| \leqq 1 . \tag{8}$$

From Lemma 4.1.3/3(i), it follows that

$$[F^{-1}a^+(\xi', \xi_n)\, Fu]\,(x', x_n)=0 \quad \text{if} \quad x_n \leqq 0 \tag{9}$$

(recall we assumed that u is smooth). On the other hand we have $\operatorname{supp}(Au-Sf) \subset \overline{R^-_n}$. Consequently, by (7) and (8) we can apply Lemma 4.1.3/3(ii) to $Au-Sf$ (instead of f) and obtain

$$\begin{aligned}[F^{-1}a^+(\xi', \xi_n)\, Fu](x', x_n) &= c[F^{-1}(a^-(\xi', \xi_n))^{-1}\, FAu](x', x_n) \\ &= c[F^{-1}(a^-(\xi', \xi_n))^{-1}\, FSf](x', x_n) \quad \text{if} \quad x_n>0 .\end{aligned} \tag{10}$$

In order to put (9) and (10) together we remark that the characteristic function χ of R^+_n is a pointwise multiplier for $F^\sigma_{p,q}(R_n)$, cf. Theorem 2.8.7(ii). Hence it follows from (9) and (10) that

$$\begin{aligned}\|F^{-1}a^+(\xi', \xi_n)\, Fu \mid F^\sigma_{p,q}(R_n)\| &= c\|\chi F^{-1}(a^-(\xi', \xi_n))^{-1}\, FSf \mid F^\sigma_{p,q}(R_n)\| \\ &\leqq c'\|F^{-1}(a^-(\xi', \xi_n))^{-1}\, FSf \mid F^\sigma_{p,q}(R_n)\| .\end{aligned} \tag{11}$$

By (4) and (8) we can apply Lemma 4.1.3/2 and Remark 4.1.3/2 to obtain

$$\|u \mid F^{\sigma+m}_{p,q}(R_n)\| \leqq c\|Sf \mid F^{\sigma-m}_{p,q}(R_n)\| . \tag{12}$$

Now (3) follows from (6) and the proof is complete.

Theorem. Let A be the properly elliptic differential operator (2). Let p, q and σ be real numbers with (1) and $\min(p, q) \geqq \min(p, 1)$. Let $s=0, 1, 2, \ldots$ Then there exists a positive number c such that

$$\begin{aligned}&\|u \mid F^{\sigma+m+s}_{p,q}(R^+_n)\| \\ &\leqq c\left[\|Au \mid F^{\sigma-m+s}_{p,q}(R^+_n)\| + \sum_{j=0}^{m-1}\left\|\frac{\partial^j u}{\partial x_n^j}(x', 0) \mid B^{\sigma+m+s-j-\frac{1}{p}}_{p,p}(R_{n-1})\right\| \right. \\ &\left. \quad +\|u \mid F^{\sigma+m+s-1}_{p,q}(R^+_n)\|\right]\end{aligned} \tag{13}$$

holds for all $u \in F^{\sigma+m+s}_{p,q}(R^+_n)$.

Proof. Step 1. First we generalize the above lemma. Let

$$u \in F^{\sigma+m+s,r}_{p,q}(R_n), \quad \frac{\partial^j u}{\partial x_n^j}(x', 0) = 0 \quad \text{if} \quad j = 0, \ldots, m-1, \tag{14}$$

$$(Fu)(\xi', \xi_n) = 0 \quad \text{if} \quad |\xi'| \leq 1, \tag{15}$$

where $s = 0, 1, 2, \ldots$ and $r \geq 0$. The boundary values in (13) and (14) make sense, cf. Theorem 2.7.2(ii) (obviously, $F^{\sigma+m+s,r}_{p,q}(R_n) \subset F^{\sigma+m+s}_{p,q}(R_n)$ if $r \geq 0$). We claim that there exists a constant c such that

$$\|u \mid F^{\sigma+m+s,r}_{p,q}(R_n^+)\| \leq c\|Au \mid F^{\sigma-m+s,r}_{p,q}(R_n^+)\| \tag{16}$$

for every function u with (14) and (15). Let $s = r = 0$ and let χ be the characteristic function of R_n^+. Then it follows from (3.4.3/5) and Remark 3.4.3/2, with R_n^+ instead of Ω, that χu satisfies also (14) and (15) with $s = r = 0$ (recall that χ depends only on x_n). We apply the above lemma to χu. Since only the values of $u(x)$ in R_n^+ play a role, we obtain (16) if $s = r = 0$. Next we prove (16) with $s = 0$ and $r > 0$. Let u be a function which satisfies (14) and (15) with $s = 0$. Let

$$g = F^{-1}(1 + |\xi'|^2)^{\frac{r}{2}} Fu.$$

Then it is easy to see that g also satisfies (14) and (15) with $s = r = 0$ (and with g instead of u). (16) yields

$$\|u \mid F^{\sigma+m,r}_{p,q}(R_n^+)\| = \|g \mid F^{\sigma+m}_{p,q}(R_n^+)\| \leq c\|Ag \mid F^{\sigma-m}_{p,q}(R_n^+)\| \tag{17}$$

$$= c\,\|F^{-1}(1 + |\xi'|^2)^{\frac{r}{2}} FAu \mid F^{\sigma-m}_{p,q}(R_n^+)\| = c\|Au \mid F^{\sigma-m,r}_{p,q}(R_n^+)\|.$$

This proves (16) if $s = 0$ and $r \geq 0$. We prove the general case by mathematical induction. Let us assume that (16) holds for a fixed s (with $s = 0, 1, 2, \ldots$) and every $r \geq 0$. Let u be a function which satisfies (14) and (15) with $s+1$ instead of s. Then it follows from Lemma 4.2.1 that

$$\|u \mid F^{\sigma+m+s+1,r}_{p,q}(R_n^+)\| \leq c(\|Au \mid F^{\sigma-m+s+1,r}_{p,q}(R_n^+)\| + \|u \mid F^{\sigma+m+s,r+1}_{p,q}(R_n^+)\|). \tag{18}$$

We apply (16) (with $r+1$ instead of r) to the last term on the right-hand side of (18). Since

$$F^{\sigma-m+s+1,r}_{p,q}(R_n^+) \subset F^{\sigma-m+s,r+1}_{p,q}(R_n^+)$$

we obtain (16) with $s+1$ instead of s. The proof of (16) is complete.

Step 2. Our next aim is to drop (15). Let u be a function which satisfies (14) with $r = 0$. Let $\varphi(t)$ be an infinitely differentiable function on $[0, \infty)$ such that

$$\varphi(t) = 1 \quad \text{if} \quad 0 \leq t \leq 1 \quad \text{and} \quad \varphi(t) = 0 \quad \text{if} \quad 2 \leq t < \infty.$$

Let

$$u_0(x) = F^{-1}\varphi(|\xi'|)\, Fu, \quad u_1(x) = F^{-1}(1 - \varphi(|\xi'|))\, Fu. \tag{19}$$

We use the fact that $\varphi(|\xi'|)$ is a Fourier multiplier for all spaces $F^{\gamma}_{\alpha,\beta}(R_n)$ with $0 < \alpha < \infty$, $0 < \beta < \infty$ and $-\infty < \gamma < \infty$ (cf. Remark 1 after the proof). Then u_0 and u_1 satisfy (14) with $r = 0$ and u_1 has the additional property (15). Consequently we have (16) with u_1 instead of u and with $r = 0$, i.e.

$$\|u_1 \mid F^{\sigma+m+s}_{p,q}(R_n^+)\| \leq c\|F^{-1}(1 - \varphi(|\xi'|))\, FAu \mid F^{\sigma-m+s}_{p,q}(R_n^+)\|, \tag{20}$$

where we have used the definition of u_1. As in the first step of the proof of Lemma 4.2.1 we have

$$\|u_1 \mid F^{\sigma+m+s}_{p,q}(R_n^+)\| \leq c\|Au \mid F^{\sigma-m+s}_{p,q}(R_n^+)\|. \tag{21}$$

In order to estimate u_0 we remark that $(1+|\xi'|^2)^{\frac{1}{2}}\varphi(|\xi'|)$ is also a Fourier multiplier for $F^{\gamma}_{\alpha,\beta}(R_n)$ with $0<\alpha<\infty$, $0<\beta<\infty$ and $-\infty<\gamma<\infty$. Hence we obtain by the same argument as above that

$$\|u_0 \mid F^{\sigma+m+s}_{p,q}(R_n^+)\| = \|F^{-1}\varphi(|\xi'|)(1+|\xi'|^2)^{\frac{1}{2}} FF^{-1}(1+|\xi'|^2)^{-\frac{1}{2}} Fu \mid F^{\sigma+m+s}_{p,q}(R_n^+)\|$$
$$\leqq c\|F^{-1}(1+|\xi'|^2)^{-\frac{1}{2}} Fu \mid F^{\sigma+m+s}_{p,q}(R_n^+)\| = c\|u \mid F^{\sigma+m+s,-1}_{p,q}(R_n^+)\|,$$

cf. also (4.2.1/7). Lemma 4.2.1 yields

$$\|u_0 \mid F^{\sigma+m+s}_{p,q}(R_n^+)\| \leqq c(\|Au \mid F^{\sigma-m+s,-1}_{p,q}(R_n^+)\| + \|u \mid F^{\sigma+m+s-1}_{p,q}(R_n^+)\|). \tag{22}$$

But the first term on the right-hand side of (22) can be estimated from above by the right-hand side of (21). This proves that

$$\|u \mid F^{\sigma+m+s}_{p,q}(R_n^+)\| \leqq c(\|Au \mid F^{\sigma-m+s}_{p,q}(R_n^+)\| + \|u \mid F^{\sigma+m+s-1}_{p,q}(R_n^+)\|) \tag{23}$$

if u satisfies (14) with $r=0$ and $s=0, 1, 2, \ldots$

Step 3. We prove (13). Let $u \in F^{\sigma+m+s}_{p,q}(R_n^+)$. Theorem 2.7.2 (ii) yields

$$\frac{\partial^j u}{\partial x_n^j}(x', 0) \in B^{\sigma+m+s-j-\frac{1}{p}}_{p,p}(R_{n-1}) \quad \text{with} \quad j=0, \ldots, m-1. \tag{24}$$

Furthermore, by the same theorem we find a function $v \in F^{\sigma+m+s}_{p,q}(R_n^+)$ such that

$$\frac{\partial^j v}{\partial x_n^j}(x', 0) = \frac{\partial^j u}{\partial x_n^j}(x', 0) \quad \text{if} \quad j=0, \ldots, m-1. \tag{25}$$

$$\|v \mid F^{\sigma+m+s}_{p,q}(R_n^+)\| \leqq c \sum_{j=0}^{m-1} \left\| \frac{\partial^j v}{\partial x_n^j}(x', 0) \mid B^{\sigma+m+s-j-\frac{1}{p}}_{p,p}(R_{n-1}) \right\|. \tag{26}$$

We apply (23) to $u-v$ and obtain

$$\|u \mid F^{\sigma+m+s}_{p,q}(R_n^+)\| \leqq c\|u-v \mid F^{\sigma+m+s}_{p,q}(R_n^+)\| + c\|v \mid F^{\sigma+m+s}_{p,q}(R_n^+)\| \tag{27}$$
$$\leqq c'\|Au-Av \mid F^{\sigma-m+s}_{p,q}(R_n^+)\| + c'\|u-v \mid F^{\sigma+m+s-1}_{p,q}(R_n^+)\| + c'\|v \mid F^{\sigma+m+s}_{p,q}(R_n^+)\|.$$

The term $\|Av \mid F^{\sigma-m+s}_{p,q}(R_n^+)\|$ can be estimated from above by the last term on the right-hand side of (27). Then it follows from (27) and (25), (26) that

$$\|u \mid F^{\sigma+m+s}_{p,q}(R_n^+)\|$$
$$\leqq c\|Au \mid F^{\sigma-m+s}_{p,q}(R_n^+)\| + c\|u \mid F^{\sigma+m+s-1}_{p,q}(R_n^+)\| + c\|v \mid F^{\sigma+m+s}_{p,q}(R_n^+)\|$$
$$\leqq c'\|Au \mid F^{\sigma-m+s}_{p,q}(R_n^+)\| + c'\|u \mid F^{\sigma+m+s-1}_{p,q}(R_n^+)\|$$
$$+c' \sum_{j=0}^{m-1} \left\| \frac{\partial^j u}{\partial x_n^j}(x', 0) \mid B^{\sigma+m+s-j-\frac{1}{p}}_{p,p}(R_{n-1}) \right\|.$$

The proof is complete.

Remark 1. In Step 2 we used the fact that $\varphi(|\xi'|)$ (and then also $(1+|\xi'|^2)^{\frac{1}{2}}\varphi(|\xi'|)$) is a Fourier multiplier for $F^{\gamma}_{\alpha,\beta}(R_n)$ if $0<\alpha<\infty$, $0<\beta<\infty$ and $-\infty<\gamma<\infty$. By the lifting property from Theorem 2.3.8 (ii) it is sufficient to prove this statement for large values of γ: the set of all Fourier multipliers for $F^{\gamma}_{\alpha,\beta}(R_n)$ is independent of γ. Let γ be large. We put

$$F^{-1}\varphi(|\xi'|)\, Ff = F'^{-1}\varphi(|\xi'|)\, F'f$$

in (2.5.10/2) with α, β, γ instead of p, q, s, respectively. Then

$$|(F^{-1}\varphi(|\xi'|)\, Ff)(x)| \leqq c \sup_{y \in R_n} \frac{|f(x-y)|}{|1+|y|^a}$$

and

$$|(\Delta_h^M F^{-1}\varphi(|\xi'|)\,Ff)\,(x)| \leqq c \min\,(|h|^M, |h|^a) \sup_{y\in R_n} \frac{|f\,(x-y)|}{1+|y|^a},$$

where we assume that $M > \gamma > a > 0$. In other words, we can estimate the quasi-norm (2.5.10/2) with $F^{-1}\varphi(|\xi'|)Ff$ instead of f from above by

$$\left\| \sup_{y\in R_n} \frac{|f\,(x-y)|}{1+|y|^a} \mid L_p(R_n) \right\|.$$

However, the latter expression can be estimated from above by $\|f \mid F^s_{p,q}(R_n)\|$, cf. the end of the proof of Corollary 2.5.9/1. Hence

$$\|F^{-1}\varphi(|\xi'|)\,Ff \mid F^{\gamma}_{\alpha,\beta}(R_n)\| \leqq c\|f \mid F^{\gamma}_{\alpha,\beta}(R_n)\|$$

if γ is large. The proof is complete.

As we mentioned at the beginning of this subsection, the above theorem has an immediate counterpart for the spaces $B^s_{p,q}$, where the limiting cases $p=\infty$ and/or $q=\infty$ are included. In this sense we replace (1) by

$$\begin{cases} \text{either} \quad 1\leqq p\leqq\infty, \quad 0<q\leqq\infty, \quad \frac{1}{p}-1<\sigma<\frac{1}{p} \\ \text{or} \quad 0<p<1, \quad 0<q\leqq\infty, \quad n\left(\frac{1}{p}-1\right)<\sigma<\frac{1}{p}, \end{cases} \tag{28}$$

(cf. Fig. 4.2.2).

Corollary. Let A be the properly elliptic differential operator (2). Let p, q, and σ be real numbers with (28). Let $s=0, 1, 2, \ldots$ Then there exists a positive number c such that

$$\|u \mid B^{\sigma+m+s}_{p,q}(R_n^+)\| \leqq c \Bigg[\|Au \mid B^{\sigma-m+s}_{p,q}(R_n^+)\| + \sum_{j=0}^{m-1} \left\| \frac{\partial^j u}{\partial x_n^j}(x',0) \mid B_{p,q}^{\sigma+m+s-j-\frac{1}{p}}(R_{n-1}) \right\| + \|u \mid B^{\sigma+m+s-1}_{p,q}(R_n^+)\| \Bigg] \tag{29}$$

holds for all $u \in S(R_n)$.

Proof. The proof follows the same line as the proof of the above theorem, cf. the Theorems 2.9.4 and 2.8.7, and in particular Remark 3.4.3/3. The limiting cases are always included. Instead of Theorem 2.5.10 in Remark 1 one can use Theorem 2.5.12.

Remark 2. In the above proofs we sometimes used the fact that the functions u are smooth. If $p<\infty$ and $q<\infty$, we can make such an assumption without restriction of generality because those functions are dense in the corresponding spaces. If $p=\infty$ and/or $q=\infty$, this is not the case. This is the reason for the restriction $u\in S(R_n)$ in the above corollary. If $p<\infty$ and $q<\infty$, then (29) holds for all $u \in B^{\sigma+m+s}_{p,q}(R_n^+)$. The most interesting case of the above corollary is $p=\infty$, because that case cannot be obtained immediately from the spaces in the above theorem by real interpolation.

4.2.3. A Priori Estimates. Part II: R_n^+, constant coefficients, general boundary problem

We follow [I, 5.3.3.], where we must replace the basic space L_p, with $1<p<\infty$, by the basic spaces $F^{\sigma}_{p,q}$, with (4.2.2/1) and $\min\,(p, q) \geqq \min\,(p, 1)$.

Theorem. Let $\{A; B_1, \ldots, B_m\}$ be a regular elliptic system in R_n^+ in the sense of Definition 4.1.2/4(ii), where

$$Au = \sum_{|\alpha|=2m} a_\alpha D^\alpha u \quad \text{and} \quad B_j u = \sum_{|\alpha|=m_j} b_{j,\alpha} D^\alpha u \tag{1}$$

with $j=1, \ldots, m$, are differential operators having only constant top-order coefficients. Let p, q, and σ be real numbers with (4.2.2/1) and $\min(p, q) \geqq \min(p, 1)$. Let $s=0, 1, 2, \ldots$ Then there exists a positive number c such that

$$\|u \mid F^{\sigma+2m+s}_{p,q}(R^+_n)\| \leqq c \Bigg[\|Au \mid F^{\sigma+s}_{p,q}(R^+_n)\| + \sum_{j=1}^{m} \|(B_j u)(x', 0) \mid B^{\sigma+2m+s-m_j-\frac{1}{p}}_{p,p}(R_{n-1})\| + \|u \mid F^{\sigma+2m+s-1}_{p,q}(R^+_n)\| \Bigg] \tag{2}$$

holds for all $u \in F^{\sigma+2m+s}_{p,q}(R^+_n)$.

Proof. Step 1. We begin with a preliminary. Let $w(x)$ be a function on R_n with classical derivatives up to order M such that

$$\sup_{x \in R_n} (1+|x|^2)^M \sum_{|\alpha| \leqq M} |D^\alpha w(x)| < \infty . \tag{3}$$

We always assume that M is sufficiently large. Let $(Aw)(x)=0$ if $x_n \geqq 0$ and

$$(F'w)(\xi', x_n)=0 \quad \text{if} \quad |\xi'| \leqq 1 \quad \text{and} \quad x_n \geqq 0 . \tag{4}$$

We recall that F' is the $(n-1)$-dimensional Fourier transform, cf. the descriptions preceding Lemma 4.1.4/2. We apply this lemma. Then it follows from the notions introduced in Lemma 4.1.4/1 that

$$\frac{\partial^l}{\partial x_n^l}(F'w)(\xi', 0) = c \sum_{j=1}^{m} c_{j,l}(\xi')(F'B_j w)(\xi', 0)$$

where $l=0, \ldots, m-1$. Consequently,

$$\begin{aligned} &\sum_{l=0}^{m-1} \left\| \frac{\partial^l w}{\partial x_n^l}(x', 0) \mid B^{\sigma+2m+s-l-\frac{1}{p}}_{p,p}(R_{n-1}) \right\| \\ &= \sum_{l=0}^{m-1} \left\| F'^{-1}\left(\frac{\partial^l}{\partial x_n^l} F'w \right)(\xi', 0) \mid B^{\sigma+2m+s-l-\frac{1}{p}}_{p,p}(R_{n-1}) \right\| \\ &\leqq c \sum_{l=0}^{m-1} \sum_{j=1}^{m} \|F'^{-1} c_{j,l}(\xi') F'B_j w(\xi', 0) \mid B^{\sigma+2m+s-l-\frac{1}{p}}_{p,p}(R_{n-1})\| . \end{aligned} \tag{5}$$

We have $(F'B_j w)(\xi', x_n)=0$ if $|\xi'| \leqq 1$ and $x_n \geqq 0$, cf. (4). This shows that we can apply Lemma 4.1.4/1 and Remark 4.1.4/1. Thus it follows that the right-hand side of (5) can be estimated from above by

$$c \sum_{l=0}^{m-1} \sum_{j=1}^{m} \|F'^{-1}(1+|\xi'|^2)^{\frac{l-m_j}{2}} F'B_j w(\xi', 0) \mid B^{\sigma+2m+s-l-\frac{1}{p}}_{p,p}(R_{n-1})\| . \tag{6}$$

Finally the lifting property from Theorem 2.3.8 shows that

$$\begin{aligned} &\sum_{l=0}^{m-1} \left\| \frac{\partial^l w}{\partial x_n^l}(x', 0) \mid B^{\sigma+2m+s-l-\frac{1}{p}}_{p,p}(R_{n-1}) \right\| \\ &\leqq c \sum_{j=1}^{m} \|(B_j w)(x', 0) \mid B^{\sigma+2m+s-m_j-\frac{1}{p}}_{p,p}(R_{n-1})\| . \end{aligned} \tag{7}$$

Step 2. We prove (2). First we remark that the formulation makes sense, cf. Theorem 2.7.2(ii). Furthermore, by a density argument it is sufficient to assume that u is smooth, e.g. the restriction of an element of $S(R_n)$ to R^+_n. Let $u \in S(R_n)$ and let u_0 and u_1 be given by (4.2.2/19). Obviously, F and F^{-1} in (4.2.2/19) can be replaced by F' and F'^{-1}, respectively. Let $S=S^N$ be the extension operator from Theorem 2.9.4(ii), where we always assume that N is sufficiently large. Let

$$v(x)=F^{-1}a^{-1}(\xi)\,F[SAu_1]\,, \tag{8}$$

where $a(\xi)=a(x,\xi)$ is defined by (4.1.2/2). The extension operator S is given by (2.9.2/10). By this explicit formula it follows that

$$\begin{aligned} v(x) &= F^{-1}a^{-1}(\xi)\,F[SF'^{-1}\,(1-\varphi(|\xi'|))\,F'Au] \\ &= F^{-1}a^{-1}(\xi)\,FF'^{-1}\,(1-\varphi(|\xi'|))\,F'SAu \\ &= F^{-1}a^{-1}(\xi)\,(1-\varphi(|\xi'|))\,FSAu \\ &= F'^{-1}\,(1-\varphi(|\xi'|))\,F'F^{-1}a^{-1}(\xi)\,\psi(\xi')\,FSAu\,, \end{aligned} \tag{9}$$

where $\psi(\xi')$ is an infinitely differentiable function in R_{n-1} which vanishes near the origin, and with $\psi(\xi')=1$ if ξ' belongs to the support of $1-\varphi(|\xi'|)$. The function SAu satisfies a condition of type (3). Then it is well-known that $FSAu$ has the same property (with different values of M). Thus it follows from (9) that $v(x)$ also has this property, where we may assume that M is as large as we want. Furthermore, by (8) and (9),

$$(Av)(x)=(Au_1)(x) \quad \text{if} \quad x_n\geqq 0\,, \tag{10}$$

$$(F'v)(\xi',x_n)=0 \quad \text{if} \quad |\xi'|\leqq 1\,. \tag{11}$$

Let $w=u_1-v$. Then we have (3), (4) and $(Aw)(x)=0$ if $x_n\geqq 0$. In other words, the hypotheses of Step 1 are satisfied, and we can apply (7). Then

$$\begin{aligned} &\sum_{l=0}^{m-1}\left\|\frac{\partial^l u_1}{\partial x_n^l}(x',0)\mid B_{p,p}^{\sigma+2m+s-l-\frac{1}{p}}(R_{n-1})\right\| \\ &\leqq c\sum_{l=0}^{m-1}\left\|\frac{\partial^l w}{\partial x_n^l}(x',0)\mid B_{p,p}^{\sigma+2m+s-l-\frac{1}{p}}(R_{n-1})\right\| \\ &\quad +c\sum_{l=0}^{m-1}\left\|\frac{\partial^l v}{\partial x_n^l}(x',0)\mid B_{p,p}^{\sigma+2m+s-l-\frac{1}{p}}(R_{n-1})\right\| \\ &\leqq c'\sum_{j=1}^{m}\|(B_jw)(x',0)\mid B_{p,p}^{\sigma+2m+s-m_j-\frac{1}{p}}(R_{n-1})\|+c'\|v\mid F_{p,q}^{\sigma+2m+s}(R_n^+)\|\,, \end{aligned} \tag{12}$$

where we have used Theorem 2.7.2(ii). Now replace w in the first term on the right-hand side of (12) by u_1-v. The terms in v we estimate also via Theorem 2.7.2(ii). Then

$$\begin{aligned} &\sum_{l=0}^{m-1}\left\|\frac{\partial^l u_1}{\partial x_n^l}(x',0)\mid B_{p,p}^{\sigma+2m+s-l-\frac{1}{p}}(R_{n-1})\right\| \\ &\leqq c\sum_{j=1}^{m}\|(B_ju_1)(x',0)\mid B_{p,p}^{\sigma+2m+s-m_j-\frac{1}{p}}(R_{n-1})\|+c\|v\mid F_{p,q}^{\sigma+2m+s}(R_n^+)\|\,. \end{aligned} \tag{13}$$

(9) yields

$$\begin{aligned} \|v\mid F_{p,q}^{\sigma+2m+s}(R_n^+)\| &\leqq c\|v\mid F_{p,q}^{\sigma+2m+s}(R_n)\| \\ &\leqq c\|F^{-1}\,(1-\varphi(|\xi'|))\,(1+|\xi|^2)^m\,a^{-1}(\xi)\,FSAu\mid F_{p,q}^{\sigma+s}(R_n^+)\|\,. \end{aligned} \tag{14}$$

Since $a(\xi)\sim|\xi|^{2m}$, the function $(1-\varphi(|\xi'|))\,(1+|\xi|^2)^m a^{-1}(\xi)$ is a Fourier multiplier for $F_{p,q}^{\sigma+s}(R_n)$, cf. Remark 4.2.2/1. Together with the extension property of S we have

$$\|v\mid F_{p,q}^{\sigma+2m+s}(R_n^+)\|\leqq c\|Au\mid F_{p,q}^{\sigma+s}(R_n^+)\|\,. \tag{15}$$

Applying Theorem 4.2.2 to u_1 (instead of u) we obtain from (13) and (15) that

$$\|u_1 \mid F_{p,q}^{\sigma+2m+s}(R_n^+)\| \leq c \Bigg[\|Au_1 \mid F_{p,q}^{\sigma+s}(R_n^+)\| + \|u_1 \mid F_{p,q}^{\sigma+2m+s-1}(R_n^+)\| + \sum_{j=1}^{m} \|(B_j u_1)(x', 0) \mid B_{p,p}^{\sigma+2m+s-m_j-\frac{1}{p}}(R_{n-1})\| + \|Au \mid F_{p,q}^{\sigma+s}(R_n^+)\|\Bigg]. \tag{16}$$

The first term on the right-hand side of (16) can be estimated by the last one in the same way as above,

$$\|Au_1 \mid F_{p,q}^{\sigma+s}(R_n^+)\| \leq \|F'^{-1}(1-\varphi(|\xi'|))\, F'SAu \mid F_{p,q}^{\sigma+s}(R_n)\| \leq c\|Au \mid F_{p,q}^{\sigma+s}(R_n^+)\|.$$

Similarly, one can replace u_1 in the second and in the third term on the right-hand side of (16) by u (estimates from above). Thus

$$\|u_1 \mid F_{p,q}^{\sigma+2m+s}(R_n^+)\| \leq c \Bigg[\|Au \mid F_{p,q}^{\sigma+s}(R_n^+)\| + \|u \mid F_{p,q}^{\sigma+2m+s-1}(R_n^+)\| + \sum_{j=1}^{m} \|(B_j u)(x', 0) \mid B_{p,p}^{\sigma+2m+s-m_j-\frac{1}{p}}(R_{n-1})\|\Bigg]. \tag{17}$$

As far as u_0 is concerned we have (4.2.2/22). This estimate with $s+m$ instead of s shows that $\|u_0 \mid F_{p,q}^{\sigma+2m+s}(R_n^+)\|$ can also be estimated from above by the right-hand side of (17). Now (2) follows from $u=u_0+u_1$. The proof is complete.

Corollary. Let $\{A; B_1, \ldots, B_m\}$ be the regular elliptic system in R_n^+ from the above theorem. Let p, q, and σ be real numbers with (4.2.2/28). Let $s=0, 1, 2, \ldots$ Then there exists a positive number c such that

$$\|u \mid B_{p,q}^{\sigma+2m+s}(R_n^+)\| \leq c \Bigg[\|Au \mid B_{p,q}^{\sigma+s}(R_n^+)\| + \|u \mid B_{p,q}^{\sigma+2m+s-1}(R_n^+)\| + \sum_{j=1}^{m} \|(B_j u)(x', 0) \mid B_{p,q}^{\sigma+2m+s-m_j-\frac{1}{p}}(R_{n-1})\|\Bigg] \tag{18}$$

holds for all $u \in S(R_n)$.

Proof. The proof follows the same line as the proof of the above theorem. Instead of Theorem 4.2.2 we use Corollary 4.2.2.

Remark. We have the same situation as in Remark 4.2.2/2. If $p<\infty$ and $q<\infty$, then (18) holds for all $u \in B_{p,q}^{\sigma+2m+s}(R_n^+)$. This follows by completion. However if either p or q equals infinity, then the restriction of $S(R_n)$ to R_n^+ is not dense in the corresponding spaces.

4.2.4. A Priori Estimates. Part III: Bounded domain, variable coefficients, general boundary problem

Let Ω be a bounded C^∞-domain in the sense of Definition 3.2.1. The balls K_j with $j=1, \ldots, N$ and the mappings $y=f^{(j)}(x)$ always have the meaning of this definition, where the special form of $y=f^{(j)}(x)$ which we described in Remark 3.2.1/1 will be of some interest for us. On the other hand, we modify the resolution of unity at the end of 3.2.1. somewhat. If the balls K_j with $j=1, \ldots, N$ have the above meaning, we introduce open balls K_j with $j=-M, \ldots, 0$ such that

$$\overline{K_j} \subset \Omega \quad \text{if} \quad j=-M, \ldots, 0 \quad \text{and} \quad \bigcup_{j=-M}^{N} K_j \supset \overline{\Omega}. \tag{1}$$

Let $\{\psi_j(x)\}_{j=-M}^{N}$ be a corresponding resolution of unity, i.e.

$$\operatorname{supp} \psi_j \subset K_j \quad \text{if} \quad j=-M, \ldots, N \quad \text{and} \quad \sum_{j=-M}^{N} \psi_j(x)=1 \quad \text{if} \quad x \in \bar{\Omega}. \tag{2}$$

Finally, the centre of K_j is denoted by x^j. We follow essentially [I, 5.3.4.].

Theorem. Let $\{A; B_1, \ldots, B_m\}$ be a regular elliptic system in Ω in the sense of Definition 4.1.2/4. Let p, q, and σ be real numbers with (4.2.2/1) and $\min(p, q) \geqq \min(p, 1)$. Let $s=0, 1, 2, \ldots$ Then there exist two positive numbers c_1 and c_2 such that

$$\begin{aligned} & c_1\|u \mid F_{p,q}^{\sigma+2m+s}(\Omega)\| \\ & \leqq \|Au \mid F_{p,q}^{\sigma+s}(\Omega)\| + \sum_{j=1}^{m} \|B_j u \mid B_{p,p}^{\sigma+2m+s-m_j-\frac{1}{p}}(\partial\Omega)\| + \|u \mid F_{p,q}^{\sigma}(\Omega)\| \\ & \leqq c_2\|u \mid F_{p,q}^{\sigma+2m+s}(\Omega)\| \end{aligned} \tag{3}$$

holds for every $u \in F_{p,q}^{\sigma+2m+s}(\Omega)$.

Proof. Step 1. The right-hand side of (3) is an immediate consequence of Theorem 3.3.2(ii) and Theorem 3.3.5(ii) (including their counterparts with $\partial\Omega$ instead of Ω) and the traces from Theorem 3.3.3.

Step 2. In order to prove the left-hand side of (3) we use the above balls K_j (centered at x^j) and the resolution of unity (2). In this step we deal with the case $j=-M, \ldots, 0$. Let $u \in F_{p,q}^{\sigma+2m+s}(\Omega)$. Then $\psi_j u \in F_{p,q}^{\sigma+2m+s}(\Omega)$, cf. Theorem 3.3.2, and after extension by zero $\psi_j u \in F_{p,q}^{\sigma+2m+s}(R_n)$. Let $a(x^j, \xi)$ be the polynomial from (4.1.2/2). We recall that $a(x^j, \xi) \sim |\xi|^{2m}$. Then it follows from Theorem 2.3.8 that

$$\begin{aligned} & \|\psi_j u \mid F_{p,q}^{\sigma+2m+s}(\Omega)\| \leqq \|\psi_j u \mid F_{p,q}^{\sigma+2m+s}(R_n)\| \\ & \leqq c\|F^{-1}a(x^j, \xi)\, F\psi_j u \mid F_{p,q}^{\sigma+s}(R_n)\| + c\|\psi_j u \mid F_{p,q}^{\sigma+s}(R_n)\| \\ & \leqq c\Big\| \sum_{|\alpha|=2m} a_\alpha(x^j)\, D^\alpha \psi_j u \mid F_{p,q}^{\sigma+s}(R_n)\Big\| + c\|\psi_j u \mid F_{p,q}^{\sigma+s}(R_n)\| \\ & \leqq c'\|A\psi_j u \mid F_{p,q}^{\sigma+s}(R_n)\| + c' \sum_{|\alpha|=2m} \|(a_\alpha(x) - a_\alpha(x^j))\, D^\alpha \psi_j u \mid F_{p,q}^{\sigma+s}(R_n)\| \\ & \quad + c' \sum_{|\alpha|<2m} \|a_\alpha(x)\, D^\alpha \psi_j u \mid F_{p,q}^{\sigma+s}(R_n)\| + c'\|\psi_j u \mid F_{p,q}^{\sigma+s}(R_n)\|\,. \end{aligned} \tag{4}$$

The third term on the right-hand side of (4) (and also the fourth one) can be estimated from above by $c\|\psi_j u \mid F_{p,q}^{\sigma+s+2m-1}(R_n)\|$, cf. Theorem 2.3.8 and Theorem 2.8.2. In order to estimate the second term on the right-hand side of (4) we use first Theorem 2.3.8(ii) (with $s=0, 1, 2, \ldots$ instead of m) and then Proposition 3.4.1/2(ii) (with σ instead of s). Then we can estimate this term from above by

$$\varepsilon\|u \mid F_{p,q}^{\sigma+2m+s}(\Omega)\| + c\|\psi_j u \mid F_{p,q}^{\sigma+2m+s-1}(R_n)\|\,, \tag{5}$$

where c is independent of ε. The number ε depends on the radius τ of the balls K_j with $j=-M, \ldots, 0$. We may assume that $\varepsilon>0$ is as small as we want. Finally, we remark that $A\psi_j u = \psi_j Au +$ lower terms. These lower terms can also be estimated by $c\|u \mid F_{p,q}^{\sigma+s+2m-1}(\Omega)\|$. Combining these estimates we obtain

$$\begin{aligned} & \|\psi_j u \mid F_{p,q}^{\sigma+2m+s}(\Omega)\| \\ & \leqq \varepsilon\|u \mid F_{p,q}^{\sigma+2m+s}(\Omega)\| + c\|Au \mid F_{p,q}^{\sigma+s}(\Omega)\| + c\|u \mid F_{p,q}^{\sigma+2m+s-1}(\Omega)\|, \end{aligned} \tag{6}$$

where $\varepsilon>0$ may be chosen arbitrarily small and c depends on the choice of ε and $j=-M, \ldots, 0$.

Step 3. In this step we deal with the case $j=1, \ldots, N$. Let j be fixed. We may assume that $x^j=0$ and that $y=f^{(j)}(x)$ has the canonical form described in Remark 3.2.1/1. In particular, the Jacobian $\frac{\partial(y)}{\partial(x)}(0)$ is the unit matrix. Let $u(x) \in F^{\sigma+2m+s}_{p,q}(\Omega)$ and let $\psi_j(x)$ be the above function. Then we transform $u(x)\,\psi_j(x)$ via $y=f^{(j)}(x)$ and denote the transformed functions by a dash, e.g. $\psi_j'(y)=\psi_j(f^{(j)^{-1}}(y))$ etc. Let

$$(A\psi_j u)'(y)=A'\psi_j'(y)\,u'(y)=\sum_{|\alpha|\leq 2m} a_\alpha'(y)\,D^\alpha\psi_j'(y)\,u'(y)\,, \tag{7}$$

$$(B_l\psi_j u)'(y)=B_l'\psi_j'(y)\,u'(y)=\sum_{|\beta|\leq m_l} b_{l,\beta}'(y)\,D^\beta\psi_j'(y)\,u'(y)\,, \tag{8}$$

$$A_0'v(y)=\sum_{|\alpha|=2m} a_\alpha'(0)\,D^\alpha v(y)\,, \tag{9}$$

$$B_{0,l}'v(y)=\sum_{|\beta|=m_l} b_{l,\beta}'(0)\,D^\beta v(y)\,. \tag{10}$$

In (7) and (9) we have $y \in f^{(j)}(\Omega\cap K_j)$ and $y \in R_n$, respectively. In (8) and (10) we assume additionally $y_n=0$; furthermore $l=1, \ldots, m$. By the above hypotheses, it follows that $a_\alpha'(0)=a_\alpha(0)$ if $|\alpha|=2m$, and $b_{l,\beta}'(0)=b_{l,\beta}(0)$ if $l=1, \ldots, m$ and $|\beta|=m_l$. In particular, $\{A_0'; B_{0,1}', \ldots, B_{0,m}'\}$ is a regular elliptic system in R_n^+ in the sense of Definition 4.1.2/4 having only constant top-order coefficients. Theorem 4.2.3 yields

$$\begin{aligned}\|\psi_j'u' \mid F^{\sigma+2m+s}_{p,q}(R_n^+)\| \leq c[\|A_0'\psi_j'u' \mid F^{\sigma+s}_{p,q}(R_n^+)\| + \|\psi_j'u' \mid F^{\sigma+2m+s-1}_{p,q}(R_n^+)\| \\ +\sum_{l=1}^{m} \|(B_{0,l}'\psi_j'u')(y',0) \mid B^{\sigma+2m+s-m_j-\frac{1}{p}}_{p,p}(R_{n-1})\|]\,.\end{aligned} \tag{11}$$

In order to replace A_0' and $B_{0,l}'$ in (11) by A' and B_l', respectively, we use the same technique as in the second step. After re-transformation we find that

$$\begin{aligned}\|\psi_j u \mid F^{\sigma+2m+s}_{p,q}(\Omega)\| \leq \varepsilon\|u \mid F^{\sigma+2m+s}_{p,q}(\Omega)\| + \varepsilon\|u \mid B^{\sigma+2m+s-\frac{1}{p}}_{p,p}(\partial\Omega)\| \\ +c\|Au \mid F^{\sigma+s}_{p,q}(\Omega)\| + c\sum_{l=1}^{m} \|B_l u \mid B^{\sigma+2m+s-m_j-\frac{1}{p}}_{p,p}(\partial\Omega)\| \\ +c\|u \mid F^{\sigma+2m+s-1}_{p,q}(\Omega)\| + c\,\|u \mid B^{\sigma+2m+s-1-\frac{1}{p}}_{p,p}(\partial\Omega)\|\,,\end{aligned} \tag{12}$$

where $\varepsilon>0$ can be chosen arbitrarily small (which depends on the diameters of $K_1, \ldots, K_N$). Cf. also Definition 3.2.2/2 and Proposition 3.2.3. By Theorem 3.3.3 the second term on the right-hand side of (12) can be estimated from above by the first one, and the last one by the last but one term. Now we combine (6) and (12) and choose ε in an appropriate way. Then

$$\begin{aligned}&\|u \mid F^{\sigma+2m+s}_{p,q}(\Omega)\| \\ &\leq c\|Au \mid F^{\sigma+s}_{p,q}(\Omega)\| + c\sum_{l=1}^{m} \|B_l u \mid B^{\sigma+2m+s-m_j-\frac{1}{p}}_{p,p}(\partial\Omega)\| + c\|u \mid F^{\sigma+2m+s-1}_{p,q}(\Omega)\|.\end{aligned} \tag{13}$$

Step 4. We claim that

$$\|u \mid F^{\sigma+2m+s-1}_{p,q}(\Omega)\| \leq \varepsilon\|u \mid F^{\sigma+2m+s}_{p,q}(\Omega)\| + c\|u \mid F^{\sigma}_{p,q}(\Omega)\|, \tag{14}$$

where $\varepsilon>0$ is an arbitrary number and c depends on ε. However, (14) is an easy consequence of the real interpolation formula (3.3.6/10) and (3.2.4/3). Now the left-hand side of (3) follows from (13) and (14). The proof is complete.

Corollary. Let $\{A; B_1, \ldots, B_m\}$ be a regular elliptic system in Ω in the sense of Definition 4.1.2/4. Let p, q, and σ be real numbers with (4.2.2/28). Let $s=0, 1, 2, \ldots$ Then there exist two positive numbers c_1 and c_2 such that

$$c_1\|u \mid B^{\sigma+2m+s}_{p,q}(\Omega)\| \leqq \|Au \mid B^{\sigma+s}_{p,q}(\Omega)\| + \sum_{j=1}^{m} \|B_j u \mid B_{p,q}^{\sigma+2m+s-m_j-\frac{1}{p}}(\partial\Omega)\| + \|u \mid B^{\sigma}_{p,q}(\Omega)\| \leqq c_2\|u \mid B^{\sigma+2m+s}_{p,q}(\Omega)\| \tag{15}$$

holds for every $u \in S(R_n)$.

Proof. The proof is the same as the proof of the above theorem. Instead of Theorem 4.2.3 we use Corollary 4.2.3.

Remark 1. The above theorem and the corollary are the main goals of this chapter. In the corollary we again restrict (15) to smooth functions u, where $u \in S(R_n)$ can be replaced by $u \in C^\infty(\Omega)$, cf. 3.2.4. If $p<\infty$ and $q<\infty$, then (15) holds for every $u \in B^{\sigma+2m+s}_{p,q}(\Omega)$.

Remark 2. The classical a priori estimates in (Hölder)-Zygmund spaces or in L_p-spaces with $1<p<\infty$ are special cases of (3) and (15). The most interesting fact of (3) and (15) is the extension of these classical a priori estimates to certain spaces with $p \leqq 1$ (and/or $q \leqq 1$). However these extensions are restricted by (4.2.2/1) and (4.2.2/28), cf. also Fig. 4.2.2. This is a consequence of our method. It would be of great interest to find modifications, or other methods, such that (9) and (15) hold without any restrictions on p and q (the restrictions on σ can be removed via real and complex interpolation, cf. the following section.)

4.3. Boundary Value Problems

4.3.1. Introduction and Hypothesis

The main results in this chapter are the a priori estimates in Theorem 4.2.4 and in Corollary 4.2.4. We have mentioned that these a priori estimates contain the classical a priori estimates in the (Hölder)-Zygmund spaces and in the L_p-spaces (including the Sobolev spaces $W^m_p(\Omega)$ where $m=1, 2, 3, \ldots$) with $1<p<\infty$ as special cases (concerning the Zygmund spaces cf. also 4.3.4.). The theory of the boundary value problems (4.1.1/3) in these spaces is based on the corresponding a priori estimates. Via the real interpolation method and the classical complex interpolation method one can extend this theory to the Besov spaces $\Lambda^s_{p,q}(\Omega)$ with $s>0$, $1<p<\infty$, $1\leqq q\leqq\infty$ and Bessel-potential spaces $H^s_p(\Omega)$ with $s>0$, $1<p<\infty$. This was the main aim of [I, Chapter 5]. We always assume that Ω is a bounded C^∞-domain in R_n, cf. Definition 3.2.1. The main goal of the theory of the boundary value problems (4.1.1/3) in (Hölder)-Zygmund spaces and in Sobolev spaces (and also of its extensions to the above spaces $\Lambda^s_{p,q}(\Omega)$ and $H^s_p(\Omega)$ in the sense of [I, Chapter 5]) is the statement that the regular elliptic system $\{A; B_1, \ldots, B_m\}$ in Ω is a Φ-operator (Noether operator, Fredholm operator, operator with index) between appropriate Banach spaces. We recall that a bounded linear operator T which maps a Banach space C_0 into a Banach space C_1 is called a Φ-operator if its kernel (null space) $N(T)$ is finite-dimensional and if its range (image) $R(T)$ is closed (in C_1) and of finite codimension. In order to give the idea, we formulate a well-known special result, which may be found e.g. in [I, 5.4.5.]. Let $\{A; B_1, \ldots, B_m\}$ be a regular elliptic system in Ω in the sense of Definition 4.1.2/4. Let $1<p<\infty$. Then T, given by $Tu=\{Au; B_1u, \ldots, B_mu\}$, is a Φ-operator

$$\text{from} \quad W^{2m}_p(\Omega) \quad \text{into} \quad L_p(\Omega) \times \prod_{j=1}^{m} \Lambda_{p,p}^{2m-m_j-\frac{1}{p}}(\partial\Omega) \tag{1}$$

(where $\Lambda_{p,p}^{2m-m_j-\frac{1}{p}}(\partial\Omega)=B_{p,p}^{2m-m_j-\frac{1}{p}}(\partial\Omega)$ in our previous notations). Of particular interest is the case $\dim N(T)=\operatorname{codim} R(T)=0$. Then T yields a one-to-one mapping from $W_p^{2m}(\Omega)$ onto the space on the right-hand side of (1). This problem is closely connected with the question concerning the spectrum and the resolvent set of T. The latter is the collection of all complex numbers λ such that

$$\dim N(T-\lambda E)=\operatorname{codim} R\,(T-\lambda E)=0$$

(where E is the unit operator). The following alternative is well-known: either the resolvent set of T is empty (i.e. the spectrum of T coincides with the complex plane) or the spectrum of T is an infinite countable set of complex numbers with no finite cluster points (cf. e.g. [I, 5.4.4.]). Of interest are additional conditions for the regular elliptic system $\{A;\,B_1,\ldots,B_m\}$ such that the resolvent set of T is not empty (then we have the second case of the above alternative). Sufficient (and almost necessary) conditions of that type have been given by S. Agmon [1] (a description may also be found in [I, Theorem 4.9.1]). Obviously, if 0 is an element of the resolvent set of T then

$$\begin{aligned}&(Au)(x)=0 \quad \text{if} \quad x\in\bar{\Omega} \quad \text{and}\\ &(B_ju)(y)=0 \quad \text{if} \quad j=1,\ldots,m \quad \text{and} \quad y\in\partial\Omega\end{aligned} \tag{2}$$

with $u(x)\in C^\infty(\bar{\Omega})$ has only the trivial solution $u(x)\equiv 0$ (the space $C^\infty(\bar{\Omega})$ has been defined at the beginning of 3.2.4.). The converse assertion is also valid: if (2), with $u(x)\in C^\infty(\bar{\Omega})$, has only the trivial solution $u(x)\equiv 0$, then 0 is an element of the resolvent set of T, cf. [I, 5.4.4.]. Of course, it would be of interest to extend the theory of boundary value problems for $\{A;\,B_1,\ldots,B_m\}$ from the (Hölder)-Zygmund spaces, the Sobolev spaces, the above Besov spaces $\Lambda_{p,q}^s(\Omega)$ and Bessel-potential spaces $H_p^s(\Omega)$ to the more general spaces $B_{p,q}^s(\Omega)$ and $F_{p,q}^s(\Omega)$ from the third chapter of this book. However, there are two restrictions. The first one is connected with the theory of Φ-operators: some techniques which we used in [I, Chapter 5] are based on Banach spaces, and there are no immediate counterparts for quasi-Banach spaces (recall that $B_{p,q}^s(\Omega)$ and $F_{p,q}^s(\Omega)$ are not Banach spaces if $\min(p,q)<1$). Perhaps one can find appropriate modifications of these techniques which allow the transfer of the full theory of boundary value problems for $\{A;\,B_1,\ldots,B_m\}$ in Ω from the above more or less classical spaces to the more general spaces $B_{p,q}^s(\Omega)$ and $F_{p,q}^s(\Omega)$. We prefer another way. We assume from the very beginning that 0 is an element of the resolvent set of the above operator T. Of course, one can replace T by $T-\lambda E$.

Hypothesis. *Let Ω be a bounded C^∞-domain, cf. Definition* 3.2.1. *Let $\{A;\,B_1,\ldots,B_m\}$ be a regular elliptic system in Ω in the sense of Definition* 4.1.2/4 *such that* (2) *with $u(x)\in C^\infty(\bar{\Omega})$ has only the trivial solution $u(x)\equiv 0$.*

Remark. Recall that $C^\infty(\bar{\Omega})$ is the restriction of $S(R_n)$ to Ω, cf. also the beginning of 3.2.4. In the same way one defines $C^\infty(\partial\Omega)$ as the restriction of $S(R_n)$ to $\partial\Omega$. Then $C^\infty(\partial\Omega)$ is the set of all complex-valued infinitely differentiable functions on $\partial\Omega$.

Reformulation. The Hypothesis means that 0 belongs to the resolvent set of T. Then it follows that

$$(Au)(x)=f(x)\in C^\infty(\bar{\Omega})\,, \tag{3}$$

$$(B_ju)(y)=g_j(y)\in C^\infty(\partial\Omega) \quad \text{if} \quad y\in\partial\Omega \quad \text{and} \quad j=1,\ldots,m\,, \tag{4}$$

has a uniquely determined solution $u(x)\in C^\infty(\bar{\Omega})$: by the above considerations there exists a uniquely determined solution $u(x)\in W_p^{2m}(\Omega)$ of (3), (4). Subseqently, the smoothness theory of regular elliptic boundary value problems proves $u \in C^\infty(\bar{\Omega})$, cf. e.g. [I, Chapter 5]. We can take this statement as a reformulaton of the Hypothesis.

We have spoken about two restrictions, the first of which we circumvented via the above Hypothesis. The second one is more serious. The later statements are based on Theorem 4.2.4 and Corollary 4.2.4. In particular, we have the restrictions (4.2.2/1) and $\min(p, q) \geqq \min(p, 1)$ in Theorem 4.2.4. and (4.2.2/28) in Corollary 4.2.4, respectively. That means $p > \frac{n-1}{n}$, cf. Fig. 4.2.2. With the exception of $n=1$, this is a serious restriction on p and with our method we have no possibility of relaxing it. As far as $\min(p, q) \geqq \min(p, 1)$ in Theorem 4.2.4 is concerned, we mentioned several times that this condition can be replaced by milder ones, e.g. by (3.4.3/2). However, in any case there remain restrictions on q if p is given. Finally, the restrictions on σ and s in Theorem 4.2.4 and Corollary 4.2.4 can be removed via real and complex interpolation (and we shall do it). The main problem of this chapter is clear: the unnatural restrictions on p and q in the theorems in 4.3.2. and 4.3.3. should be removed.

4.3.2. The Basic Theorem

Our first task is to remove the terms $\|u \mid F^\sigma_{p,q}(\Omega)\|$ and $\|u \mid B^\sigma_{p,q}(\Omega)\|$ in Theorem 4.2.4 and Corollary 4.2.4, respectively if Hypothesis 4.3.1 is satisfied. This can be done by routine arguments, but we sketch a proof.

Theorem. Let $\{A; B_1, \ldots, B_m\}$ be the regular elliptic system in Ω from Hypothesis 4.3.1. Let p, q, and σ be real numbers with (4.2.2/1) and $\min(p, q) \geqq \min(p, 1)$. Let $s = 0, 1, 2, \ldots$ Then there exist two positive numbers c_1 and c_2 such that

$$c_1\|u \mid F^{\sigma+2m+s}_{p,q}(\Omega)\| \leqq \|Au \mid F^{\sigma+s}_{p,q}(\Omega)\| + \sum_{j=1}^{m} \|B_j u \mid B^{\sigma+2m+s-m_j-\frac{1}{p}}_{p,p}(\partial\Omega)\| \leqq c_2\|u \mid F^{\sigma+2m+s}_{p,q}(\Omega)\| \tag{1}$$

holds for every $u \in F^{\sigma+2m+s}_{p,q}(\Omega)$.

Proof. The right-hand side of (1) follows immediately from (4.2.4/3). In order to prove the left-hand side of (1), we assume that there does not exist a positive constant c_1 with the property claimed. Then we find a sequence $\{u_l\}_{l=1}^\infty \subset F^{\sigma+2m+s}_{p,q}(\Omega)$ with $\|u_l \mid F^{\sigma+2m+s}_{p,q}(\Omega)\| = 1$ and

$$\|Au_l \mid F^{\sigma+s}_{p,q}(\Omega)\| + \sum_{j=1}^{m} \|B_j u_l \mid B^{\sigma+2m+s-m_j-\frac{1}{p}}_{p,p}(\partial\Omega)\| \to 0 \quad \text{if} \quad l \to \infty. \tag{2}$$

In Remark 1 below we prove that the embedding from $F^{\sigma+2m+s}_{p,q}(\Omega)$ into $F^\sigma_{p,q}(\Omega)$ is compact. Hence, we may assume without restriction of generality that $\{u_l\}_{l=1}^\infty$ is a fundamental sequence in $F^\sigma_{p,q}(\Omega)$. Then it follows from (4.2.4/3) and (2) that $\{u_l\}_{l=1}^\infty$ is also a fundamental sequence in $F^{\sigma+2m+s}_{p,q}(\Omega)$. Let u be the limiting element. Then

$$\|u \mid F^{\sigma+2m+s}_{p,q}(\Omega)\| = 1 \tag{3}$$

and (4.3.1/2). In Remark 2 below we prove that $u \in C^\infty(\bar{\Omega})$. Then it follows from Hypothesis 4.3.1 that $u(x) \equiv 0$, which contradicts (3). The proof is complete.

Corollary. Let $\{A; B_1, \ldots, B_m\}$ be the regular elliptic system in Ω from

Hypothesis 4.3.1. Let p, q, and σ be real numbers with (4.2.2/28). Let $s=0, 1, 2, \ldots$ Then there exist two positive numbers c_1 and c_2 such that

$$c_1\|u \mid B^{\sigma+2m+s}_{p,q}(\Omega)\| \leqq \|Au \mid B^{\sigma+s}_{p,q}(\Omega)\| + \sum_{j=1}^{m} \|B_j u \mid B_{p,q}^{\sigma+2m+s-m_j-\frac{1}{p}}(\partial\Omega)\| \leqq c_2\|u \mid B^{\sigma+2m+s}_{p,q}(\Omega)\| \tag{4}$$

holds for every $u \in S(R_p)$.

Proof. The proof is the same as the proof of the above theorem if we use Corollary 4.2.4 instead of Theorem 4.2.4.

Remark 1. Let $0<p\leqq\infty$, $0<q\leqq\infty$, $-\infty<s<\infty$, and $\varepsilon>0$. We prove that the embedding from $B^{s+\varepsilon}_{p,q}(\Omega)$ into $B^{s}_{p,q}(\Omega)$ and from $F^{s+\varepsilon}_{p,q}(\Omega)$ into $F^{s}_{p,q}(\Omega)$ is compact (with $p<\infty$ in the latter case). By (3.2.4/3) and the counterparts of (2.3.2/7) and (2.3.2/8) with Ω instead of R_n it is sufficient to prove that the embedding from $B^{s+\varepsilon}_{p,q}(\Omega)$ into $B^{s}_{p,q}(\Omega)$ is compact. If ε is a large positive number, it follows from Theorem 3.3.1, that the embedding from $B^{s+\varepsilon}_{p,q}(\Omega)$ into the Zygmund space $\mathscr{C}^{s+\delta}(\Omega)=B^{s+\delta}_{\infty,\infty}(\Omega)$ is continuous, where $s+\delta>1$, $1<\delta<\varepsilon-\frac{n}{p}$. We use the well-known fact that the embedding from $\mathscr{C}^{s+\delta}(\Omega)$ into $\mathscr{C}^{s+\delta-1}(\Omega)$ is compact (this is essentially the theorem of Arzelà-Ascoli). Furthermore, by Theorem 3.3.1 the embedding from $\mathscr{C}^{s+\delta-1}(\Omega)$ into $B^{s}_{p,q}(\Omega)$ is continuous. Hence, the embedding from $B^{s_1}_{p,q}(\Omega)$ into $B^{s_0}_{p,q}(\Omega)$ is compact if s_1-s_0 is large. Let $s=(1-\Theta)\,s_0+\Theta s_1$ with $0<\Theta<1$. Then it follows from the real interpolation formula (3.3.6/9) that

$$\|u \mid B^{s}_{p,q}(\Omega)\| \leqq c\|u \mid B^{s_0}_{p,q}(\Omega)\|^{1-\Theta}\|u \mid B^{s_1}_{p,q}(\Omega)\|^{\Theta} \tag{5}$$

holds for every $u\in B^{s_1}_{p,q}(\Omega)$, where c is independent of u. If $\{u_l\}_{l=1}^{\infty}$ is a bounded sequence in $B^{s_1}_{p,q}(\Omega)$ which converges in $B^{s_0}_{p,q}(\Omega)$, then it follows from (5) that it converges also in $B^{s}_{p,q}(\Omega)$. The proof is complete.

Remark 2. We prove the following statement. Let p, q, and σ be real numbers satisfying (4.2.2/1) and let $u\in F^{\sigma+2m}_{p,q}(\Omega)$ satisfy (4.3.1/2). We claim that $u(x)\equiv 0$. First we assume that $1<p<\infty$. Then $u\in H^{\sigma-\varepsilon+2m}_{p}(\Omega)$, where $\varepsilon>0$ is an arbitrary number (recall that $H^{\sigma-\varepsilon+2m}_{p}(\Omega)$ is a Bessel-potential space). Hence, the desired assertion is essentially a consequence of [I, Theorem 5.7.1]. If $0<p\leqq 1$, it follows from (3.3.1/8)| that $u\in F^{\tilde{\sigma}+2m}_{\tilde{p},q}(\Omega)$ with $1<\tilde{p}<\infty$ such that $\tilde{p}$, q, and $\tilde{\sigma}$ satisfy (4.2.2/1), too. We have the situation previously treated and consequently $u(x)\equiv 0$.

4.3.3. The Main Theorem

We always assume that Ω is a bounded C^∞-domain in R_n, cf. Definition 3.2.1.

Theorem. Let $\{A; B_1, \ldots, B_m\}$ be the regular elliptic system in Ω from Hypothesis 4.3.1.

(i) Let

$$\begin{cases} \text{either} \quad 1\leqq p<\infty, \quad 0<q\leqq\infty, \quad s>\frac{1}{p}-1 \\ \text{or} \quad \frac{n-1}{n}<p<1, \quad 0<q\leqq\infty, \quad s>n\left(\frac{1}{p}-1\right). \end{cases} \tag{1}$$

Then $\{A; B_1, \ldots, B_m\}$ yields an isomorphic mapping

$$\text{from} \quad B^{s+2m}_{p,q}(\Omega) \quad \text{onto} \quad B^{s}_{p,q}(\Omega)\times\prod_{j=1}^{m} B_{p,q}^{s+2m-m_j-\frac{1}{p}}(\partial\Omega). \tag{2}$$

(ii) Let

$$\begin{cases} \text{either} \quad 1 \leqq p < \infty, \quad 1 \leqq q < \infty, \quad s > \frac{1}{p} - 1 \\ \text{or} \quad \frac{n-1}{n} < p < 1, \quad p \leqq q < \infty, \quad s > n\left(\frac{1}{p} - 1\right). \end{cases} \tag{3}$$

Then $\{A; B_1, \ldots, B_m\}$ yields an isomorphic mapping

$$\text{from} \quad F^{s+2m}_{p,q}(\Omega) \quad \text{onto} \quad F^s_{p,q}(\Omega) \times \prod_{j=1}^{m} B^{s+2m-m_j-\frac{1}{p}}_{p,p}(\partial\Omega). \tag{4}$$

Proof. Step 1. Let p, q and σ be real numbers with (4.2.2/1) and $\min(p, q) \geqq \min(p, 1)$. Let $k = 0, 1, 2, \ldots$ Then it follows from Theorem 4.3.2 and the reformulation of Hypothesis 4.3.1 that $\{A; B_1, \ldots, B_m\}$ yields an isomorphic mapping

$$\text{from} \quad F^{\sigma+2m+k}_{p,q}(\Omega) \quad \text{onto} \quad F^{\sigma+k}_{p,q}(\Omega) \times \prod_{j=1}^{m} B^{\sigma+k+2m-m_j-\frac{1}{p}}_{p,p}(\partial\Omega). \tag{5}$$

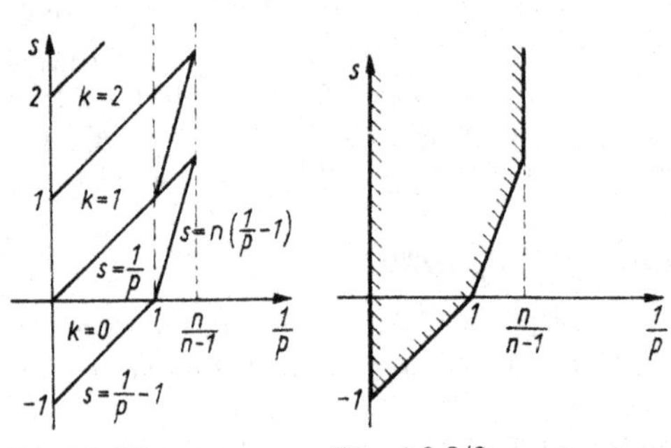

Fig. 4.3.3/1 Fig. 4.3.3/2

The restrictions on p and q are the same as in (3). We wish to apply the complex interpolation formula (3.3.6/8) with $p_0 = p_1 = p$, $q_0 = q_1 = q$, $s_0 = \sigma$ and $s_1 = \sigma + k$ with $k = 1, 2, 3, \ldots$, where p, q, and σ have the above meaning. In other words, the interpolation of the areas in Fig. 1 which are characterized by the different values of k, yields the shaded area in Fig. 2, which coincides with (3). In Remark 1 below we prove that both $\{A; B_1, \ldots, B_m\}$ and its inverse $\{A; B_1, \ldots, B_m\}^{-1}$ have the interpolation property (cf. Remark 3.3.6/4). Hence (5), (3.3.6/8), and Proposition 3.3.6(i) (including the claimed interpolation property) prove part (ii) of the theorem.

Step 2. Part (i) of the theorem follows from part (ii), the real interpolation formula (3.3.6/10) and its counterpart with $\partial\Omega$ instead of Ω in the sense of Proposition 3.3.6(i) (for the real interpolation method there are no problems with the interpolation property). The proof is complete.

Remark 1. We outline a proof that both $T = \{A; B_1, \ldots, B_m\}$ and T^{-1} have the interpolation property for the complex interpolation method. First, were mark that an arbitrary differential operator $\tilde{A}u = \sum_{|\alpha| \leqq M} \tilde{a}_\alpha(x) D^\alpha u$ on R_n with C^∞-coefficients $\tilde{a}_\alpha(x)$ (all derivatives are bounded on R_n) has the interpolation property. For derivatives D^α this is obvious, cf. the properties of an $S'(R_n)$-analytic function $f(z)$ in 2.4.4., which are preserved for $D^\alpha f(z)$. Concerning multiplications with $\tilde{a}_\alpha(x)$ cf. Step 1 of the proof of Corollary 2.8.2. It follows that $\tilde{A}$ has the interpolation property for the complex interpolation of spaces on R_n. The corresponding assertion for spaces on Ω follows from Definition 3.3.6 and Remark 3.3/6/1. Similarly for spaces on $\partial\Omega$, cf. the end of 3.3.6., e.g. (3.3.6/17). This proves that the above operator T has the interpolation property for the complex interpolation method. Via the definitions in 3.3.6. and the notion of an $S'(R_n)$-analytic function $f(z)$ in 2.4.4. one can introduce $D'(\Omega)$-analytic functions and $D'(\partial\Omega)$-analytic functions. The procedure is quite clear and so we shall not go into details. Again $f(z)$ is defined in the strip $\{z \mid 0 \leqq \operatorname{Re} z \leqq 1\}$. On the basis of Theorem 4.3.2, it is not complicated to prove that T yields a one-to-one mapping of $D'(\Omega)$-analytic functions $f(z)$ with

$$f(it) \in F^{\sigma+2m}_{p,q}(\Omega), \quad f(1+it) \in F^{\sigma+2m+k}_{p,q}(\Omega)$$

($k=0, 1, 2, \ldots$ and uniformly bounded quasi-norms with respect to $t \in R_1$) onto a corresponding subset of $D'(\Omega) \times D'(\partial\Omega) \times \ldots \times D'(\partial\Omega)$-analytic functions $g(z)$ with

$$g(it) \in F^{\sigma}_{p,q}(\Omega) \times \prod_{j=1}^{m} B_{p,p}^{\sigma+2m-m_j-\frac{1}{p}}(\partial\Omega), \quad t \in R_1, \tag{6}$$

$$g(1+it) \in F^{\sigma+k}_{p,q}(\Omega) \times \prod_{j=1}^{m} B_{p,p}^{\sigma+2m+k-m_j-\frac{1}{p}}(\partial\Omega), \quad t \in R_1. \tag{7}$$

Since every $D'(\Omega) \times D'(\partial\Omega) \times \ldots . \times D'(\partial\Omega)$-analytic function $g(z)$ satisfying (6) and (7) can be represented in this way, i.e. $g(z) = Tf(z)$, it follows that T^{-1} also has the interpolation property with respect to the complex interpolation method.

Remark 2. We proved part (i) of the theorem by real interpolation of part (ii). Another possibility would be the direct application of Corollary 4.3.2. However, in that case we must exclude $q=\infty$ in (1), because (4.3.2/4) is restricted to smooth functions, which are not dense in $B^s_{p,\infty}(\Omega)$. The interpolation method has the advantage that it includes non-separable spaces in which $C^\infty(\bar{\Omega})$ is not dense.

4.3.4. Boundary Value Problems in Zygmund Spaces

What about the case $p=\infty$ in Theorem 4.3.3? In 2.3.4. we introduced the spaces $F^s_{\infty,q}(R_n)$ with $1<q\leqq\infty$. We proved in Theorem 2.5.8/2 that $F^0_{\infty,2}(R_n)=bmo(R_n)$ (the local space of all functions with bounded mean oscillation). However, with few exceptions, we restricted the study of the spaces $F^s_{p,q}$ to the case $p<\infty$. One can hope (and conjecture) that the theory of the spaces $F^s_{p,q}$ with $p<\infty$ can be carried over to the spaces $F^s_{\infty,q}$ (with $1<q\leqq\infty$). In this sense one can expect that Theorem 4.3.3(ii) can be extended in some way to these spaces. However, there are no results at present. More interesting for us (and within our considerations) is the question of whether Theorem 4.3.3(i) can be extended to $p=\infty$. It will be sufficient to deal with the case $p=q=\infty$, since other values of q can be included afterwards via the real interpolation formula (3.3.6/9) and its counterpart with $\partial\Omega$ instead of Ω in the sense of Proposition 3.3.6(i). Let

$$\mathcal{C}^s(\Omega) = B^s_{\infty,\infty}(\Omega) \quad \text{and} \quad \mathcal{C}^s(\partial\Omega) = B^s_{\infty,\infty}(\partial\Omega) \quad \text{if} \quad -\infty<s<\infty. \tag{1}$$

If $s>0$, we have the usual Zygmund spaces (or Hölder spaces if $0<s\neq$ integer). cf. (2.5.12/22) (including the given references) and (3.4.2/2).

Theorem. Let $\{A; B_1, \ldots, B_m\}$ be the regular elliptic system in Ω from Hypothesis 4.3.1. Let $s>-1$. Then $\{A; B_1, \ldots, B_m\}$ yields an isomorphic mapping

$$\text{from} \quad \mathcal{C}^{s+2m}(\Omega) \quad \text{onto} \quad \mathcal{C}^s(\Omega) \times \prod_{j=1}^{m} \mathcal{C}^{s+2m-m_j}(\partial\Omega). \tag{2}$$

Proof. Let $c^s(\Omega)$ be the completion of $C^\infty(\bar{\Omega})$ in the norm $\|\cdot \mid B^s_{\infty,\infty}(\Omega)\|$ (recall that $C^\infty(\bar{\Omega})$ is the restriction of $S(R_n)$ to Ω). Similarly $c^s(\partial\Omega)$. It follows from Corollary 4.3.2 (with $p=q=\infty$) and from the reformulation of the hypothesis in 4.3.1. that $\{A; B_1, \ldots, B_m\}$ yields an isomorphic mapping

$$\text{from} \quad c^{s+2m}(\Omega) \quad \text{onto} \quad c^s(\Omega) \times \prod_{j=1}^{m} c^{s+2m-m_j}(\partial\Omega) . \tag{3}$$

In H. Triebel [10, pp. 1155/1156] we proved the following real interpolation formula: if $-\infty < s_0 < s_1 < \infty$, $0 < \Theta < 1$ and $s = (1-\Theta)\, s_0 + \Theta s_1$, then

$$\left(c^{s_0}(\Omega),\, c^{s_1}(\Omega)\right)_{\Theta,\infty} = \mathcal{C}^s(\Omega) \tag{4}$$

and a corresponding formula with $\partial\Omega$ instead of Ω. Application of (4) to (3) yields the desired assertion.

Remark 1. The above-mentioned real interpolation proves that $\{A; B_1, \ldots, B_m\}$ yields also an isomorphic mapping

$$\text{from} \quad \mathring{B}^{s+2m}_{\infty,q}(\Omega) \quad \text{onto} \quad B^s_{\infty,q}(\Omega) \times \prod_{j=1}^{m} B^{s+2m-m_j}_{\infty,q}(\partial\Omega)$$

if $s > -1$ and $0 < q < \infty$.

Remark 2. This application was the main reason for the formulation of Corollary 4.3.2 and its forerunners in the preceding subsections. Once the above theorem has been established it is clear that the a priori estimate (4.3.2/4) holds for every $u \in B^{\sigma+2m+s}_{p,q}(\Omega)$ including the limiting cases $\max(p, q) = \infty$.

Remark 3. The most interesting aspect of the above proof is the following: it is sufficient to prove a priori estimates in the sense of Corollary 4.3.2 for smooth functions (and not to be forced to fight with a lot of technical difficulties for non-smooth functions). The rest is done by interpolation.

5. Homogeneous Function Spaces

5.1. Definitions and Basic Properties

5.1.1. Introduction (to Part II)

A we mentioned in the Preface, Part II of this book deals with further types of function spaces which are more or less closely related to the spaces $B^s_{p,q}(R_n)$ and $F^s_{p,q}(R_n)$ from Chapter 2 and their counterparts on domains in Chapter 3. There are many possible modifications and we discuss a few of them. We restrict ourselves to formulations, comments, and occasionally to outlines of some basic ideas. As far as detailed proofs are concerned the reader is asked to consult the quoted references. In many cases the omitted proofs are more or less obvious modifications of corresponding proofs in Chapter 2. We shall give hints. This chapter deals with homogeneous spaces, which are very close to the "non-homogeneous" spaces from the second chapter. The periodic counterparts of these spaces are studied in Chapter 9. Chapters 7 and 8 deal with weighted counterparts. The necessary preparations for these two chapters will be given in Chapter 6. It deals with weighted spaces of entire analytic functions, and is a generalization and modification of the first chapter. Finally, in Chapter 10 we list very briefly further possibilities.

5.1.2. The Spaces $Z(R_n)$ and $Z'(R_n)$

The Euclidean n-space R_n, the Schwartz space $S(R_n)$ and the space $S'(R_n)$ of tempered distributions have the same meaning as in 1.2.1. Let $x=(x_1, \ldots, x_n)$ be the general point of R_n. We recall that $S(R_n)$ is a complete locally convex space, where the topology is generated by the norms (1.2.1/1). The Fourier transform F and its inverse F^{-1} have the same meaning as in (1.2.1/2) and (1.2.1/3), respectively. They are extended in the usual way to $S'(R_n)$. Let

$$Z(R_n)=\{\varphi \mid \varphi\in S(R_n),\ (D^\alpha F\varphi)(0)=0 \quad \text{for every multi-index } \alpha\}. \tag{1}$$

We consider $Z(R_n)$ as a subspace of $S(R_n)$, including the topology. Thus $Z(R_n)$ is a complete locally convex space (recall that F and F^{-1} yield isomorphic mappings from $S(R_n)$ onto itself). Let $Z'(R_n)$ be the topological dual of $Z(R_n)$, i.e. the set of all continuous linear functionals on $Z(R_n)$. If $f\in S'(R_n)$, the restriction of f to $Z(R_n)$ belongs to $Z'(R_n)$. Furthermore, if $P(x)$ is an arbitrary polynomial in R_n, then

$$(f+P(x))(\varphi)=f(\varphi) \quad \text{if} \quad \varphi\in Z(R_n). \tag{2}$$

Conversely, if $f\in Z'(R_n)$, then f can be extended linearly and continuously from $Z(R_n)$ to $S(R_n)$, i.e. to an element of $S'(R_n)$. If f_1 and f_2 are two different extensions of f, then $\operatorname{supp} F(f_1-f_2)=\{0\}$, i.e. f_1-f_2 is a polynomial. In other words, we may identify $Z'(R_n)$ with the factor space $S'(R_n)/P(R_n)$, where $P(R_n)$ is the collection of all polynomials $\sum a_\alpha x^\alpha$ in R_n, where $x^\alpha=\prod_{k=1}^{n} x_k^{\alpha_k}$, and a_α are complex numbers (finite sum). Subspaces of $Z'(R_n)$ are considered as subspaces of $S'(R_n)$ modulo $P(R_n)$.

5.1.3. Definitions

This subsection is the counterpart of 2.3.1.

Definition 1. Let $\dot{\Phi}(R_n)$ be the set of all systems $\varphi=\{\varphi_j(x)\}_{j=-\infty}^{\infty}\subset S(R_n)$ such that:

$$\operatorname{supp}\varphi_j\subset\{x \mid 2^{j-1}\leqq|x|\leqq 2^{j+1}\} \quad \text{if } j \text{ is an integer}, \tag{1}$$

H. Triebel, *Theory of Function Spaces*, Modern Birkhäuser Classics,
DOI 10.1007/978-3-0346-0416-1_5, © Birkhäuser Verlag 1983

for every multi-index α there exists a positive number c_α such that

$$2^{j|\alpha|}|D^\alpha \varphi_j(x)| \leqq c_\alpha \quad \text{for all integers } j \text{ and all } x \in R_n \,, \tag{2}$$

and

$$\sum_{j=-\infty}^{\infty} \varphi_j(x) \equiv 1 \quad \text{for every} \quad x \in R_n - \{0\} \,. \tag{3}$$

Remark 1. This is the counterpart of Definition 2.3.1/1. Remark 2.3.1/1 shows that there exist systems with the desired properties.

Definition 2. Let $-\infty < s < \infty$ and $0 < q \leqq \infty$. Let $\varphi = \{\varphi_j(x)\}_{j=-\infty}^{\infty} \in \dot{\Phi}(R_n)$.
(i) If $0 < p \leqq \infty$, then

$$\dot{B}^s_{p,q}(R_n) = \left\{ f \mid f \in Z'(R_n), \quad \|f \mid \dot{B}^s_{p,q}(R_n)\|^\varphi = \left(\sum_{j=-\infty}^{\infty} 2^{jsq} \|F^{-1}\varphi_j F f \mid L_p(R_n)\|^q \right)^{\frac{1}{q}} < \infty \right\} \tag{4}$$

(modification if $q=\infty$).
(ii) If $0 < p < \infty$, then

$$\dot{F}^s_{p,q}(R_n) = \left\{ f \mid f \in Z'(R_n), \quad \|f \mid \dot{F}^s_{p,q}(R_n)\|^\varphi = \left\| \left(\sum_{j=-\infty}^{\infty} 2^{sjq} |(F^{-1}\varphi_j F f)(\cdot)|^q \right)^{\frac{1}{q}} \mid L_p(R_n) \right\| < \infty \right\} \tag{5}$$

(modification if $q=\infty$).

Remark 2. These are the "homogeneous" counterparts of $B^s_{p,q}(R_n)$ and $F^s_{p,q}(R_n)$ from Definition 2.3.1/2. One asks the same basic questions as in the second chapter: are these spaces independent of the choice of φ? Are these spaces complete, i.e. are they quasi-Banach spaces? Do these spaces have nice properties and nice equivalent quasi-norms? One can try to develop the theory of these spaces parallel to Chapter 2. It turns out that some properties of the spaces $B^s_{p,q}(R_n)$ and $F^s_{p,q}(R_n)$ have immediate counterparts for the spaces $\dot{B}^s_{p,q}(R_n)$ and $\dot{F}^s_{p,q}(R_n)$, respectively (with minor technical changes in the corresponding proofs), e.g. maximal inequalities, Fourier multiplier theorems, lifting properties, some equivalent quasi-norms, some embedding theorems, and some interpolation formulas. We formulate a few (but not all) of these counterparts later on. Spaces of this type have been studied in a far more general setting (which covers the spaces from Chapter 2 and the above ones) in [F, Chapter 2]. We refer also to [S, Chapter 3]. The spaces $\dot{B}^s_{p,q}(R_n)$, in particular with $p \geqq 1$ and $q \geqq 1$, have been studied rather extensively by J. Peetre, cf. J. Peetre [8] including the references given there, and J. Bergh, J. Löfström [1, Chapter 6]. As we mentioned some basic assertions for the spaces $B^s_{p,q}(R_n)$ and $F^s_{p,q}(R_n)$ have more or less immediate counterparts for the spaces $\dot{B}^s_{p,q}(R_n)$ and $\dot{F}^s_{p,q}(R_n)$, respectively. On the other hand, however, some of the most striking features of the spaces $B^s_{p,q}(R_n)$ and $F^s_{p,q}(R_n)$ have no counterparts. The pointwise multipliers, cf. e.g. Theorem 2.8.2, and the diffeomorphism property from Theorem 2.10.2 are typical examples. The reason seems to be quite clear. If $g(x)$ is an infinitely differentiable function on R_n such that all derivatives of $g(x)$ are uniformly bounded on R_n, then $f(x) \to g(x)\, f(x)$ maps $S(R_n)$ into itself, but in general does not map $Z(R_n)$ into itself. Similarly, if $\psi(x)$ is as in 2.10.1., then $f(x) \to f(\psi(x))$ maps $S(R_n)$ onto itself, but in general does not map $Z(R_n)$ onto (or into) itself. This shows that at least in general one cannot expect to find counterparts of Theorem 2.8.2 and of Theorem 2.10.2 for the spaces $\dot{B}^s_{p,q}(R_n)$ and $\dot{F}^s_{p,q}(R_n)$. However, these two theorems (and also the other assertions for pointwise multipliers) played a crucial role in the theory of the chapters 2–4. In particular, they were the basis for the definitions of the spaces $B^s_{p,q}(\Omega)$ and $F^s_{p,q}(\Omega)$ on bounded C^∞-domains in R_n. This explains why we prefered to study the spaces $B^s_{p,q}(R_n)$ and $F^s_{p,q}(R_n)$ and not the spaces $\dot{B}^s_{p,q}(R_n)$ and $\dot{F}^s_{p,q}(R_n)$.

Remark 3. The very first question in connection with (4) and (5) is whether these spaces make sense as subspaces of $Z'(R_n)$. We always interpret $Z'(R_n)$ as $S'(R_n)/P(R_n)$ (i. e. $S'(R_n)$ modulo polynomials). If $P(x)$ is a polynomial and $f\in S'(R_n)$, it follows immediately that

$$\|f+P(x) \mid \dot{B}^s_{p,q}(R_n)\|^\varphi=\|f \mid \dot{B}^s_{p,q}(R_n)\|^\varphi ;$$

similarly for $\|f \mid \dot{F}^s_{p,q}(R_n)\|^\varphi$. This shows that Definition 2 makes sense.

Remark 4. We denoted these spaces as "homogeneous". The reason is the following. If $-\infty<s<\infty$, $0<q\leqq\infty$, and $0<p\leqq\infty$ are given numbers (with $p<\infty$ in the case of the spaces $\dot{F}^s_{p,q}(R_n)$), then there exists a positive constant c such that

$$\|f(\lambda\cdot) \mid \dot{B}^s_{p,q}(R_n)\|^\varphi\leqq c\lambda^{n-\frac{s}{p}}\|f \mid \dot{B}^s_{p,q}(R_n)\|^\varphi \tag{6}$$

holds for all $\lambda>0$ and all $f\in\dot{B}^s_{p,q}(R_n)$, and

$$\|f(\lambda\cdot) \mid \dot{F}^s_{p,q}(R_n)\|^\varphi\leqq c\lambda^{s-\frac{n}{p}}\|f \mid \dot{F}^s_{p,q}(R_n)\|^\varphi \tag{7}$$

holds for all $\lambda>0$ and all $f\in\dot{F}^s_{p,q}(R_n)$. The proof is a simple consequence of (3.4.1/3), in particular it follows that the inequalities (6) and (7) are equalities if $\lambda=2^l$, where l is an arbitrary integer. The corresponding assertion for the "non-homogeneous" spaces $B^s_{p,q}(R_n)$ and $F^s_{p,q}(R_n)$ is more complicated, cf. Proposition 3.4.1/1.

5.1.4. The Spaces $\dot{F}^s_{\infty,q}(R_n)$

We introduce the homogeneous counterpart of the spaces $F^s_{\infty,q}(R_n)$ from 2.3.4.

Definition. Let $-\infty<s<\infty$, $1<q<\infty$ and $\varphi=\{\varphi_k(x)\}_{k=-\infty}^{\infty}\in\dot{\Phi}(R_n)$. Then

$$\dot{F}^s_{\infty,q}(R_n)=\Big\{f \mid f\in Z'(R_n),\ \exists\{f_k(x)\}_{k=-\infty}^{\infty}\subset L_\infty(R_n)\ \text{such that}$$
$$f=\sum_{k=-\infty}^{\infty}F^{-1}\varphi_k Ff_k\ \text{in}\ Z'(R_n)\ \text{and} \tag{1}$$
$$\Big\|\Big(\sum_{k=-\infty}^{\infty}2^{skq}\,|f_k(x)|^q\Big)^{\frac{1}{q}} \mid L_\infty(R_n)\Big\|<\infty\Big\}.$$

Remark 1. This is an extension of Definition 5.1.3/2(ii) and the homogeneous counterpart of Definition 2.3.4. As in 2.3.4., one introduce the norm

$$\|f \mid \dot{F}^s_{\infty,q}(R_n)\|^\varphi=\inf\Big\|\Big(\sum_{k=-\infty}^{\infty}2^{skq}\,|f_k(x)|^q\Big)^{\frac{1}{q}} \mid L_\infty(R_n)\Big\|, \tag{2}$$

where the infimum is taken over all admissible representations of f in the sense of (1). If f is represented by (1) in $Z'(R_n)$ and if $P(x)$ is a polynomial, then $f+P(x)$ has the same representation (with the same $f_k s$). Consequently, $\dot{F}^s_{\infty,q}(R_n)$ makes sense as a subspace of $Z'(R_n)$. Another question is whether $\dot{F}^s_{\infty,q}(R_n)$ is independent of the choice of φ. In the next subsection we formulate a corresponding result for the spaces $\dot{F}^s_{p,q}(R_n)$ with $0<p<\infty$, and $0<q\leqq\infty$. Then it is not hard to see that Proposition 2.3.4/2 has a homogeneous counterpart. In particular, we have

$$(\dot{F}^s_{1,q}(R_n))'=\dot{F}^{-s}_{\infty,q'}(R_n)\,,\quad 1\leqq q<\infty\,,\quad \frac{1}{q}+\frac{1}{q'}=1$$

in the sense of dual spaces (appropriate interpretations, as in 2.11.1. (with $Z(R_n)$ and $Z'(R_n)$ instead of $S(R_n)$ and $S'(R_n)$, respectively). The last formula explains our restrictions on q in the above definition. It follows that the spaces from the above definition are independent of φ. Of particular interest are the spaces $\dot{F}^s_{\infty,2}(R_n)$. On the basis of Theorem 5.2.4, it follows that these spaces coincide with the spaces $I_s(BMO)$ in R. S. Strichartz [3, 4].

Remark 2. Proposition 2.3.4/1 can be carried over to the homogeneous spaces, cf. also [S, 3.1.2.].

5.1.5. Basic Properties

The notion of a quasi-Banach space has been introduced in 1.2.2. We recall that $Z'(R_n)$ is equipped with the strong topology. If A is a quasi-normed space then $Z(R_n) \subset A$ always means continuous (topological) embedding. Similarly $A \subset Z'(R_n)$.

Theorem. (i) $\dot{B}^s_{p,q}(R_n)$ is a quasi-Banach space if $-\infty < s < \infty$, $0 < p \leqq \infty$ and $0 < q \leqq \infty$ (Banach space if $1 \leqq p \leqq \infty$ and $1 \leqq q \leqq \infty$), and the quasi-norms $\|f \mid \dot{B}^s_{p,q}(R_n)\|^\varphi$ with $\varphi \in \dot{\Phi}(R_n)$ and $\|f \mid \dot{B}^s_{p,q}(R_n)\|^\psi$ with $\psi \in \dot{\Phi}(R_n)$ are equivalent to each other. Furthermore,

$$Z(R_n) \subset \dot{B}^s_{p,q}(R_n) \subset Z'(R_n) \tag{1}$$

for $-\infty < s < \infty$, $0 < p \leqq \infty$, and $0 < q \leqq \infty$. If $-\infty < s < \infty$, $0 < p < \infty$ and $0 < q < \infty$, then $Z(R_n)$ is dense in $\dot{B}^s_{p,q}(R_n)$.

(ii) $\dot{F}^s_{p,q}(R_n)$ is a quasi-Banach space if $-\infty < s < \infty$ and if either $0 < p < \infty$, $0 < q \leqq \infty$, or $p = \infty$, $1 < q \leqq \infty$ (Banach space if $\min(p, q) \geqq 1$), and the quasi-norms $\|f \mid \dot{F}^s_{p,q}(R_n)\|^\varphi$ with $\varphi \in \dot{\Phi}(R_n)$ and $\|f \mid \dot{F}^s_{p,q}(R_n)\|^\psi$ with $\psi \in \dot{\Phi}(R_n)$ are equivalent to each other. Furthermore,

$$Z(R_n) \subset \dot{F}^s_{p,q}(R_n) \subset Z'(R_n) \tag{2}$$

for $-\infty < s < \infty$, $0 < p < \infty$, $0 < q \leqq \infty$ or $-\infty < s < \infty$, $p = \infty$, $1 < q \leqq \infty$. If $-\infty < s < \infty$, $0 < p < \infty$ and $0 < q < \infty$, then $Z(R_n)$ is dense in $\dot{F}^s_{p,q}(R_n)$.

Remark. This is the counterpart of Theorem 2.3.3, including the considerations in 2.3.4. The proof is the same.

Convention. We shall not distinguish between equivalent quasi-norms of a given quasi-Banach space. In this sense we write $\|f \mid \dot{B}^s_{p,q}(R_n)\|$ instead of $\|f \mid \dot{B}^s_{p,q}(R_n)\|^\varphi$ with $\varphi \in \dot{\Phi}(R_n)$. Similarly $\|f \mid \dot{F}^s_{p,q}(R_n)\|$.

5.2. Further Properties

5.2.1. Maximal Inequalities

The maximal inequalities from Theorem 2.3.6 played a crucial role in the study of the spaces $B^s_{p,q}(R_n)$ and $F^s_{p,q}(R_n)$. Thus it is quite natural to ask for a homogeneous counterpart. There is no serious difficulty in transferring the considerations from 2.3.6. to the homogeneous case. First, we introduce the class $\dot{\mathfrak{K}}_L(R_n)$, as in Definition 2.3.6/1, as the collection of all systems $\varphi = \{\varphi_j(x)\}_{j=-\infty}^{\infty} \subset S(R_n)$ of functions with compact supports in $R_n - \{0\}$, such that

$$C(\varphi) = \sup\, (|x|^L + |x|^{-L}) \sum_{|\alpha| \leqq L} |D^\alpha \varphi_j(2^j x)| < \infty\,, \tag{1}$$

where the supremum is taken over all $x \in R_n - \{0\}$ and all integers j, cf. also [F, 2.2.2.]. In [F, 2.3.2.], we proved a maximal inequality in a far more general setting. We formulate its specialization to our situation. If $f \in Z'(R_n)$, we use the notation $(\varphi_j^* f)(x)$ of a maximal function from (2.3.6/2), where j is an integer.

Theorem. Let $-\infty<s<\infty$.

(i) Let $0<p\leqq\infty$ and $0<q\leqq\infty$. If $a>\frac{n}{p}$ in (2.3.6/2) and if L is a sufficiently large natural number, then there exists a positive number c such that

$$\left(\sum_{j=-\infty}^{\infty} 2^{jsq}\, \|\varphi_j^* f \mid L_p(R_n)\|^q\right)^{\frac{1}{q}} \leqq cC(\varphi)\|f \mid \dot{B}^s_{p,q}(R_n)\|^\psi \tag{2}$$

holds for all systems $\psi\in\dot{\Phi}(R_n)$, all systems $\varphi\in\dot{\mathcal{R}}_L(R_n)$, and all $f\in\dot{B}^s_{p,q}(R_n)$ (modification if $q=\infty$).

(ii) Let $0<p<\infty$ and $0<q\leqq\infty$. If $a>\frac{n}{\min(p,q)}$ in (2.3.6/2) and if L is a sufficiently large natural number, then there exists a positive number c such that

$$\left\|\left(\sum_{j=-\infty}^{\infty} 2^{jsq}|(\varphi_j^* f)(\cdot)|^q\right)^{\frac{1}{q}} \Big| L_p(R_n)\right\| \leqq cC(\varphi)\|f \mid \dot{F}^s_{p,q}(R_n)\|^\psi \tag{3}$$

holds for all systems $\psi\in\dot{\Phi}(R_n)$, all systems $\varphi\in\dot{\mathcal{R}}_L(R_n)$ and all $f\in\dot{F}^s_{p,q}(R_n)$ (modification if $q=\infty$).

Remark. This is the homogeneous counterpart of Theorem 2.3.6 and a special case of [F, Theorem 2.3.2]. In contrast to Theorem 2.3.6 we did not consider a family of systems φ^τ. The reason is the following. First, we have a direct reference to [F, 2.3.2.]. Second, we needed the formulation in Theorem 2.3.6 (with a family φ^τ instead of a single system) in order to prove the diffeomorphism property for the spaces $B^s_{p,q}(R_n)$ and $F^s_{p,q}(R_n)$, cf. Theorem 2.10.2. In general, there is no hope of proving a counterpart for the homogeneous spaces $\dot{B}^s_{p,q}(R_n)$ and $\dot{F}^s_{p,q}(R_n)$, cf. Remark 5.1.3/2. In other words, a generalization of the above theorem to families φ^τ seems to be possible but is not very interesting to us.

5.2.2. Fourier Multipliers

Theorem 2.3.7 is a simple consequence of Theorem 2.3.6. In the same way one can prove a Fourier multiplier theorem for the homogeneous spaces $\dot{B}^s_{p,q}(R_n)$ and $\dot{F}^s_{p,q}(R_n)$.

Theorem. Let $-\infty<s<\infty$ and $0<q\leqq\infty$. Let A be either $\dot{B}^s_{p,q}(R_n)$ with $0<p\leqq\infty$ or $\dot{F}^s_{p,q}(R_n)$ with $0<p<\infty$. If the natural number N is sufficient large, then there exists a positive constant c such that

$$\|F^{-1}mFf \mid A\| \leqq c\,\left(\sup_{|\alpha|\leqq N}\ \sup_{x\in R_n} |x|^{|\alpha|}\,|D^\alpha m(x)|\right)\|f \mid A\| \tag{1}$$

holds for all infinitely differentiable functions $m(x)$ and all $f\in A$.

Remark 1. This theorem is a simple consequence of Theorem 5.2.1, cf. also the proof of Theorem 2.3.7 and Remark 2.3.7/1. Furthermore, in the same way as in Remark 2.3.7/4 one can extend the above theorem to the spaces $\dot{F}^s_{\infty,q}(R_n)$ with $1<q\leqq\infty$.

Remark 2. One can ask for improvements of the above theorem, E.g. it should be possible to find a counterpart of Theorem 2.4.8 (including the underlying complex interpolation formulas). Furthermore, the considerations in Section 2.6. have a homogeneous counterpart. We do not go into detail, cf. [F, 2.4.3.] and also J. Peetre [8].

5.2.3. Equivalent Norms

In Chapter 2 we studied very extensively equivalent quasi-norms, representations etc. for the spaces $B^s_{p,q}(R_n)$ and $F^s_{p,q}(R_n)$. The problem arises whether some of these considerations can be carried over to the homogeneous spaces. We introduce the lifting operator $\dot{I}_\sigma$,

$$\dot{I}_\sigma f=F^{-1}|x|^\sigma Ff, \quad f\in Z'(R_n), \tag{1}$$

where $-\infty<\sigma<\infty$. It is not hard to see that $\dot{I}_\sigma$ makes sense, and maps $Z'(R_n)$ onto itself, cf. also (2.3.8/1).

Theorem 1. (i) Let $-\infty<s<\infty$, $-\infty<\sigma<\infty$, and $0<q\leqq\infty$. If $0<p\leqq\infty$, then $\dot{I}_\sigma$ maps $\dot{B}^s_{p,q}(R_n)$ isomorphically onto $\dot{B}^{s-\sigma}_{p,q}(R_n)$, and if $0<p<\infty$, then $\dot{I}_\sigma$ maps $\dot{F}^s_{p,q}(R_n)$ isomorphically onto $\dot{F}^{s-\sigma}_{p,q}(R_n)$.
(ii) If $1<p<\infty$, then

$$\dot{F}^0_{p,2}(R_n)=L_p(R_n)\,. \tag{2}$$

(iii) If $1<p<\infty$ and $m=0, 1, 2, \ldots$, then

$$\sum_{|\alpha|=m}\|D^\alpha f\mid L_p(R_n)\|=\|f\mid \dot{W}^m_p(R_n)\| \tag{3}$$

is an equivalent norm on $\dot{F}^m_{p,2}(R_n)$.

Remark 1. Part (i) is the counterpart of Theorem 2.3.8, including the proof. We must interpret (2). All spaces in this chapter are considered as subspaces of $Z'(R_n)$, i. e. as elements of $S'(R_n)$ modulo polynomials. However, it is obvious that for any $f\in S'(R_n)$ we have

$$\|f\mid L_p(R_n)\|=\inf\|f+P\mid L_p(R_n)\|\,, \tag{4}$$

where the infimum is taken over all polynomials. In this sense, $L_p(R_n)$ can be interpreted as a subspace of $Z'(R_n)$. The proof of (2) is the same as in Theorem 2.5.6, cf. (2.5.6/5). The interpretation of (3) is the same as in (4) (modulo polynomials). In this sense one can introduce homogeneous Sobolev spaces via

$$\dot{W}^m_p(R_n)=\{f\mid f\in Z'(R_n)\,,\ \ \|f\mid \dot{W}^m_p(R_n)\|<\infty\}\,. \tag{5}$$

Then (iii) can be reformulated as $\dot{F}^m_{p,2}(R_n)=\dot{W}^m_p(R_n)$, where $1<p<\infty$ and $m=0, 1, 2, \ldots$ The proof is similar to that in 2.5.6.

Theorem 2. Let $1\leqq p\leqq\infty$, $1\leqq q\leqq\infty$ and $s>0$. If M is an integer such that $M>s$, then

$$\left(\int_{R_n}|h|^{-sq}\sup_{\substack{|\varrho|\leqq|h|\\ \varrho\in R_n}}\|\Delta^M_\varrho f\mid L_p(R_n)\|^q\frac{dh}{|h|^n}\right)^{\frac{1}{q}} \tag{6}$$

and

$$\left(\int_{R_n}|h|^{-sq}\|\Delta^M_h f\mid L_p(R_n)\|^q\frac{dh}{|h|^n}\right)^{\frac{1}{q}} \tag{7}$$

are equivalent norms in $\dot{B}^s_{p,q}(R_n)$ (modification if $q=\infty$).

Remark 2. This is the counterpart of Theorem 2.5.12 (with the restriction $\min(p, q)\geqq 1$), in particular, the above notations have the same meaning as in 2.5.12. The proof that (6) is an equivalent norm follows the same lines as in 2.5.12., cf. also J. Bergh, J. Löfström [1, p. 147]. From that one obtains in the same way as in the third step of the proof of Theorem 2.5.12 that (7) is also an equivalent norm in $\dot{B}^s_{p,q}(R_n)$. The interpretation of the norms in (6) and (7) is similar to that in (4). Cf. also S. G. Samko [1, 2].

Remark 3. In connection with (2), and also with (6) and (7), it is quite natural to compare the homogeneous spaces from this chapter with the non-homogeneous spaces from Chapter 2. We have e.g.

$$B^s_{p,q}(R_n)=L_p(R_n)\cap\dot{B}^s_{p,q}(R_n)$$

if $s>0$, $1\leqq p\leqq\infty$ and $1\leqq q\leqq\infty$, cf. J. Bergh, J. Löfström [1, p. 148]. Another possibility has been discovered by R. Johnson [5, 6]. He studies subspaces of homogeneous spaces which are invariant under multiplication by e^{ixh} for all $h\in R_n$. Let

$$\pi\dot{B}^s_{p,q}(R_n)=\{f\mid e^{ixh}f\in\dot{B}^s_{p,q}(R_n)\text{ for all }h\in R_n\}\,.$$

He proved that $\pi\dot{B}^s_{p,q}(R_n)=B^s_{p,q}(R_n)$ if $s>0$, $1\leq p\leq\infty$, and $1\leq q\leq\infty$. Cf. R. Johnson [5, 6] and P. Nilsson [1] for further results and more details.

Remark 4. There is a full counterpart of Theorem 2.12.2: (i) If $0<p\leq\infty$, $0<q\leq\infty$ and $-\infty<s<\infty$, and if m and k are as in Theorem 2.12.2(i), then (2.12.2/8) without $\|f\mid L_p(R_n)\|$ and (2.12.2/9) without $\|f\mid L_p(R_n)\|$ are equivalent quasi-norms on $\dot{B}^s_{p,q}(R_n)$. (ii) If $0<p<\infty$, $0<q\leq\infty$ and $-\infty<s<\infty$, and if m is a sufficiently large natural number, then (2.12.2/10) without $\|f\mid L_p(R_n)\|$ and (2.12.2/11) without $\|f\mid L_p(R_n)\|$ are equivalent quasi-norms on $\dot{F}^s_{p,q}(R_n)$. A proof (and also more general assertions) may be found in H. Triebel [24].

5.2.4. The Hardy Spaces $H_p(R_n)$, the Spaces $BMO(R_n)$

In 2.5.8. we proved that $h_p(R_n)=F^0_{p,2}(R_n)$ and $bmo(R_n)=F^0_{\infty,2}(R_n)$, where $h_p(R_n)$ is the local Hardy space and $bmo(R_n)$ is the local version of $BMO(R_n)$. In this subsection we formulate corresponding statements for the Hardy spaces $H_p(R_n)$ with $0<p<\infty$ and for $BMO(R_n)$. The spaces $H_p(R_n)$ and $BMO(R_n)$ have been briefly mentioned in Remark 2.2.2/2. In order to compare these spaces with $\dot{F}^0_{p,2}(R_n)$ we must interpret them as subspaces of $Z'(R_n)$. Let $\varphi(x)$ be an infinitely differentiable function on R_n with compact support and $\varphi(x)=1$ in a neighbourhood of the origin. Let again $\varphi_t(x)=\varphi(tx)$ with $t>0$ and $x\in R_n$. If $0<p<\infty$ and $f\in Z'(R_n)$, we put

$$\|f\mid H_p(R_n)\|^\varphi=\inf\|\sup_{t>0}|(F^{-1}\varphi_tF(f+P))(\cdot)|\,L_p(R_n)\|\,, \tag{1}$$

where the infimum is taken over all polynomials $P(x)$. If $f=f(x)$ is a locally integrable function on R_n, then f_Q is the mean value (2.2.2/13), where Q is a cube in R_n. For such functions we put

$$\|f\mid BMO(R_n)\|=\inf\sup_Q\frac{1}{|Q|}\int_Q|f(x)+P(x)-f_Q-P_Q|\,dx\,, \tag{2}$$

where the supremum is taken over all cubes in R_n and the infimum over all polynomials $P(x)$. If $f\in Z'(R_n)$ is not locally integrable on R_n, then we put $\|f\mid BMO(R_n)\|=\infty$. Let us discuss (1) and (2). If $P(x)$ is a polynomial, then

$$(F^{-1}\varphi_tFP)(x)=P(x),\quad t>0\,.$$

In particular, because $p<\infty$

$$\|\sup_{t>0}|(F^{-1}\varphi_tFP)(\cdot)|\mid L_p(R_n)\|^\varphi<\infty\quad\text{if and only if}\quad P(x)\equiv0\,. \tag{3}$$

Furthermore, as has been proved by C. Fefferman, E. M. Stein [2, pp. 141/142] we have

$$\inf_C\int_{R_n}\frac{|f(x)-C|}{1+|x|^{n+1}}\,dx\leq c\sup_Q\frac{1}{|Q|}\int_Q|f(x)-f_Q|\,dx\,, \tag{4}$$

where c is independent of f. If $f(x)=P(x)$ such that the right-hand side of (4) is finite, it follows that $P(x)\equiv c$ is a constant. These considerations shed some light on (1) and (2). In particular, (1) and (2) are the usual definitions in the sense of C. Fefferman, E. M. Stein [2] modulo polynomials. As has been shown in C. Fefferman, E. M. Stein [2], the quasi-norms (1) for different choices of φ (and fixed p) are equivalent to each other. In this sense we omit φ in $\|f\mid H_p(R_n)\|^\varphi$.

Definition. If $0<p<\infty$, then

$$H_p(R_n)=\{f\mid f\in Z'(R_n),\ \|f\mid H_p(R_n)\|<\infty\}$$

and

$$BMO(R_n)=\{f\mid f\in Z'(R_n),\ \|f\mid BMO(R_n)\|<\infty\}\,.$$

Remark 1. Modulo polynomials these are the usual Hardy spaces $H_p(R_n)$ and the space $BMO(R_n)$ of all functions of bounded mean oscillation on R_n in the sense of C. Fefferman, E. M. Stein [2]. We remark that $Z(R_n)$ is dense in $H_p(R_n)$, cf. A. P. Calderón, A. Torchinsky [1, II, Theorem 1.8(v)]. More details may be found in [S, 3.1.3. and 3.1.4.].

Theorem. If $0<p<\infty$, then

$$H_p(R_n)=\dot{F}^0_{p,2}(R_n) \tag{5}$$

(equivalent quasi-norms) and

$$BMO(R_n)=\dot{F}^0_{\infty,2}(R_n)\,. \tag{6}$$

Remark 2. Formula (5) is due to J. Peetre [5, 6]. An independent proof has also been given in [S, 3.2.1.]. An extension of this result to anisotropic Hardy spaces in the sense of A. P. Calderón, A. Torchinsky [1, II] has been given in H. Triebel [13]. Formula (6) may be found in [S, 3.2.2.]. The above theorem is the homogeneous counterpart of the considerations in 2.5.8., cf. also the references given there. (6) shows that $\dot{F}^s_{\infty,2}(R_n)$ coincides with the spaces $I_s(BMO)$ (which have been studied recently by R. S. Strichartz [3].

5.2.5. Miscellaneous Properties

As has been mentioned in Remark 5.1.3/2, a few of the most striking features of the spaces $B^s_{p,q}(R_n)$ and $F^s_{p,q}(R_n)$ have no homogeneous counterparts and the homogeneous spaces are unsuitable for the study of elliptic differential equations in domains. On the other hand, besides the above statements, further properties for the spaces from Chapter 2 can be carried over to the homogeneous spaces $\dot{B}^s_{p,q}(R_n)$ and $\dot{F}^s_{p,q}(R_n)$. The real and complex interpolation formulas from Theorem 2.4.2 and Theorem 2.4.7 have homogeneous counterparts. The complex interpolation formulas for the homogeneous spaces can be proved on the basis of Theorem 2.4.9/1. The duality theory of Section 2.11. has a homogeneous counterpart. As far as embedding theorems are concerned one must be a little more careful. On the one hand, (2.3.2/5), (2.3.2/6), (2.3.2/9), and (probably) Theorem 2.7.1 have homogeneous counterparts, but not (2.3.2/7) and (2.3.2/8). In any case if one wishes to deal with homogeneous spaces, then one should carefully examine the corresponding theorems and proofs for the non-homogeneous spaces.

6. Ultra-Distributions and Weighted Spaces of Entire Analytic Functions

6.1. Ultra-Distributions

6.1.1. Introduction (to Chapters 6 and 7)

In Chapter 6 we continue the studies on entire analytic functions which we began in the first chapter. A description of the problems treated has been given in Section 1.1. The goal is the proof of estimates of type (1.1/5) and related maximal inequalities. Chapter 1 dealt essentially with the unweighted case. This was not only convenient but completely sufficient for the main purpose of this book: the study of unweighted spaces of type $B^s_{p,q}$ and $F^s_{p,q}$ on R_n and on domains, and their applications to elliptic boundary value problems in Part I. On the other hand, weighted spaces of several types have attracted much attention in recent times. So it is quite natural to ask for weighted spaces of type $B^s_{p,q}$ and $F^s_{p,q}$ on R_n and on domains. If one wishes to study weighted spaces on R_n in the spirit of Chapter 2, then one needs assertions for weighted spaces of entire analytic functions in the sense of Chapter 1. As has been mentioned in Section 1.1., there are no serious difficulties in extending the considerations in the first chapter to weights of at most polynomial growth, e.g. $\varrho(x)=(1+|x|)^\varkappa$, where $\varkappa$ is a real number. But weights of polynomial growth do not represent the true limit of admissible weight functions, which is very roughly $e^{|x|}$ from above and $e^{-|x|}$ from below. In order to deal with weights which grow more rapidly than any polynomial, the usual distributions are not sufficient. One must switch over from the usual distributions to the more natural but unfortunately also more complicated ultra-distributions. Chapter 6 deals with weighted spaces of entire analytic functions. We begin with a description of ultra-distributions and then follow the treatment given in Chapter 1. In this sense, Chapter 6 is an appendix to Chapter 1: we restrict ourselves to formulations and comments. Furthermore, Chapter 6 is more or less a summary of some parts of [F, Chapter 1], where the interested reader may find detailed proofs and also a number of additional remarks and comments. The vector-valued spaces $L^\Omega_p(l_q)$ from Section 1.6. played a crucial role in Chapter 2. Hence, we are forced to look for a weighted counterpart. Few difficulties arise and we postpone these problems to Chapter 7. In Section 7.1. we deal with maximal inequalities and Fourier multipliers theorems for weighted spaces of $L_p(l_q)$-type. On that basis we introduce in Section 7.2. weighted spaces of type $B^s_{p,q}$ and $F^s_{p,q}$ and describe a few basic assertions. In this sense, Chapter 7 is the direct continuation of Chapter 6 and a very modest counterpart of Chapter 2.

6.1.2. Definitions

Subsections 6.1.2.–6.1.4. coincide essentially (besides a few modifications) with [F, 1.2.1.–1.2.3.]. On the other hand we follow closely G. Björck [1]. We use the notations from Chapter 1, cf. in particular 1.2.1.

Definition 1. (i) $\mathfrak{M}$ is the collection of all real-valued functions $\omega(x)$ on R_n such that $\omega(x)=\sigma(|x|)$, where $\sigma(t)$ is an increasing continuous concave function on $[0,\infty)$ with

$$\sigma(0)=0\,, \tag{1}$$

$$\int_0^\infty \frac{\sigma(t)}{1+t^2}\,dt<\infty\,, \tag{2}$$

$$\sigma(t)\geqq c+d\log(1+t) \quad \text{if} \quad t\geqq 0\,, \tag{3}$$

where c is an appropriate real number and d is an appropriate positive number.

H. Triebel, *Theory of Function Spaces*, Modern Birkhäuser Classics,
DOI 10.1007/978-3-0346-0416-1_6, © Birkhäuser Verlag 1983

(ii) $\mathfrak{M}_E$ is the collection of all real-valued functions $\omega(x)$ on R_n such that $\omega(x)=\sigma(|x|)$, where $\sigma(t)$ is a continuous non-negative, not identically vanishing function on $[0, \infty)$ satisfying (1), (2) and

$$e^{\sigma(t)} = \sum_{k=0}^{\infty} a_k t^k \quad \text{if} \quad t \geqq 0, \tag{4}$$

with $a_k \geqq 0$ and

$$(k+l)!\, a_{k+l} \leqq k!\, a_k l!\, a_l$$

for $k=0, 1, 2, \ldots$ and $l=0, 1, 2, \ldots$

Remark 1. As far as part (ii) is concerned we refer to G. Björck [1, pp. 364/365], cf. also Y. Domar [1]. We have $\mathfrak{M}_E \subset \mathfrak{M}$; in other words, $\mathfrak{M}_E$ is a subclass of $\mathfrak{M}$. In this chapter we work with the class $\mathfrak{M}$. The considerations in Section 7.1. are based on $\mathfrak{M}_E$.

Remark 2. It is not hard to see that $\omega(x) \in \mathfrak{M}$ is subadditive, i.e.

$$\omega(x+y) \leqq \omega(x) + \omega(y) \quad \text{if} \quad x \in R_n \quad \text{and} \quad y \in R_n .$$

Typical examples of functions $\omega(x) \in \mathfrak{M}$ are given by

$$\omega(x) = \log(1+|x|)^d \quad \text{with} \quad d > 0, \tag{5}$$

$$\omega(x) = |x|^\beta \quad \text{with} \quad 0 < \beta < 1. \tag{6}$$

Definition 2. Let $\omega(x) \in \mathfrak{M}$.
(i) D_ω is the collection of all complex-valued integrable functions $\varphi(x)$ on R_n with compact support such that

$$\|\varphi\|_{\omega,\lambda} = \int_{R_n} |F\varphi(x)|\, e^{\lambda\omega(x)} dx < \infty \quad \text{for every} \quad \lambda > 0 . \tag{7}$$

(ii) S_ω is the collection of all complex-valued integrable functions $\varphi(x)$ on R_n such that both $\varphi(x)$ and $(F\varphi)(x)$ are infinitely differentiable and

$$p_{\alpha,\lambda}(\varphi) = \sup_{x \in R_n} e^{\lambda\omega(x)} |D^\alpha \varphi(x)| < \infty, \tag{8}$$

$$\pi_{\alpha,\lambda}(\varphi) = \sup_{x \in R_n} e^{\lambda\omega(x)} |D^\alpha (F\varphi)(x)| < \infty, \tag{9}$$

for every multi-index α and every $\lambda > 0$.

Remark 3. We use the same notations as in Chapter 1, cf. 1.2.1. In particular, F and F^{-1} denote the Fourier transform and its inverse on R_n, respectively. Furthermore, as in the first chapter we omit R_n in our notations, e.g. we write S_ω instead of $S_\omega(R_n)$. There is no danger of confusion because all spaces in this chapter are defined on R_n.

Remark 4. If $\omega(x)$ is given by (5), then $S_\omega = S$ is the usual space of tempered distributions, and $D_\omega = D$ is the usual space of all infinitely differentiable functions on R_n with compact support.

Remark 5. S_ω is endowed with a topology which is generated by the semi-norms (8) and (9). Of course, (8), (9) is the counterpart of (1.2.1/1.) In any case, S_ω is a complete locally convex space, and we have

$$S_\omega \subset S \tag{10}$$

(continuous embedding). Furthermore, S_ω is dense in S, cf. [F, Remark 1.2.2/3].

Definition 3. Let $\omega(x) \in \mathfrak{M}$. Then S'_ω is the topological dual of S_ω (equipped with the strong topology).

Remark 6. We recall that the topological dual of S_ω is the collection of all continuous linear functionals on S_ω. By (10) and the density property mentioned above we have

$$S' \subset S'_\omega \tag{11}$$

in the usual interpretation. This justifies denoting the elements of S'_ω as tempered ultra-distributions or as ω-tempered distributions.

Definition 4. Let $\omega(x)\in\mathfrak{M}$ and $f\in S'_\omega$. Then the Fourier transform Ff and the inverse Fourier transform $F^{-1}f$ of f are given by

$$(Ff)(\varphi)=f(F\varphi)\,,\quad (F^{-1}f)(\varphi)=f(F^{-1}\varphi)\,, \tag{12}$$

respectively, for all $\varphi\in S_\omega$.

Remark 7. It follows from the definition of S_ω that both F and F^{-1} yield isomorphic mappings from S_ω onto itself. This shows that the last definition is meaningful. Furthermore, F and F^{-1} yield also isomorphic mappings from S'_ω onto itself. (11) shows that (12) is a generalization of the Fourier transform on S'.

Remark 8. As we mentioned above, we follow essentially the approach given by G. Björck [1], who based his considerations on A. Beurling [1]. More precisely, A. Beurling [1] introduced the ultra-distributions D'_ω, the topological dual of D_ω. Other systematic treatments of ultra-distributions are given by C. Roumieu [1, 2], I. M. Gelfand, G. E. Schilow [1], and H. Komatsu [1]. Further references and historical remarks may be found in I. M. Gelfand, G. E. Schilow [1], and J.-L. Lions, E. Magenes [2, III, pp. 22–24].

Remark 9. What about the assumptions in Definition 1? It is not necessary to deal with radial functions $\omega(x)$, there being a corresponding theory for more general functions, cf. G. Björck [1]. Condition (3) ensures that D_ω and S_ω are subspaces of D and S, respectively. (1) is a technical condition. The main hypothesis is (2), which is due to A. Beurling [1]. He proved the following assertion: if $\omega(x)=\sigma(|x|)$, where $\sigma(t)$ is an increasing continuous concave function on $[0,\infty)$ with $\sigma(0)=0$ and if D_ω is defined as above, then D_ω is non-trivial if and only if (2) holds (non-trivial means that there exists a function $\varphi\in D_\omega$ which does not vanish identically), cf. G. Björck [1, p. 359] (also for generalizations of this statement). This shows that Beurling's condition (2) is near to the Denjoy-Carleman non-quasi-analyticity condition.

6.1.3. Basic Properties

We list some properties of D_ω and S_ω which are similar to those for D and S (recall that D and S are special cases of D_ω and S_ω, respectively). As far as proofs are concerned we refer to G. Björck [1] and [F, 1.2.2.].

Proposition 1. Let $\omega(x)\in\mathfrak{M}$.
(i) D_ω is the collection of all functions of S_ω with compact support. D_ω is dense in S_ω.
(ii) (Resolution of unity). Let Ω be a compact subset and let $K_1,\ldots,K_N$ be open balls with $\bigcup_{j=1}^{N} K_j\supset\Omega$. There exist real functions $\varphi_1(x),\ldots,\varphi_N(x)$ with $\varphi_j(x)\in D_\omega$,

$$\operatorname{supp}\varphi_j\subset K_j\,,\quad 0\leqq\varphi_j(x)\leqq 1\quad\text{if}\quad x\in R_n\,, \tag{1}$$

$$\sum_{j=1}^{N}\varphi_j(x)=1\quad\text{if}\quad x\in\Omega\,. \tag{2}$$

(iii) S_ω is a topological algebra under pointwise multiplication and under convolution.
(iv) $\varphi(x)\to x^\alpha\varphi(x)$ and $\varphi(x)\to D^\alpha\varphi(x)$ are continuous operators on S_ω.
(v) The translation $\varphi(x)\to\varphi(x-\tau)$, the dilatation $\varphi(x)\to\varphi(ax)$, and the multiplication $\varphi(x)\to e^{ix\tau}\varphi(x)$ are continuous operators on S_ω, where $\tau\in R_n$ and the real number $a\neq 0$ are given.

Remark 1. We recall that $x^\alpha=x_1^{\alpha_1}\ldots x_n^{\alpha_n}$ with $x=(x_1,\ldots,x_n)\in R_n$ and the multi-index α. Obviously, (ii)–(v) are familiar assertions in D and S. The resolution of unity in part (ii) of the above proposition is of crucial importance. It is the basis of the theory of ω-tempered distributions. This statement and Remark 6.1.2/9 show that (6.1.2/2) is very natural. In this sense ultra-distributions are more natural than the usual distributions.

Remark 2. With the help of the resolution of unity one can define the support of $f \in S'_\omega$ in the usual way. Furthermore, by standard procedures the multiplications by $\varphi(x) \in S_\omega$, x^α and $e^{ix\tau}$, the translation, dilatation and differentiation can be extended from S_ω to S'_ω. We formulate these assertions in the following proposition, together with a few properties which we mentioned in the preceding subsection and an approximation procedure.

Proposition 2. Let $\omega(x) \in \mathfrak{M}$.
(i) S_ω and S'_ω are complete locally convex spaces (with respect to the indicated topologies). The Fourier transform F and its inverse F^{-1} are isomorphic mappings from S_ω onto itself and from S'_ω onto itself.
(ii) The multiplications by $\varphi(x) \in S_\omega$, x^α (with the multi-index α), and $e^{ix\tau}$ (with $\tau \in R_n$), the translation, the dilatation and the differentiation are continuous operators in S'_ω.
(iii) Let $\varphi(x) \in S_\omega$ with $\varphi(0)=1$ and $\operatorname{supp} F\varphi \subset \{y \mid |y| \leq 1\}$.
Then

$$\varphi(\varepsilon x)\,\psi(x) \to \psi(x) \quad \text{in} \quad S_\omega \quad \text{if} \quad \varepsilon \downarrow 0 \quad \text{and} \quad \psi \in S_\omega\,, \tag{3}$$

$$\varphi(\varepsilon\,\cdot)\,f \to f \quad \text{in} \quad S'_\omega \quad \text{if} \quad \varepsilon \downarrow 0 \quad \text{and} \quad f \in S'_\omega\,. \tag{4}$$

Furthermore, if $Ff \in S'_\omega$ has compact support and if $\varepsilon > 0$, then $\varphi(\varepsilon\,\cdot)\,f \in S_\omega$.

Remark 3. The approximation formulas (3) and (4) can be reformulated as

$$F^{-1}\varphi(\varepsilon\,\cdot)F\psi \to \psi \quad \text{in} \quad S_\omega \quad \text{if} \quad \varepsilon \downarrow 0 \quad \text{and} \quad \psi \in S_\omega\,, \tag{5}$$

$$F^{-1}\varphi(\varepsilon\,\cdot)\,Ff \to f \quad \text{in} \quad S'_\omega \quad \text{if} \quad \varepsilon \downarrow 0 \quad \text{and} \quad f \in S'_\omega\,. \tag{6}$$

Regular ω-tempered distributions. If μ is a Borel measure on R_n such that

$$\varphi \to \int_{R_n} \varphi(x)\, d\mu$$

is an element of S'_ω for some $\omega \in \mathfrak{M}$, then we identity μ with the generated element of S'_ω. This makes sense because

$$\int_{R_n} \varphi(x)\, d\mu = 0 \quad \text{for all} \quad \varphi \in S_\omega$$

has only the trivial solution $\mu=0$, cf. [F, pp. 16/17]. This is an extension of the well-known procedure where some functions and measures are identified with corresponding distributions (of course, functions which differ only on a set of Lebesgue measure 0 must be identified). A typical example is

$$e^{\lambda\omega(x)} \in S'_\omega \tag{7}$$

for all real numbers λ. If e.g. $\omega(x)=|x|^\beta$ with $0<\beta<1$ and $\lambda>0$, then such a function cannot be interpreted as a usual tempered distribution. This shows the advantage of ω-tempered distributions. In particular, the Fourier transform of $e^{\lambda|x|^\beta}$ makes sense.

6.1.4. Paley-Wiener-Schwartz Theorems for Ultra-Distributions

The Paley-Wiener-Schwartz theorems from 1.2.1 for S and S' have been of great use to us, so it is quite natural to ask for corresponding statements for S_ω and S'_ω, respectively. We use the same notations and interpretations as those preceding Theorem 1.2.1/1 and in Remark 1.2.1.

Theorem 1. Let $\omega(x) \in \mathfrak{M}$. The following two assertions are equivalent to each other:
(i) $\varphi \in S_\omega$ and $\operatorname{supp} F\varphi \subset \{y \mid |y| \leq b\}$,
(ii) $\varphi(z)$ is an entire analytic function of n complex variables and for any $\lambda > 0$ and any $\varepsilon > 0$ there exists a constant $c_{\lambda,\varepsilon}$ such that

$$|\varphi(z)| \leq c_{\lambda,\varepsilon} e^{(b+\varepsilon)|y| - \lambda\omega(x)} \tag{1}$$

holds for all $z = x + iy$ with $x \in R_n$ and $y \in R_n$.

Theorem 2. Let $\omega(x)\in\mathfrak{M}$. The following two assertions are equivalent to each other:
(i) $f\in S'_\omega$ and $\operatorname{supp} Ff\subset\{y \mid |y|\leqq b\}$,
(ii) $f(z)$ is an entire analytic function of n complex variables and for an appropriate real number λ and for any $\varepsilon>0$ there exists a constant c_ε such that

$$|f(z)|\leqq c_\varepsilon e^{(b+\varepsilon)|y|+\lambda\omega(x)} \tag{2}$$

holds for all $z=x+iy$ with $x\in R_n$ and $y\in R_n$.

Remark 1. If $\omega(x)=\log(1+|x|)$, cf. (6.1.2/5), then the two theorems coincide with the corresponding theorems in 1.2.1.

Remark 2. The above theorems are due to G. Björck [1, pp. 365 and 379] (concerning the modified notations cf. [F, Remark 1.2.3/1]). Theorems of the above type are known. We refer to L. Hörmander [2, Theorem 1.7.7 and Lemma 5.7.2], C. Roumieu [2, pp. 158 and 184], H. Komatsu [1, I, p. 81], H. Komatsu [1, II, p. 610], and H. Triebel [3, II, Appendix].

6.2. Inequalities of Plancherel-Polya-Nikol'skij Type

6.2.1. Admissible Weight Functions

We use the same notations as in the first chapter. In particular, $L_{p,\mu}$ has the meaning of 1.2.2., where μ is a Borel measure on R_n and $0<p\leqq\infty$ (we recall that we omit "R_n" because all spaces in this chapter are defined on R_n). Furthermore, the cubes Q_k^h and the set K_h, with $h>0$, of admissible Borel measures in R_n have been introduced in Definition 1.2.4. We carry over these notations in the sequel. In the framework of ultra-distributions we are in a better position, and can essentially extend the class of admissible Borel measures, cf. also the examples in Remark 1.2.4.

Definition. (i) Let $\omega(x)\in\mathfrak{M}$. Then $R(\omega)$ denotes the collection of all Borel-measurable real functions $\varrho(x)$ on R_n such that there exists a positive constant c with

$$0<\varrho(x)\leqq c\varrho(y)\,e^{\omega(x-y)} \tag{1}$$

for every $x\in R_n$ and every $y\in R_n$.
(ii) Let $\omega(x)\in\mathfrak{M}$, $\varrho(x)\in R(\omega)$ and $\mu\in K_h$. Then $K_h(\varrho,\mu)$ denotes the collection of all Borel-measurable real functions $\varkappa(x)$ on R_n with the following two properties: (a) there exists a positive number c with

$$0\leqq\varkappa(x)\leqq c\varrho(x) \tag{2}$$

for every $x\in R_n$, and (b) there exist two positive numbers δ and c', and a Borel-measurable subset G of R_n with $\mu(G\cap Q_k^h)\geqq\delta$ for every cube Q_k^h and

$$\varkappa(x)\geqq c'\varrho(x)\quad\text{for every}\quad x\in G. \tag{3}$$

Remark 1. One may assume without restriction of generality that

$$\mu(Q_k^h)=1\quad\text{for every cube}\quad Q_k^h. \tag{4}$$

Then the above definition coincides with [F, Definition 1.3.1].

Remark 2. We discuss the above definition. Let $\varrho(x)\in R(\omega)$. From (6.1.3/7) and (1) with $y=0$ it follows that

$$\varrho(x)\in S'_\omega. \tag{5}$$

Furthermore, (1) yields

$$c e^{-\omega(x)}\leqq\varrho(x)\leqq c' e^{\omega(x)},\quad x\in R_n, \tag{6}$$

where c and c' are appropriate positive numbers. This shows that the growth of $\varrho(x)$ is not only restricted from above but also from below. If $\omega_1(x)\in\mathfrak{M}$, $\omega_2(x)\in\mathfrak{M}$ and $\lambda>0$, then $\lambda\omega_1(x)\in\mathfrak{M}$

and $\omega_1(x)+\omega_2(x)\in\mathfrak{M}$, cf. Definition 6.1.2/1. Furthermore, if $\varrho_1(x)\in R(\omega_1)$ and $\varrho_2(x)\in R(\omega_2)$, then

$$\begin{cases} \lambda\varrho_1(x)\in R(\omega_1)\,, & \varrho_1^{\lambda}(x)\in R(\lambda\omega_1)\,,\\ \varrho_1(x)+\varrho_2(x)\in R(\omega_1+\omega_2)\,, & \varrho_1^{-1}(x)\in R(\omega_1)\,,\\ \varrho_1(x)\,\varrho_2(x)\in R(\omega_1+\omega_2)\,, & \dfrac{\varrho_1(x)}{\varrho_2(x)}\in R(\omega_1+\omega_2)\,. \end{cases} \tag{7}$$

Remark 3. We describe a few exemples. If $\omega(x)\in\mathfrak{M}$, then

$$\varrho(x)=\mathrm{e}^{\omega(x)}\in R(\omega)\,. \tag{8}$$

This follows from the subadditivity of $\omega(x)$, cf. Remark 6.1.2/2. In particular,

$$\varrho(x)=(1+|x|)^d\in R\,(\log\,(1+|x|)^d)\,,\quad d>0\,, \tag{9}$$

$$\varrho(x)=\mathrm{e}^{|x|^\beta}\in R(|x|^\beta)\,,\qquad 0<\beta<1\,, \tag{10}$$

cf. Remark 6.1.2/2.

Remark 4. The subclass $\mathfrak{M}_E$ of $\mathfrak{M}$ from Definition 6.1.2/1 is sufficiently large. Let $d>0$ and $0<\beta<1$. Then one can find appropriate functions $\omega_1(x)$ and $\omega_2(x)$ such that

$$\omega_1(x)\in\mathfrak{M}_E\,,\quad \varrho(x)=(1+|x|)^d\in R(\omega_1)\,, \tag{11}$$

$$\omega_2(x)\in\mathfrak{M}_E\,,\quad \varrho(x)=\mathrm{e}^{|x|^\beta}\in R(\omega_2)\,. \tag{12}$$

These are the two most interesting concrete examples. The first one, i.e. $\varrho(x)=(1+|x|)^d$ with $d>0$ can be treated in the framework of the usual distributions, but not the second one, i.e. $\varrho(x)=\mathrm{e}^{|x|^\beta}$ with $0<\beta<1$. This class leads to the so-called Gevrey ultra-distributions.

Remark 5. Essentially, the measures $\mu_0\in K_h$ have a lattice structure, in particular if (4) is satisfied. Of special interest are Lebesgue measure $\mu_0=\mathrm{d}x$ and atomic measures with $\mu_0(Q_k^h)=\mu_0(\{x^k\})=1$, where $x^k\in Q_k^h$. Our goal is to extend inequalities of type (1.3.4/1) and (1.4.1/3) to measures of the type $\nu=\varkappa(x)\,\mu_0$ with $\mu_0\in K_h$ and $\varkappa(x)\in K_h(\varrho,\mu_0)$ in the sense of the above definition. In other words, the general measure ν is divided into a measure μ_0 with lattice-structure and a weight function $\varkappa(x)$, which on the one hand may grow almost as rapidly as an exponential function and on the other may be zero on a set of infinite Lebesgue measure. Such general measures ν include e.g. weight functions of type $\prod_{j=1}^{n}|x_j|^{d_j}$ with $d_j>0$ and $|x|^d$ with $d>0$.

6.2.2. Some Inequalities

Our next aim is the extension of Theorem 1.3.1, Proposition 1.3.2, and Proposition 1.3.3. to the weighted case. With a few technical modifications the proofs can be carried over from the unweighted to the weighted case. For details we refer to [F, 1.3.2.–1.3.4.]. If $\omega(x)\in\mathfrak{M}$ and if Ω is a compact subset of R_n, then

$$S_\omega^\Omega=\{\varphi\mid\varphi\in S_\omega\,,\quad \operatorname{supp} F\varphi\subset\Omega\}\,, \tag{1}$$

cf. (1.3.1/1). We recall that M is the Hardy-Littlewood maximal function from 1.2.3.

Theorem. Let $\omega(x)\in\mathfrak{M}$ and $\varrho(x)\in R(\omega)$. Let Ω be a compact subset of R_n.
(i) Let $0<r<\infty$. There exist two positive numbers c_1 and c_2 such that

$$\sup_{z\in R_n}\varrho(x-z)\,\frac{|\nabla\varphi(x-z)|}{1+|z|^{\frac{n}{r}}}\le c_1\sup_{z\in R_n}\varrho(x-z)\,\frac{|\varphi(x-z)|}{1+|z|^{\frac{n}{r}}}\le c_2[(M|\varrho\varphi|^r)(x)]^{\frac{1}{r}} \tag{2}$$

holds for all $\varphi\in S_\omega^\Omega$ and all $x\in R_n$.

(ii) Let $0<p\leqq q\leqq\infty$ and let α be a multi-index. There exists a positive constant c such that

$$\|\varrho D^\alpha\varphi \mid L_q\|\leqq c\|\varrho\varphi \mid L_p\| \tag{3}$$

holds for all $\varphi\in S^\Omega_\omega$.

(iii) Let $0<p\leqq\infty$. There exist three positive numbers h_0, c_1, and c_2, such that

$$c_1\Big(\sum_{k\in Z_n}|\varrho(x^k)\,\varphi(x^k)|^p\Big)^{\frac{1}{p}}\leqq h^{-\frac{n}{p}}\|\varrho\varphi \mid L_p\|\leqq c_2\Big(\sum_{k\in Z_n}|\varrho(x^k)\,\varphi(x^k)\,|^p\Big)^{\frac{1}{p}} \tag{4}$$

holds for all h with $0<h<h_0$, all sets $\{x^k\}_{k\in Z_n}$ with $x^k\in Q^h_k$ and all $\varphi\in S^\Omega_\omega$ (Modification if $p=\infty$).

Remark. As in the first chapter, (2) plays a key role. In order to prove inequalities of type (3) (at least if $p=q$) we apply the scalar case of the maximal inequality from Theorem 1.2.3 for the unweighted L_p-spaces. Another possibility would be to use the unweighted counterpart (1.3.1/2) of (2) and to generalize (the scalar case of) Theorem 1.2.3 to weighted spaces. We formulate a result in this direction, which may be found in H. Triebel [17, p. 102], (cf. also K. F. Andersen, R. T. John [1]). It is a combination of some fundamental ideas in R. R. Coifman, C. Fefferman [1], and C. Fefferman, E. M. Stein [1]. First we recall Muckenhoupt's A_p-condition, cf. B. Muckenhoupt [1], and R. R. Coifman, C. Fefferman [1]. Let $1<p<\infty$ and let $\omega(x)>0$ a. e. be a locally integrable function on R_n such that $\omega^{-1/(p-1)}(x)$ is also locally integrable on R_n. Then $\omega(x)$ satisfies the A_p-condition if and only if

$$\sup\left(\frac{1}{|Q|}\int_Q\omega(x)\,\mathrm{d}x\right)\left(\frac{1}{|Q|}\int_Q\omega^{-1/(p-1)}(x)\,\mathrm{d}x\right)^{p-1}<\infty, \tag{5}$$

where the supremum is taken over all cubes Q in R_n. Let $\|f_k \mid L_{p,\mu}(l_q)\|$ be given by (1.2.2/5) where the Lebesgue measure $\mathrm{d}x$ is replaced by $\mathrm{d}\mu=\omega(x)\,\mathrm{d}x$, cf. (1.2.2/1). Then it is clear what is meant by $L_{p,\mu}(l_q)$. The extension of Theorem 1.2.3. now reads as follows. Let $1<p<\infty$ and $1<q<\infty$. Then there exists a positive constant c such that

$$\|Mf_k \mid L_{p,\mu}(l_q)\|\leqq c\|f_k \mid L_{p,\mu}(l_q)\| \tag{6}$$

holds for all sequences $\{f_k(x)\}^\infty_{k=0}\in L_{p,\mu}(l_q)$ if and only if $\omega(x)$ satisfies the A_p-condition. Of course one can apply the scalar case of this extension of Theorem 1.2.3 to (2). Then one obtains generalizations of (3). We do not go into details and restrict ourselves to the formulations in the above theorem. In this context we refer also to B. Jawerth [3]. Another generalization of some assertions of the theorem for weighted mixed L_p-quasi-norms has been given by B. Stöckert [1].

6.2.3. The Basic Inequality

We generalize Theorem 1.3.4, cf. also [F, 1.3.4.]. $\|\cdot \mid L_{p,\mu}\|$ has been defined in 1.2.2.

Theorem. Let $\omega(x)\in\mathfrak{M}$ and $\varrho(x)\in R(\dot\omega)$. Let Ω be a compact subset of R_n and let $0<p\leqq q\leqq\infty$. There exists a positive number h_0 (which depends only on $\omega(x)$, $\varrho(x)$, Ω, p, and q) with the following property: if $0<h<h_0$, $\mu_1\in K_h$, $\mu_2\in K_h$, $\varkappa_1(x)\in K_h(\varrho,\mu_1)$ and $\varkappa_2(x)\in K_h(\varrho,\mu_2)$, then there exists for every multi-index α a positive number c such that

$$\|\varkappa_1 D^\alpha\varphi \mid L_{q,\mu_1}\|\leqq c\|\varkappa_2\varphi \mid L_{p,\mu_2}\| \tag{1}$$

holds for all $\varphi\in S^\Omega_\omega$.

Remark. A proof may be found in [F, pp. 28/29]. This is a rather general inequality of Plancherel-Polya-Nikol'skij type and the problem arises whether these restrictions are natural. That the number h_0 must be sufficiently small follows from the representation formula (1.3.5/2). It is not hard to see that h_0 tends to zero if $\Omega=\{y \mid |y|\leqq b\}$ and $b\to\infty$. In [F, 1.3.5.] we discussed the other conditions in the above theorem. For exemple, one cannot expect inequalities of type (1) with $p=q$ if μ_1 and μ_2 are Lebesgue measure and if $\varkappa_1(x)=\varkappa_2(x)=\mathrm{e}^{-|x|^\beta}$ with $\beta>1$. Also the "limiting" cases $\varkappa_1(x)=\varkappa_2(x)=\mathrm{e}^{|x|}$ or $\varkappa_1(x)=\varkappa_2(x)=\mathrm{e}^{-|x|}$ cause trouble. Furthermore, (6.1.2/2) is rather natural. For details we refer to [F, 1.3.5.].

6.3. L_p-Spaces of Analytic Functions

6.3.1. Definition and Main Inequalities

This is the weighted counterpart of 1.4.1., cf. also [F, 1.4.1. and 1.4.2.]. $\|\cdot \mid L_{p,\mu}\|$ has the same meaning as in 1.2.2.

Definition. Let $\omega(x)\in\mathfrak{M}$ and $\varrho(x)\in R(\omega)$. Let Ω be a compact subset of R_n and let $0<p\leqq\infty$. Let $\mu\in K_h$ and $\varkappa(x)\in K_h(\varrho, \mu)$. Then

$$L^{\Omega}_{p,\mu}(\varkappa(x))=\{f \mid f\in S'_\omega, \operatorname{supp} Ff\subset\Omega, \|\varkappa f \mid L_{p,\mu}\|<\infty\}\,. \tag{1}$$

If μ is Lebesgue measure, we put $L^{\Omega}_{p,\mu}(\varkappa(x))=L^{\Omega}_p(\varkappa(x))$.

Remark 1. We assume that $\omega(x)$ and $\varrho(x)$ are fixed once and for all. For that reason we did not indicate $\omega(x)$ and $\varrho(x)$ on the left-hand side of (1). Furthermore, we are only interested in spaces $L^{\Omega}_{p,\mu}(\varkappa(x))$ for which $\|\varkappa f \mid L_{p,\mu}\|$ is a quasi-norm. This will be ensured if $0<h<h_0$, where h_0 has the meaning of Theorem 6.2.3. Under this assumption we shall see that $L^{\Omega}_{p,\mu}(\varkappa(x))=L^{\Omega}_p(\varrho(x))$. Furthermore, one can prove that $L^{\Omega}_{p,\mu}(\varkappa(x))$ with $h<h_0$ is independent of the choice of $\omega(x)$, cf. [F, p. 39] for details. In other words, the spaces $L^{\Omega}_{p,\mu}(\varkappa(x))$ depend only on Ω, p and $\varrho(x)$ (if h is sufficiently small).

Theorem. Let $\omega(x)\in\mathfrak{M}$ and $\varrho(x)\in R(\omega)$. Let Ω be a compact subset of R_n.
(i) Let $0<p\leqq\infty$. If $0<r<p$, then there exists a positive constant c such that

$$\left\| \sup_{z\in R_n} \varrho(\cdot - z)\frac{|f(\cdot - z)|}{1+|z|^{\frac{n}{r}}} \,\middle|\, L_p\right\| \leqq c\|\varrho f \mid L_p\| \tag{2}$$

holds for all $f\in L^{\Omega}_p(\varrho(x))$.
(ii) Let $0<p\leqq q\leqq\infty$. Let h_0 be the positive number from Theorem 6.2.3. Let $0<h<h_0$, $\mu_1\in K_h$, $\mu_2\in K_h$, $\varkappa_1(x)\in K_h(\varrho, \mu_1)$ and $\varkappa_2(x)\in K_h(\varrho, \mu_2)$. If α is a multi-index, then there exists a positive constant c such that

$$\|\varkappa_1 D^\alpha f \mid L_{q,\mu_1}\|\leqq c\|\varkappa_2 f \mid L_{p,\mu_2}\| \tag{3}$$

holds for all $f\in L^{\Omega}_{p,\mu_2}(\varkappa_2(x))$.

Remark 2. This is the weighted counterpart of Theorem 1.4.1 and is an extension of Theorem 6.2.3 from S^{Ω}_ω to $L^{\Omega}_{p,\mu_2}(\varkappa_2(x))$, cf. [F, 1.4.2]. The inequality (3) is the main goal of this chapter. It shows that inequalities of type (1.1/5) hold for a large variety of measures. Furthermore, at least in our approach, it is convenient to split the measure from (1.1/5) into a μ-part and a $\varkappa$-part in the sense of (3), where the μ-measure has a lattice structure and the $\varkappa(x)$-function is responsible for the growth.

6.3.2. Basic Properties

Here is the weighted counterpart of 1.4.2., cf. also [F, 1.4.3.]. We recall that $\subset$ always stands for continuous embedding.

Theorem. Let $\omega(x)\in\mathfrak{M}$ and $\varrho(x)\in R(\omega)$. Let Ω be a compact subset of R_n and let $0<p\leqq\infty$. Let h_0 be the positive number from Theorem 6.2.3. (with $q=p$) and let $0<h<h_0$.
(i) If $\mu\in K_h$ and $\varkappa(x)\in K_h(\varrho, \mu)$, then $L^{\Omega}_{p,\mu}(\varkappa(x))$ is a quasi-Banach space (Banach space if $1\leqq p\leqq\infty$) and

$$S^{\Omega}_\omega\subset L^{\Omega}_{p,\mu}(\varkappa(x))\subset S'_\omega$$

(topological embedding).

(ii) If $\mu_1 \in K_h$, $\mu_2 \in K_h$, $\varkappa_1(x) \in K_h(\varrho, \mu_1)$ and $\varkappa_2(x) \in K_h(\varrho, \mu_2)$, then

$$L^{\Omega}_{p,\mu_1}(\varkappa_1(x)) = L^{\Omega}_{p,\mu_2}(\varkappa_2(x)) . \tag{1}$$

Remark 1. In particular, we have $L^{\Omega}_{p,\mu_1}(\varkappa_1(x)) = L^{\Omega}_p(\varrho(x))$, i.e. the spaces $L^{\Omega}_{p,\mu}(\varkappa(x))$ from Definition 6.3.1 are independent of μ and $\varkappa(x)$ if h is sufficiently small.

Remark 2. Further properties of these spaces may be found in [F, 1.4.4. and 1.4.5.].

Remark 3. In Section 1.5. we studied Fourier multipliers for the spaces L^{Ω}_p. It is natural (and also possible) to extend these considerations to the weighted spaces $L^{\Omega}_p(\varrho(x))$. However we shall not go into detail and refer the interested reader to [F, pp. 50–67]. The needed Fourier multiplier assertions for the spaces $L^{\Omega}_p(\varrho(x))$ can be obtained from the scalar case of Theorem 7.1.1, which is the weighted counterpart of Theorem 1.6.3 (and Theorem 1.6.2).

7. Weighted Function Spaces on R_n

7.1. Maximal Inequalities, Fourier Multipliers and Littlewood-Paley Theorems

7.1.1. The Spaces $L_p^\Omega(R_n, \rho(x), l_q)$

In Chapters 1 and 6 of this book we omitted R_n in order to characterize the domain on which the considered spaces are defined. In all other chapters we indicate the underlying domain, such as R_n, Ω etc. Subsection 7.1.1. deals with the weighted counterpart of the spaces from Section 1.6. First we generalize the notation (1.2.2/5) by

$$\|f_k \mid L_p(R_n, \varrho(x), l_q)\| = \left(\int_{R_n} \varrho^p(x) \left(\sum_{k=0}^{\infty} |f_k(x)|^q \right)^{\frac{p}{q}} dx \right)^{\frac{1}{p}}, \tag{1}$$

where $0<p<\infty$ and $0<q\leqq\infty$ (modification if $q=\infty$), and $\varrho(x)$ is a positive Borel-measurable weight function. Furthermore, we specify the sequence $\Omega=\{\Omega_k\}_{k=0}^{\infty}$ of compact subsets of R_n from Definition 1.6.1 by

$$\Omega_k = \{y \mid y \in R_n, |y| \leqq 2^k\}, \quad k=0, 1, 2, \ldots \tag{2}$$

Let $\mathfrak{M}$ and $R(\omega)$ be the classes from Definition 6.1.2/1 and Definition 6.2.1(i). Furthermore, $D_\omega = D_\omega(R_n)$, $S_\omega = S_\omega(R_n)$, and $S'_\omega = S'_\omega(R_n)$ have the same meaning as in 6.1.2. If $\varphi = \{\varphi_k(x)\}_{k=0}^{\infty} \subset D_\omega(R_n)$, then we put

$$A_\lambda^\varphi = \sup_k \|\varphi_k(2^k \cdot)\|_{\omega,\lambda}, \tag{3}$$

where λ is a positive number and $\|\cdot\|_{\omega,\lambda}$ has been defined in (6.1.2/7) (we assume in the sequel that $\omega(x)\in\mathfrak{M}$ is fixed once and for all).

Definition. Let $\omega(x)\in\mathfrak{M}$, $\varrho(x)\in R(\omega)$, $0<p<\infty$ and $0<q\leqq\infty$. Let $\Omega=\{\Omega_k\}_{k=0}^{\infty}$ be given by (2). Then

$$L_p^\Omega(R_n, \varrho(x), l_q) = \{f \mid f=\{f_k\}_{k=0}^{\infty} \subset S'_\omega(R_n), \quad \operatorname{supp} Ff_k \subset \Omega_k \quad \text{if} \quad k=0, 1, 2, \ldots, \quad \text{and} \quad \|f_k \mid L_p(R_n, \varrho(x), l_q)\| < \infty\}. \tag{4}$$

Remark 1. These are the weighted counterparts of the spaces $L_p^\Omega(l_q) = L_p^\Omega(R_n, l_q)$ from Definition 1.6.1 and the vector-valued generalizations of the spaces from Definition 6.3.1 (cf. Remark 6.3.2/1). It is not complicated to prove that $L_p^\Omega(R_n, \varrho(x), l_q)$ is a quasi-Banach space (Banach space if $\min(p, q)\geqq 1$), cf. Theorem 1.6.1 and Theorem 6.3.2.

Theorem. Let $\omega(x)\in\mathfrak{M}$, $\varrho(x)\in R(\omega)$, $0<p<\infty$, $0<q\leqq\infty$ and $0<r<\min(p, q)$. There exist two positive numbers c and λ such that

$$\begin{aligned} &\|F^{-1}\varphi_k F f_k \mid L_p(R_n, \varrho(x), l_q)\| \\ &\leqq \left\| \sup_{z\in R_n} \varrho(\cdot - z) \frac{|(F^{-1}\varphi_k F f_k)(\cdot - z)|}{1+|2^k z|^{\frac{n}{r}}} \,\Big|\, L_p(R_n, l_q) \right\| \\ &\leqq c A_\lambda^\varphi \|f_k \mid L_p(R_n, \varrho(x), l_q)\| \end{aligned} \tag{5}$$

holds for all $\varphi=\{\varphi_k(x)\}_{k=0}^{\infty} \subset D_\omega(R_n)$ and all $\{f_k\}_{k=0}^{\infty} \in L_p^\Omega(R_n, \varrho(x), l_q)$.

Remark 2. The first inequality in (5) is trivial, the second is a generalization of Theorem 1.6.2, Theorem 1.6.3, and Theorem 6.3.1. It is a maximal inequality and also a Fourier multiplier theorem. A proof may be found in H. Triebel [17, Theorem 1]. Of interest is the number A_λ^φ. It can be improved, cf. H. Triebel [17, formula (13)]. From this improvement it follows that

H. Triebel, *Theory of Function Spaces*, Modern Birkhäuser Classics,
DOI 10.1007/978-3-0346-0416-1_7, © Birkhäuser Verlag 1983

in the unweighted case, i.e. $\varrho(x)\equiv 1$, the right-hand side of (5) coincides essentially with the right-hand side of (1.6.3/1) (with $\varphi_l(2^l\,\cdot)$ instead of $M_l(d_l\,\cdot)$). Furthermore, if $\varrho(x)=(1+|x|)^d$ with $d>0$ (or more generally $\omega(x)=\log\,(1+|x|)^d$ and $\varrho(x)\in R(\omega)$), then we have (1.6.3/1) with $L_p(R_n, \varrho(x), l_q)$ instead of $L_p(l_q)$ and with $\varkappa>\frac{n}{2}+\frac{n}{\min\,(p,q)}+d$, cf. H. Triebel [17, p. 93].

7.1.2. The Spaces $L_p(R_n, \varrho(x), l_q)$

The l_q-valued version of the famous Michlin-Hörmander Fourier multiplier theorem for the classical spaces $L_p(R_n)$ with $1<p<\infty$ has been formulated in (2.4.8/7). If $q=2$, one has the more general version of Proposition 2.5.6, which was the basis for proving Littlewood-Paley theorems for $L_p(R_n)$ and more generally for the Bessel-potential spaces $H^s_p(R_n)$. We wish to extend these considerations to the weighted case.

Definition. Let $0<p<\infty$ and $0<q\leqq\infty$. Let $\varrho(x)>0$ be a Borel-measurable function on R_n. Then

$$L_p(R_n, \varrho(x), l_q)=\{f \mid f=\{f_k(x)\}_{k=0}^{\infty}\subset L_p(R_n), \|f_k \mid L_p(R_n, \varrho(x), l_q)\|<\infty\}\,.$$

Remark 1. Obviously, $L_p(R_n, \varrho(x), l_q)$ is a quasi-Banach space (Banach space if $\min\,(p, q)\geqq 1$). If one wishes to find Fourier multiplier theorems for these spaces, then it is reasonable to restrict p and q by $1<p<\infty$ and $1<q<\infty$. We deal with the spaces $L_p(R_n, \varrho(x), l_q)$ in the framework of ultra-distributions. This suggests $\omega(x)\in\mathfrak{M}$ and $\varrho(x)\in R(\omega)$, where $\mathfrak{M}$ and $R(\omega)$ have the same meaning as in the previous subsection, cf. Definition 6.1.2/1 and Definition 6.2.1(i). However this is not sufficient. We follow the treatment given in H. Triebel [17] (cf. also H. Triebel [4]), in which we reduced the weighted cases to the unweighted cases in (2.4.8/7) and in Proposition 2.5.6. The main idea is to approximate $\omega(x)$ by polynomials, and this is exactly the point where the subclass $\mathfrak{M}_E$ of $\mathfrak{M}$ from Definition 6.1.2(ii) comes in, cf. also Remark 6.1.2/1 and Remark 6.2.1/4. In particular, the important weight functions $\varrho(x)=(1+|x|)^d$ with $d>0$, $\varrho(x)=e^{|x|^\beta}$ and $\varrho(x)=e^{-|x|^\beta}$ with $0<\beta<1$ belong to $R(\omega)$ with appropriate functions $\omega(x)\in\mathfrak{M}_E$. We formulated Proposition 2.5.6 for arbitrary matrices. However, only three types of matrices are of interest: diagonal, vertical and horizontal ones. The matrix $(\varphi_{k,j}(x))_{k,j=0}^{\infty}$ is called a diagonal matrix if

$$\begin{cases}\varphi_{k,k}(x)=\varphi_k(x) \quad \text{and}\\ \varphi_{k,j}(x)=0 \quad \text{for} \quad k=0,1,2,\ldots \quad \text{and} \quad j=0,1,2,\ldots \quad \text{with} \quad j\neq k,\end{cases} \tag{1}$$

a vertical matrix if

$$\varphi_{k,0}(x)=\varphi_k(x) \quad \text{and} \quad \varphi_{k,j}(x)=0 \quad \text{for} \quad k=0,1,2,\ldots \quad \text{and} \quad j=1,2,3,\ldots, \tag{2}$$

and a horizontal matrix if

$$\varphi_{0,k}(x)=\varphi_k(x) \quad \text{and} \quad \varphi_{j,k}(x)=0 \quad \text{for} \quad k=0,1,2,\ldots \quad \text{and} \quad j=1,2,3,\ldots \tag{3}$$

In all three cases we put $\varphi=\{\varphi_k(x)\}_{k=0}^{\infty}$ and A_λ^φ has the meaning of (7.1.1/3). Finally, we recall that $D_\omega=D_\omega(R_n)$ has been defined in 6.1.2.

Theorem. Let $\omega(x)\in\mathfrak{M}_E$, $\varrho(x)\in R(\omega)$ and $1<p<\infty$. Let $\{\varphi_{k,j}(x)\}_{k,j=0}^{\infty}\subset D_\omega(R_n)$ be either a diagonal, vertical, or horizontal matrix in the above sense with

$$\operatorname{supp}\varphi_0\subset\{y \mid |y|\leqq 2\},\ \operatorname{supp}\varphi_l\subset\{y \mid 2^{l-1}\leqq|y|\leqq 2^{l+1}\}$$

if $l=1, 2, 3, \ldots$

(i) There exist two positive numbers c and λ such that

$$\left\|\sum_{j=0}^{\infty}F^{-1}\varphi_{k,j}Ff_j \mid L_p(R_n, \varrho(x), l_2)\right\|\leqq cA_\lambda^\varphi\,\|f_k \mid L_p(R_n, \varrho(x), l_2)\| \tag{4}$$

holds for all matrices $(\varphi_{k,j}(x))_{k,j=0}^{\infty}$ (with $\varphi=\{\varphi_k\}_{k=0}^{\infty}$) of the above type and all systems $\{f_k\}_{k=0}^{\infty}\in L_p(R_n, \varrho(x), l_2)$.

(ii) Let $1<q<\infty$. There exist two positive numbers c and λ such that

$$\|F^{-1}\varphi_k F f_k \mid L_p(R_n, \varrho(x), l_q)\| \leq c A_\lambda^\varphi \|f_k \mid L_p(R_n, \varrho(x), l_q)\| \tag{5}$$

holds for all diagonal matrices of the above type (with $\varphi=\{\varphi_k\}_{k=0}^\infty$) and all systems $\{f_k\}_{k=0}^\infty \in L_p(R_n, \varrho(x), l_q)$.

Remark 2. A proof of this theorem may be found in H. Triebel [17]. The proof goes through in the diagonal cases of (i) and (ii). However for the horizontal and vertical case of (i) it is not correct[1]). We sketch a proof. First we consider the vertical case of (4) with $p=2$. Even in that case (4) is not obvious, but can be proved by rather elementary calculations. Afterwards one interpolates this estimate (i.e. the vertical case of (4) with $p=2$) with the unweighted version of the vertical case of (4) (i.e. $1<p<\infty$, $\varrho(x)=1$), which is valid. This interpolation proves the vertical case of (4) for all $1<p<\infty$ and all admissible weights $\varrho(x)$. Finally, the horizontal case is a matter of duality.

Remark 3. $\varrho(x)$ is of at most polynomial growth, then we remarked at the end of 7.1.1. that A_λ^φ can be replaced by $\sup_k \|\varphi_k(2^k\cdot) \mid H_2^\varkappa(R_n)\|$. This is also valid for the above theorem (with the same $\varkappa$ as at the end of 7.1.1.). In particular, only a finite number of derivatives of φ_k come in. However if $\varrho(x)$ grows more rapidly than any polynomial, then all derivatives of φ_k are needed.

Remark 4. The admissible weight functions in the above theorem have no local singularities, such as tending to zero or infinity at points or on subsets of R_n. The question arises whether the above theorem can be extended to weight functions $\varrho(x)$ with such (sufficiently tame) singularities. In particular, it would be of interest to study the case where $\varrho(x)$ satisfies the A_p-condition from Remark 6.2.2 and then to combine such weights with those from the theorem. In connection with Fourier multiplier theorems for weighted L_p-spaces we refer to S. I. Agalarov [1], D. S. Kurtz [1], D. S. Kurtz, R. L. Wheeden [1, 2], B. Muckenhoupt, R. L. Wheeden, Wo-Sang Young [1], and Wo-Sang Young [1].

7.1.3. The Spaces $L_p(R_n, \varrho(x))$

Theorem 7.1.2(i) provides a possibility of proving a Littlewood-Paley theorem for weighted L_p-spaces if the underlying system of functions is smooth enough. For that purpose we introduce the counterpart of the class $\Phi(R_n)$ from Definition 2.3.1/1.

Definition 1. Let $\omega(x)\in\mathfrak{M}$. Then $\Phi_\omega(R_n)$ denotes the collection of all systems $\varphi=\{\varphi_j(x)\}_{j=0}^\infty \subset D_\omega(R_n)$ such that

$$\begin{cases} \operatorname{supp}\varphi_0 \subset \{x \mid |x|\leq 2\} \\ \operatorname{supp}\varphi_j \subset \{x \mid 2^{j-1}\leq |x| \leq 2^{j+1}\} \quad \text{if} \quad j=1,2,3,\ldots, \end{cases} \tag{1}$$

$$A_\lambda^\varphi = \sup_k \|\varphi_k(2^k\cdot)\|_{\omega,\lambda} < \infty \quad \text{for every} \quad \lambda>0, \tag{2}$$

$$\sum_{j=0}^\infty \varphi_j(x) = 1 \quad \text{for every} \quad x\in R_n. \tag{3}$$

Remark 1. We recall that $D_\omega(R_n)=D_\omega$ and $\|\cdot\|_{\omega,\lambda}$ have been introduced in Definition 6.1.2/2(i). Furthermore, we have $D_\omega(R_n)=S_\omega(R_n)\cap D(R_n)$, cf. Proposition 6.1.3/1. The numbers A_λ^φ played an important role in Theorem 7.1.1 and Theorem 7.1.2. The class $\Phi_\omega(R_n)$ is not-empty. This can be proved in the same way as in Remark 2.3.1/1. In particular, we may assume that $\varphi_0\in D_\omega$, $\varphi_1\in D_\omega$ and $\varphi_j(x)=\varphi_1(2^{-j+1}x)$ if $j=2,3,4,\ldots$

Definition 2. Let $0<p<\infty$. Let $\varrho(x)>0$ be a Borel-measurable function on R_n. Then

$$L_p(R_n, \varrho(x)) = \{f \mid \varrho f \in L_p(R_n)\}. \tag{4}$$

[1]) I am indebted to Dr. H.-J. Schmeisser for pointing this out to me.

Remark 2. This is the scalar case of the spaces $L_p(R_n, \varrho(x), l_q)$ from Definition 7.1.2, quasi-normed by $\|f \mid L_p(R_n, \varrho(x))\| = \|\varrho f \mid L_p(R_n)\|$. Obviously, $L_p(R_n, \varrho(x))$ is a quasi-Banach space (Banach space if $1 \leqq p < \infty$).

Theorem. Let $\omega(x) \in \mathfrak{M}_E$, $\varrho(x) \in R(\omega)$ and $1 < p < \infty$. Let $\{\varphi_k(x)\}_{k=0}^\infty \in \Phi_\omega(R_n)$. Then there exist two positive constants c_1 and c_2 such that

$$c_1\|f \mid L_p(R_n, \varrho(x))\| \leqq \|F^{-1}\varphi_k Ff \mid L_p(R_n, \varrho(x), l_2)\| \leqq c_2\|f \mid L_p(R_n, \varrho(x))\| \tag{5}$$

holds for all $f \in L_p(R_n, \varrho(x))$.

Remark 3. This is a theorem of Littlewood-Paley type. It follows easily from the horizontal and vertical cases of Theorem 7.1.2, cf. H. Triebel [17, pp. 98/99] (the unweighted counterpart of this proof can also be found in Step 2 of the proof of Theorem 2.5.6).

7.2. Weighted Spaces of Type $B^s_{p,q}$ and $F^s_{p,q}$

7.2.1. Definition

This is the weighted counterpart of 2.3.1. We use the above notations, in particular $\mathfrak{M}_E$ and $R(\omega)$ from Definition 6.1.2/1(ii) and Definition 6.2.1(i).

Definition. Let $\omega(x) \in \mathfrak{M}_E$ and $\varrho(x) \in R(\omega)$. Let $-\infty < s < \infty$, $0 < q \leqq \infty$ and $\varphi = \{\varphi_k(x)\}_{k=0}^\infty \in \Phi_\omega(R_n)$.
(i) If $0 < p \leqq \infty$, then

$$B^s_{p,q}(R_n, \varrho(x)) = \{f \mid f \in S'_\omega(R_n), \tag{1}$$
$$\|f \mid B^s_{p,q}(R_n, \varrho(x))\|^\varphi = \|\varrho 2^{js} F^{-1}\varphi_j Ff \mid l_q(L_p(R_n))\| < \infty\}.$$

(ii) If $0 < p < \infty$, then

$$F^s_{p,q}(R_n, \varrho(x)) = \{f \mid f \in S'_\omega(R_n), \tag{2}$$
$$\|f \mid F^s_{p,q}(R_n, \varrho(x))\|^\varphi = \|\varrho 2^{js} F^{-1}\varphi_j Ff \mid L_p(R_n, l_q)\| < \infty\}.$$

Remark 1. If $\varrho(x) \equiv 1$, then (1) and (2) coincide with (2.3.1/5) and (2.3.1/6), respectively. Furthermore, introducing the weighted counterparts of $F^s_{\infty,q}(R_n)$ from Definition 2.3.4 we can ask the same questions as for the unweighted spaces and try to develop a theory of weighted spaces parallel to Chapter 2. We shall not be concerned with this comprehensive task but restrict ourselves instead to a few examples.

Remark 2. The above weight functions $\varrho(x)$ have no local singularities. It would be desirable (and probably also possible) to extend the class of admissible weight functions $\varrho(x)$ in the same way as has been described in Remark 7.1.2/4. Weighted spaces of $F^s_{p,q}$-type, where the weights satisfy Muckenhoupt's A_p-condition (cf. Remark 6.2.2) have been studied by V. M. Kokilašvili [1–6] and Bui Huy Qui [4]. One can expect that weighted spaces of type $F^0_{p,2}$ with $0 < p < \infty$ are near to corresponding weighted Hardy spaces. As far as the latter spaces are concerned, we cite a few articles, where further references can be found: Bui Huy Qui [1, 3], J. Garcia-Cuerva [1], B. Muckenhoupt, R. L. Wheeden [1], J.-O. Strömberg, A. Torchinsky [1], J.-O. Strömberg, R. L. Wheeden [1], R. L. Wheeden [1], H. P. Heinig, R. Johnson [1].

7.2.2. Basic Properties

Theorem. Let $\omega(x) \in \mathfrak{M}_E$ and $\varrho(x) \in R(\omega)$. Let $-\infty < s < \infty$ and $0 < q \leqq \infty$.
(i) If $0 < p \leqq \infty$, then $B^s_{p,q}(R_n, \varrho(x))$ is a quasi-Banach space (Banach space if $1 \leqq p \leqq \infty$ and $1 \leqq q \leqq \infty$) and the quasi-norms $\|f \mid B^s_{p,q}(R_n, \varrho(x))\|^\varphi$ with $\varphi \in \Phi_\omega(R_n)$ and $\|f \mid B^s_{p,q}(R_n, \varrho(x))\|^\psi$ with $\psi \in \Phi_\omega(R_n)$ are equivalent to each other.

(ii) If $0<p<\infty$, then $F^s_{p,q}(R_n, \varrho(x))$ is a quasi-Banach space (Banach space if $1\leq p<\infty$ and $1\leq q\leq\infty$) and the quasi-norms $\|f \mid F^s_{p,q}(R_n, \varrho(x))\|^\varphi$ with $\varphi\in\Phi_\omega(R_n)$ and $\|f \mid F^s_{p,q}(R_n, \varrho(x))\|^\psi$ with $\psi\in\Phi_\omega(R_n)$ are equivalent to each other.

Remark. The proof is an easy consequence of Theorem 7.1.1. In particular, Definition 7.2.1 makes sense. The above theorem is the weighted counterpart of Theorem 2.3.3. The other assertions of that theorem can also be transfered. Hence it can be shown that

$$S_\omega(R_n)\subset B^s_{p,q}(R_n, \varrho(x))\subset S'_\omega(R_n)\,,$$

$$S_\omega(R_n)\subset F^s_{p,q}(R_n, \varrho(x))\subset S'_\omega(R_n)$$

for all admissible values of s, p and q. Furthermore, $S_\omega(R_n)$ is dense in $B^s_{p,q}(R_n, \varrho(x))$ and in $F^s_{p,q}(R_n, \varrho(x))$ if $-\infty<s<\infty$, $0<p<\infty$ and $0<q<\infty$.

7.2.3. Weighted Sobolev Spaces

In 2.5.6. we proved that the classical Sobolev spaces $W^m_p(R_n)$ can be identified with $F^m_{p,2}(R_n)$, where $m=0, 1, 2, \ldots$ and $1<p<\infty$. If $\omega(x)\in\mathfrak{M}_E$, $\varrho(x)\in R(\omega)$ and $1<p<\infty$, then

$$F^0_{p,2}(R_n, \varrho(x))=L_p(R_n, \varrho(x)) \tag{1}$$

is a simple reformulation of Theorem 7.1.3. On the basis of Theorem 7.1.1, it is not complicated to extend this assertion to the spaces $F^m_{p,2}(R_n, \varrho(x))$, cf. H. Triebel [17].

Definition. Let $\omega(x)\in\mathfrak{M}_E$, $\varrho(x)\in R(\omega)$, $1<p<\infty$ and $m=0, 1, 2, \ldots$ Then

$$W^m_p(R_n, \varrho(x))=\{f \mid D^\alpha f\in L_p(R_n, \varrho(x)) \quad\text{if}\quad 0\leq|\alpha|\leq m\}\,. \tag{2}$$

Remark. $W^m_p(R_n, \varrho(x))$ is a subspace of $S'_\omega(R_n)$. Furthermore, when endowed with the norm

$$\|f \mid W^m_p(R_n, \varrho(x))\| = \sum_{|\alpha|\leq m} \|D^\alpha f \mid L_p(R_n, \varrho(x))\|$$

it is a Banach space. Obviously, $W^0_p(R_n, \varrho(x))=L_p(R_n, \varrho(x))$.

Theorem. Let $\omega(x)\in\mathfrak{M}_E$, $\varrho(x)\in R(\omega)$, $1<p<\infty$ and $m=0, 1, 2, \ldots$ Then

$$W^m_p(R_n, \varrho(x))=F^m_p(R_n, \varrho(x)) \tag{3}$$

(equivalent norms). Furthermore,

$$\|f \mid L_p(R_n, \varrho(x))\| + \sum_{k=1}^n \left\| \frac{\partial^m f}{\partial x_k^m} \mid L_p(R_n, \varrho(x)) \right\|$$

is an equivalent norm in $W^m_p(R_n, \varrho(x))$.

7.2.4. Characterizations by Approximation

One can expect that many properties of the spaces $B^s_{p,q}(R_n)$ and $F^s_{p,q}(R_n)$ from the second chapter have weighted counterparts. We describe an example: the characterization of the weighted Sobolev spaces $W^m_p(R_n, \varrho(x))$ from the previous subsection by approximations in the sense of 2.5.3. There is little doubt that the results can be extended to other spaces $B^s_{p,q}(R_n, \varrho(x))$ and $F^s_{p,q}(R_n, \varrho(x))$. However, in order to give the idea, the theorem below will be sufficient. First, we introduce the weighted counterpart of the class $\mathfrak{A}_p(R_n)$ from (2.5.3/4). If $\omega(x)\in\mathfrak{M}_E$, $\varrho(x)\in R(\omega)$ and $0<p<\infty$, then $\mathfrak{A}_p(R_n, \varrho(x))$ denotes the collection of all systems

$a=\{a_j(x)\}_{j=0}^{\infty}$ with

$$a_k(x)\in S'_\omega(R_n)\cap L_p(R_n,\varrho(x))\,,$$
$$\operatorname{supp} Fa_k\subset\{y\mid |y|\leq 2^{k+1}\}\quad\text{if}\quad k=0,1,2,\ldots \tag{1}$$

If $1\leq p<\infty$, then $L_p(R_n,\varrho(x))\subset S'_\omega(R_n)$, which simplifies (1).

Theorem. If $1<p<\infty$ and $m=1,2,3,\ldots$, then

$$W^m_p(R_n,\varrho(x))=\{f\mid f\in S'_\omega(R_n),\ \exists a=\{a_k(x)\}_{k=0}^{\infty}\in\mathfrak{A}_p(R_n,\varrho(x))$$
$$\text{such that}\quad f=\lim_{k\to\infty}a_k\quad\text{in}\quad S'_\omega(R_n)\,,$$
$$\|f\mid W^m_p(R_n,\varrho(x))\|^a=\|\varrho a_0\mid L_p(R_n)\|+\|2^{mk}\varrho\,(f-a_k)\mid L_p(R_n,l_2)\|<\infty\}\,.$$

Furthermore,

$$\|f\mid W^m_p(R_n,\varrho(x))\|^{\times}=\inf\|f\mid W^m_p(R_n,\varrho(x))\|^a\,,$$

where the infimum is taken over all admissible systems $a\in\mathfrak{A}_p(R_n,\varrho(x))$, is an equivalent norm in $W^m_p(R_n,\varrho(x))$.

Remark 1. If we use Theorem 7.2.3, then we can carry over the proof of Theorem 2.5.3, cf. also H. Triebel [17, p. 101].

Remark 2. Recall that $\varrho(x)=\mathrm{e}^{|x|^\beta}$ and $\varrho(x)=\mathrm{e}^{-|x|^\beta}$ with $0<\beta<1$ are admissible weights (Gevrey classes). On the other hand, the above theorem fails if $\varrho(x)=\mathrm{e}^{|x|}$ or $\varrho(x)=\mathrm{e}^{-|x|}$. Corresponding remarks may be found in [F, pp, 34/35] and in H. Triebel [17, p. 101].

8. Weighted Function Spaces on Domains and Degenerate Elliptic Differential Equations

8.1. Weighted Function Spaces on Domains

8.1.1. Introduction and Definitions

In [I, 3.2.3.] we introduced the weighted function spaces $B^s_{p,q}(\Omega, \varrho^\mu(x), \varrho^\nu(x))$ and $H^s_p(\Omega, \varrho^\mu(x), \varrho^\nu(x))$, where Ω is an arbitrary domain on R_n and $\varrho(x)$ is a weight function which degenerates near the boundary $\partial\Omega$ of Ω and at infinity (if Ω is unbounded). Furthermore $1<p<\infty, 1\leqq q\leqq\infty$ and $s\geqq 0$. We treated these spaces in some detail and used them to study degenerate elliptic differential equations in [I, Chapter 6]. In H. Triebel [11] we extended these considerations to the spaces $B^s_{p,p}(\Omega, \varrho^\mu(x), \varrho^\nu(x))$ with $0<p<\infty$. It is the aim of this chapter to give an idea of these spaces and their applications to elliptic differential equations. For that purpose, we restrict ourselves to model cases, both for the spaces (in particular for the domain Ω) and the elliptic differential operators. Essentially, we follow the summaries given in [S, R. 4 and R. 6]. A reader who is interested in proofs and generalizations is asked to consult [I] and H. Triebel [11].

Let $\Omega=\{x \mid x\in R_n, |x|<1\}$ be the open unit ball in R_n. Let $\Omega_0=\left\{x \mid |x|<\frac{1}{2}\right\}$ and

$$\Omega_j=\{x \mid 1-2^{-(j-1)}<|x|<1-2^{-(j+1)}\} \quad \text{if} \quad j=1, 2, 3, \ldots$$

Let $\bar{\Omega}_k$ be covered by a finite number of balls $K_{k,l}=\{x \mid |x-x_{k,l}|<2^{-k-2}\}$ with $x_{k,l}\in\Omega_k$, i.e.

$$\bar{\Omega}_k \subset \bigcup_{l=1}^{N_k} K_{k,l} \quad \text{with} \quad k=0, 1, 2, \ldots$$

It is assumed that there exists a natural number L such that at most L balls $K_{k,l}$ (where $k=0,1,2,\ldots$ and $l=1,\ldots,N_k$) have a non-empty intersection. Let $\varphi=\{\varphi_{k,l}\}_{\substack{k=0,1,2,\ldots\\ l=1,\ldots,N_k}}$ be a smooth resolution of unity with respect to the balls $K_{k,l}$, i.e. $\varphi_{k,l}(x)$ are infinitely differentiable real functions on R_n with

$$0\leqq\varphi_{k,l}(x)\leqq 1\,, \quad \operatorname{supp}\varphi_{k,l}\subset K_{k,l}\,,$$

$$\sum_{k=0}^{\infty}\sum_{l=1}^{N_k}\varphi_{k,l}(x)=1 \quad \text{if} \quad x\in\Omega\,,$$

$$|D^\gamma\varphi_{k,l}(x)|\leqq c_\gamma 2^{k|\gamma|} \quad \text{if} \quad x\in R_n\,,$$

or every multi-index γ, where the c_γ's are appropriate numbers independent of k and l. Let $\Phi(\Omega)$ be the collection of all such systems φ. We specialize the above-mentioned weight function $\varrho(x)$ by $\varrho(x)=(d(x))^{-1}$, where $d(x)$ is the distance of a point $x\in\Omega$ from the boundary of Ω, i.e.

$$d(x)=\inf_{|y|=1}|x-y|, \quad |x|<1\,.$$

In [I, 3.2.3.] we introduced the spaces

$$B^s_{p,q}(\Omega, \varrho^\mu(x), \varrho^\nu(x))=\left\{f \mid f\in D'(\Omega), \|f \mid B^s_{p,q}(\Omega, \varrho^\mu(x), \varrho^\nu(x))\|\right. \tag{1}$$

$$\left.=\left(\sum_{k=0}^{\infty}\sum_{l=1}^{N_k}(2^{k\mu}\|\varphi_{k,l}f \mid B^s_{p,q}(R_n)\|^p+2^{k\nu}\|\varphi_{k,l}f \mid L_p(R_n)\|^p\right)^{\frac{1}{p}}<\infty\right\},$$

where $1<p<\infty, 1\leqq q\leqq\infty$, $s>0$, μ and ν are real numbers with $\nu\geqq\mu+sp$, and $\varphi=\{\varphi_{k,l}\}\in\Phi(\Omega)$. In [I, 3.2.3. etc.], we treated spaces of this type in arbitrary domains Ω and for a large class of weight functions. $D'(\Omega)$ is the collection of all complex-valued distributions on Ω, and

H. Triebel, *Theory of Function Spaces*, Modern Birkhäuser Classics,
DOI 10.1007/978-3-0346-0416-1_8, © Birkhäuser Verlag 1983

$\varphi_{k,l} f$ is extended by zero outside of $K_{k,l}$. In [I, 3.2.4], we proved that $B^s_{p,q}(\Omega, \varrho^\mu(x), \varrho^\nu(x))$ is independent of $\varphi \in \Phi(\Omega)$. Furthermore, it is a Banach space. In H. Triebel [11] we considered corresponding spaces with $p=q$ and $0<p<\infty$. The main problem was to find an appropriate substitute for the basic spaces $L_p(R_n)$. It turns out that the spaces $B^\sigma_{p,p}(R_n)$ with $0<p<\infty$ and $n\left(\frac{1}{p}-1\right)<\sigma<\frac{n}{p}$ can be taken as basic spaces.

Definition. Let $0<p<\infty$ and $\varphi=\{\varphi_{k,l}\}_{\substack{k=0,1,2,\dots\\ l=1,\dots,N_k}} \in \Phi(\Omega)$.

(i) Let $n\left(\frac{1}{p}-1\right)<\sigma<\frac{n}{p}$ and let $\varkappa$ be a real number. Then

$$B^\sigma_p(\Omega, \varkappa)=\Big\{f \mid f\in D'(\Omega), \|f \mid B^\sigma_p(\Omega,\varkappa)\|^\varphi = \left(\sum_{k=0}^\infty \sum_{l=1}^{N_k} 2^{k\varkappa}\|\varphi_{k,l}f \mid B^\sigma_{p,p}(R_n)\|^p\right)^{\frac{1}{p}}<\infty\Big\}. \tag{2}$$

(ii) Let $n\left(\frac{1}{p}-1\right)<\sigma<\frac{n}{p}$ and $s\geqq\sigma$. Let $\varkappa$ and ζ be two real numbers with $\varkappa+\sigma p\geqq\zeta+sp$. Then

$$B^{s,\sigma}_p(\Omega, \zeta, \varkappa)=\Big\{f \mid f\in D'(\Omega), \|f \mid B^{s,\sigma}_p(\Omega,\zeta,\varkappa)\|^\varphi = \left(\sum_{k=0}^\infty \sum_{l=1}^{N_k} (2^{k\zeta}\|\varphi_{k,l}f \mid B^s_{p,p}(R_n)\|^p+2^{k\varkappa}\|\varphi_{k,l}f \mid B^\sigma_{p,p}(R_n)\|^p)\right)^{\frac{1}{p}}<\infty\Big\}. \tag{3}$$

Remark. Obviously, under the above specializations $B^s_{p,q}(\Omega, \varrho^\mu(x), \varrho^\nu(x))$ with $p=q$ and $B^{s,\sigma}_p(\Omega, \zeta, \varkappa)$ are related to each other. Now we have not such a standard basic space as $L_p(R_n)$. Its place is taken by $B^\sigma_{p,p}(R_n)$ and so there is the additional index σ. Of course, we have

$$B^{\sigma,\sigma}_p(\Omega, \varkappa, \varkappa)=B^\sigma_p(\Omega, \varkappa) . \tag{4}$$

8.1.2. Basic Properties

In connection with the spaces from 8.1.1, several questions arise. First, it must be ensured that the spaces $B^{s,\sigma}_p(\Omega, \zeta, \varkappa)$ are independent of $\varphi\in\Phi(\Omega)$. Furthermore, one is interested in dense subsets, topological embeddings and nice equivalent quasi-norms. In the next subsection we add some remarks about the latter problem. Let $D(\Omega)$ be the usual space of all test functions on Ω (where Ω is the open unit ball), i.e. the collection of all complex-valued infinitely differentiable functions in Ω with compact support, equipped in the usual way with a locally convex topology. Let

$$D(\bar\Omega)=\{\varphi \mid \varphi\in D(R_n), \operatorname{supp}\varphi\subset\bar\Omega\} .$$

The topology in $D(\bar\Omega)$ is generated by the semi-norms $\sup_{x\in\Omega} |D^\alpha\varphi(x)|$, where α is an arbitrary multi-index.

Theorem. All notations have the same meaning as in Definition 8.1.1. In particular, Ω is the open unit ball in R_n.

(i) $B^{s,\sigma}_p(\Omega, \zeta, \varkappa)$ is a quasi-Banach space (Banach space if $1\leqq p<\infty$). It is independent of the choice of the balls $K_{k,l}$ and of $\varphi\in\Phi(\Omega)$. Furthermore, $D(\Omega)$ and $D(\bar\Omega)$ are dense in $B^{s,\sigma}_p(\Omega, \zeta, \varkappa)$ and

$$D(\Omega)\subset D(\bar\Omega)\subset B^{s,\sigma}_p(\Omega, \zeta, \varkappa) , \tag{1}$$

(topological embedding).

(ii) If η is an arbitrary real number, then $f \to d^{\eta}(x)f$ is an isomorphic mapping from $B_p^{\sigma}(\Omega, \varkappa - \eta p)$ onto $B_p^{\sigma}(\Omega, \varkappa)$.

Remark. A proof may be found in H. Triebel [11]. By (8.1.1/4) the spaces $B_p^{\sigma}(\Omega, \varkappa)$ are included in part (i) of the theorem.

8.1.3. Inner Descriptions

The problem is to find explicit nice equivalent quasi-norms for the spaces $B_p^{\sigma}(\Omega, \varkappa)$ and $B_p^{\sigma,s}(\Omega, \zeta, \varkappa)$ from Definition 8.1.1. We are looking for the weighted counterparts of (3.4.2/6) and (3.4.2/8). Let $1 \leqq p < \infty$ and let $\sigma > 0$ with $\sigma = [\sigma] + \{\sigma\}$, where $[\sigma]$ is an integer and $0 < \{\sigma\} < 1$ (in particular, σ is not an integer). Then it follows from (3.4.2/6) that

$$\|f \mid B_{p,p}^{\sigma}(\Omega)\| = \left(\int_{\Omega} |f(x)|^p \, dx + \sum_{|\alpha| \leqq [\sigma]} \int_{\Omega \times \Omega} \frac{|D^{\alpha}f(x) - D^{\alpha}f(y)|^p}{|x-y|^{n+\{\sigma\}p}} \, dx \, dy \right)^{\frac{1}{p}} \tag{1}$$

is an equivalent norm in $B_{p,p}^{\sigma}(\Omega)$. By Proposition 3.4.2 this assertion can be extended to $B_{p,p}^{\sigma}(\Omega)$ with $0 < p < 1$ if

$$n\left(\frac{1}{p} - 1\right) < \sigma < 1 , \tag{2}$$

cf. also Fig. 3.4.2 (equivalent quasi-norm).

Theorem 1. Let

$$\begin{cases} \text{either} \quad 1 < p < \infty , \quad 0 < \sigma < \frac{n}{p}, \quad \sigma \neq \text{integer} , \quad \sigma - \frac{1}{p} \neq \text{integer} , \\ \text{or} \quad 0 < p \leqq 1 , \quad n\left(\frac{1}{p} - 1\right) < \sigma < 1 . \end{cases} \tag{3}$$

Then (1) is an equivalent quasi-norm in $B_p^{\sigma}(\Omega, 0)$.

Remark 1. In particular $B_p^{\sigma}(\Omega, 0)$ is the completion of $D(\Omega)$ in the quasi-norm (1), provided (3) is satisfied. In other words, if $\mathring{B}_{p,p}^{\sigma}(\Omega)$ is the completion of $D(\Omega)$ in the quasi-norm $\|\cdot \mid B_{p,p}^{\sigma}(\Omega)\|$, then $\mathring{B}_{p,p}^{\sigma}(\Omega) = B_p^{\sigma}(\Omega, 0)$ provided (3) is satisfied. If $1 < p < \infty$ and $\sigma > 0$, put

$$\tilde{B}_{p,p}^{\sigma}(\Omega) = \{f \mid f \in B_{p,p}^{\sigma}(R_n), \operatorname{supp} f \subset \bar{\Omega}\} .$$

We have $\tilde{B}_{p,p}^{\sigma}(\Omega) = \mathring{B}_{p,p}^{\sigma}(\Omega)$ if $1 < p < \infty$ and $\sigma - \frac{1}{p} \neq$ integer, cf. [I, Theorem 4.3.2/1]. Now one can extend the statement of the above theorem in the following way: if $1 < p < \infty$ and $0 < \sigma < \frac{n}{p}$, then

$$B_p^{\sigma}(\Omega, 0) = \tilde{B}_{p,p}^{\sigma}(\Omega) . \tag{4}$$

This includes the cases of either σ or $\sigma - \frac{1}{p}$ is an integer. Now it follows from (3.4.2/6) that we have also equivalent norms in $B_p^{\sigma}(\Omega, 0)$ if $1 < p < \infty$ and if σ is an integer with $0 < \sigma < \frac{n}{p}$ (in that case higher differences are needed).

In the following theorem, $d(x)$ again denotes the distance of a point $x \in \Omega$ from the boundary $\partial\Omega = \{y \mid |y| = 1\}$, cf. 8.1.1.

Theorem 2. Let

$$\begin{cases} \text{either} \quad 1<p<\infty \quad \text{and} \quad 0<\sigma<\dfrac{n}{p} \\ \text{or} \quad 0<p\leqq 1 \quad \text{and} \quad n\left(\dfrac{1}{p}-1\right)<\sigma<1 . \end{cases} \tag{5}$$

Let $\varkappa$ be a real number, $\nu>0$ and $\mu>\nu+2m$ with $m=1, 2, 3, \ldots$ Then $B_p^\sigma(\Omega, \varkappa)$ is the completion of $D(\Omega)$ in the quasi-norm $\|d^{-\frac{\varkappa}{p}} f \mid B_p^\sigma(\Omega, 0)\|$ and $B_p^{\sigma+2m,\sigma}(\Omega, \varkappa+p\mu, \varkappa+p\nu)$ is the completion of $D(\Omega)$ in the quasi-norm

$$\sum_{|\alpha|=2m} \|d^{-\frac{\varkappa}{p}-\mu} D^\alpha f \mid B_p^\sigma(\Omega, 0)\| + \|d^{-\frac{\varkappa}{p}-\nu} f \mid B_p^\sigma(\Omega, 0)\| .$$

Remark 2. The combination of these two theorems produces many explicit quasi-norms.

8.2. Degenerate Elliptic Differential Equations

8.2.1. Definition and A Priori Estimate

Let again $\Omega=\{x \mid |x|<1\}$ be the open unit ball in R_n with the boundary $\partial\Omega=\{y \mid |y|=1\}$ Let $d(y)=\inf\limits_{y\in\partial\Omega} |x-y|$ be the distance of a point $x\in\Omega$ from $\partial\Omega$.

Definition. Let $m=1, 2, 3, \ldots$ and let ν and μ be two real numbers with $\nu>0$ and $\nu>\mu+2m$. Then A,

$$Af=d^{-\mu}(x)\,(-\Delta)^m f+d^{-\nu}(x) f , \quad f\in D'(\Omega) , \tag{1}$$

denotes a degenerate elliptic differential operator.

Remark. This is a special case of a large class of degenerate elliptic differential operators which have been studied in [I, Chapter 6]. In H. Triebel [11], we extended this theory to the above weighted spaces. In order to give the idea, we formulate an a priori estimate in this subsection and, as a consequence, a mapping property in the next subsection.

Theorem. Let $0<p<\infty$ and $n\left(\frac{1}{p}-1\right)<\sigma<\frac{n}{p}$. Let $\varkappa$ be a real number. Then there exist a real number λ_0 and two positive numbers c_1 and c_2 such that

$$\begin{aligned} & c_1 \|Af+\lambda f \mid B_p^\sigma(\Omega, \varkappa)\| \\ & \leqq \|f \mid B_p^{\sigma+2m,\sigma}(\Omega, \varkappa+p\mu, \varkappa+p\nu)\| + |\lambda|\, \|f \mid B_p^\sigma(\Omega, \varkappa)\| \\ & \leqq c_2 \|Af+\lambda f \mid B_p^\sigma(\Omega, \varkappa)\| \end{aligned}$$

holds for all complex numbers λ with Re $\lambda=\lambda_0$ and all $f\in D(\Omega)$.

8.2.2. A Mapping Property

The operator A has the same meaning as in the previous subsection. Let E be the identity.

Theorem. Let $0<p<\infty$ and $n\left(\frac{1}{p}-1\right)<\sigma<\frac{n}{p}$. Let $\varkappa$ be a real number. Then there exists a real number λ_0 such that for all complex numbers λ with Re$\lambda\geqq\lambda_0$, the operator $A+\lambda E$ yields an isomorphic mapping from $B_p^{\sigma+2m,\sigma}(\Omega, \varkappa+p\mu, \varkappa+p\nu)$ onto $B_p^\sigma(\Omega, \varkappa)$.

Remark. A proof has been given in H. Triebel [11].

9. Periodic Function Spaces

9.1. Introduction and Definitions

9.1.1. Introduction

Let T_n be the n-torus. As usual, we represent T_n in R_n by the cube

$$T_n = \{x \mid x \in R_n,\ x = (x_1, \ldots, x_n),\ |x_j| \leq \pi \text{ with } j = 1, \ldots, n\}, \tag{1}$$

where opposite points are identified. More precisely, $x \in T_n$ and $y \in T_n$ are identified if and only if $x - y = 2\pi k$, where $k = (k_1, \ldots, k_n)$ is a vector with the integer-valued components $k_1, \ldots, k_n$. We denote the lattice of such vectors k with integer-valued components by Z_n. The problem is to introduce and to study spaces of type $B^s_{p,q}$ and $F^s_{p,q}$ on T_n. One can try to develop such a theory parallel to Chapter 2, where the Euclidean n-space R_n is replaced by the n-torus T_n. In principle, such an approach should be possible. Another possibility is to extend distributions on T_n periodically to R_n and to compare spaces on T_n with corresponding spaces on R_n. The disadvantage of the latter approach is quite clear. If one extends e.g. a function $f(x) \in L_p(T_n)$ with $1 < p < \infty$ periodically to R_n, then the extended function does not belong to $L_p(R_n)$. One can try to correct this inconvenience if one interprets the extended functions as elements of $L_p(R_n, \varrho(x))$ where $\varrho(x)$ is an appropriate weight function on R_n. We follow this approach and identify the spaces $B^s_{p,q}(T_n)$ and $F^s_{p,q}(T_n)$ on T_n with subspaces of appropriate spaces $B^s_{p,q}(R_n, \varrho(x))$ and $F^s_{p,q}(R_n, \varrho(x))$ from Chapter 7, respectively, although we shall not give a systematic treatment. Our aim is to introduce the spaces $B^s_{p,q}(T_n)$ and $F^s_{p,q}(T_n)$, to give affirmative answers for some basic questions and to formulate a theorem about Fourier multipliers and maximal inequalities for trigonometrical polynomials, which is of self-contained interest and which may be considered as the main result of this chapter. Further details about this approach may be found in H. Triebel [21].

9.1.2. Periodic Distributions on R_n

We use the notations from 1.2.1., in particular the Euclidean n-space R_n, the Schwartz space $S(R_n)$, and the space of tempered distributions $S'(R_n)$. The Fourier transform F and its inverse F^{-1} have been defined in (1.2.1/2) and (1.2.1/3), respectively. They are extended in the usual way to $S'(R_n)$. We recall that

$$Z_n = \{k \mid k = (k_1, \ldots, k_n) \in R_n,\ k_j \text{ integers}\}.$$

Definition. $f \in S'(R_n)$ is called a periodic distribution on R_n if

$$f(\varphi(\cdot)) = f(\varphi(\cdot + 2\pi k)) \tag{1}$$

holds for all $k \in Z_n$ and all $\varphi \in S(R_n)$.

Remark 1. Of course, $2\pi k = (2\pi k_1, \ldots, 2\pi k_n)$ if $k = (k_1, \ldots, k_n)$.

Proposition. $f \in S'(R_n)$ is a periodic distribution on R_n if and only if it can be represented as

$$f(\varphi) = \sum_{k \in Z_n} a_k (F\varphi)(k), \quad \varphi \in S(R_n), \tag{2}$$

where $\{a_k\}_{k \in Z_n}$ is a sequence of complex numbers of at most polynomial growth, i.e. there exist two positive constants c and N such that

$$|a_k| \leq c\,(1 + |k|)^N \quad \text{for all} \quad k \in Z_n. \tag{3}$$

H. Triebel, *Theory of Function Spaces*, Modern Birkhäuser Classics,
DOI 10.1007/978-3-0346-0416-1_9, © Birkhäuser Verlag 1983

Remark 2. The proof is not very complicated, incluing the following reformulation: $f \in S'(R_n)$ is a periodic distribution on R_n if and only if it can be represented as

$$f = \sum_{k \in Z_n} a_k e^{ikx} \quad (\text{convergence in } S'(R_n)), \tag{4}$$

where $\{a_k\}_{k \in Z_n}$ is a sequence of at most polynomial growth.

9.1.3. Function Spaces on T_n

Let $D(T_n)$ be the collection of all complex-valued infinitely differentiable functions on T_n. The n-torus T_n is given by (9.1.1/1), where opposite points are identified, in particular, implicid in $f \in D(T_n)$ is that $f(x) = f(y)$ if $x \in T_n$, $y \in T_n$ and $x - y = 2\pi k$ (in other words, the function $f(x) \in D(T_n)$ is 2π-periodic). The locally convex topology in $D(T_n)$ is generated by the semi-norms $\sup_{x \in T_n} |D^\alpha f(x)|$, where α is an arbitrary multi-index. Let $D'(T_n)$ be the topological dual of $D(T_n)$. We call the elements of $D'(T_n)$ periodic distributions on T_n. It is well-known that any $f \in D'(T_n)$ can be represented as

$$f = \sum_{k \in Z_n} a_k e^{ikx} \quad (\text{convergence in } D'(T_n)), \tag{1}$$

where $\{a_k\}_{k \in Z_n}$ is a sequence of complex numbers with at most polynomial growth. $\Phi(R_n)$ has the meaning of Definition 2.3.1/1. If $0 < p \leqq \infty$, then $L_p(T_n)$ has the usual meaning, in particular

$$\|f \mid L_p(T_n)\| = \left(\int_{T_n} |f(x)|^p \, dx \right)^{\frac{1}{p}} \quad \text{for} \quad 0 < p < \infty$$

and

$$\|f \mid L_\infty(T_n)\| = \operatorname{ess\,sup}_{x \in T_n} |f(x)| .$$

Furthermore, if $0 < p \leqq \infty$ and $0 < q \leqq \infty$, then $\|f_k \mid L_p(T_n, l_q)\|$ and $\|f_k \mid l_q(L_p(T_n))\|$ have the same meaning as in (1.2.2/5) and (1.2.2/6) with T_n instead of R_n.

Definition. Let $-\infty < s < \infty$ and $0 < q \leqq \infty$. Let $\varphi = \{\varphi_j(x)\}_{j=0}^{\infty} \in \Phi(R_n)$.

(i) If $0 < p \leqq \infty$, then

$$B^s_{p,q}(T_n) = \{f \mid f \in D'(T_n) \quad \text{is given by (1)},$$
$$\|f \mid B^s_{p,q}(T_n)\|^\varphi = \|2^{ks} \sum_{l \in Z_n} \varphi_k(l)\, a_l e^{ilx} \mid l_q(L_p(T_n))\| < \infty\} . \tag{2}$$

(ii) If $0 < p < \infty$, then

$$F^s_{p,q}(T_n) = \{f \mid f \in D'(T_n) \quad \text{is given by (1)},$$
$$\|f \mid F^s_{p,q}(T_n)\|^\varphi = \|2^{ks} \sum_{l \in Z_n} \varphi_k(l)\, a_l e^{ilx} \mid L_p(T_n, l_q)\|^\varphi < \infty\} . \tag{3}$$

Remark 1. This is the periodic counterpart of Definition 2.3.1/2. The sum $\sum_{l \in Z_n} \varphi_k(l)\, a_l e^{ilx}$ is finite for every k, i.e. it is a trigonometrical polynomial. At first glance, the definition looks somewhat more complicated than Definition 2.3.1/2. However, if we identify (1) with (9.1.2/4) (and we shall do this in the sequel), then we have

$$F^{-1}\varphi_k F\left(\sum_l a_l e^{ilx}\right) = \sum_l \varphi_k(l)\, a_l e^{ilx} . \tag{4}$$

This shows that the structure of the spaces in (2) and (3) is the same as in Definition 2.3.1/2.

Remark 2. Periodic Sobolev spaces (cf. 9.2.3.) and Besov spaces $B^s_{p,q}(T_n)$ with $\min(p, q) \geqq 1$ have been studied for a long time, cf. S. M. Nikol'skij [3], where also one can find many references. We refer also to P. I. Lizorkin [6].

Remark 3. Recently P. Oswald [4] has given a different definition of spaces of type $B^s_{p,q}(T_1)$ and $F^s_{p,q}(T_1)$. He extends the classical approach of the Hardy spaces via analytic functions in the unit disc in the complex plane. In particular, the classical Hardy spaces are special cases of his spaces of $F^s_{p,q}$ type. Cf. also Ė. A. Storoženko [2, 3].

9.1.4. Periodic Function Spaces on R_n

Formula (9.1.3/4) shows that it would be desirable to identify the spaces from Definition 9.1.3 with spaces of periodic distributions on R_n of type $B^s_{p,q}(R_n)$ and $F^s_{p,q}(R_n)$. However, one needs weight functions, for otherwise the corresponding quasi-norms in (9.1.3/2) and (9.1.3/3) with R_n instead of T_n would be rather meaningless. We use the spaces $B^s_{p,q}(R_n, \varrho(x))$ and $F^s_{p,q}(R_n, \varrho(x))$ from Definition 7.2.1 with $\varrho(x)=(1+|x|)^{-\varkappa}$, where $\varkappa$ is a positive number. In that case it is not necessary to deal with ultra-distributions, the usual distributions being sufficient. In other words, if $\varrho(x)=(1+|x|)^{-\varkappa}$, we can replace $S'_\omega(R_n)$ in (7.2.1/1) and (7.2.1/2) by $S'(R_n)$.

Definition. Let $-\infty<s<\infty$, $0<q\leqq\infty$ and $\varrho(x)=(1+|x|)^{-\varkappa}$ with $\varkappa>0$.
(i) Let $0<p\leqq\infty$. Then $B^s_{p,q,\pi}(R_n, \varrho(x))$ is the collection of all periodic distributions on R_n which belong to $B^s_{p,q}(R_n, \varrho(x))$.
(ii) Let $0<p<\infty$. Then $F^s_{p,q,\pi}(R_n, \varrho(x))$ is the collection of all periodic distributions on R_n which belong to $F^s_{p,q}(R_n, \varrho(x))$.

Remark. $B^s_{p,q,\pi}(R_n, \varrho(x))$ is considered as a subspace of $B^s_{p,q}(R_n, \varrho(x))$ with the same quasi-norm. Similarly $F^s_{p,q,\pi}(R_n, \varrho(x))$. They are closed subspaces. Furthermore, they are non-trivial if $\varkappa p>n$. Periodic distributions on R_n have been introduced in 9.1.2.

9.2. Properties

9.2.1. The Main Theorem

We have a one-to-one relation between the periodic distributions on T_n and those on R_n, cf. (9.1.2/4) and (9.1.3/1). In this sense we can interpret the spaces $B^s_{p,q,\pi}(R_n, \varrho(x))$ and $F^s_{p,q,\pi}(R_n, \varrho(x))$ from Definition 9.1.4 as subspaces of $D'(T_n)$ and in this sense the spaces from Definitions 9.1,3 and 9.1.4 are comparable. The statement of the following theorem must be understood in this way.

Theorem. Let $-\infty<s<\infty$ and $0<q\leqq\infty$.
(i) If $0<p\leqq\infty$, then $B^s_{p,q}(T_n)$ is a quasi-Banach space (Banach space if $1\leqq p\leqq\infty$ and $1\leqq q\leqq\infty$) and the quasi-norms $\|f \mid B^s_{p,q}(T_n)\|^\varphi$ with $\varphi\in\Phi(R_n)$ and $\|f \mid B^s_{p,q}(T_n)\|^\psi$ with $\psi\in\Phi(R_n)$ are equivalent to each other. If $\varrho(x)=(1+|x|)^{-\varkappa}$ with $\varkappa p>n$, then

$$B^s_{p,q}(T_n)=B^s_{p,q,\pi}(R_n, \varrho(x)) \tag{1}$$

(in the sense of the above interpretation).
(ii) If $0<p<\infty$, then $F^s_{p,q}(T_n)$ is a quasi-Banach space (Banach space if $1\leqq p<\infty$ and $1\leqq q\leqq\infty$) and the quasi-norms $\|f \mid F^s_{p,q}(T_n)\|^\varphi$ with $\varphi\in\Phi(R_n)$ and $\|f \mid F^s_{p,q}(T_n)\|^\psi$ with $\psi\in\Phi(R_n)$ are equivalent to each other. If $\varrho(x)=(1+|x|)^{-\varkappa}$ with $\varkappa p>n$, then

$$F^s_{p,q}(T_n)=F^s_{p,q,\pi}(R_n, \varrho(x)) \tag{2}$$

(in the sense of the above interpretation).

Remark. The proof is an easy consequence of (9.1.3/4) and Theorem 7.2.2. Upon proving (1) and (2), the rest is obvious.

9.2.2. Fourier Multipliers and Maximal Inequalities

Once Theorem 9.2.1 has been established, assertions for the weighted spaces of type $B^s_{p,q}$ and $F^s_{p,q}$ can be spezialized to the periodic situation. We formulate the counterpart of Theorem 7.1.1.

Theorem. Let $0<p<\infty$, $0<q\leqq\infty$ and $0<r<\min(p,q)$. Let $\lambda>\frac{n}{2}+\frac{n}{p}+\frac{n}{\min(p,q)}$. There exists a positive number c such that

$$\begin{aligned}&\Big\|\sum_{k\in Z_n} a^l_k\varphi_l(k)\,\mathrm{e}^{ikx}\mid L_p(T_n,l_q)\Big\|\\&\leqq\left\|\sup_{z\in T_n}\frac{\big|\sum_{k\in Z_n} a^l_k\varphi_l(k)\,\mathrm{e}^{ik(x-z)}\big|}{1+|2^kz|^{\frac{n}{r}}}\;\Big|\;L_p(T_n,l_q)\right\|\\&\leqq c\sup_l\|\varphi_l(2^l\cdot)\mid H^\lambda_2(R_n)\|\,\Big\|\sum_{k\in Z_n}a^l_k\mathrm{e}^{ikx}\mid L_p(T_n,l_q)\Big\|\end{aligned}\tag{1}$$

holds for all trigonometrical polynomials $\sum_{k\in Z_n} a^l_k\mathrm{e}^{ikx}$ with $a^l_k=0$ if $|k|>2^l$, and all systems $\{\varphi_l(x)\}^\infty_{l=0}\subset S(R_n)$.

Remark. We recall that $H^\lambda_2(R_n)$ denotes the usual Bessel-potential spaces ($H^\lambda_2(R_n)=W^\lambda_2(R_n)$ is a Sobolev space if λ is a natural number). The summation in (1), as far as the l_q-quasi-norm is concerned, is taken with respect to l. The above theorem follows essentially from Theorem 7.1.1 and Remark 7.1.1/2 (where d must be replaced by $|d|$ if $d<0$). The restrictions for λ are not natural. One would try to improve these restrictions to $\lambda>\sigma_{p,q}$, where $\sigma_{p,q}$ has the meaning of (2.4.8/1), cf. Theorem 2.4.9/2.

9.2.3. Periodic Sobolev Spaces

One can try to find explicit equivalent norms and quasi-norms for the spaces $B^s_{p,q}(T_n)$ and $F^s_{p,q}(T_n)$. For instance, the method from 2.5.7. can be extended immediately to the periodic case. As a result, one obtains equivalent norms for the spaces $B^s_{p,q}(T_n)$ with $s>0$, $1\leqq p\leqq\infty$ and $1\leqq q\leqq\infty$ which look like (2.2.2/6), (2.2.2/9.) and (2.2.2/10) with T_n instead of R_n. In this connection there arises a problem about the periodic Sobolev spaces $W^m_p(T_n)$ with $1<p<\infty$ and $m=0,1,2,\ldots$ which can be defined as follows. $L_p(T_n)$ with $1<p<\infty$ has the usual meaning and

$$W^m_p(T_n)=\{f\mid f\in D'(T_n),\ \|f\mid W^m_p(T_n)\|=\sum_{|\alpha|\leqq m}\|D^\alpha f\mid L_p(T_n)\|<\infty\}.\tag{1}$$

Let $L_p(T_n)=W^0_p(T_n)$. The counterpart of Theorem 2.5.6 is an easy consequence of (9.2.1/2) and Theorem 7.2.3. It reads as follows.

Theorem. If $1<p<\infty$ and $m=0,1,2,\ldots$, then $W^m_p(T_n)=F^m_{p,2}(T_n)$, and $\|f\mid W^m_p(T_n)\|$ from (1) is an equivalent norm in $F^m_{p,2}(T_n)$.

9.2.4. The Problem of Strong Summability

The problem of strong summability for one-dimensional trigonometric series can be described as follows. Let

$$f(x)=\sum_{k=-\infty}^{\infty}a_k\mathrm{e}^{ikx}\in D'(T_1)\quad\text{and}\quad s_N(x)=\sum_{k=-N}^{N}a_k\mathrm{e}^{ikx},\tag{1}$$

where $s_N(x)$ are the usual partial sums. We have

$$s_N(x) \to f(x) \quad \text{if} \quad N \to \infty \tag{2}$$

at least in $D'(T_1)$. If $f(x)$ has additional properties, e.g. $f(x) \in L_p(T_1)$ with $1 < p < \infty$, then (2) holds in those spaces. What about the degree of approximation of $|f(x) - s_N(x)|$? Of course, one can measure that degree of approximation by several means. One of them would be to question whether

$$\left(\sum_{N=1}^{\infty} N^{\gamma q} \, | f(x) - s_N(x)|^q \right)^{\frac{1}{q}} \quad \text{belongs to } L_p(R_n)\,, \tag{3}$$

where $0 < q \leqq \infty$, $0 < p \leqq \infty$ and γ are given real numbers. This is the problem of strong summability. At least the case $p = \infty$ has attracted some attention in recent times. There is a long series of papers by L. Leindler, cf. the two survey papers by L. Leindler [1, 2] and the references given there. If f belongs to certain function spaces, e.g. Hölder-Zygmund spaces, what can be said about (3)? Conversely, if (3) holds what can be said about the smoothness properties of f? Of course, (3) can be modified in several directions. Of interest seems to be the replacement of $s_N(x)$ by somewhat smoother approximations, e.g. by $\sum_{k=-\infty}^{\infty} a_k \varphi_N(k) \, e^{ikx}$ where $\{\varphi_N(x)\}_{N=1}^{\infty}$ are (smooth) functions on R_1. The partial sums $s_N(x)$ can be obtained in this way if $\varphi_N(x)$ is identifies with appropriate characteristic functions. Let us assume that we can replace $|f(x) - s_N(x)|$, with $2^k \leqq N < 2^{k+1}$, in the question (3) (e.g. via Fourier multiplier theorems) by $|f(x) - s_{2^k}(x)|$. Then (3) can essentially be reformulated as

$$\left(\sum_{k=0}^{\infty} 2^{k\gamma q} 2^k |f(x) - s_{2^k}(x)|^q \right)^{\frac{1}{q}} \in L_p(T_n) \ ? \tag{4}$$

But this is very close to the one-dimensional periodic counterpart of the approximation procedures of the spaces $F^s_{p,q}(R_n)$ from 2.5.3. with $s = \gamma + \frac{1}{q}$, cf. also 2.2.5. In other words, there is some hope that the problem of strong summability can be treated (at least partly) via approximation procedures for the periodic spaces $F^s_{p,q}(T_1)$. Afterwards one can use embedding theorems for $F^s_{p,q}(T_1)$ into e.g. Hölder spaces. This problem can be generalized from the one-dimensional case to the n-dimensional case. First results in this direction have been obtained recently by H.-J. Schmeisser, W. Sickel [1].

10. Further Types of Function Spaces

10.1. Anisotropic Functions Spaces

Let $1<p<\infty$ and $\bar{m}=(m_1, \ldots, m_n)$, where m_k are non-negative integers. Then

$$W_p^{\bar{m}}(R_n)=\left\{f \mid f\in L_p(R_n), \|f \mid W_p^{\bar{m}}(R_n)\|=\|f \mid L_p(R_n)\|+\sum_{k=1}^{n}\left\|\frac{\partial^{m_k}f}{\partial x_k^{m_k}} \mid L_p(R_n)\right\|<\infty\right\} \tag{1}$$

is an anisotropic Sobolev space on the Euclidean n-space R_n. If $m_1=\ldots=m_n=m$, then $W_p^{\bar{m}}(R_n)=W_p^m(R_n)$ is the usual (isotropic) Sobolev space from (2.2.2/7), cf. also Theorem 2.5.6. In contrast to the usual Sobolev spaces, the smoothness properties of an element of an anisotropic Sobolev space depend in general on the chosen direction in R_n. Similarly, one can introduce anisotropic Besov spaces $\Lambda_{p,q}^{\bar{s}}(R_n)$, where $\bar{s}=(s_1, \ldots, s_n)$, $1<p<\infty$ and $1\leqq q\leqq\infty$. Spaces of that type on R_n and on domains in R_n have been studied in great detail in S. M. Nikol'skij [3] and in O. V. Besov, V. P. Il'in, S. M. Nikol'skij [1]. Cf. also A. Kufner, O. John, S. Fučik [1, 8.5] and [I, 2.13]. A treatment of these spaces from the standpoint of interpolation theory can be found in H.-J. Schmeisser, H. Triebel [1] and H.-J. Schmeisser [1]. The problem arises of studying these spaces in the spirit of Chapter 2. In other words, we are looking for anisotropic spaces $B_{p,q}^{\bar{s}}(R_n)$ and $F_{p,q}^{\bar{s}}(R_n)$ with $\bar{s}=(s_1, \ldots, s_n)$, $0<q\leqq\infty$ and $0<p\leqq\infty$ ($p<\infty$ in the case of the spaces $F_{p,q}^{\bar{s}}(R_n)$) which coincide with $B_{p,q}^{s}(R_n)$ and $F_{p,q}^{s}(R_n)$, respectively, if $s_1=\ldots=s_n=s$. The first steps in this direction have been made in [F, 2.5.2.] and in B. Stöckert, H. Triebel [1, Section 4]. There is little doubt that a full counterpart of Chapter 2 exists, although such a theory is not yet worked out in detail. We restrict ourselves to the definition of these spaces and to a few comments.

The first task is the generalization of the set $\Phi(R_n)$ from Definition 2.3.1/1. Let

$$a=(a_1, \ldots, a_n)\in R_n \quad \text{with} \quad a_j>0 \quad \text{and} \quad \sum_{j=1}^{n} a_j=n. \tag{2}$$

We introduce the anisotropic distance of $x\in R_n$ from the origin,

$$|x|_a=\left(\sum_{k=1}^{n}|x_k|^{\frac{2}{a_k}}\right)^{\frac{1}{2}} \quad \text{where} \quad x=(x_1, \ldots, x_n)\in R_n .$$

The isotropic case is given by $a_1=\ldots=a_n=1$, where $|x|_a=|x|$. The counterpart of $\Phi(R_n)$ reads as follows: $\Phi^a(R_n)$ is the collection of all systems $\varphi=\{\varphi_j(x)\}_{j=0}^{\infty}\subset S(R_n)$ such that

$$\begin{cases}\operatorname{supp}\varphi_0\subset\{x \mid |x|_a\leqq 2\}, \\ \operatorname{supp}\varphi_j\subset\{x \mid 2^{j-1}<|x|_a<2^{j+1}\} \quad \text{if} \quad j=1, 2, 3, \ldots ;\end{cases} \tag{3}$$

for every multi-index α there exists a positive number c_α such that

$$2^{ja\alpha}|D^\alpha\varphi_j(x)|\leqq c_\alpha \quad \text{for any} \quad j=0, 1, 2, \ldots \quad \text{and any} \quad x\in R_n; \tag{4}$$

and

$$\sum_{j=0}^{\infty}\varphi_j(x)=1 \quad \text{for any} \quad x\in R_n . \tag{5}$$

Of course, $a\alpha=\sum_{k=1}^{n} a_k\alpha_k$ is the scalar product. In the isotropic case with $a_1=\ldots=a_n=1$, we have $a\alpha=|\alpha|$.

Definition. Let $-\infty<s<\infty$ and $0<q\leqq\infty$. Let a be given by (2) and $\bar{s}=\left(\frac{s}{a_1}, \ldots, \frac{s}{a_n}\right)$. Let $\varphi=\{\varphi_j(x)\}_{j=0}^{\infty}\in\Phi^a(R_n)$.

H. Triebel, *Theory of Function Spaces*, Modern Birkhäuser Classics,
DOI 10.1007/978-3-0346-0416-1_10,© Birkhäuser Verlag 1983

(i) If $0<p\leqq\infty$, then

$$B^{\bar{s}}_{p,q}(R_n)=\{f\mid f\in S'(R_n),\ \|f\mid B^{\bar{s}}_{p,q}(R_n)\|^{\varphi}=\|2^{sj}F^{-1}\varphi_jFf\mid l_q(L_p(R_n))\|<\infty\}\,. \tag{6}$$

(ii) If $0<p<\infty$, then

$$F^{\bar{s}}_{p,q}(R_n)=\{f\mid f\in S'(R_n),\ \|f\mid F^{\bar{s}}_{p,q}(R_n)\|^{\varphi}=\|2^{sj}F^{-1}\varphi_jFf\mid L_p(R_n,l_q)\|<\infty\}\,. \tag{7}$$

Remark 1. This is the anisotropic generalization of Definition 2.3.1/2. One can also introduce the spaces $F^{\bar{s}}_{\infty,q}(R_n)$ with $1<q\leqq\infty$ by generalizing Definition 2.3.4. Furthermore, there is no difficulty in introducing homogeneous anisotropic spaces $\dot{B}^{\bar{s}}_{p,q}(R_n)$ and $\dot{F}^{\bar{s}}_{p,q}(R_n)$ by generalizing Definition 5.1.3/2 and Definition 5.1.4.

Remark 2. There is a full counterpart of Theorem 2.3.3. Furthermore, we have

$$W^{\bar{m}}_p(R_n)=F^{\bar{m}}_{p,2}(R_n)\quad\text{if}\quad 1<p<\infty,\quad \bar{m}=(m_1,\ldots,m_n)\,,$$

where the m_k's are natural numbers. A. P. Calderón, A. Torchinsky [1, II] introduced anisotropic Hardy spaces, which generalize the spaces $H_p(R_n)$ from Definition 5.2.4 (which are essentially the Hardy spaces in the sense of C. Fefferman, E. M. Stein [2]), cf. also J. C. Peral, A. Torchinsky [1]. Let us denote these spaces by $H_{p,a}(R_n)$, where a has the above meaning. We proved in H. Triebel [13] that

$$H_{p,a}(R_n)=\dot{F}^{\bar{o}}_{p,2}(R_n)\quad\text{if}\quad 0<p<\infty$$

and $\bar{o}=(o,\ldots,o)=\left(\frac{o}{a_1},\ldots,\frac{o}{a_n}\right)$.

Remark 3. If $\bar{s}=(s_1,\ldots,s_n)$ with $s_j>0$ is given, then the above numbers s and $a=(a_1,\ldots,a_n)$ can be calculated as follows:

$$\frac{1}{s}=\frac{1}{n}\sum_{k=1}^{n}\frac{1}{s_k}\quad\text{and}\quad a_k=\frac{s}{s_k}$$

(similarly if $s_j<0$). In other words, s is something like the mean smoothness and a measures the anisotropy.

10.2. Generalizations

We dealt with several types of spaces $B^s_{p,q}$ and $F^s_{p,q}$. As far as unweighted spaces on R_n are concerned, we mention the spaces $B^s_{p,q}(R_n)$ and $F^s_{p,q}(R_n)$ from Chapter 2, their homogeneous counterparts $\dot{B}^s_{p,q}(R_n)$ and $\dot{F}^s_{p,q}(R_n)$ from Chapter 5, and the anisotropic generalizations $B^{\bar{s}}_{p,q}(R_n)$ and $F^{\bar{s}}_{p,q}(R_n)$ from Section 10.1. (including their homogeneous counterparts). One can ask whether a unified treatment exists which covers all these cases. The considerations in Chapters 2 and 5 and in Section 10.1. where based on Chapter 1, i.e. on the study of l_q-valued L_p-spaces of entire analytic functions on R_n. One can ask whether these basic spaces $L^{\Omega}_p(R_n,l_q)$ from Section 1.6. can be replaced by more general spaces. Let $\bar{p}=(p_1,\ldots,p_n)$ with $0<p_k\leqq\infty$ if $k=1,\ldots,n$. One can try to replace $L_p(R_n)$ by

$$L_{\bar{p}}(R_n)=L_{p_n}(R_1,L_{p_{n-1}}(R_1,\ldots,L_{p_1}(R_1)\ldots) \tag{1}$$

and similarly l_q by $l_{\bar{q}}=l_{q_n}(l_{q_{n-1}}(\ldots l_{q_1}\ldots)$ with $\bar{q}=(q_1,\ldots,q_n)$. We add a few remarks about these possibilities.

General Coverings. The first question is related to different types of coverings of R_n and $R_n-\{0\}$: coverings by "dyadic" balls (or differences of such balls) in the sense of (2.3.1/1) and (5.1.3/1) and their anisotropic generalizations in (10.1/3). In [F, 2.2], we studied such dyadic coverings in a rather general setting and introduced related spaces of type $B^s_{p,q}$ and $F^s_{p,q}$. The spaces from Chapters 2 and 5 and from Section 10.1. are special cases of such a general approach. We do not go into detail, but we wish to mention that other types of spaces also can be obtained in that way as special cases, e.g. spaces with dominating mixed smoothness properties. The prototype of such a space is given by

$$S_p^m W(R_n) = \{f \mid f \in S'(R_n),\ \|f \mid S_p^m W(R_n)\| = \sum_{k \in \{m\}} \|D^k f \mid L_p(R_n)\| < \infty\} \tag{2}$$

with $1 < p < \infty$. Furthermore, $m = (m_1, \ldots, m_n)$ is a given multi-index with the natural numbers $m_1, \ldots, m_n$ as its components, and $\{m\}$ is the collection of all multi-indices $(\Theta_1 m_1, \ldots, \Theta_n m_n)$, where the numbers Θ_j are (independently) either 0 or 1. Spaces of this type have been introduced by S. M. Nikol'skij [2], cf. also P. Grisvard [1], G. Sparr [1], and T. I. Amanov [1]. $S_p^m W(R_n)$ is a space of Sobolev type. Similarly, one can introduce corresponding spaces of type $B_{p,q}^s$ and $F_{p,q}^s$ with dominating mixed smoothness properties, cf. H. Triebel [3, III–V]. All these spaces and also those from Chapters 2 and 5 and from Section 10.1. are covered by the approach given in [F, Chapter 2]. That approach is based on dyadic coverings of R_n or subsets of R_n. However, if one looks at the theory in Section 1.6. (which was the basis for many considerations, at least as far as basic questions are concerned), then it is quite clear that one can start with rather arbitrary coverings of R_n or of subsets of R_n. For exemple, it would be of interest to study spaces of type $B_{p,q}^s$ and $F_{p,q}^s$ on R_n which are based on coverings of R_n by congruent cubes, cf. H. Triebel [25] and the references given there to the work of H. G. Feichtinger, in particular H. G. Feichtinger [1]. In connection with more general coverings we refer also to G. A. Kaljabin [1, 3] and M. L. Gol'dman [1, 2, 4, 5].

Mixed Quasi-Norms. We discuss briefly the second question mentioned above. We wish to replace $l_q(L_p(R_n))$ and $L_p(R_n, l_q)$ in the theory in Section 1.6. and afterwards in Definition 2.3.1/2 by e.g. $l_{\bar{q}}(L_{\bar{p}}(R_n))$ and $L_{\bar{p}}(R_n, l_{\bar{q}})$, where $l_{\bar{q}}$ and $L_{\bar{p}}(R_n)$ have the above meaning. Of course there are a lot of other possibilities, e.g.

$$L_{p_n}(R_1, l_{q_n}(L_{p_{n-1}}(R_1, l_{q_{n-1}}, \ldots) \ldots)) . \tag{3}$$

The corresponding spaces are related to the above-mentioned spaces with dominating mixed smoothness properties and corresponding spaces of type

$$B_{p_2,q_2}^{s_2}(R_{n_2}, B_{p_1,q_1}^{s_1}(R_{n_1})) \quad \text{or} \quad F_{p_2,q_2}^{s_2}(R_{n_2}, F_{p_1,q_1}^{s_1}(R_{n_1})) . \tag{4}$$

These spaces (on domains) are of interest for a detailed study of mapping properties of integral operators, cf. A. Pietsch [2, 3] and H. Triebel [20] (cf. also A. Pietsch [1, Ch. 22] for the general background). This connection with integral operators was the motivation for a rather extensive study of spaces of type $B_{p,q}^s$ and $F_{p,q}^s$ on the basis of quasi-norms of type (3) (and modifications) by H.-J. Schmeisser [2–5]. We refer also to B. Stöckert [1, 2], where corresponding basic spaces in the sense of Chapters 1 and 6 have been treated.

10.3. Abstract Spaces and Spaces Related to Orthogonal Expansions

In Chapter 9 we treated the spaces $B_{p,q}^s(T_n)$ and $F_{p,q}^s(T_n)$ on the n-torus T_n. These spaces are based on the complete (with respect to $L_2(T_n)$) orthonormal system

$$\{(2\pi)^{-\frac{n}{2}} e^{ikx}\}_{k \in Z_n} . \tag{1}$$

There arise two questions. First one can ask whether the orthonormal system (1) can be replaced by other classical orthonormal systems, such as the system of Hermite functions, Laguerre functions, Jacobi functions, respectively, or the system of spherical harmonics. The second question is to find an abstract approach which covers not only the above classical orthogonal expansions but also the spaces from Chapter 2, Section 10.1. etc. There is no complex answer, at least as far as the full abstract counterpart of the spaces $B_{p,q}^s(R_n)$ and $F_{p,q}^s(R_n)$ for all admissible values of the parameters s, p and q is concerned. However, for spaces of type $B_{p,q}$ with $s > 0$, $1 \leq p \leq \infty$ and $1 \leq q \leq \infty$, there is an abstract counterpart, which has been developed independently by J. Peetre [8, Chapter 10] and in [F, Chapter 3]; cf. also H. Triebel [22]. We describe in very rough terms the basic idea and than add a few comments concerning concrete applications.

Abstract Spaces. Let H be a separable complex Hilbert space and let A be a self-adjoint positive-definite operator in H. Let

$$\Lambda a = \int_0^\infty t \,\mathrm{d}E(t)\, a \tag{2}$$

be the spectral decomposition of Λ. Let $R_{l,\varrho}$ with

$$R_{l,\varrho} a = \int_0^\varrho \left(1 - \frac{t}{\varrho}\right)^l \mathrm{d}E(t)\, a\,, \quad a \in H\,, \tag{3}$$

be the Riesz means. Let A be a Banach space. We assume that both H and A are continuously embedded in a common Hausdorff space. Furthermore, let $A \cap H$ be dense in H and dense in A. The main hypothesis of the theory can be formulated as follows: there exist a natural number l and a positive number c such that

$$\sup_{\varrho > 0} \|R_{l,\varrho} a \mid A\| \leq c \|a \mid A\| \quad \text{for all} \quad a \in H \cap A\,. \tag{4}$$

After extension by continuity, $\{R_{l,\varrho}\}_{\varrho > 0}$ is a family of uniformly bounded continuous linear operators on A. We assume that $R_{l,\varrho} a \to a$ if $\varrho \to \infty$ (convergence in A) for every $a \in A$. Let Φ be the collection of all systems $\varphi = \{\varphi_j(x)\}_{j=0}^\infty$ of infinitely differentiable functions on R_1 such that

$$\operatorname{supp} \varphi_0 \subset (-2, 2),\ \operatorname{supp} \varphi_j \subset (2^{j-1}, 2^{j+1}) \quad \text{with} \quad j = 1, 2, 3, \ldots;$$

for every non-negative integer k there exists a positive constant c_k such that

$$2^{jk} |\varphi_j^{(k)}(x)| \leq c_k \quad \text{for all} \quad j = 0, 1, 2, \ldots \text{ and all } x \in R_1;$$

and

$$\sum_{j=0}^\infty \varphi_j(x) = 1 \qquad \text{for all} \quad x \geq 0$$

$\left(\varphi_j^{(k)} \text{ means } \frac{\mathrm{d}^k \varphi_j}{\mathrm{d}x^k}\right)$. Φ is the counterpart of $\Phi(R_n)$ from Definition 2.3.1/1. In the usual way, we construct the operators $\varphi_j(\Lambda)$,

$$\varphi_j(\Lambda)\, a = \int_0^\infty \varphi_j(t)\, \mathrm{d}E(t)\, a\,.$$

Let $s > 0$, $1 \leq q \leq \infty$ and $\varphi \in \Phi$. Then the spaces $B_q^s(A)$ are defined by

$$B_q^s(A) = \left\{ a \mid a \in A,\ \|a \mid B_q^s(A)\|^\varphi = \left(\sum_{j=0}^\infty 2^{jsq} \|\varphi_j(\Lambda)\, a \mid A\|^q \right)^{\frac{1}{q}} < \infty \right\}. \tag{5}$$

These are the abstract Besov spaces which have been studied in the literature cited above. In other words, we define spaces via spectral decompositions. A similar procedure has been described in 2.2.4. as a motivation for the introduction of the spaces $B_{p,q}^s(R_n)$ and $F_{p,q}^s(R_n)$. What about concrete cases if we specialize H, Λ and A? Let

$$H = L_2(R_n)\,, \quad \Lambda = -\Delta + E\,, \quad A = L_p(E_n)\,, \tag{6}$$

with $1 \leq p < \infty$, where E is the unit operator. Then one can prove that the above hypotheses, in particular (4), are satisfied. In that case, $B_q^s(A)$ coincides with the space $B_{p,q}^{2s}(R_n)$ from Definition 2.3.1/2. In other words, the spaces $B_{p,q}^s(R_n)$ and $F_{p,q}^s(R_n)$ from Definition 2.3.1/2 can be obtained via spectral decompositions of the Laplacian. For details and further examples we refer to [F, 3.5].

Orthogonal Expansions. If we replace R_n in (6) by T_n, then we obtain the spaces $B_{p,q}^{2s}(T_n)$ from Chapter 9 (under the above restrictions for the parameters s, p and q). In particular, (4) is satisfied in that case, too. One can ask whether other classical orthogonal expansions fit in that framework. For instance, if

$$H = L_2(R_1)\,, \quad (\Lambda f)(x) = -f''(x) + x^2 f(x)\,, \quad A = L_p(R_1) \tag{7}$$

with $1 \leq p < \infty$, then A is a positive-definite self-adjoint operator with pure point spectrum. The corresponding complete orthonormal system $\{H_k(x)\}_{k=0}^{\infty}$ of eigenfunctions is given by the Hermite functions

$$H_k(x) = c_k e^{\frac{x^2}{2}} \frac{d^k}{dx^k} (e^{-x^2}) .$$

The hypothesis (4) is satisfied and we obtain Besov spaces which look similar to those in (9.1.3/2) if the system (1) is replaced by $\{H_k(x)\}_{k=0}^{\infty}$. Similar one can deal with Laguerre functions, Jacobi functions and spherical harmonics, cf. [F, 3.5.5.] for details and references. Recently, T. Runst [1] has made a thorough investigation of such spaces. We refer also to T. Runst, W. Sickel [1], where the problem of strong summability in the sense of 9.2.4. for Jacobi expansions has been studied.

References

Adams, R. A.
1. Sobolev Spaces. Academic Press; New York, San Francisco, London, 1975.

Agalarov, S. I.
1. A Littlewood – Paley theorem and Fourier multipliers in weighted spaces with mixed norms. (Russian) Trudy Mat. Inst. Steklov 140 (1976), 3–13.

Agmon, S.
1. On the eigenfunctions and on the eigenvalues of general elliptic boundary value problems. Comm. Pure Appl. Math. 15 (1962), 119–147.

Agmon, S.; Douglis, A.; Nirenberg, L.
1. Estimates near the boundary for solutions of elliptic partial differential equations satisfying general boundary conditions. I. Comm. Pure Appl. Math. 12 (1959), 623–727.

Amanov, T. I.
1. Spaces of Differentiable Functions with Dominating Mixed Derivatives. (Russian) Nauka Kaz. SSR, Alma-Ata 1976.

Andersen, K. F.; John, R. T.
1. Weighted inequalities for vector-valued maximal functions and singular integrals. Studia Math. 69 (1980), 19–31.

Arkeryd, L.
1. On the L^p estimates for elliptic boundary problems. Math. Scand. 19 (1966), 59–76.

Aronszajn, N.
1. Boundary values of functions with finite Dirichlet integral. Techn. Report 14, Univ. of Kansas 1955, 77–94.

Aronszajn, N.; Smith, K. T.
1. Theory of Bessel potentials. I. Ann. Inst. Fourier (Grenoble) 11 (1961), 385–476.

Bagby, R. J.
1. Parabolic potentials with support on a half space. Illinois J. Math. 18, 2 (1974), 213–222.

Bennett, C.; Gilbert, J. E.
1. Homogeneous algebras on the circle: II. Multipliers, Ditkin conditions. Ann. Inst. Fourier (Grenoble), 22, 3 (1972), 21–50.

Berezanskij, Ju. M.
1. Decomposition in Eigenfunctions of Self-Adjoint Operators. (Russian) Naukova dumka, Kiev, 1965.

Bergh, J.; Löfström, J.
1. Interpolation Spaces. An Introduction. Berlin, Heidelberg, New York: Springer-Verlag 1976.

Besov, O. V.
1. On a family of function spaces. Embedding theorems and applications. (Russian) Dokl. Akad. Nauk SSSR 126 (1959), 1163–1165.

H. Triebel, *Theory of Function Spaces*, Modern Birkhäuser Classics,
DOI 10.1007/978-3-0346-0416-1,

2. On a family of function spaces in connection with embeddings and extensions. (Russian) Trudy Mat. Inst. Steklov 60 (1961) 42–81.
3. On the growth of the mixed derivative of functions of $C^{(l_1,l_2)}$. (Russian) Mat. Zametki 15 (1974), 355–362.

Besov, O. V.; Il'in, V. P.; Nikol'skij, S. M.
1. Integral Representations of Functions and Embedding Theorems. (Russian) Moskva: Nauka 1975 [English translation: Scripta Series in Math., Washington, Halsted Press, New York, Toronto, London: V. H. Winston & Sons 1978/79].

Beurling, A.
1. Quasi-analyticity and general distributions. Lectures 4 and 5, A. M. S. Summer Institute, Stanford 1961.

Björck, G.
1. Linear partial differential operators and generalized distributions. Ark. Mat. 6 (1966), 351–407.

Boas, R. P.
1. Entire Functions. New York: Academic Press, Inc. Publishers 1954.

Boman, J.
1. Supremum norm estimates for partial derivatives of functions of several real variables. Illinois J. Math. 16 (1972), 203–216.

Bui Huy Qui
1. Some aspects of weighted and non-weighted Hardy spaces. Kôkyûroku Res. Inst. Math. Sci. 383 (1980), 38–56.
2. Results on Besov-Hardy-Triebel spaces. Techn. Report.
3. Weighted Hardy spaces. Math. Nachr. 103 (1981), 45–62.
4. Weighted Besov and Triebel spaces: Interpolation by the real method. Hiroshima Math. J. (to appear)

Burenkov, V. I.; Gol'dman, M. L.
1. On the extension of functions of L_p. (Russian) Trudy Mat. Inst. Steklov 150 (1979), 31–51.

Calderón, A. P.
1. Lebesgue spaces of functions and distributions. "Part. Diff. Equations", Proc. Symp. Math. 4, Amer. Math. Soc., 1961, 33–49.
2. Intermediate spaces and interpolation, the complex method. Studia Math. 24 (1964), 113–190.

Calderón, A. P.; Torchinsky, A.
1. Parabolic maximal functions associated with a distribution. I, II. Advances in Math. 16 (1975), 1–64 and 24 (1977), 101–171.

Calderón, C. P.; Fabes, E. B.; Rivière, N. M.
1. Maximal smoothing operators. Indiana Univ. Math. J. 23 (1974), 889–898.

Carleson, L.
1. An explicit unconditional basis in H^1. Bull. Sci. Math. (2) 104 (1980), 405–416.

Chodak, L. B.
1. On the approximation of functions by algebraic polynomials in the metric L_p, where $0<p<1$. (Russian) Mat. Zametki 30 (1981), 321–332.

Ciesielski, Z.
1. Constructive function theory and spline systems. Studia Math. 53 (1975), 277–302.
2. Equivalence, unconditionality and convergence a. e. of spline bases in L_p spaces. Banach Center Publications 4, "Approximation Theory", Warszawa 1979, 55–68.
3. Equivalence and shift property of spline bases in L_p spaces. Proc. Intern. Conference in Blagoevgrad 1977, "Constructive Function Theory", Sofia 1980, 281–291.

Ciesielski, Z.; Domsta, J.
1. Construction of an orthonormal basis in $C^m(I^d)$ and $W_p^m(I^d)$. Studia Math. 41 (1972), 211–224.

Ciesielski, Z.; Figiel, T.
1. Construction of Schauder bases in function spaces on smooth compact manifolds. Proc. Intern. Conf. "Approximation and Function Spaces", Gdansk 1979. Amsterdam: North Holland 1981, 217–232.
2. Spline approximation and Besov spaces on compact manifolds. Preprint, Warszawa 1981.

Coifman, R. R.; Fefferman, C.
1. Weighted norm inequalities for maximal functions and singular integrals. Studia Math. 51 (1974), 241–250.
Domar, Y.
1. Harmonic analysis based on certain commutative Banach algebras. Acta Math. 96 (1956), 1–66.
Dunford, N.; Schwartz, J. T.
1. Linear Operators, I, II. New York, London: Interscience Publishers 1958, 1963.
Duren, P. L.; Romberg, B. W.; Shields, A. L.
1. Linear functionals on H^p spaces with $0<p<1$. J. Reine Angew. Math. 238 (1969), 32–60.
Edmunds, D. E.
1. Embeddings of Sobolev spaces. Proc. Spring School "Nonlinear Analysis, Function Spaces, and Applications", Teubner-Texte Math. 19, 38–58. Leipzig: Teubner 1979.
Edwards, R. E.
1. Functional Analysis. New York, Chicago, San Francisco, Toronto, London: Holt, Rinehardt and Winston 1965.
Edwards, R. E.; Gaudry, G. I.
1. Littlewood-Paley and Multiplier Theory. Berlin, Heidelberg, New York: Springer-Verlag 1977.
Fefferman, C.
1. Characterizations of bounded mean oscillation. Bull. Amer. Math. Soc. 77 (1971), 587–588.
2. The multiplier problem for the ball. Ann. of Math. 94 (1971), 330–336.
Fefferman, C.; Rivière, N. M.; Sagher, Y.
1. Interpolation between H^p spaces. The real method. Trans. Amer. Math. Soc. 191 (1974), 75–81.
Fefferman, C.; Stein, E. M.
1. Some maximal inequalities. Amer. J. Math. 93 (1971), 107–115.
2. H^p spaces of several variables. Acta Math. 129 (1972), 137–193.
Feichtinger, H. G.
1. Modulation spaces on locally compact abelian groups. (to appear)
Flett, T. M.
1. Lipschitz spaces of functions on the circle and the disc. J. Math. Anal. Appl. 39 (1972), 125–158.
Gagliardo, E.
1. Proprietà di alcune classi di funzioni in più variabili. Ricerche Mat. 7 (1958), 102–137.
Garcia-Cuerva, J.
1. Weighted H^p spaces. Dissertationes Math. 162 (1979).
Gelfand, I. M.; Schilow, G. E.
1. Verallgemeinerte Funktionen (Distributionen). II. Berlin: VEB Deutscher Verlag der Wissenschaften 1962.
Goldberg, D.
1. A local version of real Hardy spaces. Duke Math. J. 46 (1979), 27–42.
2. Local Hardy spaces. Proc. Symp. Pure Math. 35, I, 1979, "Harmonic Analysis in Euclidean Spaces", 245–248.
Gol'dman, M. L.
1. On the traces of functions of a generalized Liouville class. (Russian) Sibirsk. Mat. Ž. 17 (1976), 1236–1255.
2. Description of the traces of an anisotropic generalized Liouville class. (Russian) Dokl. Akad. Nauk SSSR 233 (1977), 273–276.
3. On the extension of functions of $L_p(R_m)$ in spaces with a greater number of dimensions. (Russian) Mat. Zametki 25 (1979), 513–520.
4. Description of the traces of some functional spaces. (Russian) Trudy Mat. Inst. Steklov 150 (1979), 99–127.
5. The decomposition method for the description of general spaces of Besov type. (Russian) Trudy Mat. Inst. Steklov 156 (1980), 47–81.
Golovkin, K. K.
1. On equivalent norms in fractional spaces. (Russian) Trudy Mat. Inst. Steklov 66 (1962), 364–383.

2. On the impossibility of some inequalities between functional norms. (Russian) Trudy Mat. Inst. Steklov 70 (1964), 5–25.
3. Parameter-normed spaces and normed classes. (Russian) Trudy Mat. Inst. Steklov 106 (1969), 3–134.

Grevholm, B.
1. On the structure of the spaces $\mathfrak{L}_k^{p,\lambda}$, Math. Scand. 26 (1970), 241–254.

Grisvard, P.
1. Commutativité de deux foncteurs d'interpolation et applications. J. Math. Pures Appl. 45 (1966), 143–290.

Heinig, H. P.; Johnson, R.
1. Weighted norm inequalities for L^r-valued integral operators and applications. (to appear)

Herz, C. S.
1. Lipschitz spaces and Bernstein's theorem on absolutely convergent Fourier transforms. J. Math. Mechanics 18 (1968), 283–324.

Hörmander, L.
1. Estimates for translation invariant operators in L_p spaces. Acta Math. 104 (1960), 93–140.
2. Linear Partial Differential Operators. Berlin, Göttingen, Heidelberg: Springer-Verlag 1963.

Irodova, I. P.
1. Properties of the scale of spaces $B_p^{\lambda\Theta}$ if $0<p<1$. (Russian) Dokl. Akad. Nauk SSSR 250 (1980), 273–275.

Ivanov, V. I.
1. Direct and inverse theorems of approximation theory in the norm L_p if $0<p<1$. (Russian) Mat. Zametki 18 (1975), 641–658.

Jawerth, B.
1. Some observations on Besov and Lizorkin-Triebel spaces. Math. Scand. 40 (1977), 94–104.
2. The trace of Sobolev and Besov spaces if $0<p<1$. Studia Math. 62 (1978), 65–71.
3. Weighted norm inequalities for functions of exponential type. Ark. Mat. 15 (1977), 223–228.

John, F.; Nirenberg, L.
1. On functions of bounded mean oscillation. Comm. Pure Appl. Math. 14 (1961), 415–426.

Johnson, R.
1. Lipschitz spaces, Littlewood-Paley spaces and convoluteurs. Proc. London Math. Soc. (3), 29 (1974), 127–141.
2. Convoluteurs of H^p spaces. Bull. Amer. Math. Soc. 81, 4 (1975), 711–714.
3. Convoluteurs on H^p spaces. Technical Report, College Park 1973.
4. Maximal subspaces of Besov spaces invariant under multiplication by characters. Trans. Amer. Math. Soc. 249 (1979), 387–407.
5. Multipliers of H^p spaces. Ark. Mat. 16 (1978), 235–249.
6. Duality methods for the study of maximal subspaces of Besov spaces invariant under multiplication by characters. Technical Report, Univ. of Maryland, 1979.

Kaljabin, G. A.
1. Embedding theorems for generalized Besov and Liouville spaces. (Russian) Dokl. Akad. Nauk SSSR 232 (1977), 1245–1248.
2. Characterization of functions of spaces of Besov-Lizorkin-Triebel type. (Russian) Dokl. Akad. Nauk SSSR 236 (1977), 1056–1059.
3. Description of the traces of anisotropic spaces of Triebel-Lizorkin type. (Russian) Trudy Mat. Inst. Steklov 150 (1979), 160–173.
4. Multiplier conditions of function spaces of Besov and Lizorkin-Triebel type. (Russian) Dokl. Akad. Nauk SSSR 251 (1980), 25–26.
5. The description of functions of classes of Besov-Lizorkin-Triebel type. (Russian) Trudy Mat. Inst. Steklov 156 (1980), 82–109.
6. Criteria of the multiplication property and the embedding in C of spaces of Besov-Lizorkin-Triebel type. (Russian) Mat. Zametki 30 (1981), 517–526.

Kokilašvili, V. M.
1. On Fourier multipliers. (Russian) Dokl. Akad. Nauk SSSR 220 (1975), 19–22.
2. On singular integrals and Fourier multipliers. (Russian) Akad. Nauk Gruzin. SSR, Trudy Tbiliss. Mat. Inst. 53 (1976), 38–61.

3. Maximal inequalities and multipliers in weighted Lizorkin-Triebel spaces. (Russian) Dokl. Akad. Nauk SSSR 239 (1978), 42–45.
4. Anisotropic Bessel-potentials. Embedding theorems for limiting indices. (Russian) Akad. Nauk Gruzin. SSR, Trudy Tbiliss. Mat. Inst. 58 (1978), 134–149.
5. On multiple Hilbert transforms and multipliers in weighted spaces. (Russian) Soobšč. Akad. Nauk Gruzin. SSR 98 (1980), 285–288.
6. On traces for weighted anisotropic Lizorkin-Triebel spaces. Proc. Intern. Conf. "Approximation and Function Spaces", Gdańsk, 1979. Amsterdam: North-Holland, 1981, 338–344.

Komatsu, H.

1. Ultradistributions: I: Structure theorems and a characterization. II: The kernel theorem and ultradistributions with support on a manifold. J. Fac. Sci. Univ. Tokyo Sect. IA Math. 20 (1973), 25–105; 24 (1977), 607–628.

Krotov, V. G.

1. On the differentiability of functions of L^p, $0<p<1$. (Russian) Mat. Sb. 117 (1982), 45–113.

Kufner, A.

1. Weighted Sobolev Spaces. Teubner-Texte Math. 31, Leipzig: Teubner 1980.

Kufner, A.; John, O.; Fučik, S.

1. Function Spaces. Prague: Academia 1977.

Kurtz, D.

1. Littlewood-Paley and multiplier theorems on weighted L^p spaces. Trans. Amer. Math. Soc. 259 (1980), 235–254.

Kurtz, D.; Wheeden, R. L.

1. Results on weighted norm inequalities for multipliers. Trans. Amer. Math. Soc. 255 (1979), 343–362.
2. A note on singular integrals with weights. Proc. Amer. Math. Soc. 81 (1981), 3, 391–397.

De Leeuw, K.; Mirkil, H.

1. A priori estimates for differential operators in L_∞-norm. Illinois J. Math. 8 (1964), 112–124.

Leindler, L.

1. On the strong summability and approximation of Fourier series. Banach Center Publications 4, "Approximation Theory", Warsaw 1979, 143–157.
2. Strong approximation of Fourier series and structural properties of functions. Acta Math. Acad. Sci. Hungar. 33 (1979), 105–125.

Lions, J. L.; Magenes, E.

1. Problèmes aux limites non homogènes. IV. Ann. Scuola Norm Sup. Pisa 15 (1961), 311–326.
2. Non-Homogeneous Boundary Value Problems and Applications. I, II, III. Berlin, Heidelberg, New York: Springer-Verlag 1972, 1973.

Littman, W.; McCarthy, C.; Rivière, N. M.

1. L^p-multiplier theorems. Studia Math. 30 (1968), 193–217.

Lizorkin, P. I.

1. On Fourier multipliers in the spaces $L_{p,\Theta}$. (Russian) Trudy Mat. Inst. Steklov 89 (1967), 231–248.
2. Operators connected with fractional derivatives and classes of differentiable functions. (Russian) Trudy Mat. Inst. Steklov 117 (1972), 212–243.
3. Generalized Hölder classes of functions in connection with fractional derivatives. (Russian) Trudy Mat. Inst. Steklov 128 (1972), 172–177.
4. Properties of functions of the spaces $\Lambda^r_{p\Theta}$. (Russian) Trudy Mat. Inst. Steklov 131 (1974), 158–181.
5. Isomorphism of Besov spaces $B^{(r)}_{p,\Theta}(R^n)$ with the space $l_\Theta(l_p)$. On zero-spaces $B^0_{p,\Theta}(R^n)$. Anal. Math. 2 (1976), 203–210.
6. On bases and multipliers for the spaces $B^r_{p,\Theta}(\Pi)$. (Russian) Trudy Mat. Inst. Steklov 143 (1977), 88–104.

Madych, W. R.

1. Anisotropic H^p, real interpolation, and fractional Riesz potentials. Trans. Amer. Math. Soc. 232 (1977), 255–263.

Madych, W. R.; Rivière, N. M.

1. Multipliers of the Hölder classes. J. Functional Analysis 21 (1976), 369–379.

Maurey, B.
1. Isomorphismes entre espaces H_1. Acta Math. 145 (1980), 79–120.
Mazja, W.
1. Einbettungssätze für Sobolevsche Räume, I, II. Teubner-Texte Math. 21, 28. Leipzig: Teubner 1979, 1980.
Maz'ja, V. G.; Šapošnikova, T. O.
1. On multipliers in function spaces with fractional derivatives. (Russian) Dokl. Akad. Nauk SSSR 244 (1979), 1065–1067.
2. On multipliers in Sobolev spaces. (Russian) Vestnik Leningrad. Univ. Mat. Meh. Astronom. 7(2) (1979), 33–40.
3. On traces and extensions of multipliers in the spaces W_p^l. (Russian) Uspechi Mat. Nauk 34 (1979), 205–206.
4. Multipliers in pairs of spaces of differentiable functions. (Russian) Trudy Moskov. Mat. Obšč. 43 (1981), 37–80.
5. Multipliers in pairs of potential spaces. (Russian) Math. Nachr. 99 (1980), 363–379.
6. Multipliers in S. L. Sobolev spaces in a domain. (Russian) Math. Nachr. 99 (1980), 165–183.
7. On sufficient conditions for belonging to classes of multipliers. (Russian) Math. Nachr. 100 (1981), 151–162.
Meyer, Y.
1. Régularité des solutions des équations aux dérivées partielles non linéaires [d' après J.-M. Bony]. Sém. Bourbaki, 32e année, 1979/80, n° 560.
Michlin, S. G.
1. On multipliers for Fourier integrals. (Russian) Dokl. Akad. Nauk SSSR 109 (1956), 701–703.
2. Fourier integrals and singular integrals of several variables. (Russian) Vestnik Leningrad. Univ. Mat. Meh. Astronom. 7 (1957), 143–155.
Miranda, C.
1. Partial Differential Equations of Elliptic Type. 2. ed. Berlin, Heidelberg, New York: Springer-Verlag 1970.
Mizuhara, T.
1. On Fourier multipliers of Lipschitz classes. Tôhoku Math. J. 24 (1972), 263–268.
2. On $A(\alpha;\, p, q)$ multiplier theorems. Bull. Yamagata Univ. Natur. Sci. 8 (1975), 491–498.
3. On Besov multipliers. Bull. Yamagata Univ. Natur. Sci. 9 (1976), 47–52.
4. On a theorem of Fourier multipliers and its applications. Bull. Yamagata Univ. Natur. Sci. 9 (1977), 217–220.
5. On some function spaces and Lipschitz spaces. J. London Math. Soc. 25 (1982), 75–87.
Muckenhoupt, B.
1. Weighted norm inequalities for the Hardy maximal function. Trans. Amer. Math. Soc. 165 (1972), 207–226.
Muckenhoupt, B.; Wheeden, R. L.
1. On the dual of weighted H^1 on the half space. Studia Math. 63 (1978), 57–79.
Muckenhoupt, B.; Wheeden, R. L.; Young, Wo-Sang
1. L^2 multipliers with power weights.
Murata, M.
1. Asymptotic behaviors at infinity of solutions of certain linear partial differential equations. J. Fac. Sci. Univ. Tokyo Sect. IA Math. 23 (1976), 107–148.
Neri, U.
1. Fractional integration on the spaces H^1 and its dual. Studia Math. 53 (1975), 175–189.
Nessel, R. J.; Wilmes, G.
1. Nikolskii-type inequalities for trigonometric polynomials and entire functions of exponential type. J. Austral. Math. Soc. 25 (1978), 7–18.
2. Über Ungleichungen vom Bernstein-Nikolskii-Riesz-Typ in Banach Räumen. Forschungsberichte des Landes Nordrhein-Westfalen 2841. Opladen: Westd. Verlag, 1979.
Neugebauer, C. J.
1. Smoothness of Bessel potentials and Lipschitz functions. Indiana Univ. Math. J. 26 (1977), 585–591.
Nikol'skij, S. M.
1. Inequalities for entire functions of finite order and their application in the theory of differ-

entiable functions of several variables. (Russian) Trudy Mat. Inst. Steklov 38 (1951), 244–278.
2. Functions with dominating mixed derivatives in Hölder classes of several variables. (Russian) Sibirsk Mat. Ž. 4 (1963), 1342–1364.
3. Approximation of Functions of Several Variables and Imbedding Theorems. Second ed. (Russian). Moskva: Nauka 1977. (English translation of the first edition: Berlin, Heidelberg, New York: Springer-Verlag 1975).

Nilsson, P.
1. Multiplicatively invariant subspaces of Besov spaces. Techn. Report, Univ. of Lund, 1980.
2. Personal communication (Jena, 1981)

Orlovskij, D. G.
1. On multipliers in the spaces $B^r_{p,\Theta}$. Anal. Math. 5 (1979), 207–218.

Ornstein, D.
1. A non-inequality for differential operators in the L_1-norm. Arch. Rat. Mech. Anal. 11 (1962), 40–49.

Oswald, P.
1. Approximation by splines in the metric L_p, $0<p<1$. (Russian) Math. Nachr. 94 (1980), 69–96.
2. Über die Existenz von Schauderbasen in Besov-Räumen for $0<p<1$ und $N=1$.
3. On inequalities for spline approximation and spline systems in the spaces L^p $(0<p<1)$. Proc. Intern. Conf. "Approximation and Function Spaces", Gdansk 1979. Amsterdam: North Holland 1981, 531–552.
4. Besov-Hardy-Sobolev-Räume für analytische Funktionen im Einheitskreis.
5. On Schauder bases in Hardy spaces.
6. Convergence of spline expansions in Hardy spaces $(0<p\leqq 1)$. TU Dresden, 1981.
7. L^p-Approximation durch Reihen nach dem Haar-Orthogonalsystem und dem Faber-Schauder-System. J. Approximation Theory 33 (1981), 1–27.

Päivärinta, L.
1. On the spaces $L^A_p(l_q)$: Maximal inequalities and complex interpolation. Ann. Acad. Sci. Fenn. Ser. AI Math. Dissertationes 25, Helsinki 1980.
2. Eine Bemerkung zu den Räumen $L^A_p(l_q)$; Monotonie der Littlewood-Paley'schen Maximal-Funktionen. Reports Dep. Math. Univ. Helsinki. "Notes on Functional Analysis", Helsinki, 1980, 13–17.
3. Equivalent quasi-norms and Fourier multipliers in the Triebel spaces $F^s_{p,q}$. Math. Nachr. 106 (1982), 101–108.

Peetre, J.
1. Applications de la théorie des espaces d'interpolation dans l'analyse harmonique. Ricerche Mat. 15 (1966), 1–34.
2. Sur le espaces de Besov. C. R. Acad. Sci. Paris. Sér. A–B 264 (1967), 281–283.
3. On the theory of $\mathfrak{L}_{p,\lambda}$ spaces. J. Functional Analysis 4 (1969), 71–87.
4. Remarques sur les espaces de Besov. Le cas $0<p<1$. C. R. Acad. Sci. Paris, Sér. A–B 277 (1973), 947–950.
5. H_p Spaces. Lecture Notes, Lund 1974.
6. On spaces of Triebel-Lizorkin type. Ark. Mat. 13 (1975), 123–130.
7. The trace of Besov space – a limiting case. Technical Report, Lund 1975.
8. New Thoughts on Besov Spaces. Duke Univ. Math. Series, Duke Univ., Durham, 1976.
9. A counter-example connected with Gagliardo's trace theorem. Comment. Math. Prace Mat., tomus specialis 2 (1979), 277–282.
10. Classes de Hardy sur les variétés. C. R. Acad. Sci. Paris. Sér. A–B 280 (1975), A439–A441.
11. Remarks on the dual of an interpolation space. Math. Scand. 34 (1974), 124–128.

Peral, J. C.; Torchinsky, A.
1. Multipliers in $H^p(R^n)$, $0<p<\infty$. Ark. Mat. 17 (1979), 225–235.

Pietsch, A.
1. Operator Ideals. Berlin: VEB Deutscher Verlag der Wissenschaften 1978.
2. Eigenvalues of integral operators. I. Math. Ann. 247 (1980), 169–178. II. (to appear)
3. Approximation spaces. J. Approximation Theory 32 (1981), 115–134.

Plancherel, M.; Polya, G.
1. Fonctions entières et intégrales de Fourier multiples. Math. Helv. 9 (1937), 224–248.
Riviere, N. M.; Sagher, Y.
1. Interpolation between L^∞ and H^1, the real method. J. Functional Analysis 14 (1973), 401–409.
Roumieu, C.
1. Sur quelques extensions de la notion de distribution. Ann. Sci. École Norm. Sup., Sér. 3, 77 (1960), 41–121.
2. Ultra-distributions définies sur R^n et sur certaines classes de variétés differentiables. J. Analyse Math. 10 (1962/63), 153–192.
Ropela, S.
1. Spline bases in Besov spaces. Bull. Acad. Polon. Sci. Sér. Sci. Math. Astronom. Phys. 24 (1976), 319–325.
2. Decomposition lemma and unconditional spline bases. Bull. Acad. Polon. Sci. Sér. Sci. Math. Astronom. Phys. 24 (1976), 467–470.
Runst, T.
1. Räume vom Typ $B^s_{p,q,\alpha,\beta}$, $H^s_{p,\alpha,\beta}$ und $W^s_{p,\alpha,\beta}$ bezüglich Jacobi-Fourier Entwicklungen. Dissertation A, Jena 1981.
Runst, T.; Sickel, W.
1. On strong summability of Jacobi-Fourier expansions and smoothness properties of functions. Math. Nachr. 99 (1980), 52–61.
Samko, S. G.
1. The spaces $L^\alpha_{p,r}(R^n)$ and hyper-singular integrals. (Russian) Studia Math. 61 (1977), 193–230.
2. The description of the image of $I^\alpha(L_p)$ of Riesz potentials. (Russian) Izv. Akad. Nauk Armjan. SSR Ser. Mat. 12 (1977), 329–334.
Schmeißer, H.-J.
1. Anisotropic spaces. II. (Equivalent norms for abstract spaces, function spaces with weights of Sobolev-Besov type). Math. Nachr. 79 (1977), 55–73.
2. On spaces of functions and distributions with mixed smoothness properties of Besov-Triebel-Lizorkin type. I: Basic properties. Math. Nachr. 98 (1980), 233–250.
3. On spaces of functions and distributions with mixed smoothness properties of Besov-Triebel-Lizorkin type. II: Fourier multipliers and approximation representations. Math. Nachr. 106 (1982), 187–200.
4. Remarks on spaces with dominating mixed smoothness properties in the sense of S. M. Nikol'skij and T. I. Amanov. Czechoslovak Math. J.
5. On Fourier multipliers and Littlewood-Paley theorems for mixed L_p-spaces. Representations of Sobolev spaces with dominating mixed derivatives. Z. Analysis Anwendungen
Schmeißer, H.-J.; Sickel, W.
1. On strong summability of multiple Fourier series and smoothness properties of functions. Anal. Math. 8 (1982), 57–70.
Schmeißer, H.-J.; Triebel, H.
1. Anisotropic spaces. I. (Interpolation of abstract spaces and function spaces). Math. Nachr. 73 (1976), 107–123.
Schwartz, L.
1. Théorie des distributions. Paris: Hermann 1973.
Semadeni, Z.
1. Banach Spaces of Continuous Functions. I. Warszawa: PWN – Polish Scientific Publ. 1971.
Shamir, E.
1. Une propriété des espaces $H^{s,p}$. C. R. Acad. Sci. Paris, Sér. A–B 255 (1962), 448–449.
Sjölin, P.; Strömberg, J.-O.
1. Basis properties of Hardy spaces. (to appear)
2. Spline systems as bases in Hardy spaces. (to appear)
Slobodeckij, L. N.
1. Generalized Sobolev spaces and their applications to boundary value problems of partial differential equations. (Russian) Leningrad. Gos. Ped. Inst. Učep. Zap. 197 (1958), 54–112.

Sobolev, S. L.
1. The Cauchy problem in a functional space. (Russian) Dokl. Akad. Nauk SSSR 3 (1935), 291–294.
2. Méthode nouvelle à resoudre le problème de Cauchy pour les équations linéaires hyperboliques normales. Mat. Sb. 1 (1936), 39–72.
3. On a theorem of functional analysis. (Russian) Mat. Sb. 4 (1938), 471–497.
4. Einige Anwendungen der Funktionalanalysis auf Gleichungen der mathematischen Physik. Berlin: Akademie-Verlag 1964.

Sparr, G.
1. Interpolation of several Banach spaces. Ann. Mat. Pura Appl. 99 (1974), 247–316.

Stein, E. M.
1. Singular Integrals and Differentiability Properties of Functions. Princeton: Princeton Univ. Press 1970.

Stein, E. M.; Weiss, G.
1. On the theory of harmonic functions of several variables. I. The theory of H^p spaces. Acta Math. 103 (1960), 25–62.
2. Introduction to Fourier Analysis on Euclidean Spaces. Princeton: Princeton Univ. Press 1971.

Stein, E. M.; Zygmund, A.
1. Boundedness of translation invariant operators on Hölder and L^p spaces. Ann. of Math. 85 (1967), 337–349.

Stöckert, B.
1. Ungleichungen vom Plancherel-Polya-Nikol'skij-Typ in gewichteten L_p^Ω-Räumen mit gemischten Normen. Math. Nachr. 86 (1978), 19–32.
2. Multiplikationen in gewichteten anisotropen L_p^Ω-Räumen ganzer analytischer Funktionen exponentiellen Typs. Math. Nachr. 86 (1978), 33–40.

Stöckert, B.; Triebel, H.
1. Decomposition methods for function spaces of $B^s_{p,q}$ type and $F^s_{p,q}$ type. Math. Nachr. 89 (1979), 247–267.

Storoženko, Ė. A.
1. On the approximation of functions of the class $L^p, 0<p<1$, by algebraic polynomials. (Russian) Izv. Akad. Nauk. SSSR Ser. Mat. 41 (1977), 652–662.
2. Approximation of functions of class H^p, $0<p\leqq 1$. (Russian) Mat. Sb. 105 (1978), 601–621.
3. On theorems of Jackson type for H^p, $0<p<1$. Izv. Akad. Nauk SSSR Ser. Mat. 44 (1980), 948–962.

Storoženko, Ė. A.; Krotov, V. G.; Oswald, P.
1. Direct and inverse theorems of Jackson type in the spaces L^p, $0<p<1$. (Russian) Mat. Sb. 98 (1975), 395–415.

Storoženko, Ė. A.; Oswald, P.
1. Jackson's theorem in the spaces $L^p(R^k)$, $0<p<1$. (Russian) Dokl. Akad. Nauk SSSR 229 (1976), 554–557.
2. Moduli of smoothness and best approximation in the spaces L^p, $0<p<1$. Anal. Math. 3 (1977), 141–150.
3. Jackson's theorem in the spaces $L^p(R^k)$, $0<p<1$. (Russian) Sibirsk. Mat. Ž. 19 (1978), 888–901.

Strichartz, R. S.
1. Multipliers on fractional Sobolev spaces. J. Math. Mechanics 16 (1967), 1031–1060.
2. The Hardy space H^1 on manifolds and submanifolds. Canad. J. Math. 24 (1972), 915–925.
3. Bounded mean oscillation and Sobolev spaces. Indiana Univ. Math. J. 29 (1980), 539–558.
4. Traces of BMO-Sobolev spaces. Proc. Amer. Math. Soc. 83, 3 (1981), 509–513.

Strömberg, J.-O.
1. A modified Franklin system and higher order spline systems on R^n as unconditional basis of Hardy spaces. (to appear)

Strömberg, J.-O.; Torchinsky, A.
1. Weighted Hardy spaces.

Strömberg, J.-O.; Wheeden R. L.

1. Relations between H_u^p and L_u^p with polynomial weights. Trans. Amer. Math. Soc. 270 (1982), 439–467.

Taibleson, M. H.

1. On the theory of Lipschitz spaces of distributions on Euclidean n-space. I, II. J. Math. Mechanics 13 (1964), 407–479; 14 (1965), 821–839.

Tran Duc Long; Triebel, H.

1. Equivalent norms and Schauder bases in anisotropic Besov spaces. Proc. Roy. Soc. Edinburgh Sect. A 84 (1979), 177–183.

Triebel, H.

F. Fourier Analysis and Function Spaces. Teubner-Texte Math. 7, Leipzig: Teubner, 1977.

I. Interpolation Theory, Function Spaces, Differential Operators. Amsterdam, New York, Oxford: North-Holland Publ. Comp. 1978, and Berlin: VEB Deutscher Verlag der Wissenschaften 1978.

S. Spaces of Besov-Hardy-Sobolev Type. Teubner-Texte Math. 15, Leipzig: Teubner 1978.

1. Höhere Analysis. Berlin: VEB Deutscher Verlag der Wissenschaften 1972.
2. Spaces of distributions of Besov type on Euclidean n-space. Duality, interpolation. Ark. Mat. 11 (1973), 13–64.
3. General function spaces. II. J. Approximation Theory 19 (1977), 154–175. III. Anal. Math. 3 (1977), 221–249. IV. Anal. Math. 3 (1977), 299–315. V. Math. Nachr. 87 (1979), 129–152.
4. Spaces of distributions with weights. Multipliers in L_p-spaces with weights. Math. Nachr. 78 (1977), 339–355.
5. Multiplication properties of Besov spaces. Ann. Mat. Pura Appl. (4) 114 (1977), 87–102.
6. Multipliers and unconditional Schauder bases in Besov spaces. Studia Math. 60 (1977), 145–156.
7. On the Dirichlet problem for regular elliptic differential operators of second order in the Besov spaces $B_{p,q}^s$: The case $0<p\leqq\infty$. Proc. Conf. "Elliptische Differentialgleichungen", Rostock 1978, 237–263.
8. On Haar bases in Besov spaces. Serdica 4 (1978), 330–343.
9. On spaces of $B_{\infty,q}^s$ type and $\mathcal{C}^s$ type. Math. Nachr. 85 (1978), 75–90.
10. On Besov-Hardy-Sobolev spaces in domains and regular elliptic boundary value problems. The case $0<p\leqq\infty$. Comm. Partial Differential Equations 3 (1978), 1083–1164.
11. Strongly degenerate elliptic differential operators in Besov spaces: the case $0<p<\infty$. Beiträge Anal. 13 (1979), 27–47.
12. Equivalent quasi-norms in spaces of Besov-Hardy-Sobolev type and smoothness properties of functions. Wiss. Z. Friedrich-Schiller-Univ. Jena, Math.-Naturwiss. Reihe (2) 29 (1980), 195–201.
13. Theorems of Littlewood-Paley type for *BMO* and for anisotropic Hardy spaces. Proc. Intern. Conf. "Constructive Function Theory '77", Sofia, 1980, 525–532.
14. On the spaces $F_{p,q}^s$ of Hardy-Sobolev type: Equivalent quasi-norms, multipliers, spaces on domains, regular elliptic boundary value problems. Comm. Partial Differential Equations 5 (1980), 245–291.
15. On $L_p(l_q)$-spaces of entire analytic functions of exponential type: Complex interpolation and Fourier multipliers. The case $0<p<\infty$, $0<q<\infty$. J. Approximation Theory 28 (1980), 317–328.
16. A remark on complex interpolation theorems by Calderón-Torchinsky and Päivärinta. Reports Dep. Math. Univ. Helsinki, "Notes on Functional Analysis", Helsinki 1980, 18–21.
17. Maximal inequalities, Fourier multipliers, Littlewood-Paley theorems, and approximations by entire analytic functions in weighted spaces. Anal. Math. 6 (1980), 87–103.
18. Spline bases and spline representations in function spaces. Arch. Math. 36 (1981), 348–359.
19. Complex interpolation and Fourier multipliers for the spaces $B_{p,q}^s$ and $F_{p,q}^s$ of Besov-Hardy-Sobolev type: The case $0<p\leqq\infty$, $0<q\leqq\infty$. Math. Z. 176 (1981), 495–510.
20. A remark on integral operators from the standpoint of Fourier analysis. Proc. Intern. Conf. "Approximation and Function Spaces", Gdansk, 1979. Amsterdam: North-Holland, 1981, 821–829.

21. Periodic spaces of Besov-Hardy-Sobolev type and related maximal inequalities. Proc. Intern. Conf. "Functions, Series, Operators", Budapest 1980.
22. Mapping properties of abstract integral operators. Beiträge Anal. 17 (1981), 7–18.
23. Über die Existenz von Schauderbasen in Sobolev-Besov-Räumen. Isomorphiebeziehungen. Studia Math. 46 (1973), 83–100.
24. Characterizations of Besov-Hardy-Sobolev spaces via harmonic functions, temperatures and related means. J. Approximation Theory 35 (1982), 275–297.
25. Modulation spaces on the euclidean n-space. Z. Analysis Anwendungen.

Uno, Y.
1. Lipschitz functions and convolution. Proc. Japan Acad. 50 (1974), 785–788.

Walsh, T.
1. The dual of $H^p(R_+^{n+1})$ for $p<1$. Canad. J. Math. 25 (1973), 567–577.

Wheeden, R. L.
1. A boundary value characterization of weighted H^1. Enseignement Math. 22 (1976), 121–134.

Wigley, N. M.
1. A Banach algebra structure for H^p. Canad. Math. Bull. 18 (4), (1975), 597–603.

Wojtaszczyk, P.
1. The Franklin system is an unconditional basis in H_1. (to appear)
2. H_p spaces, $p\leq 1$ and spline systems. (to appear)

Young, Wo-Sang
1. Weighted norm inequalities for multipliers. Proc. Symp. Pure Math. 35, Part 1, "Harmonic Analysis in Euclidean Spaces", 1979, 133–139.

Yosida, K.
1. Functional Analysis. Berlin, Göttingen, Heidelberg: Springer-Verlag 1965.

Zolésio, J.-L.
1. Multiplication dans les espaces de Besov. Proc. Roy. Soc. Edinburgh, Sect. A 78 (1977), 113–117.

Zygmund, A.
1. Smooth functions. Duke Math. J. 12 (1945), 47–76.
2. On the preservation of classes of functions. J. Math. Mechanics 8 (1959), 889–895.

Index